Stochastic Modeling in Economics and Finance

Applied Optimization

Volume 75

Series Editors:

Panos M. Pardalos
University of Florida, U.S.A.

Donald Hearn
University of Florida, U.S.A.

Stochastic Modeling in Economics and Finance

by

Jitka Dupačová

Jan Hurt

and

Josef Štěpán

Department of Probability and Mathematical Statistics,
Faculty of Mathematics and Physics,
Charles University, Prague

KLUWER ACADEMIC PUBLISHERS

DORDRECHT / BOSTON / LONDON

A C.I.P. Catalogue record for this book is available from the Library of Congress.

ISBN 1-4020-0840-6

Published by Kluwer Academic Publishers,
P.O. Box 17, 3300 AA Dordrecht, The Netherlands.

Sold and distributed in North, Central and South America
by Kluwer Academic Publishers,
101 Philip Drive, Norwell, MA 02061, U.S.A.

In all other countries, sold and distributed
by Kluwer Academic Publishers,
P.O. Box 322, 3300 AH Dordrecht, The Netherlands.

Printed on acid-free paper

Printed in the Netherlands.

To my husband Václav

To Jarmila, Eva, and in memory of my parents

To my wife Iva

CONTENTS

PREFACE

The three authors of this book are my colleagues (moreover, one of them is my wife). I followed their work on the book from initial discussions about its concept, through disputes over notation, terminology and technicalities, till bringing the manuscript to its present form. I am honored by having been asked to write the preface.

The book consists of three Parts. Though they may seem disparate at first glance, they are purposively tied together. Many topics are discussed in all three Parts, always from a different point of view or on a different level.

Part I presents basics of financial mathematics including some supporting topics, such as utility or index numbers. It is very concise, covering a surprisingly broad range of concepts and statements about them on not more than 100 pages. The mathematics of this Part is undemanding but precise within the limits of the chosen level. Being primarily an introductory text for a beginner, Part I will be useful to the enlightened reader as well, as a manual of notions and formulas used later on.

The more extensive Part II deals with stochastic decision models. Multistage stochastic programming is the main methodology here. The scenario-based approach is adopted with special attention to scenarios generation and via scenarios approximation. The output analysis is discussed, i.e. the question how to draw inference about the true problem from the approximating one. Numerous applications of the presented theory vary from portfolio optimal control to planning electric power generation systems or to managing technological processes. A case study on a bond investment problem is reported in detail. A survey of numerical techniques and available software is added. Mathematics of Part II is still of standard level but the application of the presented methods may be laborious.

The final Part III requires from the reader higher mathematical education including measure-theoretical probability theory. In fact, Part III is a brief textbook on stochastic analysis oriented to what is called diffusion financial mathematics. The apparatus built up in chapters on martingales and on stochastic integration leads to a precise formulation and to rigorous proving of many results talked about already in Part I. The author calls his proofs honest; indeed, he does not facilitate his task by unnecessarily simplifying assumptions or by skipping laborious algebra.

The audience of the book may be diverse. Students in mathematics interested in applications to economics and finance may read with benefit all Parts I,II,III and then study deeper those topics they find most attractive. Students and researchers in economics and finance may learn from the book of using stochastic methods in their fields. Specialists in optimization methods or in numerical mathematics will get acquainted with important optimization problems in finance and economics and with their numerical solution, mainly through Part II of the book. The probabilistic Part III can be appreciated especially by professional mathematicians; otherwise, this Part will be a challenge to the reader to raise his/her mathematical culture. After all, a challenge is present in all three Parts of the book through numerous unsolved exercises and through suggestions for further reading given in bibliographical notes.

I wish the book many readers with deep interest.

V. Dupač

ACKNOWLEDGMENTS

This volume could not come into being without support of several institutions and a number of individuals. We wish to express our sincere gratitude to every one of them.

First of all we thank to Ministry of Education of Czech Republic[1], Grant Agency of Czech Republic[2] and Directorate General III (Industry) of the European Commission[3] who supported the scientific and applied projects listed below that substantially influenced the contents and form of the text. We gratefully acknowledge the financial support from the companies NEWTON Investment Ltd and ALAX Ltd and appreciate the particularly helpful technical assistance provided by the Czech Statistical Office.

The authors are very indebted to Pavel Popela from the Brno University of Technology who, using his extensive experience with the numerical solutions to the problems in the field of Stochastic Programing, wrote Chapter II.8. Horrand I. Gassmann from the Dalhousie University read very carefully this Chapter and offered some valuable proposals for improvements. We thank also Marida Bertocchi from the University of Bergamo whose effective cooperation within the project (3) influenced the presentation of results in Chapter II.6.

We have to say many thanks, indeed, to our colleagues and friends Václav Dupač and Josef Machek who agreed to read the text. They expended a great effort using their extensive knowledge both of Mathematics and English to make many invaluable suggestions, pressing for higher clarity and consistency of our presentation. Further, we are particularly grateful to Jaromír Antoch for his invaluable help in the process of technical preparation of the book. The authors are also indebted to their present and former PhD students at the Charles University of Prague: Alena Henclová and Petr Dobiáš deserve credits for their efficient and swift technical assistance. Part III owes much to Petr Dostál, Daniel Hlubinka, Karel Janeček and Petr Ševčík who, cruelly tried out as the first readers, have then become enthusiastic and respected critics.

Finally, we thank our publisher *Kluwer Academic Publishers* and, above all, the senior editor John R. Martindale for publishing the book.

J. Dupačová, J. Hurt, and J. Štěpán

---

[1] MSM 1132000008 Mathematical Methods in Stochastics
[2] GAČR 402/99/1136, GAČR 201/99/0264, GAČR 201/00/0770
[3] INCO'95, HPC/Finance Project, no. 951139

Part I

FUNDAMENTALS

I.1   MONEY, CAPITAL, AND SECURITIES

money, capital, investment, interest, cash flows, financing business, securities, financial market, financial institutions, financial system

1.1   Money and Capital

Money is the means which facilitates the exchange of goods and services. Commonly, money appears in forms like banknotes, coins, and bank deposits. There are three functions ascribed to money: (i) a medium of exchange, (ii) a unit of value, expressing the value of goods and services in terms of a single unit of measure (Czech Krones, e.g.), (iii) a store of wealth. Money is, no doubts, better means for trade than *barter* (direct exchange of goods or services without monetary consideration), but still insufficient for more complicated and/or sophisticated financial operations like investment.

Capital is wealth (usually unspent money) or better to say accumulated money which is used to produce or generate more wealth via an economic activity.

1.2   Investment

Individuals or companies face the problem how to handle their income. They can either spend it immediately, or save it, or partly spend and partly save. In either of the mentioned possibilities, they must decide how to spend and how to save. In the latter case (saving), they postpone their immediate consumption in favour of investment. In that case, they become *investors* and *investment* may therefore be defined as postponed consumption. Usually, the consumption–investment decision is made so as to maximize the expected utility (level of satisfaction) of the investor. While the immediate consumption is sure (up to certain limits), the result of an investment is almost always uncertain. Investments (or assets) can be classified into two classes; *real* and *financial*. A *real asset* is a physical commodity like land, a building, a car. A (financial) *security* or a *financial asset* represents a claim (expressed in money terms) on some other economic unit. (see [143], e.g.).

1.3   Interest

The reward for both postponed consumption and the uncertainty of investment is usually paid in the form of interest. Interest is a time dependent quantity depending on, roughly speaking, time to the postponed consumption. *Interest* in wider sense is either a charge for borrowed money that is generally a percentage of the amount borrowed or the return received by capital on its investment. Simply, interest is the price of deferred consumption paid to ultimate savers. Note that the actual allocation of savings in a reasonably functioning economy is accomplished through interest rates, see next Section. In other words, capital in a free economy is allocated through a certain price system and the interest rate expresses the cost of money.

1.4   Cash Flows

A *cash flow* is a stream of payments at some time instances generated by the investment or business involved. The *inflows* to the investor have plus sign while the *outflows* have minus sign. In accounts, the inflows are called *black figures* while outflows are called *red* or *bracket figures* since they appear either in red color or in brackets. As a rule, *net cash flows* are considered; it means that at any time instant all inflows and outflows are summed up and only the resulting sum is displayed. See I.3 for a more detailed analysis of cash flows.

1.4.1   Cash Flows Example. An investor buys an equipment for USD 90000 today. After one year he or she still is not in black figures and the loss is USD 15200. In the successive years 2, 3, 4, 5, 6 the profits (in USD) are 45000, 60000, 25000, 22000, 12000, respectively. At the end of the sixth year the investor sells the equipment for the salvage value USD 15000. The net cash flow for years $0, \ldots, 6$ is (-90000, -15200, 45000, 60000, 25000, 22000, 27000=12000+15000). Graphical illustration is given in Figure 1.

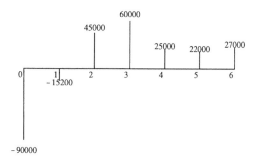

Figure 1: A cash flow

1.5   Financial and Real Estate Investment

Since handling money and capital itself is a rather complicated task, there are *financial intermediaries* and other financial institutions which should, in principle, handle money and capital efficiently. *Financial institutions* are business firms with assets in the form of either financial assets or claims like stocks, bonds, and loans. Financial institutions make loans and offer a variety of financial services (investment, life and general insurance, savings, pensions, credits, mortgages, leasing, real estates, etc.).

1.5.1   Financing the Business – Description

Almost every economic activity (of an individual, firm, bank, city, government) must be financed. In principle, there are two possibilities how to realize it; either from own funds or from outside sources (creditors, debt financing). Own funds of a company may be increased by issuing stocks resulting in the increase of equity while the debt financing usually takes form of either bank credit or issuing the debt instruments like corporate bonds. The better the expected performance of the firm is, the cheaper funds (money) are available. The financial public look on the performance of a firm through the *ratings* and the *prices* of financial instruments already issued by the firm on the market (mainly *Stock Exchange*). The most important corporations providing rating are Moody's Investor Service (shortly *Moody's*) and Standard & Poor's Corporation (shortly *Standard & Poor's*). Both the rating and price are important signals to the investors.

1.5.2 Financing the Business – Summary

We have seen that there are three main possible ways of financing; by equity (issuing stocks), and two ways of debt financing, i.e., by issuing the debt instruments like corporate bonds or just by acquiring a bank credit. A modern firm uses all the above possibilities and it is the task of financial managers to balance them. It is not so surprising that some very prospective American companies have debt to equity ratio about 70 per cent. The idea is simple; if you borrow at some 7 per cent and gain 11 per cent from the business, you are better off.

The fully self-financed company seems to be rather old-fashioned now. The tradition of the European family business may serve as an example. There are rare exceptions still surviving in these days, even among big firms in Europe. Nevertheless, the prosperous debt financed firm makes usually more profit than a comparable self-financed company.

1.6 Securities

Security (in what follows here we mean a *financial security*) is a medium of investment in the money market or capital market like shares (English) or stocks (American), bonds, options, mortgages, etc. Security is a kind of *financial asset* (everything which has a value or earning power). Speaking in accounting terms, the holder (purchaser) of it has an *asset* while the issuer or borrower (seller) has a *liability*. Security usually takes the form of an agreement (contract) between the seller and the purchaser providing an evidence of debt or of property. The holder of a security is called to be in a *long position* while the issuer is in a *short position*. Security usually gives the holder some of the following rights:

(1) returning back money or property
(2) warranted reward
(3) share on the profit generated by money provided
(4) share on the property
(5) right on decision making concerning the use of money provided.

But a security may also be an agreement between two parties (often called *Party* and *Counterparty*) on a financial or real transaction between the two. This is the case of swaps, partly the case of forwards and futures. It is difficult to say who is the issuer and who is the holder, in this case.

The basic types of securities and their forms are listed below. See [143], [138], [105], [172] for more details.

1.6.1   Fixed-Income Securities

Fixed-income securities are debt instruments characterized by a specified maturity date (the date of payoff the debt) and a known schedule of repaying the principal and interest.

1.6.1.2 Demand Deposits

Commercial banks and saving societies offer to their clients *checking accounts* or *demand deposits* which are interest bearing but the interest is usually very small. A better situation is with *savings accounts*, a type of *time deposit*. Here money is saved for a prescribed period of time and any early withdrawal is subject to penalty which usually does not exceed the interest for the period involved. The interest is higher than that of applied to demand deposits and sometimes may vary.

1.6.1.3 Certificates of Deposit

Very popular, particularly for the institutional investors, are the *Certificates of Deposit*, shortly *CD's*, mainly issued by commercial banks in large denominations. They also take the form of time deposits with fixed interest but the early withdrawal is severely penalized. CD's are usually issued on the *discounted base*, at a discount from their face value. Roughly spoken, if you want to buy a CD of the face CZK 1000000, say, payable after one year, you buy it for some CZK 920000. Remember that the return in this case is not 8 per cent.

1.6.1.4 Treasury Bills

A typical money market securities issued by the central bank are *Treasury bills, T-bills*. Their main purpose is to finance the government or their fiancées. They have maturities typically varying from weeks to one year and are also issued on the discount base.

There is one interesting point in issuing securities of the above type. A careful government (even the Czech one, now) issues T-bills through the *auction*. Prior to each auction, the central bank (representing the government, in many countries behaving independently of the government) announces the par (face) value of the security and the upper limit of the bid expressed in terms of the interest rate. Also the intended volume (*total face value*) is announced.

For example, the issuer (the bank in this case), announces that the accepted offers are up to 8 per cent p.a. It means that the issuer will only accept the offers below this rate. The submitted bids are collected and ranked according to the offers with respect to the volume and interest rate. Since the offer of the issuer is competitive, the investors who wish to catch the offer must carefully choose both the offered interest and the volume. The strategy of the issuer is the question of allocation, the problem which will be discussed later.

Note that similar policy or technique (auction) is also often used by commercial banks as well as by highly rated firms (rated as blue chips, AAA, in Standard & Poor's rating scale).

For a detailed analysis including a discussion of auctions see [143].

1.6.1.5 Coupon Bonds

A *coupon bond* is the long-term (usually from 5 to 30 years) financial instrument issued by either central or local governments (municipals), banks, and corporations. It is a debt security in which the issuer promises the holder to repay the *principal*, *par value, face value, redemption value*, or *nominal value* F at the *maturity date* and to pay (periodically, at equally spaced dates up to and including the maturity date) a fixed amount of *interest* C called *coupon* for historical reasons. The ratio $c = C/F$ is called *coupon rate*, sometimes simply *interest*. A typical period for the coupon payment is semiannual, rarely annual, but both the coupon and coupon rate are expressed on the annual base. The number of periods in a year is called *frequency*. In case of semiannually paid coupon, the frequency is 2. The bond is usually valued at a time instant between the issuing date and the maturity date. So that more important for the valuation purposes is the length of time to the maturity date called *maturity of the bond*. Maturity differs from the whole life of the bond in that only remaining payments of coupons and principal are considered. A cash flow coming from a coupon bond is illustrated in Figure 2.

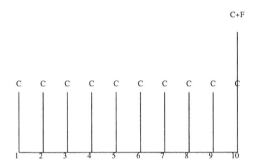

Figure 2: Cash flow of a bond

1.6.1.6 Callable Bonds

The simple coupon bond described above has an obvious disadvantage for the issuer; if the interest rates fall during the bond life, it is often possible for the issuer to get cheaper funds, for instance by issuing bonds with lower coupon. The security which partly gets rid of this feature is *callable bond*. The situation is the same as with the usual coupon bonds but in this case, the issuer has the right to buy some or all issued bonds prior to the original maturity date or *to call* them, in other words. Since the earlier repayment of the face value may cause an inconvenience to the bondholder (particularly with the reinvestment at lower interest than the coupon), the issuer should pay a reward to the bondholder in the form of *call premium*. The *call dates* and *call premiums* are stated in the offering statement. For example, if the bond is called one year before the maturity date, the payment is 101 per cent of the par value, if two years before, the payment is 102 per cent, etc. The call premium generally decreases with the date of call closer to the maturity date. Strictly speaking, the callable bond is not a fixed-income security since the payments coming from it are uncertain and depend both on the issuer policy and market interest rates.

1.6.1.7 Zero Coupon Bonds

A *zero coupon bond*, shortly *zero*, or *discount bond* pays only the face value at maturity. It is issued at discount to par value (like CD) and it pays par value at maturity. One reason for issuing such a type of bonds is that in some countries (like USA) the issuer may deduct the yearly accrued interest from taxes even though the payment is not made in cash. The bondholder (purchaser) must calculate interest income in the same way as the issuer calculates the tax deduction and should pay either corporate or personal tax even though no cash has been received. However, if the purchaser is a tax-exempt entity, like a pension fund or an individual who buys the bond for its individual retirement account, it pays no tax from the accrued interest. See [25] p. 578 for more details.

A coupon bond may be considered as a series of zero coupon bonds, all but last with face value equal to the coupon payment, and the last with the face value equal to the coupon payment plus the face value of the underlying coupon bond. This is not only a theoretical construction; the coupon components and face value of US Treasury bonds may be traded separately and such securities are called *STRIPS* – Separate Trading of Registered Interest and Principal of Securities. There are also *derivative zero coupon bonds*; a brokerage house buys usual coupon bonds, strips the coupons, and resells the stripped securities as zero coupon bonds.

1.6.1.8 Mortgage-Backed Securities

A lending institution that loans money for mortgages combines a large group of mortgages and thus creates a *pool*. The *mortgage-backed (pass-through)* security is then a long term (15 to 30 years) instrument that is collateralized by the pool of mortgages. As the homeowners make their (usually monthly) payments of the principal and interest to the lending institution, these payments are then "passed through" to the security holders in the form of coupon payments and the principal. The coupon is naturally less than the interest paid by homeowners, but the level of default is low. First, there is a warranty in real estate, second, there is a large pool of loans which diversifies the default risk.

See [143] for more details.

1.6.2   Floating-Rate Securities

Floating-rate securities' payments are not fixed in advance and rather depend on some underlying asset. The reason for issuing such securities is to reduce the interest rate risk for both the seller and the buyer. Typical examples are *floating-rate bonds* and *notes* with a coupon or interest periodically adjusted according on the underlying instrument (base rate) like LIBOR, PRIBOR, discount rate of the central bank etc. or they are simply tied to some interest rate like prime rate of a commercial bank (the interest rate for highly rated clients of the bank).

Note that LIBOR (*London InterBank Offered Rate*) is the daily published interest rate for leading currencies (GBP, EUR, USD, JPY, ...) with a variety of maturities (one day or overnight, 7 days, 14 days, 1 month, 3 months, 6 months, 1 year). LIBOR is calculated as the trimmed average (two smallest and two largest values are not considered) of the interest rates on large deposits among 8 leading banks in Great Britain. Similarly PRIBOR is an abbreviation for *Prague Interbank*

Offered Rate and is calculated in a similar way like LIBOR. Usually the calendar Actual/360 applies to all transactions.

Typically, the actual coupon rate is the interest rate of the underlying asset plus *margin* (*spread*). If the underlying instrument is LIBOR, e.g., the actual coupon rate may be actual LIBOR plus 100 basis points or actual LIBOR plus 3 per cent. The floating rates may be reset more than once a year leading to *short-term floating rates* while in the opposite case we speak of *long-term floating rates*. We also speak about *adjustable-rate securities* or *variable-rate securities*, see [60], [61].

1.6.2.1 Example (I bonds). *I bonds* are U.S. Treasury *inflation-indexed saving bonds* introduced in September 1998 with maturity on September 2028 in denominations varying from USD 50 to USD 10000. The rate – currently 6.49% p.a. – consists of two components; a *fixed rate* 3.6% which applies for the life of the bond, and *inflation rate* measured by the Consumer Price Index which can change every six months. I bonds earnings are added every month (coupon is added to the principal) and the interest is compounded semiannually. Only Federal income tax applies to the earnings. Investors cashing before 5 years are subject to a 3-month earnings penalty.

1.6.3 Corporate Stocks

Issuing stocks is a very popular method of financing business and further development of a company (corporation, firm). The most important types of stocks are common and preferred stocks. A *common stock* (US), *ordinary share* (UK) is the security that represents an ownership in a company. The equity of a company is the property of the common stock holders, hence these stocks are often called *equities*. For the investors, the stock is a piece of paper or a record in the computer giving him or her the right to engage in the decision processes concerning the company policy according to the share on common stock (voting right). Also it entitles the owner to *dividends* which consist of the amount of company's profit distributed to stockholders. This amount equals earnings less *retained earnings* (the part of earnings intended for reserves and reinvestment).

A *preferred stock* gives the holder priority over common stockholders. Preferred stockholders receive their dividend prior to common stockholders. Usually the dividend does not depend on the company's earnings and often is constant, thus resembling a coupon bond. In case of bankruptcy, the preferred stockholders have higher chance to see their claims to be satisfied. On the other hand, often they do not have voting right.

Stocks have another feature which is called *limited liability* that means that their value cannot be negative in any case.

1.6.4 Financial Derivatives

Financial derivative securities or *contingent claims* are the instruments where the payment of either party depends on the value of an *underlying asset* or assets. The underlying assets in question may be of a rather general form, e.g. stocks, bonds, commodities, currencies, stock exchange indexes, interbank offer rates, and even derivatives themselves. The underlying assets thus fall into two main groups;

commodity assets and *financial assets*. The derivatives are now traded in enormous volumes all over the world. Estimated figure for options only at 1996 was about $35 trillion. The most common derivatives are forwards, futures, options, and swaps.

1.6.4.1 Forwards and Futures

A *forward contract* is an agreement between two parties, a buyer and a seller, such that the seller undertakes to provide the buyer with a fixed amount of the currency or commodity at a fixed future date called *delivery date* for a fixed price called *delivery price* agreed today, at the beginning of the contract. For both parties this agreement is an obligation. By fixing the price today the buyer is protected against price increase while the seller is protected against price decrease. Forward is typically a privately negotiable agreement and it is not traded on exchanges. The forward contract is a risky investment from two reasons, at least. First reason is obvious; since the spot price of the underlying asset generally differs from the delivery price, the loss of one party equals the profit of the counterparty and vice versa. The second reason is the default risk in which case the seller is not willing to provide the buyer with delivery. There are also nonnegligible costs in finding a partner for this contract and fair delivery price. Therefore, the forward contracts are usually realized between reliable, highly rated parties. No money changes hands prior to delivery.

A simple example is a forward contract between a miller and a farmer producing corn. Today, April 11, 2001, they agree that the farmer will deliver 1000 bushels of corn for the delivery price USD 2.5 per bushel on September 30, 2001, the delivery date. Both parties consider these conditions of the contract as good. Assume that the spot price of corn on the delivery date would increase to USD 3 per bushel. Without the forward contract, the miller would have to buy for this price which might cause problems to him. On the other hand, with the spot price decrease to USD 2 per bushel on delivery date, the farmer who would have to sell for this price might have to go to the bankruptcy.

A *futures contract* shortly *futures*, is of a similar form as the forward but it has additional features. The futures is *standardized* (specified quality and quantity, prescribed delivery dates dependent on the type of the underlying asset). The futures are traded (they are marketable instruments) on exchanges. One of the most popular is Chicago Board of Trade (CBT). To reduce the default risk to minimum, both parties in a futures must pay so called *margins*. These margins serve as reserves and the account of any party in the contract is daily recalculated according to the actual price of the futures, the *futures price*. Such a procedure is called *marking to market*. The *initial margin* must be paid by both parties at the initiation of the contract and usually takes values between 5 to 10 per cent of the contract volume. The *maintenance margin* is a prescribed amount below the initial margin. If the account falls below this margin, it must be recovered to the initial margin by an additional payment called a *variation margin*. The contractors' accounts bring interest. The futures exchange also imposes a *daily price limit* which restricts price movements within one business day, ± 10 per cent, say. The responsibility for default is transferred to a *clearing house* that is also responsible for the clients' accounts, see [25] and [143].

The reports on futures prices in financial press provide the daily opening, highest, lowest, and closing price, the percentage change, the highest and lowest price during the lifetime of the contract, and the total number of currently outstanding contracts called *open interest*.

1.6.4.2 Options

An *option* is a contract giving its owner (*holder*, *buyer*) the right to buy or sell a specified underlying asset at a price fixed at the beginning of the contract (today) at any time before or just on a fixed date. The seller of an option is also called *writer*. It must be emphasized that an option contract gives the holder a **right** and not an **obligation** as it was the case of futures. For the writer, the contract has a potential obligation. He **must** sell or buy the underlying asset accordingly to the holder's decision. We distinguish between a *call option* (CALL) which is the right to buy and a *put option* (PUT) which is the right to sell. The fixed date of a possible delivery is called *expiry* or *maturity* date. The price fixed in the contract is called *exercise* or *strike price*. If the right is imposed we say that the option is *exercised*. If the option may be exercised at any time up to expiry date, we speak of an *American option* and if the option may be exercised **only** on expiry date, we speak of a *European option*. These are the simplest forms of options contracts and in literature such options are called *vanilla options*.

The right to buy/sell has a value called an *option premium* or *option price* which must be paid to the seller of the contract. It must be stressed that the option price is **different** from the exercise price!

Like futures, options are mostly *standardized* contracts and are traded on exchanges since 1973. The first such exchange was the Chicago Board Options Exchange (CBOE). Most common underlying assets are common stocks, stock market indexes, fixed-income securities, and foreign currencies. Options are usually short-term securities with typical maturities 3, 6, and 9 months. At any time there are options with different maturities and different strike prices available on the market. An example (taken from [143]) shows how the long term options are quoted in financial press on January 15, 1992, is in the following table:

Option	Expiry	Strike	Last
ATT	Jan93	25	$16\frac{1}{2}$
ATT	Jan93	35	$7\frac{1}{2}$
ATT	Jan93	35p	$1\frac{1}{4}$
ATT	Jan93	40	$4\frac{1}{4}$
ATT	Jan93	40p	$2\frac{3}{4}$
ATT	Jan93	50	$1\frac{3}{16}$.

This is an example of American options with different strike prices with the underlying asset AT & T common stock and with the same expiry date, the third Friday January 1993. "*p*" standing at strike price means a PUT option, the others are CALLs. "Last" means the closing price.

Another type of options are *exotic* or *path-dependent* options. These options (if exercised) pay the holder the amount dependent on the history of the underlying asset. Despite their "exotic" features, they are successfully used for hedging purposes. Since the creativity and fantasy of the developers of such products is practically unbounded, we only give some examples. Note that most of the mentioned options may be either of European or American type. For more details see [172] and [105], e.g.

A *binary* or *digital* option pays the holder a fixed amount of money if the value of the underlying asset rises above or falls below the exercise price. The payoff is independent of how far from the exercise price the asset value was at the exercise time.

A *barrier option* is a usual vanilla option but it may only be exercised if either the asset value does not cross a certain value – an *out-barrier*, or if the asset price crosses a certain value – an *in-barrier* during the life of the option contract. There are four possible cases:

(1) *up-and-in*; the option pays only if the barrier is reached from below,
(2) *down-and-in*; the option pays only if the barrier is reached from above,
(3) *up-and-out*; the option pays only if the barrier is not reached from below,
(4) *down-and-out*; the option pays only if the barrier is not reached from above.

A *compound option* is simply an option where the underlying asset is another option. If we consider only plain vanilla options, we have four possibilities again. For brevity, we describe the mechanism of a *call-on-a-call* European type compound option. Such an option gives the holder the right to buy a call option for the price K_1 at the expiry T_1. The second call option is on an underlying asset with the exercise K_2 and the expiry $T_2 > T_1$.

A *chooser option* or *as-you-like-it option* is an option which gives the holder the right to buy or sell either a call or a put option. We give an example of a *call-on-a-call-or-put*. Such a chooser option gives the holder the right to purchase for the exercise price K_1 at expiry time T_1 either a call or a put with exercise price K_2 at time T_2.

An *Asian option* is a path-dependent option with payoffs dependent on the average price of the underlying asset during the life time of the option. Such an average plays the role of the exercise price. Thus, the *average strike call* pays the holder the difference between the asset price at expiry and the average of the asset prices over some period of time, if positive, and zero otherwise. The problems arise from the proper definition of the average involved, continuous or discrete sampling, if discrete, then from prices sampled hourly or from closing prices, etc.

A *lookback option* has a payoff which also depends on maximum or minimum reached by the underlying asset over some period prior to expiry. Such a maximum or minimum plays the role of the exercise price.

1.6.4.3 Swaps

Swaps, like forwards, are mostly individual contracts between two highly rated, reliable parties which well fit the needs of both. Although the swaps are individual contracts, in practise they often follow the recommendations of the *International Swaps and Derivatives Association* (ISDA). A *swap* may be briefly characterized as an agreement on exchange of cash flows in future times with a prescribed schedule. There are two main categories of swaps; interest rate swaps and currency swaps.

In practise, the two are often combined. Swaps are used to manage interest rate exposure or uncertainty concerning the future exchange rates.

An *interest rate swap* is a contract between two parties to exchange interest streams with different characteristics based on a principal, notional amount, sometimes called the volume of a swap. The interest rates may be either fixed or floating in the same or different currencies.

A pure *currency swap* is a forward contract on the exchange of different currencies on some future date (maturity) in amounts fixed today. Another type of a currency swap is a *cross-currency swap* that consists of the initial exchange of fixed amounts of currencies and reverse final exchange of the same amounts at maturity. One or both parties may pay interest during the lifetime of the swap.

1.6.4.4 Example (Combined swap). Notional amount: CZK 34,500,000

Fixed amounts:

Initial exchange: Party A pays EUR 1,000,000 to party B, party B pays CZK 34,500,000 to party A. Maturity 10 years.

Final exchange (after 10 years): Party B pays EUR 1,000,000 to party A, party A pays CZK 34,500,000 to party B.

Floating amounts:

Party A pays to party B semiannually E6M - 3.5 per cent (spread or margin) from notional amount based on the floating rate day count fraction Actual/360, i.e., CZK $((E6M-3.5)/100) \cdot (182/360) \cdot 34,500,000$. Here E6M stands for LIBOR interest rate on EUR with maturity 6 months.

1.6.5 Miscellaneous Securities

Here we briefly mention a sample of other types of derivatives met in financial practise.

A *warrant* is a derivative security which gives the holder the right to buy a specified number of common stocks for a fixed price called *exercise price* at any time during the lifetime of the warrant. Such a security resembles a CALL option but there are two differences. First, warrant is a long-term security, 10 years say, while options have maturities up to two years. Second, perhaps a more important feature of the warrant is, that it is issued by the same company which issues the underlying stock while options are traded among investors.

Another type of security with an option is a *convertible bond*. Such a bond gives the bondholder the right to exchange the bond for another security, typically the common stock issued by the same company or just to sell back the bond to the issuing company. This is an example of a convertible bond with *put option*. Firms usually add the conversion option to lower the coupon rate. On the other hand, the issuer may reserve the right to *call back* the bonds and upon call, the bondholder either converts the bond into stocks or redeems it at the call price (convertible bond with *call option*). In this case, the coupon rate must be higher than that of usual coupon bond. In both cases we speak of *conversion premiums*.

Let us turn to floating-rate bonds (see 1.6.2). Most issuers cap their obligations to ensure that the floating coupon rate does not rise above a prespecified rate called *cap*. Thus if the face value of a bond is F, the floating rate r (say LIBOR on EUR with maturity 6 month + 3 percent) and the cap r_c, then the payment is $F \cdot \min(r, r_c)$. On the other hand, some issuers offer buyers an interest rate below which the coupon rate will not decline; such a rate is called *floor*. If the floor is r_f, then the payment is

$$F \cdot \max(\min(r, r_c), r_f) = F \cdot \min(\max(r, r_f), r_c).$$

Usually caps and floors take the form of consequent payments called caplets and floorlets, respectively.

1.7   Financial Market

Financial market consists of money market and capital market. *Money market* is a market with short-term assets or funds, up to one year say, like bills of exchange, Treasury bills (*T-bills*), and Certificates of Deposit (*CD's*). *Capital market* is a market which deals with longer-term loanable funds mainly used by industry and commerce for investment and acquisition. Usually capital markets handle securities which are related to the time horizon longer than one year.

1.8   Financial Institutions

The role of financial institutions is simple. Financial intermediaries (commercial banks, insurance companies, pension funds, e.g.) acquire debts issued by borrowers (IOU – the abbreviation for "I Owe You") and at the same time sell their own IOUs to savers. Every bank (with rare exceptions in the Czech Republic) is happy to accept your savings and handle them. It is a debt which is used by the bank in the form of loans and investments. Examples of other financial institutions are security brokers (bringing buyers and sellers of securities together), dealers, who – like brokers – intermediate but moreover purchase securities for their own accounts. There are investment bankers, mortgage bankers, and other miscellaneous financial institutions in this category, as well.

1.9   Financial System

In a civilized country, all the activities mentioned above go through the *financial system* which can be simply illustrated by the following scheme:

Ultimate borrowers, savings-deficit units $\longleftrightarrow$

$\longleftrightarrow$ Financial Intermediaries $\longleftrightarrow$

$\longleftrightarrow$ Ultimate savers, savings-surplus units.

The needs or wishes of borrowers and savers are different, of course. The borrowers need long-term loans, acceptance of significant risk by the lenders, and larger amounts of credit. Perhaps the highest priority of the lenders is *liquidity*, which means the availability of the funds (money) at the moment when these are requested. The natural needs of the savers are *safety of funds* and, particularly for small investors, accessibility of the securities in small denominations.

I.2   INTEREST RATE

interest rate, compounding, present value, future value, calendar convention, determinants of the interest rate, term structure, continuous compounding

2.1   Simple and Compound Interest

Interest rate (also *rate of interest*) is a quantitative measure of interest expressed as a proportion of a sum of money in question that is paid over a specified time period. So if the initial amount of money is PV (also called *principal* or *present value*) and the interest rate is i for the given time period, then the interest paid at the end of the period is $PV \cdot i$ and the accumulated amount of money at the end of the period (called *future value* or *terminal value*) is

$$(1) \qquad\qquad FV = PV(1+i).$$

Alternatively, the interest rate is quoted *per cent*. It will be clear from the context where $i = 0.13$ means $i = 13\%$ and vice versa. Note that r is another frequently used symbol for the rate of interest, particularly if speaking of the rate of return.

Let us consider more than one time period, say T periods, with T not necessarily integer, and the same interest rate i for one period. There are two approaches how to handle interests after each period. Under *simple interest* model, only interest from principal is received at any period. Thus the future value after T periods is

$$(2) \qquad\qquad FV_T = PV(1+iT).$$

Under *compound interest* model, the interest after each period is added to the previous principal and the interest for the next period is calculated from this increased value of the principal. The corresponding future value is

$$(3) \qquad\qquad FV_T = PV(1+i)^T.$$

In the context of the compound interest model, the process of going from present values to future values is called *compounding*.

2.1.1   Remark (Mixed Simple and Compound Interest)

Some banks or saving companies use a combination of simple and compound interest if T is not an integer. Let $T = \lfloor T \rfloor + \{T\}$ where $\lfloor \rfloor$ denotes the entire part and $\{\}$ denotes the fractional part of the argument. Then the future value is calculated as

$$(4) \qquad\qquad FV_T = PV(1+i)^{\lfloor T \rfloor}(1+i\{T\}).$$

2.1.2   Exercise. Decide what is better for the saver: future value of the savings calculated from (3) or (4).

Speaking of interest rates, it is important to state clearly the corresponding *unit of time*. In most cases, the interest rate is given as the *annual interest rate*, often stressed by the abbreviation *p.a.* (per annum). The usual notation is $i = 0.13$ p.a. or equivalently $i = 13\%$ p.a. Rarely, interest rates are given semiannually (*p.s.*, per semestre), quarterly (*p.q.*, per quartale), monthly (*p.m.*, per mensem), daily (*p.d.*, per diem). The *period of compounding* is similarly one year, six months, three months, one month, or one day. If the unit of time for the given interest rate differs from the period of compounding (which is often the case), it is very important to emphasize that we consider $i\%$ p.a. interest rate compounded semiannually, say. In this case it means that the interest rate i is so called *nominal interest rate*, and for every six month's period the actual interest rate is $i/2$. Generally, let $i^{(m)}$ be the nominal rate of interest per unit time compounded m times within the unit time so that there are m periods, each of length $1/m$, and the interest rate is $i^{(m)}/m$ per period. We also say that the nominal interest rate $i^{(m)}$ is *payable m*thly. Thus the future value of PV after T periods is

$$(5) \qquad\qquad FV_T = PV\left(1 + \frac{i^{(m)}}{m}\right)^T.$$

Of course, the actual interest rate per unit time i_{eff}, called *effective rate of interest* is *not* equal to the nominal rate of interest. Obviously,

$$(6) \qquad\qquad 1 + i_{\text{eff}} = \left(1 + \frac{i^{(m)}}{m}\right)^m.$$

2.1.3   Exercise. Compare the effective rates of interests if $i^{(m)} = 0.13$ p.a. for $m = 1, 2, 12, 365$ and comment the result.

2.2   Calendar Conventions

Assume the unit time is one year. If the number of periods n is not an integer, there are different methods to count the difference between two dates. Consider two dates, $DATE_1$, $DATE_2$, say, expressed in the form $DATE_j = YYYY_j MM_j DD_j$, $j = 1, 2$. January 13, 2013, is therefore expressed as 20130113. The most frequent conventions:

Calendar 30/360 or Euro-30/360. Under this convention all months have 30 days and every year has 360 days. The number of periods T is calculated as

$$T = \frac{1}{360}(360(YYYY_2 - YYYY_1) + 30(MM_2 - MM_1) +$$
$$\min(DD_2, 30) - \min(DD_1, 30)).$$

Calendar US-30/360. In this case, all dates ending on the 31st are changed to the 30th with the following exception: if $DD_1 < 30$ and $DD_2 = 31$ then $DATE_2$ is changed to the first of the next month.

Calendar Actual/Actual. This convention assumes the actual number of days between two dates with the actual number of days in the year.

Calendar Actual/360. The actual number of days in each month but 360 days in the year are considered. As a result, the number of periods within one year can exceed one.

Calendar Actual/365. The actual number of days in each month and 365 days in each year are considered. The leap year assumes 365 days.

Most computer systems are equipped with calendar functions, particularly with the function which returns the number of days between two dates. For example, *Mathematica* offers the function DaysBetween[*date2, date1*] which returns the actual number of days between two dates. The arguments date takes the form {year,month,day} so that March 14, 2001 is {2001,3,14} in this notation. In financial packages, the same *Mathematica* function has option DayCountBasis either "Actual/Actual"or "30/360" .

2.2.1   Exercise. Analyze the effect of the calendar conventions on savings from the point of view of a saver or a borrower.

2.3   Determinants of the Interest Rate

In a free economy, interest rates, as a price of money, are mainly determined by market supply and demand, and partly mastered by the government or central bank via money supply policy. Interest rates vary with economic environment, market position, used financial instrument, and time. The economic units which are willing to pay higher interest rates for the *funds* (=borrowed money in this case) expect higher returns on their investments. The returns are usually measured by the *rate of return* defined by:

$$\text{Rate of return} = \frac{\text{Ending price} + \text{Cash income} - \text{Beginning price}}{\text{Beginning price}},$$

sometimes quoted in per cent.

Every investment should be valued from the point of view of return, risk, inflation, and liquidity. The firm with higher return will pay higher interest for funds (money). With the rate of return 25 per cent the firm pays 20 per cent interest with pleasure. Another firm, with the rate of return 20 per cent, would not pay 20 per cent interest since then it would not have reason to develop any activity. More risky investment should be more expensive than an investment with (almost) certain return, in terms of interest rates. Inflation also makes funds more expensive. If the inflation is high, the funds may be not accessible. Short term funds (money borrowed for short time) are usually cheaper than long term funds (money borrowed for long time). Short term interest rates more or less reflect the actual state of the economy while long term interest rates reflect expectations, rational or less rational. Situation is more complicated, however, see the concept of yield curves next in this part. Denote r the rate of interest comprising all the factors mentioned above. In this context, r is also called *cost of capital*.

2.3.1   Remark (Taxation)

Almost all incomes coming from investment are subject to taxes. The few exemptions are returns on some government or municipal bonds, e.g. Thus the taxation reduces the returns. Moreover, the taxes are often different for various types of investment and sometimes are progressive, i.e., the higher the return, the higher the taxes. Thus any investment should be carefully valued with respect to tax consideration.

2.4   Decomposition of the Interest Rate

Taking into account all the factors which affect the so called *quoted* or *nominal* *interest rate r*, we can write

$$(7) \qquad 1 + r = (1 + r_0)(1 + r_{\text{infl}})(1 + r_{\text{default}})(1 + r_{\text{liquid}})(1 + r_{\text{mat}})$$

where r_0 denotes the *risk free* interest rate if we do not consider inflation, r_{infl} (*inflation premium*) is the **expected** rate of inflation, r_{default} (*default risk premium*) is the premium charged for the default risk, that is the risk that the debtor will not pay either principal or interest or both. Sometimes it is called *credit risk*. The term r_{liquid} (*liquidity premium*) stands for the risk that an asset in question is not readily convertible into cash without considerable cost. Finally, r_{mat} (*maturity risk premium*) is the premium for the risk produced by possible changes of interest rates during the life of an asset. There are two types of the maturity risk. Consider bonds, e.g. For long-term bonds, it is the *interest rate risk*; if the market interest rate rises, the prices of bonds go down. This kind of premium rises when the interest rates are more volatile. For short-term bonds, it is the *reinvestment rate risk*; if these bills become due and the actual interest rates are low, the reinvestment will result in interest income loss.

Sometimes the decomposition is given in additive form (see [25], e.g.)

$$(8) \qquad r = r_0 + r_{\text{infl}} + r_{\text{default}} + r_{\text{liquid}} + r_{\text{mat}}$$

which is a good approximation of (7) if the components of r are sufficiently small since the cross-factors of type $r_0 r_{\text{infl}}$ are small of twice higher order than the original components.

In real world, there is practically no riskless investment. For simplicity, however, the government bonds are usually considered riskless. In this case, the offered return also includes the expected rate of inflation, so that the risk free rate with a premium for expected inflation is

$$(9) \qquad (1 + r_0)(1 + r_{\text{infl}}) - 1.$$

In what follows, without further notice we will consider the riskless rate with the inflation premium.

It is also necessary to note that the above decomposition depends on the time period involved. So if we consider the one-year quoted interest rate, the corresponding expected inflation is a one-year inflation, and the risk free-rate is derived from one-year T-bills rates and the maturity risk premium has a negligible influence on the nominal rate in a stable economy. For a ten-years' quoted interest rate we should take ten-years yields of the government bonds for the riskless rate and carefully consider the other factors affecting the nominal interest rate; default, liquidity, and maturity premium in this case.

2.4.1 Remark (Rating)

Useful guides to credit risk evaluation for corporate bonds are conducted by recognized agencies like *Standard and Poor's* and *Moody's*. Based on an analysis of the firms they provide a classification into rating categories. According to Standard and Poor's, AAA is the highest rating reflecting extremely strong capacity to pay interest and to repay principal, AA means very strong capacity, A may be effected by economic conditions, etc. Further categories are BBB, BB, B, CCC, CC, C, D. Categories below BBB are sometimes considered as speculative or junk bonds. Refinement may be made by adding + or − signs. Similar categories provided by Moody's are Aaa, Aa, A, Baa, Ba, B, Caa, Ca, C, D.

2.4.2 Example.

In January 1991 the quoted interest rates for U.S. T-bonds, AAA, AA, and A were 8.0, 8.9, 9.1, and 9.4 per cent, respectively. See [25], p. 109. All these bonds had similar maturity, liquidity, and other features. So the only difference is in the default risk premium. Using formula $r_{\text{default}} = \frac{1+r}{1+r_0} - 1$ for the default premium risk we get $r_{\text{default}}(AAA) = 0.83$, $r_{\text{default}}(AA) = 1.02$, and $r_{\text{default}}(A) = 1.30$, respectively.

2.4.3 Real Return

If r is the nominal rate of interest on deposits and r_{infl} is the rate of inflation, then the *real return* on deposits is sometimes expressed in terms of the *real rate of interest* r_{real} which can be calculated from the obvious relation

$$1 + r = (1 + r_{\text{infl}})(1 + r_{\text{real}})$$

or

(10)
$$r_{\text{real}} = \frac{r - r_{\text{infl}}}{1 + r_{\text{infl}}}.$$

For small values of the components appearing in the last formula, we can use the approximation $r_{\text{real}} = r - r_{\text{infl}}$. Moreover, let r_{tax} be the tax rate imposed on the earned interest from deposits. Then we get

(11)
$$1 + r(1 - r_{\text{tax}}) = (1 + r_{\text{infl}})(1 + r_{\text{real}})$$

for the real return.

2.4.4 Example. In the Czech Republic, year 1997, the inflation rate was 0.10 (official source), r could have been taken as 0.11 (an over-optimistic value at some banks), and tax on the return on deposits was 0.15 (by law). Then we obtain the negative real return -0.6 per cent. In April 2001, the yearly inflation has been estimated as 4.1 per cent and one year term deposits net yield was about 3 per cent. So again we get the negative real return at about -1.1 per cent.

2.4.5 Exercise. Derive the corresponding relation for the *real percentage increase* in purchasing power if the percentage increase in salaries is r, the inflation rate is r_{infl}, and r_{tax} is the tax rate.

2.4.6 Example. Let us consider two investments, A and B, say, with gross returns r_A and r_B, subject to taxes i_A and i_B, respectively. The two investments provide the same net yield if

$$r_A(1 - i_A) = r_B(1 - i_B)$$

holds.

2.5   Term Structure of Interest Rates

All the interest rates in this Section relate to the equal time periods. Suppose $_tR_n$ is the actual rate of interest at time t on an n-period investment called *spot interest rate* and $_{t+1}r_{1t}$, $_{t+2}r_{1t}$, $\ldots$, $_{t+n-1}r_{1t}$ are the one-period interest rates on an investment beginning at times $t + 1$, $t + 2$, $\ldots$, $t + n - 1$, respectively, called *forward rates for one period implied in the term structure at time t.* At time t we know spot rates $_tR_1$, $_tR_2$, $\ldots$, $_tR_n$. Obviously, we can put $_tr_{1t} = {_tR_1}$. We have

$$(12) \qquad (1 + {_tR_k})^k = (1 + {_tR_1}) \prod_{j=1}^{k-1} (1 + {_{t+j}r_{1t}}), \qquad k = 1,\ldots,n.$$

From this formula we can simply obtain the forward rates

$$(13) \qquad 1 + {_{t+j}r_{1t}} = \frac{(1 + {_tR_{j+1}})^{j+1}}{(1 + {_tR_j})^j}, \qquad j = 1,\ldots,n.$$

The one-period forward rates may simply span any desired length of time. Thus, *j-period forward rate beginning at time $t + k$ implied in the term structure at time t is*

$$(14) \qquad {_{t+k}r_{jt}} = \left(\frac{(1 + {_tR_{k+j}})^{k+j}}{(1 + {_tR_k})^k} \right)^{1/j} - 1.$$

Due to the liquidity premium the relations between spot and forward rates are rarely fulfilled exactly in practise. Instead of $_{t+k}r_{1t}$ we should consider $_{t+k}r_{1t} + {_{t+k}L_{1t}}$ where the L's are the liquidity premiums embodied in the forward rates. Usually the liquidity premiums are increasing:

$$0 < {_{t+1}L_{1t}} < \cdots < {_{t+n-1}L_{1t}}.$$

2.6   Continuous Compounding

In theory, continuous compounding plays a crucial role. The idea of continuous compounding comes from the usual concept of compounding for the number of compounding periods approaching to infinity. In this case, we consider the nominal interest rate $i^{(\infty)} =: \delta$ (δ called the *force of interest* or often *interest rate* in the continuous financial mathematics) per unit time so that the future value FV of the initial investment PV (at time $t = 0$) after time T becomes

(15)
$$FV_T = PV \lim_{m \to \infty} \left(1 + \frac{i^{(\infty)}}{m}\right)^{Tm} = PV e^{\delta T}.$$

In other words, the future value grows exponentially with time according to

(16)
$$\frac{1}{FV_T}\frac{\partial FV_T}{\partial T} = \frac{\partial \ln FV_T}{\partial T} = \delta.$$

This formula is often presented in the form

(17)
$$\frac{dFV_T}{FV_T} = \delta dT.$$

If the investment is taken at time t instead of $t = 0$ (usually $t < T$), and is represented by the present value PV_t then

(18)
$$FV_T = PV_t\, e^{\delta(T-t)}.$$

So far, we have considered the force of interest to be a constant. But, the above formulation allows us to simply extend it to the case of variable force of interest δ_t depending on time t. The accumulation factor then becomes $\int_t^T \delta_s ds$ instead of $\delta(T - t)$ and the future value at time T of the unit investment at time t therefore is

(19)
$$FV_T = PV_t\, e^{\int_t^T \delta_s ds}.$$

Analogously, the expression of the present value in terms of the future value and the time dependent force of interest reads:

(20)
$$PV_t = FV_T\, e^{-\int_t^T \delta_s ds}.$$

The function

(21)
$$v(t,T) = e^{-\int_t^T \delta_s ds}$$

is called *discount function* and for $t = 0$ it is abbreviated to $v(T)$ so that $v(T) = v(0,T)$.

2.6.1   Example (Stoodley's Formula). A flexible model has been suggested by Stoodley. In spite of the fact that this model is mainly of theoretical interest, it is useful for giving a sight of a possible behavior of the time development of the interest rate. The *Stoodley's formula* says that

$$(22) \qquad\qquad \delta_t = p + \frac{s}{1 + rse^{st}}$$

where p, r, and s are properly chosen or estimated parameters.

2.6.2   Exercise. Study the behavior of the force of interest following the Stoodley's formula dependent on the parameters appearing in the formula.

2.6.3   Example (Discount Function of the Stoodley's Force of Interest). The calculation needs some algebra. Write t instead of T in the formula for $v(T)$. Then

$$(23) \quad v(t) = \exp\left\{ -\int_0^t \left(p + \frac{s}{1 + re^{sy}} \right) dy \right\} =$$

$$\exp\left\{ -\int_0^t \left(p + s - \frac{rse^{sy}}{1 + re^{sy}} \right) dy \right\} =$$

$$\exp\{ -(p+s)t + \ln(1 + re^{sy})|_0^t \} =$$

$$\exp\{ -(p+s)t \} \frac{1 + re^{st}}{1 + r} = \frac{1}{1+r} e^{-(p+s)t} + \frac{r}{1+r} e^{-pt}.$$

If we put $v_1 := e^{-(p+s)}$, $v_2 := e^{-p}$, we get

$$(24) \qquad\qquad v(t) = \frac{1}{1+r} v_1^t + \frac{r}{1+r} v_2^t.$$

From this formula it follows that the discount function can be expressed as the weighted average of the present values with **constant interest rates**.

I.3   MEASURES OF CASH FLOWS

present value, future value, annuities, equation of value, internal rate of return, duration, convexity, investment projects, payback method, yield curves

Consider first the sums (payments) $CF_0, \ldots, CF_T$ related to the equally spaced time instants $0, \ldots, T$. The interest rate for one period i will alternatively mean the *cost of capital*, the *opportunity cost rate*, i.e., the rate of return that can be earned on an alternative investment. Sometimes it is called *valuation interest rate*. The formulas below are formally valid for $i > -1$ but the case $i \geq 0$ is the only realistic one. The vector $CF = (CF_0, \ldots, CF_T)^\top$ represents a *cash flow*. Values $CF_t > 0$ are *inflows* (amounts received) and $CF_t < 0$ are *outflows* (amounts paid, deposits, costs, etc.) Define the *discount factor* v corresponding to the interest rate i by $v := 1/(1 + i)$, the *discount* by $d = 1 - v$, and the *force of interest* δ by the relation $e^\delta = 1 + i$ or $\delta = \ln(1 + i)$. **Beware** of the fact that here symbol d is different from the same symbol d used from notational reasons in Part III where d will mean the discount function, or more generally the discount process. Summary of the notation:

$$v = \frac{1}{1 + i} \qquad d = 1 - v = \frac{i}{1 + i} \qquad e^\delta = 1 + i \qquad \delta = \ln(1 + i)$$

3.1   Present Value

One of the most important characteristics of a cash flow CF is its *present value*, PV, also called *net present value*, NPV. "Net" means that inflows and outflows at the same time t are added together and thus represented by a single number CF_t. If needed, the dependence of PV on CF and either i, v, or δ will be stressed:

$$(1) \quad PV(CF, i) := PV(CF, v) := PV(CF, \delta) := \sum_{t=0}^{T} \frac{CF_t}{(1 + i)^t} =$$

$$\sum_{t=0}^{T} CF_t v^t = \sum_{t=0}^{T} CF_t e^{-\delta t}.$$

Note that the present value is expressed in *currency units* like USD or CZK.

Let $\mathbb{L}^{T+1}$ be the linear vector space of cash flows, i.e., the space of finite sequences of maximum length $T + 1$. If the actual length of a cash flow is less than $T + 1$, we complete it by zeros. The present value is a linear function on $\mathbb{L}^{T+1}$ in the following sense: if α, $\beta \in \mathbb{R}$, CF_A, $CF_B \in \mathbb{L}^{T+1}$ then

$$(2) \qquad PV(\alpha CF_A + \beta CF_B, i) = \alpha PV(CF_A, i) + \beta PV(CF_B, i).$$

Let us consider the payments $CF_0, \ldots, CF_T$ at equally spaced time instants $0, \ldots, T$, again, but with different interest rates in the compounding periods

$i := (i_1, \ldots, i_T)$ where i_t is the interest rate applied in the period $(t - 1, t)$, $t = 1, \ldots, T$. Then the present value of the given cash flow is

$$(3) \quad PV(\textbf{CF}, i) = CF_0 + \frac{CF_1}{1 + i_1} + \cdots + \frac{CF_T}{(1 + i_1) \ldots (1 + i_T)} = \sum_{t=0}^{T} \frac{CF_t}{\prod_{j=1}^{t}(1 + i_j)}$$

where $\prod_{j=1}^{0} := 1$, by definition.

Finally, let us assume that the payments $CF_{t_1}, \ldots, CF_{t_T}$ take place in some general time instants $0 < t_1 < \cdots < t_T$ and the corresponding discount factor is v. Then

$$(4) \qquad\qquad PV(\textbf{CF}, v) = CF_{t_1} v^{t_1} + \ldots CF_{t_T} v^{t_T}.$$

This formula may be generalized to the case of an arbitrary starting (or valuating) date t_0. The present value related to this date is then

$$(5) \qquad\qquad PV(\textbf{CF}, v) = CF_{t_1} v^{t_1 - t_0} + \ldots CF_{t_T} v^{t_T - t_0}.$$

One must be careful with proper interpretation of time in this case, however.

3.1.1   Example. Consider the calendar convention Actual/360 and a cash flow $CF_{t_1}, \ldots, CF_{t_T}$ where the t_j's now represent dates, the compounding is annual with the discount factor $v = 1/(1 + i)$ and the starting date is t_0. Let $d(t_j, t_k)$ denote the number of days between the dates t_j, t_k. Then

$$(6) \qquad PV(\textbf{CF}, v) = CF_{t_1} v^{d(t_1, t_0)/360} + \cdots + CF_{t_T} v^{d(t_T, t_0)/360}.$$

With daily compounding with the interest rate $i^{(360)} = i$ p.a., the formula for the present value reads

$$(7) \quad PV(\textbf{CF}, i^{(360)}) = \frac{CF_{t_1}}{(1 + i^{(360)}/360)^{d(t_1, t_0)}} + \cdots + \frac{CF_{t_T}}{(1 + i^{(360)}/360)^{d(t_T, t_0)}}.$$

A cash flow often represents an investment opportunity. The dependence of the net present value of such a cash flow is of vital importance for investment decision making. For the first insight, the graphical representation of the dependence of the present value on the cost of capital (valuation interest rate) is of interest.

3.1.2   Example. Let us consider the cash flow from 1.4.1

$$(-90000, -15200, 45000, 60000, 25000, 22000, 270000)$$

at times $t = 0, \ldots, 6(= T)$. The PV of this cash flow in dependence on the interest rate i is plotted in Figure 3. Such a type of graph is called the *present value profile*.

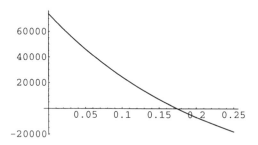

Figure 3: Present value of cash flow

3.1.3 Continuous Case

Speaking of interest rates, we were speaking of present values and future values with constant present values (investments) and a continuously varying force of interest. Here we deal with the case when even the respective cash flow changes continuously. For the sake of simplicity let us suppose that the starting point of time is set to 0 and the time at which the cash flow comes (received or paid, inflows or outflows) is t. Let us denote the cash flow coming for the period $(0, t)$ as $CF(t)$. It means that the net income for the corresponding period will be $CF(t)$, either with plus or minus sign. So the total payment made between (t_1, t_2) is $CF(t_2) - CF(t_1)$. Suppose that CF is differentiable so that the derivative $cf(t) = CF'(t)$ exists. Then the increment in income may be expressed as

$$(8) \qquad CF(t_2) - CF(t_1) = \int_{t_1}^{t_2} cf(s)ds.$$

Now we have to consider the time value of money. Between the time instants $t, t+dt$, dt being small enough, the total income is approximately $cf(t) \cdot dt$. Therefore, the present value of money received during the time interval $t, t + dt$ is $v(t)cf(t)dt$. So the present value of the cash flow over the whole period (t_1, t_2) is

$$(9) \qquad PV(CF, t_1, t_2) = \int_{t_1}^{t_2} v(t)cf(t)dt.$$

3.2 Annuities

Consider a series of T payments, each of amount 1 at times $1, \ldots, T$. Such a stream of payments is called *annuity immediate* (with payments at the end of the period). The present value of this cash flow for $i > 0$ is

$$(10) \qquad a_{\overline{T}|} := v + \cdots + v^T = \frac{1 - v^T}{i} = \frac{1 - (1+i)^{-T}}{i}$$

and often it is also called the *Present Value Interest Factor of an Annuity* abbreviated as $PVIFA_{i:T}$. For $i = 0$ we have $a_{\overline{T}|} = T$. Sometimes the interest rate is attached to symbol a: $a_{\overline{T}|i}$ or $a_{\overline{T}|i\%}$.

Consider again a series of T payments, each of amount 1 but now at times $0, \ldots, T - 1$. Such a stream of payments is called *annuity due* (with payments at the beginning of the period). The present value of this cash flow for $i > 0$ is

$$(11) \qquad \ddot{a}_{\overline{T}|} = 1 + v + \cdots + v^{T-1} = \frac{1 - v^T}{1 - v} = \frac{(1+i)(1 - (1+i)^{-T})}{i}.$$

Clearly, $\ddot{a}_{\overline{T}|} = T$ for $i = 0$. Further,

$$(12) \qquad \ddot{a}_{\overline{T}|} = (1+i)a_{\overline{T}|}, \qquad \ddot{a}_{\overline{T}|} = 1 + a_{\overline{T-1}|} \quad \text{for } T \geq 2.$$

For an infinite stream of constant payments of amount 1, the annuity is called *perpetuity* and if it is immediate or due, its present value is

$$(13) \qquad a_{\overline{\infty}|} = \frac{1}{i} \qquad \text{or} \qquad \ddot{a}_{\overline{\infty}|} = \frac{1+i}{i},$$

respectively.

3.3   Future Value

Let us consider the valuation date T, a cash flow $CF_0, \ldots, CF_T$, and the above interest rate characteristics i, v, δ. Then the future value is

$$(14) \quad FV(\boldsymbol{CF}, i) := CF_T + CF_{T-1}(1+i) + CF_{T-2}(1+i)^2 + \cdots + CF_0(1+i)^T =$$
$$\sum_{t=0}^{T} CF_t(1+i)^{T-t},$$

alternatively

$$(15) \qquad FV(\boldsymbol{CF}, v) = \sum_{t=0}^{T} CF_t v^{t-T}, \qquad FV(\boldsymbol{CF}, \delta) = \sum_{t=0}^{T} CF_t \, e^{\delta(T-t)}.$$

Obviously, $FV(\boldsymbol{CF}, i) = (1+i)^T PV(\boldsymbol{CF}, i)$ in this case.

In the case of varying interest rates we have

$$FV(\boldsymbol{CF}, i) = CF_T + CF_{T-1}(1 + i_T) + CF_{T-2}(1 + i_T)(1 + i_{T-1}) + \cdots +$$
$$CF_0(1 + i_T)(1 + i_{T-1}) \cdot \ldots \cdot (1 + i_1)$$

or

$$(16) \qquad FV(\boldsymbol{CF}, i) = \sum_{t=0}^{T} CF_t \prod_{j=t+1}^{T} (1 + i_j)$$

with $\prod_{j=T+1}^{T} := 1$.

In case of general time instants (see (4)) and a constant interest rate i we immediately get the obvious relationship

$$(17) \qquad FV(\boldsymbol{CF}, i) = (1 + i)^{t_T} PV(\boldsymbol{CF}, i).$$

3.3.1   Exercise. Modify the last result to the case of the calendar convention Actual/365.

Let us turn to the annuity immediate of an amount 1 and $i > 0$. The future value of this annuity is

(18) $s_{\overline{T}|} := 1 + (1+i) + \cdots + (1+i)^{T-1} = \dfrac{(1+i)^T - 1}{i} = (1+i)^T a_{\overline{T}|}.$

Analogously, for an annuity due, the future value is

(19) $\ddot{s}_{\overline{T}|} = \dfrac{(1+i)^T - 1}{d} = (1+i)^T \ddot{a}_{\overline{T}|}.$

Both $s_{\overline{T}|}$ and $\ddot{s}_{\overline{T}|}$ are equal to T for $i = 0$.

3.3.2   Exercise. Verify the following relations:

(20) $\ddot{s}_{\overline{T}|} = (1+i)s_{\overline{T}|} \qquad s_{\overline{T+1}|} = 1 + \ddot{s}_{\overline{T}|}.$

3.3.3   Remark

Other useful and frequently used relations:

(21) $\boxed{1 = i a_{\overline{T}|} + v^T \quad 1 = d\ddot{a}_{\overline{T}|} + v^T \quad (1+i)^T = i s_{\overline{T}|} + 1 \quad (1+i)^T = d\ddot{s}_{\overline{T}|} + 1}$

3.3.4   Exercise. Verify and give the interpretation of the preceeding formulas. (Hint: the first formula may be explained as the present value of a loan of amount 1 over the period $0, 1, \ldots, T$).

3.3.5   Remark

If the regular payments are all equal to PMT (abbreviation for PayMenT), then the corresponding present and future values are simply multiples by PMT of the corresponding a's and s's.

3.3.6   Remark (Equation of Value)

Due to technical and accounting reasons, the strict convention on the signs (inflows plus, outflows minus) leads to the following relations between the five variables involved, i.e., the present value PV, the future value FV, the interest rate i, the annuity PMT, and the number of periods T:

Annuity of amount PMT immediate.

(22) $$PV + PMT\, a_{\overline{T}|} + \frac{FV}{(1+i)^T} = 0$$

Annuity of amount PMT due.

$$(23) \qquad PV + PMT\, \ddot{a}_{\overline{T}|} + \frac{FV}{(1+i)^T} = 0$$

In the introductory courses, such a type of formulas is known as the *equation of value*. This approach is often used on financial calculators or in spread sheets. The user should carefully input the data with proper plus or minus signs for inflows and outflows, respectively.

3.3.7 Example (Installment Savings). Consider the investment of CZK 5000 in installment savings for 3 years at 3.6 per cent p.a., compounded monthly, so that $i^{(12)} = 0.036$. What will be the total of principal and interest at the end? Reasonably, installment savings represent an annuity due (payments at the beginning of the period) so that the equation of value (23) applies with $PV = 0$, $PMT = -5000$, $i = i^{(12)}/12 = 0.003$, $T = 36$. We have $\ddot{a}_{\overline{36}|} = 38.069$ so that $FV = 190349$. Compare this result with the case of 3 installment savings CZK 60000 at the beginning of every year with yearly compounding at the interest rate $i = 3.6$ per cent p.a. This results in the total savings $FV = 193273$. Give an explanation as an exercise.

3.3.8 Example and Exercise (Loans). Suppose you are able to repay CZK 5000 monthly for a 3 years' loan at $i^{(12)} = 7.2$ per cent p.a., compounded monthly. The question is, how much you can borrow under these conditions. Reasonably, the payments represent an annuity immediate (payments at the end of period) so that (22) applies to loan borrowing power $PV = ?$, $PMT = -5000$, $i = i^{(12)}/12 = 0.006$, $FV = 0$, $T = 36$. Since $a_{\overline{36}|} = 32.29$, you can borrow $PV = 161454$. In case you are able to pay CZK 60000 at the end of each year at the same interest but compounded yearly, you will obtain from (22) with $PMT = -60000$, $i = 0.072$, $T = 3$ that your loan borrowing power will decrease to $PV = 156885$. As an exercise, calculate PV under the same conditions if your balance (= remaining debt) is compounded monthly.

3.4 Internal Rate of Return (IRR)

In a simple Example 3.1.2 we have seen that depending on the interest rate the present value of a cash flow takes either positive or negative values. So the critical point is the value of the interest rate that equates the present value to zero. Consequently, we are motivated to define an *internal rate of return* (shortly *IRR*) as a solution to the equation

$$(24) \qquad PV(\boldsymbol{CF}, IRR) = \sum_{t=0}^{T} \frac{CF_t}{(1+IRR)^t} = 0.$$

In other words, *IRR* is defined as the interest rate (or the cost of capital) which equates the present value of inflows (incomes) to the present value of outflows (costs):

$$(25) \qquad \sum_{t:CF_t>0} \frac{CF_t}{(1+IRR)^t} = -\sum_{t:CF_t<0} \frac{CF_t}{(1+IRR)^t}.$$

The equivalent problem is to find a discount factor v such that

$$(26) \qquad PV(\boldsymbol{CF}, v) = \sum_{t=0}^{T} CF_T v^t = 0.$$

If $CF_T \neq 0$ then the last equation is an algebraic equation of degree T and hence it has T roots. Therefore, by the above definition, we have T internal rates of return. All the solutions can be easily obtained by standard numerical methods. Only real roots greater than -1 may have an economic meaning, however. Some authors define IRR as a positive solution to (24). But it can be simply demonstrated that some (rather strange) cash flows possess only positive IRR's with difficult economic interpretation. The cash flow $(-1000, 3600, -4310, 1716)$ has IRR's $0.1, 0.2, 0.3$, e.g. Nevertheless for "well-behaved" cash flows we have the following theorem:

3.4.1   Theorem. *Let $A_j = \sum_{t=0}^{j} CF_t$, $j = 0, 1 \ldots, T$, $A_0 \neq 0$, $A_T \neq 0$. Suppose that in the sequence $A_0, \ldots, A_T$ with zeros excluded the sign changes just once. Then there is exactly one positive IRR.*

Proof. We have the equation

$$\sum_{t=0}^{T} CF_t e^{-\delta t} = 0$$

with $e^{\delta} = 1 + i$. Since $CF_t = A_t - A_{t-1}$, $(A_{-1} := 0)$, the equation reads

$$(27) \qquad A_0 + \sum_{t=1}^{T} (A_t - A_{t-1}) e^{-\delta t} = 0.$$

Further,

$$\sum_{t=1}^{T} (A_t - A_{t-1}) e^{-\delta t} = \sum_{t=1}^{T} A_t e^{-\delta t} - \sum_{t=1}^{T} A_{t-1} e^{-\delta t} =$$

$$\sum_{t=1}^{T} A_t e^{-\delta t} - \sum_{t=0}^{T-1} A_t e^{-\delta(t+1)} = \sum_{t=1}^{T} A_t e^{-\delta t} - e^{-\delta} \sum_{t=0}^{T-1} A_t e^{-\delta t} =$$

$$(1 - e^{-\delta}) \sum_{t=1}^{T-1} A_t e^{-\delta t} + A_T e^{-\delta T} - e^{-\delta} A_0.$$

Thus (27) may be written as

$$(28) \qquad (1 - e^{-\delta}) \sum_{t=0}^{T-1} A_t e^{-\delta t} + A_T e^{-\delta T} = 0.$$

Without loss of generality suppose that $A_0 > 0$. Then there exists an index k such that $A_t \geq 0$, $t = 1, \ldots, k-1$, $A_k > 0$, $A_t \leq 0$, $t = k+1, \ldots T-1$, $A_T < 0$. Hence (28) becomes

$$(1 - e^{-\delta}) \left[\sum_{t=0}^{k} A_t e^{-\delta t} - \sum_{t=k+1}^{T-1} |A_t| e^{-\delta t} \right] - |A_T| e^{-\delta T} = 0$$

and after multiplication by $e^{\delta k}$ we get

$$(1 - e^{-\delta}) \left[\sum_{t=0}^{k} A_t e^{-\delta(t-k)} - \sum_{t=k+1}^{T-1} |A_t| e^{-\delta(t-k)} \right] - |A_T| e^{-\delta(T-k)} = 0$$

or

$$g_1(\delta) \left[g_2(\delta) - g_3(\delta) \right] - g_4(\delta) = 0,$$

say. All the g_i's are continuous, g_1, g_2 increasing, g_3, g_4 decreasing. Thus $g = g_1 [g_2 - g_3] - g_4$ is continuous and increasing. Moreover,

$$\lim_{\delta \to 0} g(\delta) = A_T < 0 \qquad \lim_{\delta \to +\infty} g(\delta) = +\infty$$

so that there is just one $\delta_0 > 0$ such that $g(\delta_0) = 0$ and $IRR = e^{\delta_0} - 1$ is the only positive IRR. $\square$

3.4.2 Remark and Example (Leasing). Financial *leasing* is an alternative form of financing. It takes a form of an agreement between two parties, the *lessee* and the leasing company called *lessor*. The lessee obtains the right to use a (usually real) asset for a period of time while the ownership of that asset remains with the lessor. At the end of the lease the ownership still remains with the lessor. But the residual (or salvage) value is usually negligible. There are many reasons for leasing, let us mention some of them. First, a company or an individual may not have money available to purchase the asset. This is often the case if the asset is too expensive like tanker or airplane. Second, there is a risk that the asset will become obsolete. Third, in most countries there exists a tax deduction advantage to promote investment. See [141], p. 512 for details. The following numerical example presents an analysis of leasing a car. The SKODA car priced CZK 227900 is leased under the following conditions: the lessee pays the sum of CZK 34185 immediately. Then the lessee pays (i) CZK 6943 =: PMT_{36} monthly for 36 months or (ii) CZK 6192 =: PMT_{42} monthly for 42 months. In both cases the payments are at the end of the month and the salvage value of the car is CZK 122. The question arises, what is the effective interest rate counted by the lessor. The IRR methodology gives the answer. We have $PV = -34185 + 227900 = 193715$, annuities with the minus sign given above, $T = 36$ and $T = 42$ months, respectively. Using a financial calculator or a spreadsheet program, we find the respective IRR's are $IRR_{36} = 1.446$ and $IRR_{42} = 1.449$ per cent monthly, so that $IRR_{36}^{12} = 17.35$ and $IRR_{42}^{12} = 17.39$ per cent p.a.

Investment projects represented by cash flows are called *normal* or *regular* if the payments change their sign just once, and are called *nonnormal* or *irregular* in the opposite case.

In the above definition of *IRR* we have implicitly supposed that the inflows from the project will be reinvested at the same interest rate, i.e., *IRR*. More often, the inflows are reinvested at the interest rate equal to the current cost of capital k, say. We can overcome this problem by a modification of the definition of *IRR* following the principle:

$$(29) \qquad PV(\text{outflows}, k) = PV(FV(\text{inflows}, k), MIRR)$$

where MIRR is called *modified rate of return*. In symbols, MIRR is defined by the equation

$$(30) \qquad -\sum_{t:CF_t<0} \frac{CF_t}{(1+k)^t} = \frac{1}{(1+MIRR)^T} \sum_{t:CF_t>0} CF_t(1+k)^{T-t}.$$

It is obvious that in this case (given k) MIRR can be expressed explicitly. Also note that for $k = IRR$ we have $MIRR = IRR$.

Another modification of IRR makes use of different interest rates for outflows and inflows, i.e., the different costs of investment and reinvestment capital, k_O and k_I, respectively. The *modified rate of return* MIRR (we use the same notation) is then defined by

$$(31) \qquad -\sum_{t:CF_t<0} \frac{CF_t}{(1+k_O)^t} = \frac{1}{(1+MIRR)^T} \sum_{t:CF_t>0} CF_t(1+k_I)^{T-t}.$$

MIRR can be explicitly calculated again.

Note that sometimes this idea is also used for the valuation of cash flows if different valuation interest rates are used for outflows and inflows. Using the above notation, the present value is expressed as

$$(32) \qquad PV(\boldsymbol{CF}, k_O, k_I) = \sum_{t:CF_t<0} \frac{CF_t}{(1+k_O)^t} + \sum_{t:CF_t>0} \frac{CF_t}{(1+k_I)^t}.$$

3.5   Duration

The *duration* is defined as the time-weighted average of the discounted payments:

$$(33) \qquad D(\boldsymbol{CF}, v) = \frac{\sum_{t=0}^{T} tCF_t v^t}{\sum_{t=0}^{T} CF_t v^t} = \frac{1}{PV(\boldsymbol{CF}, v)} \sum_{t=0}^{T} tCF_t v^t.$$

Duration is expressed in time units. So if the payments are semiannual, for instance, the duration is expressed in halves of year. It is also called *discounted mean term of the cash flow*. We have

$$\frac{\partial PV(\boldsymbol{CF}, v)}{\partial v} = \frac{1}{v} \sum_{t=0}^{T} tCF_t v^t = \frac{1}{v} D(\boldsymbol{CF}, v) PV(\boldsymbol{CF}, v)$$

and thus the duration may be expressed as

$$(34) \qquad D(\boldsymbol{CF}, v) = \frac{v}{PV(\boldsymbol{CF}, v)} \frac{\partial PV(\boldsymbol{CF}, v)}{\partial v}.$$

In economics the last expression is known as elasticity so that we may interpret the duration as an *elasticity of the net present value with respect to the discount factor*. An alternative formula for the duration expressed in terms of the interest rate reads

$$(35) \qquad D(\boldsymbol{CF}, i) = -\frac{1+i}{PV(\boldsymbol{CF}, i)} \frac{\partial PV(\boldsymbol{CF}, i)}{\partial i}.$$

From the above expressions it follows that the duration may serve either as a measure of the sensitivity of the cash flow to the interest rate or as the duration of the corresponding investment project. The first interpretation will become clear if we write the first few terms of the Taylor expansion of the relative increment of the present value of the given cash flow as a function of the interest rate; the derivatives are taken with respect to the second argument:

$$(36) \qquad \frac{PV(\boldsymbol{CF}, i + \Delta i) - PV(\boldsymbol{CF}, i)}{PV(\boldsymbol{CF}, i)} =$$
$$\frac{PV'(\boldsymbol{CF}, i)}{PV(\boldsymbol{CF}, i)} \Delta i + \frac{1}{2} \frac{PV''(\boldsymbol{CF}, i)}{PV(\boldsymbol{CF}, i)} (\Delta i)^2 + \dots \quad \approx -\frac{1}{1+i} D(\boldsymbol{CF}, i) \Delta i.$$

Note that duration, unlike the present value, **is not** a linear function of the $\boldsymbol{CF}$'s. To overcome this disadvantage sometimes the *dollar duration* is used:

$$(37) \qquad D_\$(\boldsymbol{CF}, v) = \sum_{t=0}^{T} t C F_t v^t.$$

In literature and applications we can also meet the *modified duration*:

$$D_{\mathrm{mod}}(\boldsymbol{CF}, v) = v \cdot D(\boldsymbol{CF}, v)$$

or, in terms of i

$$D_{\mathrm{mod}} = -\frac{1}{PV(\boldsymbol{CF}, i)} \frac{\partial PV(\boldsymbol{CF}, i)}{\partial i}.$$

3.6 Convexity

A finer measure of the sensitivity of a cash flow to the interest rate is the *convexity*:

$$(38) \qquad C(\boldsymbol{CF}, v) = \frac{\sum_{t=0}^{T} t(t+1) C F_t v^t}{\sum_{t=0}^{T} C F_t v^t}.$$

Convexity is expressed in squared time units. If the payments are accomplished semiannually, convexity is expressed in [year^2/4], e.g. Taking into account that

$$PV''(\boldsymbol{CF}, i) = \frac{1}{(1+i)^2} C(\boldsymbol{CF}, i) PV(\boldsymbol{CF}, i)$$

we can substitute in (36) and get a more precise formula for the relative increment of the present value in the form

(39) $\dfrac{PV(\boldsymbol{CF}, i + \Delta i) - PV(\boldsymbol{CF}, i)}{PV(\boldsymbol{CF}, i)} \approx$

$$-\frac{1}{1+i} D(\boldsymbol{CF}, i)\Delta i + \frac{1}{2(1+i)^2} C(\boldsymbol{CF}, i)(\Delta i)^2 .$$

In literature we can find a slightly different definition of the convexity, as an analogue to the modified duration:

$$C_{\mathrm{mod}}(\boldsymbol{CF}, i) = \frac{PV''(\boldsymbol{CF}, i)}{PV(\boldsymbol{CF}, i)}.$$

Then the equation for the relative change of the present value may be expressed in terms of the modified measures as

(40) $\dfrac{PV(\boldsymbol{CF}, i + \Delta i) - PV(\boldsymbol{CF}, i)}{PV(\boldsymbol{CF}, i)} \approx -D_{\mathrm{mod}}(\boldsymbol{CF}, i)\Delta i + \frac{1}{2} C_{\mathrm{mod}}(\boldsymbol{CF}, i)(\Delta i)^2.$

3.7   Comparison of Investment Projects

As usual, investment projects will be represented by the corresponding expected cash flows. Since the future cash flows are uncertain, the results of decision making process are also uncertain. The detailed qualified analysis may reduce uncertainty, however.

We will deal with a set of competing projects. The decision maker may accept one or more projects and may even decide not to accept any. The projects are said to be *mutually exclusive* if at most one of the involved projects can be accepted. And they are said to be *independent* if an arbitrary number of the competing projects (including none of them) can be accepted.

There are two broad classes of investment projects that often arise in practise. In the first case, the investors use their own capital for the initial investment and they obtain incomes generated by the initial investment in successive periods. Such projects are characterized by negative payments in the initial period(s) and positive ones afterwards. Call them *class I projects*. In the second case, the investors take a loan at the beginning, make an investment, and then they acquire the benefits and also should pay back the loan. Such projects are characterized by positive payments in the initial period(s) and negative ones afterwards. Call them *class II projects*.

There is a variety of methods for decision making and we will mention only some of the principles. All the methods start with a careful analysis of the expected stream of payments including dividends, interest obtained or paid, salvage value of the assets at the end of the project's life, etc. The cost of capital (the valuation interest rate) should take into account the riskness (uncertainty) of the project.

3.7.1 Profitability Index

A simple indicator for a class I project $\boldsymbol{CF} = (CF_0, CF_1, \ldots, CF_T)$ is the *profitability index* defined by

$$(41) \qquad PI(\boldsymbol{CF}, i) = -\frac{1}{CF_0} PV((CF_1, \ldots, CF_T), i).$$

This measure seems to be trivial but in fact it is, in some sense, equivalent to the measures based on the present value profile as we will see later. Among competing projects we select those with highest profitability indexes greater than one; we select none of them if all *PI*'s are less than one.

3.7.2 Payback Method

Another simple and rough method is the *payback method* applied again to class I projects. It is based on the *payback period* that is the number of periods required to recover the initial outflows. Formally, let us keep assumptions of Theorem 3.4.1. For a class I project we have $A_0 < 0$. Let k be the first index such that $A_k > 0$. Then the *payback period* is defined by

$$(42) \qquad PB(\boldsymbol{CF}) = k - 1 - \frac{A_{k-1}}{CF_k}.$$

Here $k - 1$ is the period just preceeding the full recovery, $-A_{k-1}$ is the uncovered cost at the beginning of this period, and CF_k (obviously positive) is the payment in the recovering period. If such a k does not exist, we set formally the payback period to infinity. Based on the payback method, we select the project(s) with the shortest payback period, or none of them if their payback periods all equal infinity.

A little better method based on this idea is the so called *discounted payback method*. Let i be a properly chosen project's cost of capital and define $A_j^{(i)} = \sum_{t=0}^{j} CF_t / (1+i)^t$. Assume again $A_0^{(i)} < 0$ and k the first index such that $A_k^{(i)} > 0$. Then the *discounted payback period* is defined as

$$(43) \qquad PB(\boldsymbol{CF}, i) = k - 1 - \frac{A_{k-1}^{(i)}}{CF_k / (1+i)^k}.$$

If such a k does not exist we set formally the discounted payback period to infinity. The decisions based on the discounted payback method are the same as in case of the usual payback method.

3.7.2.1 Exercise and Problem. Analyze and try to prove the following conjecture. For a class I project of length T ($A_T > 0$), the discounted payback period approaches T as the interest rate approaches the internal rate of the project, $i \to IRR$.

3.7.3 Methods Based on the Present Value Profile

Typically, for class I projects the present value is a decreasing (and often also convex) function of the valuation interest rate i and the opposite is true for class II projects; the present value is an increasing (and often concave) function of i. However, this is not the rule as shown in the following counterexample.

3.7.3.1 A Counterexample

Consider an artificial cash flow $CF_Z = (-6, -10, -4, -8, -3, -5, 18.5, 18.5)$. The assumptions of Theorem 3.4.1 are fulfilled. The only IRR is 0.006372. But $PV(CF_Z, i)$ is decreasing for $i < 0.39$ and increasing for $i > 0.39$.

Hence the investor should take care of the individual present value profile, i.e., the graph of the present value in dependence on the interest rate involved.

The leading rule is simple; for a given i accept the project if its present value at this interest rate i is positive:

$$\boxed{\text{Accept if} \quad PV(CF, i) > 0} .$$

For class I projects, the criterion of positive present value is equivalent to $PI(CF, i) > 1$. In case of independent projects we select all the projects with the positive present values at the given interest rate. If the projects are mutually exclusive we select that with the highest present value. If we investigate a set of projects which are mutually exclusive dependent on the valuation interest rate we should select the project that is determined by the upper envelope of the present value profiles.

For one project, the critical point is IRR. If PV is a decreasing function of i then we accept the project if the valuation interest rate is less than IRR and reject it otherwise. Analogously, if PV is an increasing function of i, we accept the project if the valuation interest rate is greater than IRR. For projects which do not possess a monotonous present value profile, we should perform a more careful analysis.

For two or more projects, the critical points are not only the IRR's of the individual projects but also their crossover rates. A *crossover rate* of two projects is such an interest rate for which the present values of the two projects are equal. Formally, let us consider two projects CF_A and CF_B. The crossover rate i_{AB} is defined as a solution to the equation

$$PV(CF_A, i_{AB}) = PV(CF_B, i_{AB}).$$

Obviously, there may be more than one solution so that we must select that one with a reasonable economic interpretation. Since the present value is a linear function on the space of cash flows, we see that the crossover rate i_{AB} is in fact the internal rate of return determined by the difference between the two projects, IRR_{A-B}:

$$PV(CF_A - CF_B, IRR_{A-B}) = 0.$$

In the neighborhood of the crossover rate the investor should take care and carefully study also the sensitivity of the present value profiles with respect to the interest rate. This is best done by looking on the duration and possibly on the convexity. Such an analysis will be better understood from the example.

3.7.3.2 Example. Let us consider five projects:

(1) A: $CF_A = (-1000, 300, 500, 200, 100)$

(2) B: $CF_B = (-1000, 47, 47, 47, 1047)$
(3) C: $CF_C = (-851.18586, 281.0005, 170.39716, 300, 200)$
(4) D: $CF_D = (-600, -500, -300, 400, 500, 600)$
(5) E: $CF_E = (1200, -400, -300, -200, -400)$.

Projects A, B, C, D are class I projects while E is a class II project. CF_B represents the cash flow of a four years coupon bond purchased for the par value 1000 giving the holder yearly coupons of 47 with redemption value 1000. The present value profiles of these projects are shown in Figure 4. Visually the present value profiles of the projects A and C coincide. The payback periods for the first four projects are

$$PB(CF_A) = 3.00 \quad PB(CF_B) = 3.82 \quad PB(CF_C) = 3.50 \quad PB(CF_D) = 4.83$$

and the discounted payback periods for two selected interest rates ($i = 0.02$, $i = 0.04$) are: $PB(CF_A, 0.02) = 3.40$, $PB(CF_B, 0.02) = 3.89$, $PB(CF_C, 0.02) = 3.70$, $PB(CF_D, 0.02) = 4.99$ and $PB(CF_A, 0.04) = 3.84$, $PB(CF_B, 0.04) = 3.97$, $PB(CF_C, 0.04) = 3.92$, $PB(CF_D, 0.04) = +\infty$. In case of independent projects, based on the discounted payback method we accept projects A, B, C, D if $i = 0$ or $i = 0.02$. For $i = 0.04$ we accept A, B, C and reject D. If the projects are mutually exclusive, we accept only A for all three values of i.

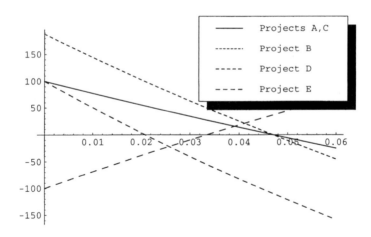

Figure 4: Present values of 5 projects

The present value is a decreasing function of i for projects A, B, C, D, and an increasing function for project E. Thus the acceptance region depends on the corresponding IRR's:

$$IRR_A = 0.0471 \quad IRR_B = 0.0470 \quad IRR_C = 0.0472 \quad IRR_D = 0.0208 \quad IRR_E = 0.0333.$$

Consider first the case of independent projects. We accept A, B, C, D for $i \leq 0.021(= IRR_D)$. For $0.021 < i \leq 0.033(= IRR_E)$ we accept A, B, C. For $0.033 < i < 0.047$ (approximately) we accept A, B, C, E; and we accept only E for

$i > 0.047.$

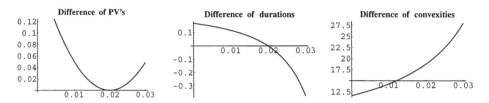

Figure 5: Characteristics of A and C

Second, consider mutually independent projects A, C, E only. Since the projects A and C have almost identical present value profiles, we must look first at the difference of their present values. In Figure 5 we have plots of $PV(\mathbf{CF}_C, i) - PV(\mathbf{CF}_A, i)$, $D(\mathbf{CF}_C, i) - D(\mathbf{CF}_A, i)$, and $C(\mathbf{CF}_C, i) - C(\mathbf{CF}_A, i)$. We see that $PV(\mathbf{CF}_C, i) \geq PV(\mathbf{CF}_A, i)$ and that the difference is negligible. We also have $PV(\mathbf{CF}_A, 0.02) = PV(\mathbf{CF}_C, 0.02)$ and $D(\mathbf{CF}_A, 0.02) = D(\mathbf{CF}_C, 0.02)$. Since the convexities fulfil the inequality $C(\mathbf{CF}_C, i) > C(\mathbf{CF}_A, i)$ we can decide in favor of project C against A. Further, the crossover rate for projects C and E is $IRR_{C-E} = 0.0391$. Thus to summarize, for $i \leq 0.0391$, we accept C and for $i > 0.0391$ we accept E, among the candidates A, C, E. If we consider all the five projects, then we obviously select B for $i < IRR_{B-E} = 0.041$ and E for greater values of i.

3.7.4   Internal Value

Suppose that the cash flow in question depends also on another variable or parameter y say, $\mathbf{CF} = \mathbf{CF}(y)$. For decision making, an important measure is the value of $y = y(i)$ such that the present value for a given interest rate i is zero. Call this value the *internal value* of the cash flow and denote it by HIV. (HIV has been introduced in [83] but we admit that such a simple indicator might have been known before.) Mathematically, HIV is defined implicitly by the relation

(44)
$$\sum_{t=0}^{T} \frac{CF_t(\mathrm{HIV}(i))}{(1+i)^t} = 0.$$

Often, the dependence of the present value on y is simple, for instance linear or quadratic. Hence, for a fixed interest rate, the analysis of the present value profile becomes more simple. Application of HIV is many-sided. Particularly, HIV is useful in valuation of all transactions where the foreign exchange rate appears, like currency swaps. In this case $y = FX$, the foreign exchange rate. The HIV can also be employed for the risk analysis of loans payable in foreign currencies or cash flows dependent on interest rates like LIBOR, etc. If more than one variable influence the cash flow involved, the above definition is still of use. The analysis is more complex in this case, however. Also a two-dimensional analysis if both the interest rate i and y vary is a rather complex task and needs a further research and analysis of particular situations.

3.8   Yield Curves

Generally, a *yield curve* plots interest rates paid on interest bearing securities against the time to maturity. Such a plot makes sense only for a class of comparable securities. Thus we may plot yield curves for government zero coupon bonds for maturities 1, 3, 6, 9, 12 months getting a completely different picture for AA rated firm's bonds for same maturities. Thus we should take into account the risk factors (cf. decomposition of interest rate) and also comparable taxation conditions.

Even for the same type of securities (like T-bills), the shape of the yield curve differs in time, i.e., the shape is different in years 2000 and 2001, say, *ceteris paribus*. This feature may be explained by many factors, like the change in spot riskless rate, inflation, and other exogenous factors. Another important feature is the internal need of the issuer for short, medium, or long financial funds.

Another problem arising with a yield curves' presentation is that the yields may be either *declared* or *actually observed* on the market. Here, by *declared yields* we mean the promised coupon rates for usual fixed coupon bonds while the *actually observed yields* are derived from the spot market price of the respective security, see I.4 for the calculation.

There is an obvious connection between the yield curve and the term structure of interest rates (cf. 2.5); for a given type of security (or a group of similar securities) with different maturities and for a given particular date t, the yield curve is the plot of the spot rates $_tR_n$, $n = 1, 2, \ldots$. The difference $_tR_n - {}_tR_1$ is called *yield spread*. Sometimes the *forward-rate curve* calculated from (2.5.13) is plotted.

A typical shape of the yield curve is *upward-sloping*, which simply means that the corresponding function is increasing and often concave. Such a yield curve is called *normal yield curve*. On the contrary, the yield curve which is *downward-sloping* (decreasing and often convex) is called *inverted yield curve*. Another shape often arising in practise is a *humped curve*; the yield curve increases at first and then decreases for longer maturities. Rarely we can meet a flat, i.e., *constant yield curve* or U-shaped curves. However, rather strange images, different from the above mentioned, can be met with in practise.

The shape and magnitude of the yield curve depend on many factors. Most important are the risk factors, the liquidity preference, and the expected inflation. Increasing risk factors (mainly default risk) cause approximately parallel upper shift of the yield curve. The higher the liquidity preference, the higher the liquidity premium for lending for longer time periods. With increasing expected inflation in future periods the longer-term rates become higher and vice versa. See 2.3 and 2.4 for explanation.

For financial decision making and also for analysis we often need yields for maturities which are not available on the market. Thus we must construct them from existing market data. To this purpose one may use purely numerical approaches like linear interpolation, e.g. Another recommended approach is based on regression models. Suppose that we have N comparable fixed or zero coupon bonds 1, $\ldots, N$ with maturities $T_1, \ldots, T_N$ and observed yields $y_1, \ldots, y_N$, respectively. The postulated parametric regression model is (see [55], e.g.)

$$(45) \qquad\qquad y_n = g(T_n; \boldsymbol{\theta}) + \varepsilon_n, \qquad n = 1, \ldots, N,$$

where the hypothetical yield curve g of a known analytical form depends on an unknown vector parameter $\boldsymbol{\theta}$ which is to be estimated, and ε_n are disturbances with zero means. The estimate $\widehat{\boldsymbol{\theta}}$ of $\boldsymbol{\theta}$ is obtained as an argument of

$$(46) \qquad \min_{\boldsymbol{\theta}} \sum_{n=1}^{N} |y_n - g(T_n; \boldsymbol{\theta})|^{\gamma}$$

for a properly chosen γ ($\gamma = 2$ for the least squares method and $\gamma = 1$ for the absolute deviation criterion, e.g.). There is also a variety of possible choices for the analytical form of g. Having the estimate $\widehat{\boldsymbol{\theta}}$, we may estimate the yield for a nonobserved maturity $T \neq T_n$, $n = 1, \ldots, N$ as

$$(47) \qquad \widehat{y}_T = g(T; \widehat{\boldsymbol{\theta}}).$$

One of the simplest forms of g is a *polynomial function* of a small degree K

$$(48) \qquad g(t; \boldsymbol{\theta}) = \sum_{k=0}^{K} \theta_k t^k$$

which leads to a polynomial regression. For $K = 3$ the corresponding function is a cubic function and 4 parameters are to be estimated. Due to bad experience with polynomial regression, other types of g are recommended.

One of the successful and recently frequently used models is the model of *cubic splines*. Assuming $T_1 < T_2 < \cdots < T_N$, we consider functions g such that (i) g is a piecewise cubic function, i.e., g equals

$$(49) \qquad g_n(t) := \alpha_n + \beta_n t + \gamma_n t^2 + \delta_n t^3 \quad \text{for } t \in [T_{n-1}, T_n], \quad n = 2, \ldots, N,$$

(ii) g is twice continuously differentiable everywhere; this is (together with (i)) equivalent to

$$g_n(T_n) = g_{n-1}(T_n), \quad g_n'(T_n) = g_{n-1}'(T_n), \quad g_n''(T_n) = g_{n-1}''(T_n), \quad n = 2, \ldots, N.$$

We then choose the function $\widehat{g}$ from this class that minimizes a combination of the residual sum of squares and the integrated squared 2nd derivative of g:

$$\widehat{g} = \operatorname*{argmin}_{g} \Big\{ \sum_{n=1}^{N} (y_n - g(T_n))^2 + \lambda \int_{T_1}^{T_N} (g''(t))^2 dt \Big\}$$

with a smoothing constant $\lambda > 0$. The resulting $\widehat{g}$ represents a compromise between fit of data and smoothness of the fitting curve. Values of the smoothing constant λ cover ordinary least squares fitting by a straight line ($\lambda \to \infty$) as one extreme, and pure numerical interpolation by a piecewise cubic functions ($\lambda = 0$) as the other one. Details of the method together with an algorithm can be found in [150].

Another flexible model has been treated by Bradley and Crane in [24] (see also Example II.5.4.4):

$$(50) \qquad g(t; \alpha, \beta, \gamma) = \alpha t^{\beta} e^{\gamma t}.$$

This model should be taken with care, however, because with wide range of observed maturities severe discrepancies may appear, see the Example below and Figure 8 in II.6. After the logarithmic transform and the reparametrization α^* the last equation becomes

$$(51) \qquad\qquad \ln g(t; \alpha^*, \beta, \gamma) = \alpha^* + \beta \ln t + \gamma t$$

which is linear in parameters and these may be simply estimated by ordinary least squares method.

Two alternative techniques of modeling the term structure of a coupon bond will be discussed in 4.1.3.

3.8.1   Example. Consider declared interest rates for term deposits of the Czech saving company as in February 1999:
Maturity (in days) 7   14   30   60   90   120   150   180 210 240 270 290
Interest rate (p.a.)   5.4 5.4 6.2 6.1 6.1 6.00 6.00 5.9 5.9 5.9 5.9 5.8 .
The yield curve is humped. Let us make a comparison of three estimating procedures: (i) fitting by a cubic function, (ii) fitting by cubic splines, (iii) fitting by (50). For (i), (iii) there are no alternatives while in case (ii), we have experimentally chosen the smoothing constant as to get the best fit from the optical point of view. The estimated curves along with the original rates are plotted in Figure 6. We see that for such a pattern it is difficult to fit the data satisfactorily by simple analytic models. Particularly, fitting by the cubic function may lead to a dangerous conclusion, i.e., that for longer maturities the yield curve rises again. This is not the only exception. Another example (not presented here) shows that even the polynomial interpolation of a very nice smooth yield curve observed at discrete times (years) 1, ... ,30, resembling a parabola, by a polynomial of the degree 29 reveals unrealistic values for some points within the intervals. We strongly recommend not to use the polynomial fitting procedure.

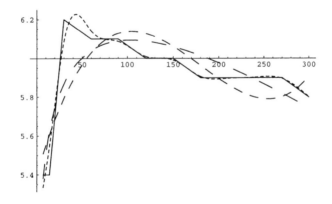

Figure 6: Fitting the yield curve
original rates (broken line)   cubic splines (- - - -)
cubic function (– – –)   Bradley-Crane (——— —)

I.4   RETURN, EXPECTED RETURN, AND RISK

return, rate of return, random walk hypothesis, Black-Scholes model, risk, parametric value at risk (VaR), nonparametric VaR

Warning

Some symbols used in the following text are very popular both in financial and financial mathematics literature, unfortunately with a different meaning. Particularly, symbols r or R may serve as typical representatives. Sometimes R or r means a return, sometimes the rate of return, sometimes the *expected rate of return*, sometimes the interest rate, etc. Also, there is an ambiguity in distinguishing between a random variable and the expected value of it. In financial literature, a random variable is often stressed by the wave, like $\widetilde{X}$, and the expected value is simply X, while in mathematics X is reserved for a random variable and EX stands for its expected value. The reader is politely asked to pay attention what the respective symbols mean.

4.1   Return

The concept of return should be considered in a dynamic setup; by *return* of a financial asset we mean the difference between the wealth (in monetary units) at the end and the beginning of the period under consideration. Consequently, this leads to the following definition of the *rate of return ROR*, say:

$$ROR = \frac{\text{wealth at the end of the period} - \text{wealth at the beginning of the period}}{\text{wealth at the beginning of the period}}.$$

Suppose that the (market) price of the underlying asset (security) at time t is P_t. Following the above idea, we may simply define the rate of return as

$$(1) \qquad\qquad R_t = \frac{P_{t+1} - P_t}{P_t}.$$

As in the case of interest rate, we will alternatively use a percentage or a decimal form of the rate of return; $R_t = 0.1$ and $R_t = 10\%$ mean the same. Taking into account the accumulation (multiplicative) effect and an analogy with the force of interest, we can define another measure $r_t^\star$ as a rate of return by

$$(2) \qquad\qquad 1 + R_t = \frac{P_{t+1}}{P_t} =: \exp(r_t^\star),$$

that is,

$$(3) \qquad r_t^\star = \ln(1 + R_t) = \ln\frac{P_{t+1}}{P_t} = \ln P_{t+1} - \ln P_t =: p_{t+1} - p_t,$$

by definition. Note that p_t's as defined above are often called *logarithmic prices*. For small values of the rate of return, $r_t^\star$ does not differ from R_t too much. By

Taylor expansion, $\ln(1 + R_t) = R_t - R_t^2/2 + \ldots$, so that the difference is of order $O(R_t^2)$. Thus, for $R_t = 0.05$ we have $r_t^\star = 0.04879$, e.g. For higher values of R_t the difference increases. The rate of return R^T for the time horizon T is then defined by the relation

$$(4) \qquad 1 + R^T = \prod_{t=1}^{T}(1 + R_t) = \exp\left(\sum_{t=1}^{T} r_t^\star\right) = \frac{P_T}{P_0}.$$

In case of securities, let us denote by R_t the rate of return for the period t, P_t the (market) price of the respective security at the end of period t, and D_t the dividend paid for the time interval $[t, t+1]$. Then

$$(5) \qquad R_t = \frac{P_{t+1} + D_{t+1} - P_t}{P_t} = \frac{D_{t+1}}{P_t} + \frac{P_{t+1} - P_t}{P_t}.$$

The first part of the rate of return, D_{t+1}/P_t, represents the so called *dividend yield*, or in case of coupon bonds, *coupon yield*, while the second part, $(P_{t+1} - P_t)/P_t$, represents the *capital yield*. Note that the dividend is usually paid rarely in comparison with the time period considered, once, or twice a year, say. For a correct expression of the rate of return we should incorporate the corresponding part of the dividend into the formula (1). If we consider the time period of one week with the yearly paid dividend D, we substitute $D_t := D/52$, e.g. In the theory, we must also distinguish between expected returns (ex ante) based on subjective probabilities and returns coming from historical data (ex post).

For an asset paying no dividends the rate of return becomes

$$(6) \qquad R_t = \frac{P_{t+1} - P_t}{P_t} = \frac{P_{t+1}}{P_t} - 1.$$

4.1.1   Random Walk Hypothesis

Under the *random walk hypothesis* the logarithmic prices follow the model

$$(7) \qquad p_{t+1} - p_t = \mu + \varepsilon_{t+1}, \quad t = 0, 1, \ldots$$

where ε_t's are either uncorrelated (weak form) or independent (strong form) identically distributed random variables (shortly iid for the latter case) with $E\,\varepsilon_t = 0$ and $\mathrm{var}\,\varepsilon_t = \sigma^2$, and μ represents a *drift* or *trend*. Next we will suppose that the ε_t's are iid. It follows that the $r_t^\star$'s are iid random variables under the random walk hypothesis. Since for $T \in \mathbb{N}$

$$p_T = p_0 + \mu T + \sum_{t=1}^{T} \varepsilon_t$$

we have $E\,p_T = p_0 + \mu T$ and $\mathrm{var}\,p_T = \sigma^2 T$. In the stationary case $\mu = 0$ there are only random fluctuations about the initial logarithmic price p_0. For the original prices P_t's we have

$$P_{t+1} = P_t e^{\mu + \varepsilon_{t+1}}$$

or

$$P_T = P_0 \exp\left(\mu T + \sum_{t=1}^{T} \varepsilon_t\right).$$

The ratios P_1/P_0, P_2/P_1, ... ,P_T/P_{T-1} are therefore iid random variables. Also the returns R_0, ... ,R_T are iid under the above assumptions. The case of normally distributed ε's will be treated in the next Section.

Sometimes it is supposed that the original price process is driven by

$$P_{t+1} - P_t = \mu + \varepsilon_{t+1}$$

with analogous assumptions on ε_t's.

Sometimes even an unrealistic assumption is made that the P_t's are independent identically distributed. However, the independence of P_t's does not generally guarantee the independence of the returns R_t's. Just look on the covariance between two successive rates of return:

(8) $\mathrm{cov}\,(R_t, R_{t-1}) = \mathrm{cov}\,(P_{t+1}/P_t, P_t/P_{t-1}) =$

$$E(P_{t+1}/P_{t-1}) - E(P_{t+1}/P_t)E(P_t/P_{t-1}) =$$

$$(\text{independence}) = EP_{t+1}E(1/P_{t-1}) - EP_{t+1}EP_tE(1/P_t)E(1/P_{t-1}) =$$

$$(\text{identically distributed}) = EP_tE(1/P_t)(1 - EP_tE(1/P_t))$$

which could hardly be zero.

4.1.2 A Simple Model for Price Development

The model presented in this Section serves as a background for more complicated models like Black-Scholes model for option valuation etc. We need only two assumptions concerning an efficient market: (i) all the past history of the price development is reflected in the present price; (ii) the response of the market on any new piece of information is immediate. Assumption (i) resembles a Markov property.

Let $\Delta t > 0$ and denote $\Delta P := P_{t+\Delta t} - P_t$, $P := P_t$ for a moment, P_0 being a starting price. In the model it is supposed that the return, $\Delta P/P$ in our case, can be decomposed into a deterministic and a stochastic part in the following way:

(9)
$$\boxed{\frac{\Delta P}{P} = \mu\Delta t + \sigma\Delta W.}$$

Here the first term $\mu\Delta t$ is the deterministic part, μ is called *drift* or a *trend coefficient* while the second part is a stochastic term with so called *volatility, standard error* or *diffusion* σ and $\Delta W := W(t + \Delta t) - W(t)$ standing for the increment of a standard Wiener process. In more general models, both μ and σ may be also functions of P and t. Recall that the *Wiener process* $\{W(t), t \geq 0\}$ is a stochastic process with continuous trajectories such that $W(0) = 0$ with probability 1, for s, t positive the distribution of $W(t) - W(s)$ is normal $N(0, |t - s|)$, and for any $0 < t_0 < t_1 < \cdots < t_n < \infty$ the random variables $W(t_0)$, $W(t_1) - W(t_0)$,

... ,$W(t_n) - W(t_{n-1})$ (the *increments*) are independent. See Part III for more details. Since the distribution of ΔW is $N(0, \Delta t)$, (9) may be written in the form

$$(10) \qquad \Delta P = \mu P \Delta t + \sigma P \varepsilon \sqrt{\Delta t}$$

where ε is an $N(0,1)$ random variable so that the return $\Delta P/P$ possesses the normal distribution $N(\mu \Delta t, \sigma^2 \Delta t)$. This formula is useful for discrete modeling and simulation. Formally, for $\Delta t \to 0$, we obtain the *stochastic differential equation* (SDE, see Theorem 12.6, p. 223 in [93])

$$(11) \qquad \boxed{\frac{dP}{P} = \mu dt + \sigma dW.}$$

This equation describes the so called *geometrical Brownian motion*, see Part III 2.2.12. We will now make use of Itô formula to characterize the development of logarithmic prices. For $f = f(P,t)$ the Itô formula reads (see Part III, Corollary 2.2.9)

$$(12) \qquad df = \left(\frac{\partial f}{\partial P} \mu P + \frac{1}{2} \frac{\partial^2 f}{\partial P^2} \sigma^2 P^2 + \frac{\partial f}{\partial t} \right) dt + \frac{\partial f}{\partial P} \sigma P dW.$$

Put $f(P) := \ln P$. The first and second derivatives of f with respect to P are $1/P$ and $-1/P^2$, respectively. After some algebra we obtain the solution to (11) for the logarithmic prices:

$$(13) \qquad d \ln P = (\mu - \tfrac{1}{2}\sigma^2)dt + \sigma dW.$$

The discrete version of the last equation is (recall that $\ln P = p$)

$$(14) \qquad \ln P_{t+\Delta t} - \ln P_t = \ln(P_{t+\Delta t}/P_t) = p_{t+\Delta t} - p_t = (\mu - \tfrac{1}{2}\sigma^2)\Delta t + \sigma \varepsilon \sqrt{\Delta t}$$

with ε distributed as $N(0,1)$ again.

The solution to the SDE for the price process with given initial value P_0 is

$$(15) \qquad P_t = P_0 \exp\{(\mu - \tfrac{1}{2}\sigma^2)t + \sigma W(t)\}$$

(see also 3.1.1 in Part III) so that

$$\mathcal{L}(p_t - p_0) = N((\mu - \tfrac{1}{2}\sigma^2)t, \sigma^2 t)$$

(see 3.1.2 in Part III) and therefore

$$(16) \qquad \mathcal{L}(P_t/P_0) = LN((\mu - \tfrac{1}{2}\sigma^2)t, \sigma^2 t),$$

where by the symbol $LN(m, s^2)$ we mean the distribution of the random variable $\exp\{N(m, s^2)\}$, the *log-normal distribution* with parameters m and s^2 which are **not** its mean and variance, respectively. The density of $LN(m, s^2)$ is

$$(17) \qquad g(x; m, s^2) = \begin{cases} \frac{1}{x\sqrt{2\pi s^2}} \exp\{-\frac{1}{2}\left(\frac{\ln x - m}{s}\right)^2\} & x > 0, \\ 0 & \text{otherwise.} \end{cases}$$

The mean of $LN(m, s^2)$ is $E\,LN(m, s^2) = \exp(m + \tfrac{1}{2}s^2)$, and the variance is $\mathrm{var}\,LN(m, s^2) = \exp(m + \tfrac{1}{2}s^2)(\exp(s^2) - 1)$.

As a consequence of (16) we can deduce that the conditional distribution of P_t given P_0 is

$$(18) \qquad \mathcal{L}(P_t|P_0) = LN(\ln P_0 + (\mu - \tfrac{1}{2}\sigma^2)t, \sigma^2 t).$$

After some algebra we obtain the conditional expectation and variance:

$$E\,(P_t|P_0) = P_0 e^{\mu t}, \qquad \mathrm{var}\,(P_t|P_0) = P_0^2 e^{2\mu t}(e^{\sigma^2 t} - 1).$$

4.1.3   Important Remark

In this Part, unless otherwise stated, by *returns* we will mean either returns or rates of returns without further specification. Either of the return defined above will be considered as a random variable denoted by ρ, ρ_t, or $\rho(t)$ for the respective time period.

4.1.4   Expected Return

Often, return is a nonnegative random variable but this is not the rule. Let us denote F the distribution function of ρ. The *expected return* of ρ is the expected value

$$r := E\rho = \int_{-\infty}^{\infty} x dF(x).$$

4.2   Risk Measurement

Here we will restrict our explanation only to cases of *quantitative measures* of risk. All of the measures discussed here are based on the variance of the random variable in question, the return in our case.

4.2.1   Standard Deviation – Volatility

Basically, the *risk* of the return is defined as the standard deviation of ρ:

$$\sigma = \sqrt{E(\rho - E\rho)^2} = \sqrt{E\rho^2 - (E\rho)^2}.$$

A *riskless asset* is an asset with $\sigma = 0$ so that the return is a constant with probability one.

In literature we can find an analogous measure based on the variance of the return, called *volatility*. This term is used either for the variance or for the standard deviation of the return or of another stochastic financial variable.

4.2.2   Example. Let us consider two assets, A and B. Suppose that the rates of return randomly depend on the state of the economy in the way showed in Figure 7. Obviously, both assets have the same expected rate of return, r_A and r_B, respectively:

$$r_A = 0.3 \cdot 100 + 0.4 \cdot 15 - 0.3 \cdot 70 = 15 \ [\%],$$

$$r_B = 0.3 \cdot 20 + 0.4 \cdot 15 + 0.3 \cdot 10 = 15 \ [\%].$$

State of Economy	Probability	Return [%]	
		A	**B**
Boom	0.3	100	20
Normal	0.4	15	15
Recession	0.3	-70	10

Figure 7: States of Economy

Their respective variances are

$$\sigma_A^2 = 0.3 \cdot 100^2 + 0.4 \cdot 15^2 + 0.3 \cdot (-70)^2 - 15^2 = 4335 \; [\%^2], \quad \sigma_B^2 = 15 \; [\%^2],$$

so that the risks are $\sigma_A = 65.84[\%]$ and $\sigma_B = 3.87[\%]$. We conclude that the investment into asset B is less risky than into A.

4.2.3 Value at Risk

Another useful and recommended measure of risk is called *Value at Risk*, shortly VaR. (Distinguish between symbols VaR – Value at Risk and var – the variance.) *Value at Risk at confidence level* $1 - \alpha$, shortly VaR$_\alpha$ is defined by the relation

$$(19) \qquad P(\rho < -\text{VaR}_\alpha) = \alpha.$$

In words, $-\text{VaR}_\alpha$ is the cut-off point under which the return will attain values only with some given (small) probability α. Thus $-\text{VaR}_\alpha$ is the 100α per cent quantile of the distribution of ρ. Different financial institutions use different levels of confidence; the Bankers Trust 99 per cent, J P Morgan 95 per cent, Citibank 95.4 per cent, e.g. Otherwise the confidence level is stated in reverse form, 1 per cent, 5 per cent, etc., but the meaning is the same; the maximum possible loss will be more than VaR with probability α or it will be less than VaR with probability $1 - \alpha$.

4.2.3.1 Parametric VaR

Let us start with the so called *parametric* VaR. Suppose that the random return possesses a distribution from a *location-scale* family of distributions. Let $G(x)$ be a distribution function free of any other parameters and suppose that the distribution function $F_{\mu,\sigma}(x)$ of the return ρ is of the form

$$(20) \qquad F_{\mu,\sigma}(x) = G\left(\frac{x - \mu}{\sigma}\right)$$

where μ is a real number called *location parameter* and $\sigma > 0$ is called *scale parameter*. In what follows, we deal with distributions for which the location parameter μ is equal to the expected return and the scale parameter σ is equal to the standard deviation. If we denote the 100α per cent quantile of the distribution function $G(x)$ as u_α, then we get for VaR$_\alpha$

$$(21) \qquad P(\rho < -\text{VaR}_\alpha) = P\left(\frac{\rho - \mu}{\sigma} < \frac{-\text{VaR}_\alpha - \mu}{\sigma}\right) = G\left(\frac{-\text{VaR}_\alpha - \mu}{\sigma}\right) = \alpha$$

and therefore

$$(22) \qquad \boxed{\text{VaR}_\alpha = -\mu - \sigma u_\alpha.}$$

Sometimes, the last quantity is called *absolute value at risk* while

$$(23) \qquad \text{VaR}_{\alpha\,\text{rel}} = -\sigma u_\alpha$$

is called *relative value at risk*. Parameters μ and σ are usually unknown (even if the analytical form of the distribution G is supposed to be known) and typically they should be estimated by their sample counterparts.

Such an approach is good if we want to calculate VaR for the return based on data coming from the respective period. If we need VaR for a subsequent period or periods, we must take into account that both the mean μ and the volatility parameter σ can change in time. There are two simple models with a theoretical background that overcome this problem.

Firstly, we assume that the mean return does not change in time, but the variance of it is proportional to time. So if we consider the prospective return after T periods after the original parameters had been obtained, we suppose that the variance is

$$(24) \qquad\qquad \sigma_T^2 = T\sigma^2.$$

It follows that the value at risk may now be computed from the formula

$$(25) \qquad\qquad \mathrm{VaR}_\alpha = -\mu - \sigma\sqrt{T}\,u_\alpha.$$

Secondly, if we suppose that the mean is proportional to time, i.e., $\mu_T = T\mu$, then the formula for VaR becomes

$$(26) \qquad\qquad \mathrm{VaR}_\alpha = -T\mu - \sigma\sqrt{T}\,u_\alpha.$$

Often it is supposed that G is the distribution function of the standard normal distribution Φ. Numerous examples show that this is not a frequent case in practise, however.

Formulas (20) to (26) relate to **return** ρ and to its characteristics μ, σ. If ρ is the **rate of return** instead, and μ, σ its characteristics, then the value at risk in (20) – (26) is expressed in terms of the initial investment P_0 as unit, in other words, the maximum possible loss (in dollars) is $-\mu P_0 - \sigma u_\alpha P_0$.

4.2.3.2 Nonparametric VaR

If only little is known about the analytical (parametric) form of the returns' distributions but a sufficient amount of (historical) data is available, then a proper method for the risk analysis may be based on a nonparametric approach. Suppose that the observed returns during a given period (one year, say) are $R_1, \ldots R_T$. For data based on daily closing prices from a stock exchange we have about $T = 250$ observations yearly, e.g. We rank the observed returns to get the ordered random sample

$$R_{(1)} \leq \cdots \leq R_{(T)}.$$

Instead of the theoretical quantile u_α in the above considerations we will use the empirical αth quantile $\widehat{u}_\alpha$ defined for $0 < \alpha < 1$ by

$$(30) \qquad \widehat{u}_\alpha = \begin{cases} R_{(\lfloor T\alpha \rfloor + 1)} & \text{if } T\alpha \neq \text{ an integer} \\ \frac{1}{2}[R_{(T\alpha)} + R_{(T\alpha+1)}] & \text{if } T\alpha \text{ is an integer} \end{cases}.$$

For a chosen confidence level α we may state that the return will not fall under $\widehat{u}_\alpha$ with probability $1 - \alpha$. Similarly, the conclusion for VaR in case of a loss follows. This may be accepted as true for a one-period prospective, in the above case for one year ahead. The extension to more than one period needs some kind of speculation, however. Some regression techniques for a trend investigation may be helpful in this case.

4.2.4   Remark (The Distribution of R and Related Quantities)

The simplest assumption in accordance with the random walk hypothesis is that the R's are iid and moreover that they are normally distributed. The empirical studies reveal that this is often not the case. Usually we meet a violation of the zero *skewness* and zero *excess* property of the supposed normal distribution. Sometimes the problem of symmetry is not too severe for returns, but an important violation may be observed with other characteristics. Concerning excess, the difference between the theoretical value for a normal random variable (equal to 0) and the actually observed values sometimes appears to be significant. In [159], p. 45, the reader may find an analysis of the excess of stock returns which shows that the distribution of the respective returns is far from normal. See also [109].

4.2.5   Stress Testing

Often it is of interest for an investor to know what will happen if the market conditions attain their extremes, either in positive or negative direction from the investor's point of view. Of course, the more unfavorable, the more important they are for the investor's decision making, and they resemble VaR (in the sense of maximum possible loss) to some extent. A possible method to see what will happen is based on a *scenario analysis*. *Stress testing* starts with a construction of scenarios covering the extreme situations involved. The scenarios may be developed either from historical experience (*historical scenarios*) or from a theoretical model of the further development of the characteristic in question (*hypothetical scenarios*).

Stress testing catches the dynamics. It is therefore a task for the decision maker to state the limits or maximum likely changes for the periods of time under investigation. There are some recommendations. For example (see [89]), the Derivatives Policy Group suggests the following guidelines for the extreme movements of the variables involved in derivative's products (all given in basic points) for a one month's period; parallel yield curve shift ±100, yield curve twisting (change in shape) ±25, stock exchange index change ±10, foreign exchange rate change ±6, volatility change ±20.

A computational problem can arise with stress testing. If the time horizon covers T periods, say, and we consider four possible outstanding values of a variable in question (typically maximum, minimum, mean, and median), we have to generate 4^T scenarios and afterwards to evaluate the desired indicator or measure. For a typical ten years' currency swap described in Example 1.6.4.4 with the interests paid semiannually we have $T = 20$, so that the total number of scenarios is $4^{20} = 1,099,511,627,776$, a pretty large number of scenarios to be analysed. Note that actually two variables affect the resulting cash flow in Example 1.6.4.4; the exchange rate and LIBOR. Hence in fact there are even more than four possibilities at every period, at least at the initial and the final period.

To avoid this trouble, usually only a few (relative to the total number) of scenarios are selected and the desired measures evaluated. The typical trajectories then cover the most optimistic and most pessimistic (worst–case) scenarios consisting of all maximum and minimum values together with the average or median trajectory.

Another reduction of the size of the problem may be reached by a careful selection from the whole set of scenarios. A useful technique of such a selection is a

Monte Carlo simulation approach. We sample a number of scenarios at random and evaluate the desired characteristics of every sampled scenario. Such characteristics create a random sample and its useful descriptive statistics can be calculated. Since these statistics are obtained from a large number of characteristics, thousands say, we may employ the standard statistical inference based on a normal distribution's assumption, using the central limit theorem's argument. Note that the Monte Carlo simulation is generally a very useful device for the risk analysis.

I.5   VALUATION OF SECURITIES

valuation of different securities (bonds, options, forwards, and futures), arbitrage, hedging, put-call parity, Black-Scholes formula, binomial model

5.1   Coupon Bonds

Consider a simple coupon bond, coupons fixed, see 1.6.1.5. For the sake of simplicity **assume that the coupons are paid annually**. The cash flow to the holder of the bond is $-P, C, C, \dots, C, C + F$, where P is the value invested into purchasing the bond. At the time of issuing, the issuer sells the bond for its face value F. If this is not the case, there is something wrong with the initial setup of the coupon rate. Usually bond valuation does not consider the initial cost of purchasing the bond, P, and rather takes into account only the future cash flow resulting from the coupon payments and the redemption of the face value at the maturity date, so that the corresponding cash flow becomes $C, C, \dots, C, C + F$. Moreover, the history of the past payments is of no interest for the holder, and he or she values the security on the basis of the expected future cash flow only.

More formally, let us suppose that the time of valuation is t while the maturity time is T, $t < T$. The coupon payments take place in times $T - \lfloor T - t \rfloor, T - \lfloor T - t \rfloor + 1,$ $\dots, T$. At time T there is the additional payment of the face value F. Altogether we have $\lfloor T - t \rfloor + 1$ payments. With the valuation interest rate i, and the corresponding discount factor $v = 1/(1 + i)$, we can express the present value of the above cash flow sometimes called the *dirty, gross, fair*, or *full price* or *value* of the bond as

$$(1) \quad PV = Cv^{T - \lfloor T - t \rfloor - t} + Cv^{T - \lfloor T - t \rfloor - t + 1} + \cdots + (C + F)v^{T - t} =$$
$$C \sum_{j=0}^{\lfloor T - t \rfloor} v^{T - \lfloor T - t \rfloor - t + j} + Fv^{T - t}.$$

This formula provides a correct expression of the present value of the bond. There is one point to be discussed, however. If $T - t$ is an integer, the above formula assumes the immediate payment of the coupon at time t. In practise this is hardly the case because the issuer states the clause of so called *ex-coupon*. It means, that after some date, called *ex-coupon date*, the bond is traded without the first forthcoming coupon and the coupon payment belongs to the former holder of the bond. Thus it is more realistic to adapt (1) to

$$(2) \qquad PV = Cv + Cv^2 + \cdots + (C + F)v^{T - t} = C \sum_{j=1}^{T-t} v^j + Fv^{T - t}.$$

Sometimes it is useful to invert the time by setting $n = T - t$. In this case, n means the time to maturity (n need not be an integer). Then

$$(3) \qquad PV = Cv^{\{n\}} + Cv^{\{n\}+1} + \cdots + (C + F)v^n = Cv^{\{n\}} \sum_{j=0}^{\lfloor n \rfloor} v^j + Fv^n$$

with the first term missing if n is an integer. The value $Cv^{\{n\}}$ is called *accrued interest*. Using the simple interest method, the accrued interest can be expressed as $C/(1+\{n\}i)$. The two values slightly differ, of course. Accrued interest is a reward to the seller of the bond compensating the loss of the next forthcoming coupon. The difference between the dirty price and the accrued interest is called *pure price, pure value, net value* of the bond which therefore takes the form

$$(5) \qquad PV_p = Cv^{\{n\}} \sum_{j=1}^{\lfloor n \rfloor} v^j + Fv^n,$$

which is also quoted in the financial press.

A very important measure of a bond is the so called yield to maturity. Let us suppose that the market price of the bond is MP. Consider the value of the bond expressed in terms of interest rate i, either from (1) or (3), $PV(i)$, *ceteris paribus*. Then the *yield to maturity*, YTM, is defined as a solution to the equation

$$(6) \qquad MP = PV(YTM).$$

Since YTM is in fact the internal rate of return and the assumptions of Theorem 3.4.1 are fulfilled, there is just one YTM.

Another very simple but frequently used measure of a bond is its *current yield*:

$$(7) \qquad \text{Current yield} = \frac{c}{MP}.$$

Note that so far we have supposed that the coupons are paid annually. We will discuss other than annual frequency of coupons later.

For further analysis it is convenient to suppose that n is an integer. Then the value of the bond (immediately after the coupon payment), now identical with the net value, becomes

$$(8) \quad PV(c,F,n,i) = C \sum_{j=1}^{n} v^j + Fv^n = C\frac{v(1-v^n)}{1-v} + Fv^n =$$

$$\frac{1}{i}F(c + (1+i)^{-n}(i-c)).$$

Note, that in ancient literature this formula is used for calculation of the net value of the bond if n is not an integer. In this case, the net value is calculated as the linear interpolation between values $P_0 = PV(c,F,\lfloor n \rfloor + 1, i)$ and $P_1 = PV(c,F,\lfloor n \rfloor, i)$. The interpolated value is $P_n = P_0 + (1-\{n\})(P_1 - P_0)$. The dirty value is calculated as $P_n + (1-\{n\})C$, the term $(1-\{n\})C$ standing for the accrued interest, without taking into account discounting.

From formula (8) we can immediately deduce that the net value of the bond at the maturity date equals its par value: $PV(c,F,0,i) = F$. Further, for the valuation interest rate i equal to the coupon rate c, $i = c$, the net value of the bond is equal to the par value independently of the time to maturity: $PV(c,F,n,c) = F$. For $i > c$, $PV(c,F,n,i)$ is a decreasing function of n and $PV(c,F,n,i) < F$ for

$n > 0$. The reverse is true for $i < c$, so that the net value is an increasing function of n and $PV(c, F, n, i) > F$. Hence, in case $i > c$, the bond is called a *discount bond* while in case $i < c$ the bond is called a *premium bond*. Thus, an increase of the interest rate will cause the value of the bond to fall, whereas a decrease of this rate will cause it to rise. As n approaches 0 (this means, to the maturity date), the net value of the bond approaches its par value F. An analysis of (8) also shows that, *ceteris paribus*, bonds with longer maturities are more sensitive to changes of i than those with shorter ones.

After some algebra we get a formula for the duration corresponding to the net value expressed in terms of the discount factor:

$$(9) \qquad D_n = \frac{1}{1-v} + \frac{n-1-nv(1+c)}{1-v+cv(v^{-n}-1)}$$

or in terms of the valuation interest rate:

$$(10) \qquad D_n = 1 + \frac{1}{i} - \frac{1+i+n(c-i)}{i+c((1+i)^n-1)}.$$

For $i = c$ the expression for the duration simplifies to

$$(11) \qquad D_n = \frac{1+c-(1+c)^{1-n}}{c}.$$

5.1.1   Example. Suppose we have a bond with par value $F = 1$ and coupon $C = 0.1$ (all in thousands CZK) so that the coupon rate is $c = 0.1$, that is 10 per cent p.a.

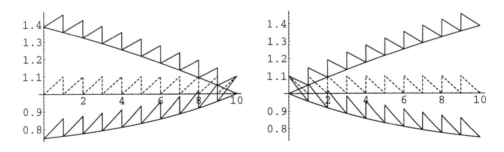

Figure 8: Value of the bond: left – normal time, right – inverted time
sawed – full price, smooth – pure price, dashed – full price for $i = c$

If the market interest rate is 0.05, the bond is a premium one, if it is 0.15, the bond is a discount one. The dependence of the price of the bond on time to maturity is graphically illustrated in Figure 8. The decreasing function on the left part of the figure corresponds to the price of the bond with the valuation interest rate $i = 0.05$, etc.

5.1.2   Exercise. Investigate modifications of the above formulas in case that the coupon payments appear semiannually, i.e., with frequency 2, which is perhaps the most frequent case.

5.1.3 Remark (Construction of the Yield Curve of Coupon Bonds)

The simplest way is to take a set of similar coupon bonds with different maturities and their calculated yields to maturity. Then some method of fitting discussed above may be applied. An alternative approach is known as *bootstrapping*. (Do not confuse with the same term used in statistics!) The idea consists in valuating every coupon payment (and also the principal) using the corresponding spot interest rate. Thus the present value formula (8) for the valuation of the bond with maturity T at time t is now

$$(12) \qquad PV = \frac{C}{1 + {}_tR_1} + \frac{C}{(1 + {}_tR_2)^2} + \cdots + \frac{C + F}{(1 + {}_tR_T)^T}.$$

Suppose for the simplicity that there are exactly T coupon bonds with maturities $1, \ldots ,T$, fixed coupons $C_1, \ldots ,C_T$, face values $F_1, \ldots ,F_T$, and (observed) market prices $MP_1, \ldots , MP_T$. For the bond j, the present value is expressed as

$$(13) \qquad PV_j = \frac{C_j}{1 + {}_tR_1} + \cdots + \frac{C_j + F_j}{(1 + {}_tR_j)^j}, \quad j = 1,\ldots,T.$$

For the first bond, the yield to maturity ${}_tR_1$ is, according to (6), calculated from the relation

$$(14) \qquad MP_1 = \frac{C_1 + F_1}{1 + {}_tR_1}.$$

For the second bond we use or bootstrap the information from the first bond (${}_tR_1$ already ascertained) using the relation

$$MP_2 = \frac{C_2}{1 + {}_tR_1} + \frac{C_2 + F_2}{(1 + {}_tR_2)^2},$$

and by recursion, having known ${}_tR_1, \ldots ,{}_tR_{j-1}$, we calculate the yield ${}_tR_j$ from the relation

$$(15) \qquad MP_j = \sum_{k=1}^{j-1} \frac{C_j}{(1 + {}_tR_k)^k} + \frac{C_j + F_j}{(1 + {}_tR_j)^j}.$$

Care should be taken if the maturities are not equally spaced. In any case, some fitting procedure is almost always necessary.

5.1.4 Callable Bonds

A callable bond means that the issuer has the right to call back the bond prior to the designed maturity. In fact, in our terminology, the issuer is the holder of a call option which has some value itself. Therefore, from the point of view of the holder of the bond, the price of the callable bond is

> price of the bond without callable feature – price of the call option.

The value of this call option may be derived by standard methods given later in 5.2.5 and Part III. Similarly, *putable bonds* can be issued and valued, see II.6.4.

5.1.5 Remark (Amortized Bonds)

An *amortized bond* is characterized by constant installment payments of the principal and interest like a loan, see 3.3.8. Suppose that the cash flow of the usual coupon bond is $CF_c = (C, \ldots, C, C + F)$ and that of the amortized bond $CF_a = (C^*, \ldots, C^*)$, both of the same length. As an exercise, find C^* such that the two bonds are equivalent in the sense of the equality of their present values under the valuation interest rate equal to the coupon rate $c = C/F$. Remind that an amortized bond is less risky than a classical coupon bond (it immediately repays the principal) and hence, in practise, the risk premium offered by the issuer is not as high as in case of the usual coupon bond. The actual C^* is less than that of calculated on the above equivalence principle.

5.1.6 Remark (Simple Bonds under Uncertainty)

Suppose that a zero coupon bond pays F with probability p and pays ηF with probability $1 - p$ at maturity, where $\eta \in [0, 1)$ is called the *recovery rate* and $\lambda := 1 - \eta$ is called the *loss rate*. The case $\eta = 0$ is equivalent to the default. Suppose that the valuation interest rate is r. The fair value of the bond is the expected present value:

$$\overline{PV} = \frac{1}{1+r}[pF + (1-p)\eta F].$$

Further, let us consider a one-period coupon bond with coupon C and par value F which sells for par F now and pays $C + F$ with probability p and $\eta(C + F)$ with probability $1 - p$ at maturity, η having the same meaning as above. The question is what is the fair value of the coupon C under the valuation interest rate r. We equate the present value and the discounted expected future value

$$F = \frac{1}{1+r}[p(C + F) + (1 - p)\eta(C + F)],$$

solve for C, and get the fair value of the coupon

$$C = \frac{F[1 + r - p - (1 - p)\eta]}{p + (1 - p)\eta}.$$

As an exercise, extend the above one-period case to a multiperiod one.

5.2 Options

We start with the valuation of options since the ideas of their pricing are general enough to be used for the valuation of other derivative securities. The key concepts, arbitrage and hedging play a crucial role in the mathematical modeling.

5.2.1 Arbitrage

All the models treated in this book assume *no-arbitrage principle*, in other words, the absence of arbitrage opportunities. By an *arbitrage opportunity* we mean any of the two situations:

(1) At the same time, the same asset is sold at different prices at different places. Nowadays, this can hardly happen in the financial world since the

information from one stock exchange is available on the stock exchange on the opposite side of the globe within a second.

(2) With zero investment at time 0 there is no probability of loss but there is a possibility of a riskless profit at time 1. More rigorously, an *arbitrage opportunity* in this case is a self-financing trading strategy with no initial investment, and a positive probability of positive profit and zero probability of negative profit later on (cf. III.3.3 and III.3.3.1).

Arbitrage opportunity is often characterized as a "money pump" and no-arbitrage principle by the slogan: "There is no such thing like a free lunch.".

5.2.2 Hedging

Hedging may be compared to insurance. It provides an insurance against unfavorable development of the market from the investor's point of view. Hedging may reduce the risk but, under no-arbitrage principle, risk cannot be fully eliminated. In principle, *hedging* consists of taking two opposite positions in the assets which are highly negatively correlated. The investor who hedges his/her position is called *hedger*. A *perfect hedge* means that the hedger combines an option and an underlying asset in such proportions that result in a riskless position and provide a riskless profit (equal to the riskless interest rate). See also III. 3.3.5. This is a rather idealized situation, however, since it does not take transaction costs into account.

5.2.3 Notation

We will assume the continuous-time world with constant riskless rate of interest (force of interest) r applied both to borrowing and lending. Symbols c and p will stand for the price of a European CALL and PUT, respectively. Analogously, symbols C and P will be used for prices of the respective American options. The price of the underlying asset (usually stock) will be denoted S and we will suppose that there are no liabilities like dividends connected with this asset during the period involved. We will also assume that S is a random variable or, more generally a stochastic process, an approach consistently adopted in Part III. Finally let K denote the strike price and T the expiry date. If necessary, we add subscript t to stress the dependence of the respective quantity on time, S_t, S_T, c_t, etc. If the option is exercised, denote the time of exercising by τ, $\tau \le T$. For Europeans, $\tau = T$.

The payoff of an exercised call option is

(16) $$(S_\tau - K)^+$$

and that of a put option is

(17) $$(K - S_\tau)^+.$$

(16) and (17) are called *terminal payoffs*. At any given $t < T$, the value $S_t - K$ for a CALL and $K - S_t$ for a PUT is called the *intrinsic value* of the respective option. This is the value which the option would have if it were exercised at time t. If the intrinsic value is positive, zero, or negative, we say that the option is *in the money*, *at the money*, or *out of money*, respectively.

5.2.4   PUT – CALL Parity

Let us consider the portfolio long one asset, long one PUT, and short one CALL. It means that we have bought one asset plus one PUT on that asset and sold one CALL on the same asset. Both the options on the asset in the portfolio are European with the same expiry date T and the same strike price K. The value of the portfolio at time $t \leq T$ is therefore

$$(18) \qquad \Pi_t = S_t + p_t - c_t.$$

Look what will happen at the expiry date. The value of the portfolio becomes

$$(19) \qquad \Pi_T = S_T + (K - S_T)^+ - (S_T - K)^+.$$

If $S_T \leq K$, then $\Pi_T = S_T + K - S_T - 0 = K$, and if $S_T > K$, then $\Pi_T = S_T + 0 - (S_T - K) = K$. We conclude that such a portfolio is riskless and leads to the certain gain K. What is the value of the portfolio at time $t < T$? Since the future value is K, the present value is $\Pi_t = Ke^{-r(T-t)}$ (for riskless investment we have used the riskless interest rate r). Thus we have obtained so called *put-call parity relation*:

$$(20) \qquad S_t + p_t = Ke^{-r(T-t)} + c_t, \quad t \leq T.$$

This is an example of risk elimination. Note that this formula cannot be applied to American options due to the early exercise feature.

5.2.5   Option Pricing

5.2.5.1 Natural Boundaries

The limited liability condition says that all option prices are non-negative. Since American options have all features like Europeans plus the right of an early exercise, they must be worth at least the Europeans:

$$C_t \geq c_t, \qquad P_t \geq p_t.$$

Further, from the put-call parity relation it follows that

$$c_t \geq (S_t - Ke^{-r(T-t)})^+.$$

For an American CALL we have

$$C_t \geq (S_t - K)^+.$$

The proof is by the contrary; suppose that $0 \leq C_t < S_t - K$. Then we can buy the CALL at C_t, immediately exercise it and thus get a riskless profit $S_t - K - C_t$ which is in a contradiction to the no-arbitrage principle.

5.2.5.2 Exercise. The quantities which are not explicitly mentioned remain constant. Prove the following propositions:
   (1) If $t_1 \leq t_2$ then $C_{t_1} \geq C_{t_2}$ and $P_{t_1} \geq P_{t_2}$.
   (2) If $K_1 \leq K_2$ then $c_t(K_1) \geq c_t(K_2)$ and $p_T(K_1) \geq p_t(K_2)$. The same holds for the Americans.
   (3) If $S_1 \leq S_2$ then $c_t(S_1) \leq c_t(S_2)$ and $p_t(S_1) \geq p_t(S_2)$. The same holds for the Americans.

5.2.5.3 The Black–Scholes Formula

Let us consider a European call option on a stock, the current price of which is known and equal to S_t. Since the payoff at the expiry date T is $(S_T - K)^+$, the present value of this payoff is

$$e^{-r(T-t)}(S_T - K)^+.$$

Next we adopt the so called *risk-neutral* valuation. Under this approach we do not consider any risk preferences of the investors. Since the higher the level of risk aversion, the higher the expected return μ will be for a risky asset, by excluding the risk preferences we conclude that the only correct risk-neutral μ is $\mu = r$, the riskless rate. At this point it is important to emphasize that by the above choice we **do not assert** that the conditional distribution of S_T given S_t is that for which $\mu = r$! It seems to be reasonable to take the conditional expected value of the discounted payoff given the current value of the underlying asset S_t as the value of the option but with the expectation taken with respect to the riskless rate r:

$$(21) \qquad c_t = e^{-r(T-t)} E^* ((S_T - K)^+ | S_t).$$

where E^* stands for the expected value in a risk-neutral world. In Black-Scholes approach we suppose that the conditional distribution of S_T given S_t adjusted for risk-neutrality (see formula (18) in 4.1.2) is log-normal

$$(22) \qquad \mathcal{L}^*(S_T | S_t) = LN(\ln S_t + (r - \frac{1}{2}\sigma^2)(T - t), \sigma^2(T - t)).$$

To evaluate (21) under assumption (22) we first calculate the expected value

$$E(m, s^2) = E(X - K)^+$$

where the random variable X possesses a log-normal distribution $LN(m, s^2)$ with the probability density function given by formula (17) in 4.1.2. After some algebra we get

$$(23) \quad E(m, s^2) = \int_K^\infty (x - K)g(x; m, s^2)dx =$$

$$e^{m+\frac{1}{2}s^2} \Phi\left(\frac{m + s^2 - \ln K}{s}\right) - K\Phi\left(\frac{m - \ln K}{s}\right),$$

where Φ stands for the distribution function of the standard normal distribution $N(0, 1)$. Substituting $m \to \ln S_t + (r - \frac{1}{2}\sigma^2)(T - t)$ and $s^2 \to \sigma^2(T - t)$ into the expression for $E(m, s^2)$ gives

$$(24) \qquad E^* ((S_T - K)^+ | S_t) = S_t e^{r(T-t)} \Phi(d_1) - K\Phi(d_2),$$

where

$$(25) \quad d_1 = \frac{\ln(S_t/K) + (r + \frac{1}{2}\sigma^2)(T - t)}{\sigma\sqrt{T - t}}, \quad d_2 = \frac{\ln(S_t/K) + (r - \frac{1}{2}\sigma^2)(T - t)}{\sigma\sqrt{T - t}}.$$

Altogether, going back to (21) we have derived the *Black-Scholes formula* for the value of a European call option:

(26)
$$c_t = S_t\Phi(d_1) - Ke^{-r(T-t)}\Phi(d_2).$$

An elementary application of the put-call parity relation provides the value of a European put option

(27)
$$p_t = Ke^{-r(T-t)}\Phi(-d_2) - S_t\Phi(-d_1),$$

where we used $\Phi(x) = 1 - \Phi(-x)$.

5.2.5.4 The Binomial Option Pricing Model

We will now assume that the stock price changes only at the equally spaced time instants t, $t + 1$, ... The time unit may be arbitrary (month, day, hour, ...). Further let us suppose that if the stock price is S_t at time t, then at time $t + 1$ it may take only one of two values, dS_t or uS_t with probability p or $1 - p$, respectively. Thus

(28)
$$P(S_{t+1} = uS_t|S_t) = p, \quad P(S_{t+1} = dS_t|S_t) = 1 - p =: q.$$

Also suppose that the changes are mutually independent and the probabilities do not depend on time. By the no-arbitrage principle we may suppose that the riskless interest rate r fulfills $d < 1 + r < u$. (Suppose on the contrary that $1 + r < d < u$, e.g. Then the investor could borrow any amount of cash at the riskless rate r, buy the stocks and sell them for at least d after one period. Such a strategy would lead to a riskless profit $d - r - 1$.) Note that the usual assumption is $d < 1 < u$ so that the price can move up and down. Next we state a relationship among p, u, d, and r in a risk-neutral world. The expected return from holding the stock should be the same as the riskless return resulting from the investment S_t at the riskless rate r. Since
$$E(S_{t+1}|S_t) = puS_t + qdS_t,$$
by the above argument we conclude that
$$puS_t + qdS_t = (1 + r)S_t$$
and this is fulfilled for

(29)
$$p = \pi := \frac{1 + r - d}{u - d}.$$

Due to our assumptions, $\pi \in (0, 1)$ is a probability called the *risk-neutral probability*. We can also obtain the risk neutral probability by the following construction. Consider the so called *replicating portfolio* consisting of A stocks and B riskless bonds which gives the same payoff as one European call option on the stock with the strike price K and expiry date $t + 1$. The terminal value of the option is
$$c_{t+1}^u := (uS_t - K)^+ \quad \text{with probability } p$$

and

$$c_{t+1}^d := (dS_t - K)^+ \quad \text{with probability } q.$$

Thus A and B must satisfy the system of equations

$$AuS_t + B(1+r) = c_{t+1}^u, \quad AdS_t + B(1+r) = c_{t+1}^d.$$

The solution to this system is

(30) $$A = \frac{c_{t+1}^u - c_{t+1}^d}{(u-d)S_t}, \quad B = \frac{uc_{t+1}^d - dc_{t+1}^u}{(1+r)(u-d)}.$$

From the obvious inequality $uc_{t+1}^d \le dc_{t+1}^u$ we observe $B \le 0$ so that the replicating portfolio always involves borrowing cash and buying the stock in the above proportions. The present value of the CALL is, after substitution from (30), given by

(31) $$c_t = AS_t + B = \frac{(1+r-d)c_{t+1}^u + (u-1-r)c_{t+1}^d}{(1+r)(u-d)} =$$

$$\frac{1}{1+r}(\pi c_{t+1}^u + (1-\pi)c_{t+1}^d),$$

where π is the risk-neutral probability defined in (29).

Up to now we have considered a one-period model. Let us look on a simple generalization for a multi-period model. After two periods we obviously have

$$P(S_{t+2} = u^2 S_t | S_t) = p^2, \quad P(S_{t+2} = udS_t | S_t) = 2pq, \quad P(S_{t+2} = d^2 S_t | S_t) = q^2.$$

Generally, after $T - t$ periods ($T > t$, T integer)

(32) $$P(S_T = u^j d^{T-t-j} S_t | S_t) = \binom{T-t}{j} p^j q^{T-t-j}, \quad j = 0, \dots, T-t.$$

This is the *binomial model* describing the probability distribution of the stock price after $T - t$ periods. By $\text{Bi}(n, p)$ we will denote the *binomial distribution* with parameters n, p, i.e., the distribution of a random variable X such that $P(X = j) = \binom{n}{j} p^j q^{n-j}$, $j = 0, 1, \dots, n$.

Consider now a European call option with strike price K and expiry date T. Using the same argument as in the derivation of (21), but with discrete discounting, the value of the option at time t is given by

(33) $$c_t = (1+r)^{t-T} E\left((S_T - K)^+ | S_t\right).$$

In a risk-neutral world we should have used the risk-neutral probability π but the option price can be expressed for an arbitrary p. We have

(34) $$c_t = (1+r)^{t-T} \sum_{j=0}^{T-t} \binom{T-t}{j} p^j q^{T-t-j} (u^j d^{T-t-j} S_t - K)^+.$$

Let J be the smallest non-negative integer such that $u^J d^{T-t-j} S_t \geq K$. Put

$$p^\star := \frac{up}{1+r} \qquad q^\star := \frac{dq}{1+r}.$$

Then

(35) $\quad c_t = S_t \sum_{j=J}^{T-t} \binom{T-t}{j} (p^\star)^j (q^\star)^{T-t-j} - K(1+r)^{t-T} \sum_{j=J}^{T-t} \binom{T-t}{j} p^j q^{T-t-j}.$

If $p = \pi$, the risk-neutral probability, then $p^\star + q^\star = 1$ so that in this case we can express (35) in the form

(36) $\quad c_t = S_t P(\mathrm{Bi}(T-t, p^\star) \geq J) - K(1+r)^{t-T} P(\mathrm{Bi}(T-t, p) \geq J).$

With the binomial model, a number of questions arise. We have seen that even under the assumption of the risk-neutral probability there are some degrees of freedom in choice of u and d. We just mention how to handle the unknown parameters appearing in the model. Some ideas are based on comparing the parameters of the discrete model to those of the continuous one. Another popular relationship between u and d is $u = 1/d$. The choice $p = \frac{1}{2}$ is also popular. Such assumptions reduce the dimension of the respective parametric space and open space for a broad discussion. See [172] and [105] for more details.

Like in classical probability theory, also here there is a close connection between the binomial model and its limiting counterpart, the normal distribution model, as a consequence of the central limit theorem for iid random variables with finite positive variances. See [144] for more details.

Since the binomial model is discrete, it enables a straightforward modeling by Monte Carlo simulation. The simulation models take the advantage of the fact that on different stages of the dynamic simulation, numerous specific features and movements of the real life problems may be incorporated. Note that some of the mentioned movements, particularly shocks, may hardly be considered in a theoretical model.

5.2.5.5 Options on Assets Paying Dividends

So far we have considered the underlying stock that does not pay any dividend. We can modify the above results also for a dividend-paying stock. If a stock pays a dividend during the life time of the option, the payment of the dividend causes the stock price to fall by an amount equal to the dividend. The *dividend yield y* is expressed as a proportion of the stock price. For the purposes of this Part, we will suppose that the dividend yield is constant and understood as continuous like the force of interest. Hence during the time interval $(t, t + \Delta t)$ the stock pays $y S_t \Delta t$. For European options we may still use Black-Scholes' type formulas (25), (26) but now with

(37) $$d_{1,2} = \frac{\ln(S_t/K) + (r - y \pm \frac{1}{2}\sigma^2)(T - t)}{\sigma\sqrt{T - t}}.$$

The formula for a PUT is given by (27) but with $d_{1,2}$ from (37).

5.2.5.6 Valuing American Options

The following widely used, but, in our opinion, questionable argument enables to value an American option on an asset which does not pay dividends. From the inequalities of 5.2.5.1 we have

$$C_t \geq c_t \geq S_t - Ke^{-r(T-t)} > S_t - K.$$

If the option is exercised at time $\tau < T$, then its value immediately becomes $S_\tau - K$ which is less than the lower bound if the option is still alive. It follows that an American call option will never be exercised prior to its expiry date and hence the value of an American CALL should be the same as that of the corresponding European CALL.

There is no such argument for American put options and/or American options on dividend paying assets. Further information on the topic can be found in [105], [116], [172], e.g.

5.2.5.7 Comparative Statics – The Greeks

In option pricing formulas there are actually five variables (also called parameters), S_t, K, r, $T - t$, and σ. The sensitivity to the option prices on these variables plays a crucial role in financial decision making and is measured by partial derivatives. Since traditionally these sensitivities are denoted by Greek letters, they are often called *Greeks*.

In what follows we will suppose that the options involved are European and that their prices are driven by the Black-Scholes formula (26) and (27). Note that the definitions in the form of derivatives given below can be used in a more general setup. Also note that the respective sensitivities for PUTs can be usually simply calculated using the put-call parity relation (20). Let V denote the value of either a CALL or a PUT.

Delta. The delta Δ is defined by

$$(38) \qquad\qquad\qquad \Delta = \frac{\partial V}{\partial S_t}.$$

After some algebra we get the delta for a call and a put option:

$$(39) \qquad\qquad \Delta_c = \Phi(d_1), \quad \Delta_p = \Delta_c - 1 = -\Phi(-d_1).$$

Obviously, since $\Delta_c > 0$, the value of a CALL is always an increasing function of S_t. The reverse is true for a PUT. The concept of delta is used for so called *delta hedging*. Suppose we are long one asset and short A call options on that asset. The value of such a portfolio is therefore $\Pi_t = S_t - Ac_t$. We wish to determine A so as to make the value of the portfolio invariant with respect to (small) changes in the asset price, i.e.,

$$\frac{\partial \Pi_t}{\partial S_t} = 0.$$

It follows that the desired $A = 1/\Delta_c$, the so called *hedge ratio*. Nevertheless, since delta is changing in time, the portfolio should be rebalanced frequently and the hedge should be a *dynamic hedge*. A dynamic hedge can be rather costly, particularly in case the transaction costs are not negligible. So dynamic hedging strategies (as well as other strategies dependent on a frequent trading in time) are good for market makers or brokers and others with low transaction costs. Observe that the lower the asset price in comparison with K, the higher the hedge ratio. A similar measure of the sensitivity is the *elasticity* of the call price with respect to the asset price defined by $E_c = \Delta_c S_t / c_t$. Note that always $E_c > 1$ (prove as an exercise). Hence the call option is more risky than the underlying asset. An analogue to delta is the duration.

Gamma. The gamma Γ is the second derivative of V with respect to the asset price:

$$(40) \qquad \Gamma = \frac{\partial^2 V}{\partial S_t^2}.$$

Gamma for CALL and PUT is the same:

$$(41) \qquad \Gamma = \Gamma_c = \Gamma_p = \frac{\varphi(d_1)}{S_t \sigma \sqrt{T-t}}$$

where φ is the probability density function of the standard normal distribution $N(0,1)$. Since $\Gamma > 0$, the values of both types, CALL or PUT, are convex functions of S_t. Observe that gamma resembles the convexity introduced in I.3.6.

Theta. The theta Θ is the time derivative of V:

$$(42) \qquad \Theta = \frac{\partial V}{\partial t}.$$

The calculation is a bit cumbersome but useful exercise (a good idea is to use some CAS (Computer Algebra System) like *Mathematica®*):

$$(43) \qquad \Theta_c = -Ke^{-r(T-t)} \left(\frac{\sigma}{2\sqrt{T-t}} \varphi(d_2) + \Phi(d_2) \right).$$

Alternatively, using the identity

$$(44) \qquad K\varphi(d_2) = S_t e^{r(T-t)} \varphi(d_1)$$

we get it in the form:

$$(45) \qquad \Theta_c = -\frac{\sigma S_t}{2\sqrt{T-t}} \varphi(d_1) - Kre^{-r(T-t)} \Phi(d_2).$$

We see that Θ_c is always negative. From the put-call parity relation we obtain

$$(46) \qquad \Theta_p = \Theta_c + Kre^{-r(T-t)}.$$

Nothing can be said about the sign of the last expression.

Rho. The rho P expresses the dependence on the riskless rate:

$$(47) \qquad\qquad P = \frac{\partial V}{\partial r};$$

for a CALL it takes the form

$$(48) \qquad\qquad P_c = K(T-t)e^{-r(T-t)}\Phi(d_2)$$

and for a PUT

$$(49) \qquad\qquad P_p = -K(T-t)e^{-r(T-t)}\Phi(-d_2).$$

Immediately we see that $P_c > 0$ and $P_p < 0$.

Vega. The vega $\mathcal{V}$ measures the sensitivity of the option price with respect to the volatility σ of the underlying asset:

$$(50) \qquad\qquad \mathcal{V} = \frac{\partial V}{\partial \sigma}.$$

For both types of options $\mathcal{V}$ is the same:

$$(51) \qquad\qquad \mathcal{V} = \mathcal{V}_c = \mathcal{V}_p = S_t\sqrt{T-t}\varphi(d_1).$$

Sometimes it is also of interest to investigate the sensitivity to the strike price but for unknown reasons the corresponding Greek is missing. Nevertheless we have

$$(52) \qquad\qquad \frac{\partial c_t}{\partial K} = -e^{-r(T-t)}\Phi(d_2)$$

which is always negative and

$$(53) \qquad\qquad \frac{\partial p_t}{\partial K} = e^{-r(T-t)}\Phi(-d_2)$$

which is always positive, both these conclusions in accordance with an intuitive insight.

5.2.5.8 Exercise. Derive formulas (41), (48), (51), (52), (53).

5.2.5.9 Volatility and Implied Volatility

The parameter σ, the volatility, is of vital importance in option pricing. Since it is difficult to speculate on its value, usually some estimates must be used.

One of the most frequently used estimates, called the *historical volatility*, is based on past data. In practise, this estimator is, in fact, the usual sample standard deviation, for instance, of quantities $\ln(S_{t+1}/S_t)$, $t = 1, \ldots, T-1$ in the Black-Scholes model. A care must be taken however: The time steps for such a calculation must be in accordance with time units in which the other quantities are measured.

More sophisticated estimation procedures are based on models of the stochastic behavior of volatility. See [26] for a review and [127] for a bootstrap estimation of volatility.

Since the other parameters in the formulas for option pricing are known at time t, and also the market value V_t^M of the option is known, then, after substituting these known values of the parameters into the Black-Scholes formula (26), we can determine the unknown volatility. The corresponding equation reads

$$(54) \qquad V_t^M = f(S_t, K, \sigma, T, t, r),$$

where f is the function resulting from the Black-Scholes formula. A solution $\hat{\sigma}$ to (54) is called *implied volatility*. Equation (54) is to be solved for an unknown σ given the values of all remaining quantities. We see that volatility cannot be explicitly expressed from (54) so that a solution must be found numerically. Moreover, it is not clear, how many solutions to the mentioned equation exist. If there are more than one, we should carefully analyze them with respect to a reasonable financial interpretation.

Modern computer algebra systems provide the users with a variety of routines and financial application libraries which can be used for the above analysis. See [147], [148], and http://www.wolfram.com for a possible approach. Some specific cases may be found in the series of papers of Benninga and Wiener: [9], [10], [11], [12], [13], [14].

5.2.5.10 Example. Let us consider 6 options, 3 CALL's and 3 PUT's, on a Volkswagen stock priced at EUR 70.72 April 23, 1999, expiring 3rd Friday, June 1999 with strike prices $K_1 = 67.5$, $K_2 = 70.0$, $K_3 = 72.5$. The actual prices for the respective CALLs were 6.31, 4.92, 3.77 and those for the PUTs 2.92, 4.08, 5.48. We have $T - t = 58/360$, $r = 0.05$. The implied volatilities computed using function FindRoot in Mathematica are 0.38, 0.38, 0.35 for CALLs and 0.41, 0.42, 0.43 for PUTs, respectively.

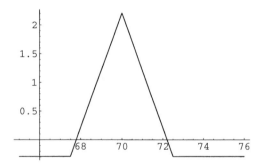

Figure 9: Payoff at expiry of a butterfly spread

Let us further consider the portfolio consisting of four the above options: long CALL with strike K_1, long PUT with strike K_3, short CALL with strike K_2, and short PUT with strike K_2. The value of that portfolio as a function of the stock price S at expiry with today's prices of the options is:

$$V(S) = -6.31 - 5.48 + 4.92 + 4.08 + (S - K_1)^+ + (K_3 - S)^+ - (S - K_2)^+ - (K_2 - S)^+.$$

This is an example of a combination of options, particularly the so called *butterfly spread*. See Figure 9 for the payoff of this portfolio.

5.2.5.11 Exercise. Examine and plot the payoffs of the following combinations of options (all on the same asset) at expiry:

(1) long CALL with strike K_1 and short CALL with strike $K_2 > K_1$ and prices $c_2 < c_1$ (*bullish spread*),

(2) long one CALL with strike K_1, long one CALL with strike $K_3 > K_1$, short two CALLs with strike $K_2 = (K_1 + K_3)/2$ and prices $c_1 > c_2 > c_3$. This is also a *butterfly spread*.

(3) long one CALL, long one PUT with the same strike K, called a *bottom straddle*,

(4) reverse of (3): short one CALL, short one PUT with the same strike K, called a *top straddle*.

Explain the motivation for the above strategies.

5.3 Forwards and Futures

Valuing both forwards and futures is practically the same from the mathematical point of view and since there is no option but obligation to deliver, it is simpler than that of valuing options. The seller must deliver the asset at time T. We can derive the forward price by the non-arbitrage principle. The seller borrows an amount S_t (=the price of the underlying asset at time t) at riskless rate and buys the asset. Hence the *forward price* must be

$$F_t = S_t e^{r(T-t)}$$

otherwise there will be a riskless profit or loss in contradiction to the nonexistence of the arbitrage. The asset may pay a dividend or need to be stored (like gold, grain, or oil) with some additional costs. If the corresponding rate is y, then the forward price becomes

$$F_t = S_t e^{(r-y)(T-t)}.$$

Note that $y > 0$ in case of dividends and $y < 0$ if there are some additional costs.

I.6   MATCHING OF ASSETS AND LIABILITIES

matching, immunization, dedicated bond portfolio, static model, dynamic model, stochastic model

6.1   Matching and Immunization

In what follows in this Chapter, by *assets* we mean the inflows and by *liabilities* the outflows of a company. The main purpose of *matching* is to balance assets and liabilities in such a way that the deficiency is either zero or as small as possible. Perhaps only of theoretical value is the case of *absolute matching*; let a_t and ℓ_t be the total assets and liabilities at time t, $t = 0, \ldots, T$, respectively. If $a_t = \ell_t$ for all t we say that assets and liabilities are *absolutely matched*. This does not sound realistic, however, so that an alternative approach is needed. The most frequent method is to match the discounted cash flows and/or other characteristics of assets and liabilities.

Suppose that the liabilities (assets) are represented by a cash flow $l \in \mathbb{L}^{T+1}$ ($a \in \mathbb{L}^{T+1}$), see 3.1. The principle of matched present values of assets and liabilities at force of interest δ is then expressed as

$$(1) \qquad\qquad PV(l, \delta) = PV(a, \delta).$$

This identity can only be satisfied for finite number of δ's with the exception of $l = a$, the absolute matching. In practise, one can choose the force of interest δ_0 which he or she believes will be most likely for the period of time under consideration. Then the matching condition for the present values is

$$(2) \qquad\qquad PV(l, \delta_0) = PV(a, \delta_0).$$

Since δ_0 is only an estimate of δ, there is a danger that for some other forces of interest, even close to the estimated one, the present value of liabilities will exceed that of assets. So it is a good idea for an investor to *immunize* his or her position by imposing further conditions expressed in terms of derived characteristics of cash flows. The condition

$$(3) \qquad\qquad D(l, \delta_0) = D(a, \delta_0)$$

requires the same duration of assets and liabilities and the condition

$$(4) \qquad\qquad C(l, \delta_0) < C(a, \delta_0)$$

guarantees that at least in a small neighborhood of δ_0, $PV(a, \delta_0) > PV(l, \delta_0)$ will hold. If we change condition (4) and require instead

$$(5) \qquad\qquad C(l, \delta_0) = C(a, \delta_0),$$

we can give an explicit solution to the problem.

Suppose $l \in \mathbb{L}^{T+1}$ is a given vector of liabilities, fixed in the sequel. Let $\mathbb{A} \subseteq \mathbb{L}^{T+1}$ be a 3-dimensional linear subspace of available assets generated by the *base assets* e_1, e_2, e_3 which form the basis of $\mathbb{A}$. Let us denote $\mathbf{E} := (e_1, e_2, e_3)$, a $(T+1) \times 3$ matrix. Thus every $a \in \mathbb{A}$ may be expressed as $a = \mathbf{E}\mathbf{x}$, where the coefficients $\mathbf{x} = (x_1, x_2, x_3)^\top$ are uniquely determined. Let us further denote

$$\mathbf{d}_j := \frac{\partial^j}{\partial \delta^j} (1, e^{-\delta}, e^{-2\delta}, \dots, e^{-T\delta})^\top \Big|_{\delta = \delta_0} ,$$

the derivatives of the discount factors taken at $\delta = \delta_0$, $j = 0, 1 \dots$, and $\mathbf{D} = (\mathbf{d}_0, \mathbf{d}_1, \mathbf{d}_2)$, a $(T+1) \times 3$ matrix. The three conditions (2), (3), and (5) may now be rewritten in the form

$$\mathbf{d}_j^\top a = \mathbf{d}_j^\top l, \qquad j = 0, 1, 2$$

or equivalently
$$\mathbf{D}^\top a = \mathbf{D}^\top l.$$

If we substitute $a = \mathbf{E}\mathbf{x}$, we get the system of three linear equations for unknown $\mathbf{x}$:
$$\mathbf{D}^\top \mathbf{E}\mathbf{x} = \mathbf{D}^\top l$$

which possesses the unique solution

(6) $$\mathbf{x} = (\mathbf{D}^\top \mathbf{E})^{-1} \mathbf{D}^\top l$$

provided the inverse exists. Exactly the same formula holds true if we, instead of three matching conditions, impose matching conditions employing higher order derivatives.

6.2 Dedicated Bond Portfolio

An important application of the idea of matching assets to liabilities is the investment strategy known as a *dedicated bond portfolio*, (see [60] e.g.) which deals with a proper selection of available bonds. In general, we may think about allocation of funds among arbitrary investment opportunities represented by their expected cash flows. A stochastic version of this problem is treated in Part II.1.2.

6.2.1 Static Model

Suppose that we have the time horizon $t = 1, \dots, T$ with investment opportunities represented by cash flows $\mathbf{CF}_1, \dots, \mathbf{CF}_N$, $\mathbf{CF}_n = (CF_{n1}, \dots, CF_{nT})^\top$, $n = 1, \dots N$ and initial acquisition costs (i.e., the cost for buying these cash flows) $\mathbf{c} = (c_1, \dots, c_N)^\top$. It means that c_n is the cost of the investment at time $t = 0$ resulting in the expected future cash flow $\mathbf{CF}_n$, $n = 1, \dots, N$. Further, let $l = (l_1, \dots, l_T)^\top$ be the expected liabilities over the considered time horizon. Let the initial wealth of the investor be $W = 1$. The objective is to create a portfolio $\mathbf{x} = (x_1, \dots, x_N)^\top$ (to find the weights in other words, $\mathbf{1}^\top \mathbf{x} = 1$) consisting of the above cash flows so as to minimize the total acquisition costs $\mathbf{c}^\top\mathbf{x}$ subject to $J+1$ matching conditions

$$(7) \qquad \sum_{n=1}^{N} x_n PV^{(j)}(\mathbf{CF}_n, \delta) = PV^{(j)}(l, \delta), \qquad j = 0, \ldots, J$$

where $PV^{(j)}$ stands for the jth derivative of the present value with respect to the force of interest δ. For $j = 0$ it means the perfect match of present values both of assets and liabilities, for $j = 1$ and 2 the perfect match of durations and convexities, respectively, etc. Further imposed conditions on portfolio may be of the type

$$(8) \qquad \mathbf{b}_L \leq \mathbf{x} \leq \mathbf{b}_U.$$

The lower limit $\mathbf{b}_L$ may represent the reasonable amounts of investment while the upper limit $\mathbf{b}_U$ may take into account some legal requirements. For example, in the Czech Republic, pension funds are not allowed to invest more than 10 per cent into one asset. In our terms, it means that the respective $x_n \leq 0.1$. We also add the natural condition $\mathbf{x} \geq \mathbf{0}$. Altogether, we have a problem of linear programming:

$$(9) \qquad \boxed{\text{Find } \min \mathbf{c}^\top\mathbf{x} \quad \text{under restrictions (7), (8), } \mathbf{x} \geq \mathbf{0}.}$$

If we abandon condition (8), the theory says that the optimal solution $\mathbf{x}^\star$ will consist of at most $J + 1$ (=number of conditions) positive weights. For $J > N - 1$ there may be no solution to the problem. However, this case is of theoretical interest only, since in practise we usually ask just for matching up to convexity, $J = 2$, in this case.

Since there is an uncertainty about the valuation force of interest δ, we usually need to solve the above problem for a set (scenarios) of expected interest rates and to discuss the solutions from the fundamental point of view.

6.2.2 Dynamic Model

In the above model, the only dynamics involved has been included via present values. In practise, the liability schedule is often determined at any time instant, $t = 1, \ldots, T$. This may be the case of obligatory balances, reserves, or solvency margins. At time t, the inflow is

$$(10) \qquad a_t = \sum_{n=1}^{N} x_n CF_{nt},$$

so that the necessary conditions to meet the liabilities at any time now read

$$(11) \qquad \sum_{n=1}^{N} x_n CF_{nt} \geq l_t, \qquad t = 1, \ldots, T.$$

It is a good idea for the investor, even under condition (11), to reinvest a possible surplus. Suppose that i_t is the short-term reinvestment interest rate for the period

$(t, t+1)$, and s_t^+ is the surplus at time t. Then the inequality condition (11) becomes the equality

$$
(12) \qquad \sum_{n=1}^{N} x_n CF_{nt} + (1 + i_{t-1}) s_{t-1}^+ - s_t^+ = l_t, \qquad t = 1, \ldots, T,
$$

with the initial surplus s_0^+, if any. Again, the optimal solution is given by solving the linear program

> Find $\min(\mathbf{c}^\top \mathbf{x} + s_0^+)$ under restrictions (8), (12), $\mathbf{x} \geq \mathbf{0}$, $\mathbf{s}^+ \geq \mathbf{0}$,

where $\mathbf{s}^+ := (s_1^+, \ldots, s_T^+)^\top$.

6.2.3 Discussion of the Restrictions

Note that if the short-terms interest rates are higher than the interest rates coming from the investment into $\mathbf{CF}$'s, the solution will naturally result in $\mathbf{x} = \mathbf{0}$ and some positive s^+'s.

6.3 A Stochastic Model of Matching

Here we give a simple stochastic version of the model given in 6.1. Suppose that the force of interest δ is now a random variable. Denote $\mathbf{d} := (1, e^{-\delta}, \ldots, e^{-T\delta})^\top$, the vector of discount factors. Note that if δ possesses a normal distribution then $e^{-j\delta}$'s posses log-normal distributions. Then the *surplus S* is also a random variable that may be expressed as

$$
(13) \qquad S = PV(\mathbf{a}, \delta) - PV(\mathbf{l}, \delta) = \mathbf{a}^\top \mathbf{d} - \mathbf{l}^\top \mathbf{d} = (\mathbf{a} - \mathbf{l})^\top \mathbf{d}
$$

and the *expected surplus* is $ES = (\mathbf{a} - \mathbf{l})^\top E\mathbf{d}$. The elements of $E\mathbf{d}$ are the moments of the log-normal random variable $e^{-\delta}$. We will find the assets $\mathbf{a}$ which minimize the *mean squared error* ES^2. Put $\mathbf{V} := E\mathbf{d}\mathbf{d}^\top$. We have then

$$
(14)\ ES^2 = E(\mathbf{a}^\top \mathbf{d} - \mathbf{l}^\top \mathbf{d})^2 = E((\mathbf{Ex} - \mathbf{l})^\top \mathbf{d}\mathbf{d}^\top (\mathbf{Ex} - \mathbf{l})) = (\mathbf{Ex} - \mathbf{l})^\top \mathbf{V}(\mathbf{Ex} - \mathbf{l}).
$$

This is obviously a convex function in $\mathbf{x}$ so that the minimum can be found by putting the gradient equal to zero:

$$
(15) \qquad \frac{\partial ES^2}{\partial \mathbf{x}} = 2\mathbf{E}^\top \mathbf{V}\mathbf{Ex} - 2\mathbf{E}^\top \mathbf{V}\mathbf{l} =: \mathbf{0}.
$$

The solution is

$$
(16) \qquad \mathbf{x} = (\mathbf{E}^\top \mathbf{V}\mathbf{E})^{-1} \mathbf{E}^\top \mathbf{V}\mathbf{l}
$$

provided the inverse exists. Thus the assets are in the form

$$
(17) \qquad \mathbf{a} = \mathbf{E}(\mathbf{E}^\top \mathbf{V}\mathbf{E})^{-1} \mathbf{E}^\top \mathbf{V}\mathbf{l}.
$$

I.7   INDEX NUMBERS AND INFLATION

construction of index numbers, properties of index numbers, stock exchange indexes, inflation, retail price index

In this Chapter we will use the following notation. Let $a = (a_1, \ldots, a_n)^\top$, $b = (b_1, \ldots, b_n)^\top \in \mathbb{R}^n$, $\alpha \in \mathbb{R}$, $\mathbf{1} := (1, 1, \ldots, 1)^\top$. The symbols "·" (more often omitted), "$*$", "$/$" will mean: $a^\top \cdot b := a^\top b := \sum_{i=1}^n a_i b_i$ (the scalar product), but $ab^\top$ is the $n \times n$ matrix with elements $a_i b_j$, $a * b := (a_1 b_1, \ldots, a_n b_n)^\top$, $a/b := (a_1/b_1, \ldots a_n/b_n)^\top$, $\alpha/a := (\alpha/a_1, \ldots, \alpha/a_n)^\top$, $a/\alpha = (a_1/\alpha, \ldots, a_n/\alpha)^\top$, $a + \alpha = (a_1 + \alpha, \ldots, a_n + \alpha)^\top$, $a^\alpha = (a_1^\alpha, \ldots, a_n^\alpha)^\top$.

7.1   Construction of Index Numbers

Index numbers (or simply indexes) serve as a means for the comparison of the same complex event either among territories at the same time (cross-section) or on the same territory in different times (time series), see [18]. Without loss of generality we will compare the events over time. The well-known indexes are *RPI* (*Retail Price Index*, used in UK; the USA equivalent of RPI is *CPI, Consumer Price Index*) and many of the stock exchange indexes like PX (Prague Stock Exchange), FTSE (Financial Times Stock Exchange), Dow Jones, Standard and Poor's (New York Stock Exchange). All of the mentioned stock exchange indexes appear in various modifications. Let us consider a complex event A, say the cost of living, which may only be observed via some particular events $A_1, \ldots, A_N$ like consumption of food, household expenditures, etc. Usually N is large so that only $n \leq N$ representatives out of $A_1, \ldots, A_N$ can be used for computation purposes. We can renumber the representatives to become $A_1, \ldots, A_n$. In period t, let the indicator of the particular event A_i be p_t^i with weight q_t^i, $t = 0, \ldots, T$, $i = 1, \ldots, n$. Denote $p_t := (p_t^1, \ldots, p_t^n)^\top$, $q_t := (q_t^1, \ldots, q_t^n)^\top$. Index t may be the index of a region or of a time period, e.g. As we note above, we will consider t as time. Similarly, the p's will usually stand for the prices while the q's for the corresponding quantities or weights. It is the goal of the theory of index numbers to find a scalar characteristics of changes of a global price level over time. The *price index* is a number which shows how the complex event A changes over time with changing prices p's, while *quantity index* measures the influence of changes in the quantities q's.

Let s be the initial (base) period and t be the current period, I_{st} the price index describing the change in price level from time s to t. From the historical point of view, the first attempt resulted in the following naive index

$$(1) \qquad I_{st} = \frac{\sum_{i=1}^n p_t^i}{\sum_{i=1}^n p_s^i} = \frac{\mathbf{1}^\top p_t}{\mathbf{1}^\top p_s},$$

which has the disadvantage that it depends on the quantity units in which the prices are given. Another index suggested by Edgeworth is simply the geometric mean of the corresponding ratios of prices:

$$(2) \qquad I_{st}^G = \sqrt[n]{\prod_{i=1}^n \left(\frac{p_t}{p_s}\right)_i}$$

which has the same unfavorable property as (1) but there is an idea behind, i.e., if the ratios p_t^i/p_s^i are random variables possessing a log-normal distribution then the geometric mean would be a good estimator. A better approach starts with the idea that the price index should be a weighted arithmetic mean of the ratios p_t^i/p_s^i with weights $w = (w_1, \ldots, w_n)^\top$:

$$(3) \qquad I_{st} = \sum_{i=1}^{n} w_i \frac{p_t^i}{p_s^i} = w^\top \frac{p_t}{p_s}.$$

A suitable (and generally accepted) choice of weights takes the form

$$(4) \qquad w = \frac{p_\sigma * q_\tau}{p_\sigma^\top q_\tau}.$$

For $\sigma = \tau = s$ we get the *Laspèyres price index*

$$(5) \qquad I_{st}^L = \frac{(p_s * q_s)^\top}{p_s^\top q_s} \frac{p_t}{p_s} = \frac{p_t^\top q_s}{p_s^\top q_s}$$

and putting $\sigma = s$, $\tau = t$ we get *Paasche price index*

$$(6) \qquad I_{st}^P = \frac{(p_s * q_t)^\top}{p_s^\top q_t} \frac{p_t}{p_s} = \frac{p_t^\top q_t}{p_s^\top q_t}.$$

The meaning of the so called *aggregate* $p_\sigma^\top q_\tau$ is clear; it is the price of a consumer's basket if he or she buys quantities q_τ for prices p_σ. The index numbers are defined as the ratios of these aggregates.

In practise, there are two ways of comparison: a) we compare the price level with the initial (base) period and afterwards we obtain $I_{01}^L, \ldots, I_{0T}^L$ if we take the weights from the base period (*base-weighted index*) or $I_{01}^P, \ldots, I_{0T}^P$ if we take the weights from the current period (*current-weighted index*), b) we create the chain of indexes $I_{01}, I_{12}, \ldots, I_{T-1,T}$, with the same meaning as in a).

The index numbers should have some desirable and natural properties: (i) $I_{tt} = 1$, (ii) $I_{st}I_{ts} = 1$ (change of time), (iii) $\prod_{t=0}^{T-1} I_{t,t+1} = I_{0T}$ (chain rule). Neither Laspèyrese nor Paasche index generally fulfil (ii) or (iii). For example,

$$I_{01}^L I_{12}^L = \frac{p_1^\top q_0}{p_0^\top q_0} \frac{p_2^\top q_1}{p_1^\top q_1} \neq I_{02}^L = \frac{p_2^\top q_0}{p_0^\top q_0}.$$

The equality in the last expression is achieved if $q_0 = q_1$ which is not too realistic since the individuals adapt to price level. The ratio

$$(7) \qquad B_{02} = \frac{I_{01} I_{12}}{I_{02}}$$

is called *bias*. Let us examine the bias for the Laspèyres index number. Put $x := p_2/p_1$, $y := q_1/q_0$, and $f = (p_1 * q_0)/p_1^\top q_0$. Obviously $1^\top f = 1$, so that f are weights. Then

$$(8) \qquad B_{02}^L = \frac{p_1^\top q_0}{p_0^\top q_0} \frac{p_2^\top q_1}{p_1^\top q_1} \frac{p_0^\top q_0}{p_2^\top q_0} = \frac{p_2^\top q_1}{p_1^\top q_0} \bigg/ \bigg(\frac{p_1^\top q_1}{p_1^\top q_0} \frac{p_2^\top q_0}{p_1^\top q_0} \bigg) = \frac{(x * y)^\top f}{x^\top f \, y^\top f}.$$

Denote further $\bar{x} := \mathbf{x}^\top \mathbf{f}$, $\bar{y} := \mathbf{y}^\top \mathbf{f}$, the weighted means, $s_x := \sqrt{\mathbf{f}^\top (\mathbf{x} - \bar{x})^2}$, $s_y := \sqrt{\mathbf{f}^\top (\mathbf{y} - \bar{y})^2}$, the weighted standard errors,

$$r_{xy} := \frac{(\mathbf{x} * \mathbf{y})^\top \mathbf{f} - \bar{x}\bar{y}}{s_x s_y},$$

the correlation coefficient, $V_x := s_x/\bar{x}$, $V_y := s_y/\bar{y}$, the coefficients of variation. Then

$$(9) \qquad B_{02}^L = \frac{s_x s_y r_{xy} + \bar{x}\bar{y}}{\bar{x}\bar{y}} = 1 + r_{xy} V_x V_y.$$

If $r_{xy} > 0$ which is the case if the demand increases, $q_1/q_0 \nearrow$, consequently the prices in the next period go up, $p_2/p_1 \nearrow$. We can conclude that the Laspèyres price index has positive bias. The reverse is true for the Paasche index. The basic ideas concerning this problem may be found in [18].

We give three examples of more sophisticated index numbers which avoid some lacks of the above indexes. The *Lowe price index* is defined by

$$(10) \qquad I_{st}^{LW} = \frac{\boldsymbol{p}_t^\top \boldsymbol{q}}{\boldsymbol{p}_s^\top \boldsymbol{q}},$$

where $\boldsymbol{q}$ are weights constant over time, possibly constructed artificially. The *Edgeworth–Marshall price index* takes the weights as the arithmetic mean of the weights of the compared periods $\frac{1}{2}(\boldsymbol{q}_s + \boldsymbol{q}_t)$:

$$(11) \qquad I_{st}^{EM} = \frac{\boldsymbol{p}_t^\top (\boldsymbol{q}_s + \boldsymbol{q}_t)}{\boldsymbol{p}_s^\top (\boldsymbol{q}_s + \boldsymbol{q}_t)}.$$

The geometric mean of the Laspèyrese and Paasche index gives the *Fisher price index number*

$$(12) \qquad I_{st}^F = \sqrt{\frac{\boldsymbol{p}_t^\top \boldsymbol{q}_s}{\boldsymbol{p}_s^\top \boldsymbol{q}_s} \frac{\boldsymbol{p}_t^\top \boldsymbol{q}_t}{\boldsymbol{p}_s^\top \boldsymbol{q}_t}}.$$

Lowe and Fisher indexes have already the desirable properties (i), (ii), and (iii).

7.2 Stock Exchange Indicators

Most of the stock exchange or market indicators are constructed in a similar way as the Laspèyres price index. There are some exceptions, however. We start with one of the oldest indicators, the *Dow Jones Industrial Average* (DJIA) which monitors 30 best stocks (called *blue chips*) traded on *the New York Stock Exchange* (NYSE). It is defined by:

$$\mathrm{DJIA}_t = \frac{1}{D_t} \sum_{i=1}^{30} p_t^i = \frac{1}{D_t} \mathbf{1}^\top \boldsymbol{p}_t,$$

where D_t is called *divisor*. Originally the divisor (in 1928) was just the number of involved stocks, $D_{1928} = 30$. Later it served to ensure continuity of the corresponding time series due to mergers, splits, replacement of the companies in the index, etc. In 1991, $D_{1991} = 0.559$. This phenomenon may be recognized as the change of representatives and the problem of continuity can be generally settled down in the following way. Let t_0 be the time of change. Let $I_1, \ldots, I_{t_0}$ be the values of the indicators based on old representatives, and $I'_{t_0}, \ldots, I'_T$ the values based on new representatives. To ensure the continuity, the following relation must hold:

$$I_{t_0} = C_{t_0} I'_{t_0}.$$

The indicators based on new representatives are afterwards multiplied by C_{t_0}, the *continuity factor*, until a further change of representatives. Hence the series will look like

$$I_1, \ldots, I_{t_0} = C_{t_0} I'_{t_0}, C_{t_0} I'_{t_0+1}, \ldots,$$

till the next change of the representatives. Most of the indicators are also adjusted (multiplied by a factor) to commence with the initial value 100 or 1000, say.

Other market indicators use the weights; the market prices p_t are weighted by the numbers of shares outstanding q_t. Therefore the value of the indicator is

$$C_{t_0} p_t^\top q_t,$$

where C_{t_0} is a proper continuity factor. A popular composite index of this type is *Standard & Poor's 500* (S & P 500) consisting of 400 industrial, 20 transportation, 40 utility, and 40 financial stocks. Another one is *NYSE Composite Index* which consists of about 1600 stocks. Finally, let us mention a sample of other frequently used indicators which are constructed similarly; *NASDAQ* (the National Association of Security Dealers Automatic Quotation), *AMEX* (American Stock Exchange) and non-American indexes *Nikkei* (Tokyo), *FTSI* (Financial Times Share Index, London), *DAX 30* (Germany), *PX 50* (Prague).

7.3   Inflation

In 2.4 we have seen that inflation has an important impact on the determination of the interest rate. *Inflation* means an increase of the general price level and, as a consequence, a decrease of the purchasing power of money. An opposite to inflation is the *deflation* which can occasionally also be observed as in the United Kingdom in the period 1920–1935. Inflation is measured by the *retail price index* (*RPI*, United Kingdom) or by the *consumer price index* (*CPI*, USA).

Usually, the retail price index is constructed as a slight modification to the Laspèyres price index by a government statistical office and its construction is a rather complex task. The weights are derived from the sample surveys of the composition of a consumer basket. They have to change from time to time. At the beginning of last century, the consumer basket consisted mainly of the essentials; in the United Kingdom the weight of food was 60 per cent in 1914; some sixty years later it was only 25 per cent and it decreased to 16 per cent in 1990.

7.3.1   Example (RPI in the Czech Republic). Thousands of goods are grouped in 10 main groups and the total of weights is 1000. The groups and their weights in 1993 were: food 327.1, housing 143.7, transport 104.8, leisure 97.5, clothing 90.9, household goods and services 77.2, other goods and services 50.5, public catering and accommodation 47.2, health care 44.2, education 16.9. Thus the importance of food in the index was 32.71 per cent. Denote these weights as q_{1993}. Let us look on the situation in August 1997 (denoted as 8/1997). The current monthly inflation is calculated from the index

$$I_{7/1997,8/1997} = \frac{p_{8/1997}^{\top} q_{1993}}{p_{7/1997}^{\top} q_{1993}} = 1.007\,,$$

i.e., 100.7 per cent with 7/1997 set to 100 per cent. Thus the monthly inflation was 0.7 per cent. In comparison to August 1996

$$I_{8/1996,8/1997} = \frac{p_{8/1997}^{\top} q_{1993}}{p_{8/1996}^{\top} q_{1993}} = 1.099$$

or 109.9 per cent. Comparison with the yearly average of 1994 is calculated as

$$I_{1994,8/1997} = \frac{p_{8/1997}^{\top} q_{1993}}{p_{1994}^{\top} q_{1993}} = 1.319\,.$$

Finally, the yearly inflation for the period September 1996 to August 1997 (9/1996 − 8/1997) compared to the same period of the past year is calculated from

$$I_{9/1995-8/1996,9/1996-8/1997} = \frac{p_{9/1996-8/1997}^{\top} q_{1993}}{p_{9/1995-8/1996}^{\top} q_{1993}} = 1.079$$

so that the current yearly inflation was 7.9 per cent.

I.8   BASICS OF UTILITY THEORY

utility function, marginal utility, risk aversion, certainty equivalent

8.1   The Concept of Utility

Utility in economic theory means a degree of satisfaction or welfare coming from an economic activity, from possession or consumption of goods. In financial world, by utility we usually mean the welfare originated from investment. Suppose that we have an N-dimensional set of investment opportunities $\mathcal{X}$. For $\mathbf{x} = (x_1, \ldots, x_N)^\top \in \mathcal{X}$, x_n will be understood as the volume of the investment into the nth investment. In *utility theory* we suppose that there is an ordering relation on $\mathcal{X} \times \mathcal{X}$ denoted by $\succsim$. If $\mathbf{x}, \mathbf{y} \in \mathcal{X}$, then $\mathbf{x} \succsim \mathbf{y}$ means that $\mathbf{x}$ is *weakly preferred* to $\mathbf{y}$. If $\mathbf{x} \succsim \mathbf{y}$ but not $\mathbf{y} \succsim \mathbf{x}$ we say that $\mathbf{x}$ is *preferred* to $\mathbf{y}$, and write $\mathbf{x} \succ \mathbf{y}$. If $\mathbf{x} \succsim \mathbf{y}$ and $\mathbf{y} \succsim \mathbf{x}$ we say that $\mathbf{x}$ is *equivalent* to $\mathbf{y}$ and write $\mathbf{x} \sim \mathbf{y}$. It is reasonable to assume that $\forall\, \mathbf{x}, \mathbf{y}, \mathbf{z} \in \mathcal{X}$ either $\mathbf{x} \succsim \mathbf{y}$ or $\mathbf{y} \succsim \mathbf{x}$ (*completeness*), $\mathbf{x} \succsim \mathbf{x}$ (*reflexivity*), and $(\mathbf{x} \succsim \mathbf{y} \wedge \mathbf{y} \succsim \mathbf{z}) \Rightarrow \mathbf{x} \succsim \mathbf{z}$ (*transitivity*).

8.2   Utility Function

If there exists a real valued function $U : \mathcal{X} \to \mathbb{R}$ such that

$$(U(\mathbf{x}) > U(\mathbf{y}) \Leftrightarrow \mathbf{x} \succ \mathbf{y}) \wedge (U(\mathbf{x}) = U(\mathbf{y}) \Leftrightarrow \mathbf{x} \sim \mathbf{y}),$$

it is called *ordinal utility function*, shortly *utility function*, and the underlying theory is known as *ordinal utility theory*. Obviously from $\mathbf{x} \succsim \mathbf{y}$ it follows that $U(\mathbf{x}) \geq U(\mathbf{y})$. For any given $c \in \mathbb{R}$, the set $\mathcal{I}_c = \{\mathbf{x} \in \mathcal{X} : U(\mathbf{x}) = c\}$ is called the *indifference set*. The corresponding plot is called the *indifference surface* or *indifference curve*. This means that from the point of view of an investor, all the investment opportunities from $\mathcal{I}_c$ provide the same degree of satisfaction and the investor cannot distinguish among them.

Finally note that for decision making the utility function contains only the information on ordering. In most cases there is no interpretation of a specific value of $U(\mathbf{x})$. If we consider any increasing function of $U(\mathbf{x})$, the conclusions remain the same. Such an *invariance property* may be an advantage in calculations.

8.2.1   Example. Let us consider the utility function ($N = 2$)

$$u(x_1, x_2) = \sqrt{x_1} + \sqrt{2x_2}.$$

The indifference curves for $c = 2,\ 2.5,\ 3$ are shown in Figure 10.

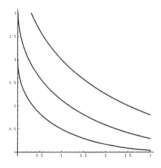

Figure 10: Indifference Curves

In the rightward direction the utility increases.

8.3   Characteristics of Utility Functions

Obviously, the utility function is increasing in the sense of preferences. The slope of the indifference curve can be expressed in terms of the respective derivative $\mathrm{d}x_j/\mathrm{d}x_i$. If $U(\mathbf{x}) = c$ then the total differential

$$\mathrm{d}U = \frac{\partial U}{\partial x_1}\mathrm{d}x_1 + \cdots + \frac{\partial U}{\partial x_N}\mathrm{d}x_N = 0$$

and if we let all $x'_k s$ but x_i and x_j constant we get

(1)
$$S_{ij} := \frac{\mathrm{d}x_j}{\mathrm{d}x_i} = -\frac{\partial U(\mathbf{x})/\partial x_i}{\partial U(\mathbf{x})/\partial x_j}.$$

S_{ij} gives the *marginal rate of substitution*; the increase by one unit of i must be compensated by the increase of S_{ij} units of j. Since usually $S_{ij} < 0$, the increase of i results in a decrease of j by $|S_{ij}|$ units. (Remark, that for unknown reasons in literature the marginal rate of substitution is defined with the opposite sign as $-\mathrm{d}x_j/\mathrm{d}x_i$.)

A simple observation of the behavior of a rational investor leads to the *law of diminishing marginal utility*:

With increasing amount of investment the additional satisfaction or utility will decrease, *ceteris paribus.*

The explanation is simple. Suppose that in situation A you invest USD 10,000 and get 10 per cent return, i.e. USD 1,000. In situation B you have already invested USD 1,000,000 and have got 10 per cent return again, i.e. USD 100,000. If you invest some additional USD 10,000 in situation B, your return will increase to USD 101,000. Surely these additional USD 1,000 in situation B will not be valued as the same amount in situation A.

Mathematically, the law of diminishing marginal utility says that the utility function describing reasonable principles of decision making is concave. We can summarize natural assumptions on a utility function:

The utility function is increasing, concave, and twice differentiable.

It should be emphasized that the invariance property (application of an increasing function to the utility function) does not preserve concavity in general.

In practise, utility functions usually depend on $\mathbf{x}$ through some aggregate functions which may be functions of another utility functions as is the following case. Let the costs of the respective investments be expressed by the *price vector* $\boldsymbol{p} = (p_1, \ldots, p_N)^\top$. Further let us suppose that the wealth of the investor is W so that he or she can choose an arbitrary $\mathbf{x}$ satisfying the condition $\boldsymbol{p}^\top \mathbf{x} \le W$. If we do not suppose an immediate consumption (which brings another problem), we may suppose $\boldsymbol{p}^\top \mathbf{x} = W$. Given a utility function $U = U(\mathbf{x})$ we can define another utility function

$$(2) \qquad U_1(W) = \max_{\mathbf{x}} \left\{ U(\mathbf{x}) : \boldsymbol{p}^\top \mathbf{x} = W \right\}$$

which, given $\boldsymbol{p}$, depends only on W.

8.4 Some Particular Utility Functions

A broad class of utility functions are *separable* in the sense

$$U(\mathbf{x}) = \sum_{n=1}^{N} w_n U_n(x_n)$$

where U_n's are utility functions and w_n's are positive weights. Such a utility function is *additively separable* but since the equivalent utility function

$$\exp\{U(\mathbf{x})\} = \prod_{n=1}^{N} \exp\{w_n U_n(x_n)\}$$

is *multiplicatively separable* we do not need to stress the kind of separability.

An example of the above is

$$(3) \qquad U_\gamma(\mathbf{x}) = \frac{1}{\gamma} \sum_{n=1}^{N} a_n x_n^\gamma$$

for positive constants a_n and $0 < \gamma \le 1$. For $\gamma = 1$ this function is linear and for $\gamma < 1$ it is concave. For $\gamma \to 0_+$ it becomes

$$(4) \qquad U_0(\mathbf{x}) = \sum_{n=1}^{N} a_n \ln x_n$$

which is always concave. This utility function is evidently related to the *Cobb-Douglas production function*

$$(5) \qquad U_{CD}(\mathbf{x}) = \prod_{n-1}^{N} x_n^{a_n}$$

but this function is both increasing and concave only for $0 < a_n < 1$. Next we review some other frequently used types of utility functions. Some of them will be analyzed in the next Section. The utility function

$$(6) \qquad U_H(W) = \frac{1-\gamma}{\gamma}\left(\frac{\beta W}{1-\gamma} + \eta\right)^\gamma$$

is defined for the values W satisfying $\beta W/(1-\gamma) + \eta > 0$, where $\beta > 0$, $\gamma \neq 1$, and η are parameters. The *exponential utility function*

$$(7) \qquad U_E(W) = -\frac{1}{\eta}e^{-\eta W}$$

is defined for $W > 0$, where $\eta > 0$ is a parameter. The already mentioned *power utility function*

$$(8) \qquad U_P(W) = \frac{1}{\gamma}W^\gamma$$

for $\gamma \to 0_+$ becomes the *logarithmic utility function*

$$(9) \qquad U_L(W) = \ln W.$$

8.5 Risk Considerations

In mean-variance portfolio theory (see Chapter 9 for details) there are two decision variables involved: μ and σ, the expected return and risk, respectively. One possible choice among various utility functions is the *quadratic utility function*

$$(10) \qquad U(\mu, \sigma; \kappa) = \mu - \kappa\sigma^2.$$

Since the objective is to maximize the utility with respect to some budget constraints, the parameter κ may be interpreted as a measure of the investor's risk aversion. The higher κ, the more adverse to risk the investor is. An analogy to (10) with parametrized expected return is

$$(11) \qquad U(\mu, \sigma; \lambda) = \lambda\mu - \sigma^2.$$

Obviously, the lower λ, the higher the risk aversion, see II.3.2.1.

So far we have not dealt with random arguments of utility functions. The above mentioned *risk aversion* may be explained as follows. The investor may decide between two investment decisions; the first one results in a fixed certain amount W while the second one results in a random amount $W + \varepsilon$, where ε is a random variable with zero mean and a positive variance, $E\varepsilon = 0$, $\operatorname{var}\varepsilon > 0$. Since we suppose increasing utility functions it follows that with probability $P(\varepsilon > 0)$ the resulting utility will fulfill $U(W + \varepsilon) > U(W)$ but with $P(\varepsilon < 0)$ it will be $U(W + \varepsilon) < U(W)$. The expected value of the resulting amount is the same in both cases, equal to W, but the *risk averse investor* will prefer certain utility to the expected one:

$$(12) \qquad EU(W + \varepsilon) < U(E(W + \varepsilon)) = U(W).$$

It is reasonable to suppose that the last inequality is true for all acceptable levels of W. By Jensen inequality, (12) is assured if U is strictly concave, i.e., $\forall \lambda \in (0,1)$ $\forall w^1, w^2$ $U(\lambda w^1 + (1-\lambda)w^2) > \lambda U(w^1) + (1-\lambda)U(w^2)$.

If the utility function of an investor is linear, then the investor is called *risk-neutral*, and if it is convex, the investor is called *risk loving* or *risk seeking*.

In what follows we will suppose that the investors are risk averse. The investor's aversion to risk can be measured in many ways but there are two measures of particular importance: absolute and relative risk aversion. Assume (see [143], p. 478) that the total investment of USD 10,000 is divided between (risky) stocks and the Treasury bills in equal proportions, USD 5,000 in stocks and USD 5,000 in T-bills. That is the decision of the investor with the initial wish to invest USD 10,000. Suppose now that there are USD 100,000 at the investor's disposal. If the investor increases the amount invested in stocks from USD 5,000 to USD 20,000, say, then he or she manifests the *decreasing absolute risk aversion*. This is the most common behavior of the investors. With increasing investor's wealth, the amount invested in risky assets also increases. Analogously, the *increasing absolute risk aversion* is characterized by the behavior of the investor who reduces the dollar investment into risky assets as his or her wealth increases. If the amount invested into stocks remains the same (not proportion but amount!), we speak of the *constant absolute risk aversion*.

A convenient measure of the absolute risk aversion based on an underlying utility function has been proposed by Arrow and Pratt and is known as the *Pratt-Arrow absolute risk aversion function*:

$$(13) \qquad\qquad A(W) = -\frac{U''(W)}{U'(W)}.$$

The related *relative risk aversion function* is defined by

$$(14) \qquad\qquad R(W) = W A(W).$$

8.5.1 Remark (HARA Utility Functions)

In (6) we have defined a class of utility functions $\{U_H\}$. The utility functions from this class are called *HARA utility functions* (abbreviation for *Hyperbolic Absolute Risk Aversion*). To find the reason just calculate the corresponding $A(W)$ from (13).

8.5.2 Exercise. Derive $A(W)$ and $R(W)$ for utility functions presented in 8.4 and comment the results.

8.6 Certainty Equivalent

We have seen that for risk averse investors the utility function is concave. A natural question arises, what certain amount is needed to achieve the same utility as the expected utility with a random wealth. In other words, let W be a random variable representing the wealth and W_c be an amount called *certainty equivalent*. By the principle, W_c is the amount that satisfies the equation

$$(15) \qquad\qquad U(W_c) = E U(W).$$

Since for strictly concave utility functions $EU(W) < U(EW)$, surely the certainty equivalent satisfies $W_c < EW$.

8.6.1   Example (Multiperiod Certainty Equivalent Model). Suppose a nonnegative random cash flow $\mathbf{CF} = (CF_1, \ldots, CF_T)$, the logarithmic utility function $U(W) = \ln(1 + W)$, and the valuation discount factor v. We are looking for a *certainty equivalent* cash flow $\mathbf{C} = (C_1, \ldots, C_T)$ which gives the holder the same utility in terms of the present value as the expected discounted random cash flow:

$$(16) \qquad \sum_{t=1}^{T} U(C_t)v^t = \sum_{t=1}^{T} EU(CF_t)v^t$$

and which is "minimal" in the following sense:

$$(17) \qquad \min_{\mathbf{C}} \sum_{t=1}^{T} C_t v^t,$$

see [100]. The solution is given by the method of Lagrange multipliers. The corresponding Lagrange function is

$$L(\mathbf{C}, \lambda) = \sum_{t=1}^{T} \left[C_t - \lambda(U(C_t) - EU(CF_t)) \right] v^t.$$

The solution is found by setting the gradient of L equal to zero

$$\frac{\partial L(\mathbf{C})}{\partial \mathbf{C}} = 0$$

which is equivalent to the system of equations

$$1 - \frac{\lambda}{C_t + 1} = 0, \quad t = 1, \ldots, T.$$

We conclude that the certainty equivalent cash flow is constant, $C_t \equiv C^*$, and C^* can be found from (16):

$$(18) \qquad C^* = \exp\left(\frac{\sum_{t=1}^{T} EU(CF_t)v^t}{\sum_{t=1}^{T} v^t} \right) - 1.$$

I.9 MARKOWITZ MEAN–VARIANCE PORTFOLIO

portfolio, efficient market, market portfolio, efficient portfolio, minimum-variance portfolio, Sharpe ratio, optimal portfolios of riskless and risky assets, separation theorem, tangency portfolio, geometry of minimum-variance portfolios

We have mentioned earlier that almost every investment is uncertain with respect to the gain obtained in the future. A natural question arises, is it possible to reduce the risk related to investment by some sophisticated procedure? The answer is yes, and the method is diversification. A very old rule says that you should divide your disposable funds (wealth) into three equal parts; one third put into deposits, one third invest into shares, and buy gold for the remaining third. This approach may seem to be naive but clearly it is a method for reducing risk. Here we deal with more exact, still elementary procedures, which give the investor hints how to diversify his or her funds. We deal with the classical topics concerning optimal portfolio selection, the rigorous treatment of which has been started by Markowitz [112]. In the explanation we will restrict ourselves to financial assets only (shares, bonds, derivatives) despite the fact that the ideas and results may be applied to real assets as well. More details and specific models are treated in Part II.

Although it is not necessary to assume too much for the purpose of the mathematical construction of an optimal portfolio, usually some reasonable and some artificial restrictive economical assumptions are made in this case, and follow Markowitz, Tobin, and Sharpe. A market is said to be the *efficient market* if it fulfills the assumptions below. (We add the comments to the assumptions in brackets: realistic – usually fulfilled, limited – may be fulfilled in most cases, restrictive – unlikely in most cases, unrealistic – hardly to be fulfilled.)

(1) The investors make decisions on their portfolios exclusively on the information based on the expected returns and covariance structure of returns, or in other words they have *homogeneous expectations* (realistic).

(2) The investors choose portfolios with the highest expected return among those with the same risk (rational behavior, realistic).

(3) The investors choose portfolios with the smallest risk among those with the same expected return (risk aversion, realistic).

(4) The assets are infinitely divisible (limited, because trading on a stock exchange is usually performed in lots – a *lot* means one hundred stocks, say – and there are extra costs for trading the fractions of lots).

(5) The investment horizon is one period of time (realistic).

(6) There are no transaction costs and taxes (limited, but the costs or taxes may partly be incorporated into the returns if they are linear functions of a traded volume).

(7) There exists just one riskless interest rate and all the investors can lend or borrow any amount of necessary funds at this riskless interest rate (unrealistic).

(8) All the assets in question are marketable (realistic).

(9) The investors can sell assets short (restrictive, mostly by legal regulations).

(10) No investor can affect the returns of the respective assets substantially (restrictive, since, in other words, it means that there is no investor with funds exceeding the other investors' funds too much).

(11) All necessary information (about means and covariances) are equally available to all the investors at the same time (restrictive).

Under these assumptions, the *market equilibrium* takes place since the investors have perfect knowledge of the market and behave in a rational way (they are risk averse).

9.1   Portfolio

By *portfolio* we mean a group of (financial) assets. A rational investor chooses his/her portfolio so as to maximize the expected return and to minimize the risk. More precisely, let us consider N assets, $1, \ldots, N$, say, and the wealth (disposable money) equal to 1. *Portfolio* is then the vector $\mathbf{x} = (x_1, \ldots, x_N)^\top$, where x_n represents the fraction of the unit wealth invested in the nth asset, $n = 1, \ldots, N$, so that $\mathbf{1}^\top \mathbf{x} = 1$. Generally, at the moment, we do not suppose $x_n \geq 0$ since the case $x_n < 0$ has an economic meaning. This is the case of *short sales*; the investor can sell a security that he or she does not own. It is equivalent to the borrowing of the respective asset, a kind of speculation. Further, let us suppose that the returns (alternatively the rates of return) of the N mentioned assets are random variables $\rho = (\rho_1, \ldots, \rho_N)^\top$ with the *expected returns* $\mathbf{r} = E\rho = (r_1, \ldots, r_N)^\top$ and the covariance matrix $\mathbf{V} = (\sigma_{ij})$ where $\sigma_{ij} = \mathrm{cov}(\rho_i, \rho_j)$, $i, j = 1, \ldots, N$. Alternatively, we will denote the diagonal elements $\sigma_i^2 := \sigma_{ii}$, the σ_i's standing for standard deviations of the returns: $\sigma_i = \sqrt{\sigma_{ii}}$. For a given portfolio p, represented by weights $\mathbf{x}$, the *expected return on the portfolio p* is

(1) $$r_p = \mathbf{r}^\top \mathbf{x}$$

and the *variance of the portfolio* (which is an abbreviation for the variance of the portfolio return) p is

(2) $$\sigma_p^2 = \mathbf{x}^\top \mathbf{V} \mathbf{x}.$$

The *risk of the portfolio p* is simply the standard deviation

(3) $$\sigma_p = \sqrt{\sigma_p^2}.$$

9.1.1   Example. Let us consider two assets A and B in the period of nine years with the corresponding returns (in per cent): 17, 13, 15, 20, 10, 16, 14, 12, 18 for the asset A and 13, 17, 15, 10, 20, 14, 16, 18, 12 for the asset B. The mean returns for both the assets based on these historical data are the same and equal to 15. The risks are also the same, 3.1225. If we invest all the unit wealth to any of the assets, we can expect a risky return 15 per cent. If we divide our unit wealth equally between the two assets, we have certain return 15 per cent over the given time interval with no risk. This is the case if the returns are perfectly negatively correlated, see Figure 11.

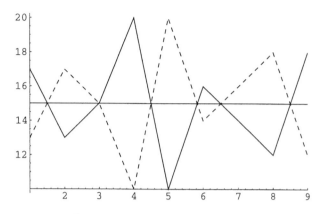

Figure 11: Returns of two assets

9.1.2 Market and Efficient Portfolio

A *market portfolio* is any portfolio in which all the assets come in the same fractions as they appear on the capital market, expressed by their capitalization (market value of the respective asset multiplied by the number of the assets). This is an abstract notion and in practise we usually substitute it by a properly chosen stock exchange index. A portfolio $\mathbf{x}^\star$ is called *efficient portfolio* if there is no other portfolio $\mathbf{x}$ such that

$$(\mathbf{r}^\top \mathbf{x}^\star < \mathbf{r}^\top \mathbf{x} \,\wedge\, \mathbf{x}^{\star\top} \mathbf{V} \mathbf{x}^\star \geq \mathbf{x}^\top \mathbf{V} \mathbf{x}) \vee (\mathbf{r}^\top \mathbf{x}^\star = \mathbf{r}^\top \mathbf{x} \,\wedge\, \mathbf{x}^{\star\top} \mathbf{V} \mathbf{x}^\star > \mathbf{x}^\top \mathbf{V} \mathbf{x}).$$

In other words, an efficient portfolio is a portfolio for which there is no other portfolio with the same or greater expected return and smaller risk.

9.2 Construction of Optimal Portfolios and Separation Theorems

There is a variety of problems concerning the choice of an "optimal portfolio". Our decisions here will be based just on the information about the expected returns and the covariance structure of the returns. This is known as *Markowitz approach*, see Part II for more details. Two basic problems appear in this context:

(i) to minimize $\frac{1}{2}\mathbf{x}^\top \mathbf{V} \mathbf{x}$ subject to $\mathbf{1}^\top \mathbf{x} = 1$, $\mathbf{r}^\top \mathbf{x} = \mu$, where μ is the prescribed expected return. In other words, the investor seeks the expected return μ with minimum risk. The corresponding portfolio is called *minimum-variance portfolio*. Note that minimizing the risk is equivalent to minimizing the variance.

(ii) to maximize the so called *Sharpe's ratio* or *Sharpe's measure of portfolio*

(4) $$\frac{\text{expected return on portfolio}}{\text{risk of portfolio}} = \frac{\mathbf{r}^\top \mathbf{x}}{\sqrt{\mathbf{x}^\top \mathbf{V} \mathbf{x}}}$$

subject to $\mathbf{1}^\top \mathbf{x} = 1$.

In any of the problems above the following cases may be considered: $\mathbf{x}$ arbitrary (short sales allowed), $\mathbf{x} \geq \mathbf{0}$ (short sales are not allowed), $\mathbf{V}$ positive definite (which implies that there is no riskless asset), or $\mathbf{V}$ just positive semidefinite. The latter case may occur if there exists a riskless asset or if the returns of two assets are perfectly correlated, e.g., in which case the matrix $\mathbf{V}$ is singular. Note that if $\mathbf{V}$ is singular then (ii) has no sense.

9.2.1 Example. Let us consider two assets with expected returns $\mathbf{r} = (8, 14)^\top$, $\sigma_{11} = 9$, $\sigma_{22} = 36$, $\sigma_{12} = \sigma_{21} = \varrho\sqrt{\sigma_{11}\sigma_{22}} = 18\varrho$ where ϱ is the correlation between returns ρ_1 and ρ_2 so that

$$\mathbf{V} = \begin{pmatrix} 9 & 18\varrho \\ 18\varrho & 36 \end{pmatrix}.$$

We will analyse the portfolio of the two assets with ϱ as a parameter. Obviously, the risks of the assets 1 and 2 are 3 and 6, respectively. Let $\mathbf{x} = (x_1, x_2)^\top$ denote a portfolio p. Since $x_1 + x_2 = 1$, we can express the expected return on the portfolio

$$r_p = 8x_1 + 14(1 - x_1) = 14 - 6x_1$$

and the variance of the portfolio

$$\sigma_p^2 = 9x_1^2 + 36\varrho x_1(1 - x_1) + 36(1 - x_1)^2.$$

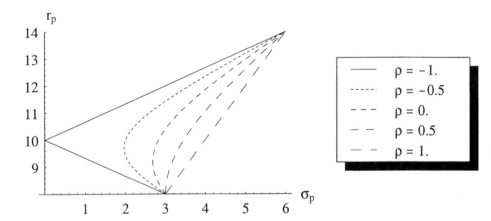

Figure 12: Efficient Frontiers

The dependence between the expected return and risk is usually plotted in the *risk – expected-return plane* or the *standard-deviation – expected-return plane* and the corresponding curves are called *efficient frontiers*. For selected values of ϱ, the dependence is illustrated in Figure 12 for $x_1 \in [0, 1]$. We see that zero risk can be attained only in the case of perfect negative correlation between the returns, $\varrho = -1$. Next we find the minimum-variance portfolio. Since for $|\varrho| < 1$ the portfolio variance σ_p^2 is a positive definite quadratic form, it suffices to find the root of the equation

$$\frac{d\sigma_p^2}{dx_1} = -18(-2\varrho + x_1(4\varrho - 5) + 4) := 0.$$

The solution to this equation is

$$x_1^\star = \frac{2(\varrho - 2)}{4\varrho - 5} \qquad x_2^\star = 1 - x_1^\star = \frac{2\varrho - 1}{4\varrho - 5}.$$

The corresponding expected return and variance of the portfolio are

$$r_p^\star = \frac{44\varrho - 46}{4\varrho - 5} \qquad \sigma_p^{2\star} = \frac{36(\varrho^2 - 1)}{4\varrho - 5}.$$

A simple analysis shows that for $\varrho \leq 0.5$ the corresponding $x_1^\star$ is in the range $[0, 1]$, while for $\varrho > 0.5$ it exceeds 1, so that to reach the minimum risk it is necessary to sale short or to borrow the asset 2. In the extreme case $\varrho = 1$, $x_1^\star = 2$, so that the necessary additional fund is obtained by selling the asset 2 of value 1 short, $x_2^\star = -1$. With zero risk, the maximum expected return is only attainable for $\varrho = -1$ in which case the expected return is 10. This can rarely happen in the real world. We can also observe that for all risks but one there are two portfolios with the same risk but two different expected returns.

9.2.2 Remark

The reader may verify that in case of two assets with $\mathbf{V} = (\sigma_{ij})$, $i, j = 1, 2$, the minimum-variance portfolio has

$$x_1^\star = \frac{\sigma_2(\sigma_2 - \varrho\sigma_1)}{\sigma_{11} - 2\varrho\sigma_1\sigma_2 + \sigma_{22}}.$$

Particularly, for $\varrho = -1$ we have $x_1^\star = \frac{\sigma_2}{\sigma_1 + \sigma_2}$, for $\varrho = 0$ $x_1^\star = \frac{\sigma_{22}}{\sigma_{11} + \sigma_{22}}$, and for $\varrho = 1$ $x_1^\star = \frac{\sigma_2}{\sigma_2 - \sigma_1}$ provided $\sigma_1 \neq \sigma_2$.

9.2.3 Example (Riskless Asset).

Suppose that we have just two assets, one riskless with return r_0, and one risky (call it A), with expected return r_A and variance σ_A^2. Let us build a portfolio from these two assets with weights x_0 standing for the riskless asset and x_A for the risky asset, $x_0 + x_A = 1$. Then the expected return on the portfolio is $r_p = (1 - x_A)r_0 + x_A r_A$ with variance $\sigma_p^2 = x_A^2\sigma_A^2$. The dependence in the *risk – expected-return* plane is linear. For $0 \leq x_A \leq 1$ it means that the investor lends the portion $1 - x_A$ of his or her money at the interest rate r_0 while for $x_A > 1$, the investor borrows at the riskless interest rate. Borrowing money at the riskless interest rate seems not to be quite realistic, however. The Government are an exception.

9.2.4   General Solution (Risky Assets, Short Sales Allowed)

Here we will solve the problem

minimize $\frac{1}{2}\mathbf{x}^\top\mathbf{V}\mathbf{x}$ subject to $\mathbf{1}^\top\mathbf{x} = 1$, $\mathbf{r}^\top\mathbf{x} = \mu$, μ prescribed.

Suppose $\mathbf{V}$ positive definite. We exclude the case $\mathbf{r} = k\mathbf{1}$ for some constant k since in this case the solution is trivial; simply take just one asset n_0 for which $n_0 = \arg\min_{1 \leq n \leq N} \sigma_n^2$. Now we can obtain the solution by the method of Lagrange multipliers. The Lagrange function for the problem is

$$L(\mathbf{x}, \lambda_1, \lambda_2) = \frac{1}{2}\mathbf{x}^\top\mathbf{V}\mathbf{x} + \lambda_1(1 - \mathbf{1}^\top\mathbf{x}) + \lambda_2(\mu - \mathbf{r}^\top\mathbf{x})$$

and the equation

$$\frac{\partial}{\partial\mathbf{x}}L = \mathbf{V}\mathbf{x} - \lambda_1\mathbf{1} - \lambda_2\mathbf{r} = 0$$

gives the optimal solution

(5)
$$\boxed{\mathbf{x}^\star = \lambda_1\mathbf{V}^{-1}\mathbf{1} + \lambda_2\mathbf{V}^{-1}\mathbf{r}.}$$

Put $A := \mathbf{1}^\top\mathbf{V}^{-1}\mathbf{1}$, $B := \mathbf{1}^\top\mathbf{V}^{-1}\mathbf{r}$, $C := \mathbf{r}^\top\mathbf{V}^{-1}\mathbf{r}$, and $\Delta := AC - B^2$. Obviously, $A > 0$, $C > 0$, $\Delta > 0$. The last inequality is a simple consequence of the Schwarz inequality since we have supposed $\mathbf{1}$ and $\mathbf{r}$ linearly independent. The constants λ_1, λ_2 can be derived from the initial conditions:

$$1 = \mathbf{1}^\top\mathbf{x} = \lambda_1 A + \lambda_2 B,$$

$$\mu = \mathbf{r}^\top\mathbf{x} = \lambda_1 B + \lambda_2 C,$$

so that

(6)
$$\boxed{\lambda_1 = \frac{C - \mu B}{\Delta} \qquad \lambda_2 = \frac{\mu A - B}{\Delta}.}$$

Now we have to distinguish the two cases:

(a) $\underline{\mathbf{1}^\top\mathbf{V}^{-1}\mathbf{r} = 0}$

First, let us note that we can hardly meet this case in practise, but, from the theoretical point of view a for given $\mathbf{V}$ we can find a subspace of $\mathbf{r}$'s of dimension $N - 1$ satisfying $\mathbf{1}^\top\mathbf{V}^{-1}\mathbf{r} = 0$. In this case

$$\lambda_1 = \frac{1}{\mathbf{1}^\top\mathbf{V}^{-1}\mathbf{1}}, \qquad \lambda_2 = \frac{\mu}{\mathbf{r}^\top\mathbf{V}^{-1}\mathbf{r}},$$

so that the minimum-variance portfolio is

$$\mathbf{x}^\star = \frac{\mathbf{V}^{-1}\mathbf{1}}{\mathbf{1}^\top\mathbf{V}^{-1}\mathbf{1}} + \frac{\mu\mathbf{V}^{-1}\mathbf{r}}{\mathbf{r}^\top\mathbf{V}^{-1}\mathbf{r}}.$$

(b) $\underline{\mathbf{1}^\top\mathbf{V}^{-1}\mathbf{r} \neq 0}$

Put

(7)
$$\mathbf{x}^1 = \frac{\mathbf{V}^{-1}\mathbf{1}}{\mathbf{1}^\top\mathbf{V}^{-1}\mathbf{1}}, \qquad \mathbf{x}^2 = \frac{\mathbf{V}^{-1}\mathbf{r}}{\mathbf{1}^\top\mathbf{V}^{-1}\mathbf{r}}.$$

The minimum-variance portfolio may now be expressed in the form (with δ's dependent on μ)

$$\mathbf{x}^\star = \delta_1\mathbf{x}^1 + \delta_2\mathbf{x}^2$$

with $\delta_1 = A(C - \mu B)/\Delta =: \delta(\mu)$ and $\delta_2 = B(\mu A - B)/\Delta = 1 - \delta(\mu)$ where $\mathbf{x}^1$, $\mathbf{x}^2$ may be considered as the basis portfolios.

Note that

$$\mathbf{1}^\top\mathbf{x}^\star = 1 = \delta_1\mathbf{1}^\top\mathbf{x}^1 + \delta_2\mathbf{1}^\top\mathbf{x}^2 = \delta_1 + \delta_2.$$

The optimal portfolio may be expressed in an alternative form:

(8)
$$\mathbf{x}^\star = \delta(\mu)\mathbf{x}^1 + (1 - \delta(\mu))\mathbf{x}^2.$$

It is important to emphasize that the basis portfolios $\mathbf{x}^1$ and $\mathbf{x}^2$ are independent of the prescribed μ but the weight $\delta(\mu)$ does depend on μ.

9.2.5 Remark (Alternative Form of the Minimum-Variance Portfolio)

Put $\mathbf{z}_1 = \frac{1}{\Delta}(C\mathbf{V}^{-1}\mathbf{1} - B\mathbf{V}^{-1}\mathbf{r})$ and $\mathbf{z}_2 = \frac{1}{\Delta}(A\mathbf{V}^{-1}\mathbf{r} - B\mathbf{V}^{-1}\mathbf{1})$, where $\mathbf{z}_1$ is a portfolio. Then (8) may be expressed in the form

(9)
$$\mathbf{x}^\star = \mathbf{z}_1 + \mu\mathbf{z}_2.$$

9.2.6 Two Funds Separation Theorem. *Let $\mathbf{x}_a$, $\mathbf{x}_b$ be two minimum-variance portfolios with expected returns r_a, r_b, respectively, $r_a \neq r_b$. Then every minimum-variance portfolio $\mathbf{x}_c$ can be expressed in the form $\mathbf{x}_c = \alpha\mathbf{x}_a + (1-\alpha)\mathbf{x}_b$ for some α. Every portfolio of the form $\mathbf{x}_c = \alpha\mathbf{x}_a + (1-\alpha)\mathbf{x}_b$ is a minimum-variance portfolio.*

Proof. Let r_c denote the expected return on the minimum-variance portfolio $\mathbf{x}_c$. Choose α such that $r_c = \alpha r_a + (1 - \alpha)r_b$, that is, $\alpha = (r_c - r_b)/(r_a - r_b)$. The coefficients λ_1, λ_2 in (6) for portfolios a, b are

$$\lambda_{1i} = \frac{1}{\Delta}(C - r_iB), \qquad \lambda_{2i} = \frac{1}{\Delta}(r_iA - B), \quad i = a,b$$

and since $\mathbf{x}_c$ is also a minimum-variance portfolio, the above relations hold for $i = c$ as well. Now

$$\mathbf{x}_c = \lambda_{1c}\mathbf{V}^{-1}\mathbf{1} + \lambda_{2c}\mathbf{V}^{-1}\mathbf{r} = \frac{1}{\Delta}(C - r_cB)\mathbf{V}^{-1}\mathbf{1} + \frac{1}{\Delta}(r_cA - B)\mathbf{V}^{-1}\mathbf{r} =$$

$$\frac{1}{\Delta}(C - \alpha r_aB - (1 - \alpha)r_bB)\mathbf{V}^{-1}\mathbf{1} + \frac{1}{\Delta}(\alpha r_aA + (1 - \alpha)r_bA - B)\mathbf{V}^{-1}\mathbf{r} =$$

$$\frac{1}{\Delta}(\alpha(C - r_aB) - (1-\alpha)(C - r_bB))\mathbf{V}^{-1}\mathbf{1} + \frac{1}{\Delta}(\alpha(r_aA - B) + (1-\alpha)(r_bA - B))\mathbf{V}^{-1}\mathbf{r} =$$

$$(\alpha\lambda_{1a} + (1 - \alpha)\lambda_{1b})\mathbf{V}^{-1}\mathbf{1} + (\alpha\lambda_{2a} + (1 - \alpha)\lambda_{2b})\mathbf{V}^{-1}\mathbf{r} =$$

$$\alpha(\lambda_{1a}\mathbf{V}^{-1}\mathbf{1} + \lambda_{2a}\mathbf{V}^{-1}\mathbf{r}) + (1 - \alpha)(\lambda_{1b}\mathbf{V}^{-1}\mathbf{1} + \lambda_{2b}\mathbf{V}^{-1}\mathbf{r}) =$$

$$\alpha\mathbf{x}_a + (1 - \alpha)\mathbf{x}_b.$$

The second assertion is obvious. $\square$

9.2.7   Remark (Covariance Between the Returns of Two Minimum-Variance Portfolios)

Let x_a, x_b be two minimum-variance portfolios with expected returns r_a, r_b, respectively. Then, after some algebra, we get the covariance

$$(10) \quad \operatorname{cov}(x_a^\top \rho, x_b^\top \rho) = x_a^\top V x_b =$$

$$(\lambda_{1a} V^{-1} 1 + \lambda_{2a} V^{-1} r)^\top V (\lambda_{1b} V^{-1} 1 + \lambda_{2b} V^{-1} r) = \frac{1}{\Delta}(A r_a r_b - B r_a - B r_b + C).$$

As a consequence, the variance of a minimum-variance portfolio x with the expected return r is

$$(11) \quad \boxed{\sigma^2(r) := \operatorname{var} x^\top \rho = \frac{1}{\Delta}(A r^2 - 2 B r + C).}$$

The *global minimum-variance portfolio* x_G is defined as the portfolio for which the variance $\sigma^2(r)$ attains its minimum. We have

$$\frac{\partial \sigma^2(r)}{\partial r} = \frac{2 A r - 2 B}{\Delta}.$$

Thus the expected return r_G corresponding to the global minimum-variance portfolio is

$$r_G = \frac{B}{A}$$

so that $\lambda_{1G} = 1/A$, $\lambda_{2G} = 0$ and

$$(12) \quad x_G = \frac{V^{-1} 1}{1^\top V^{-1} 1}$$

and the variance of the portfolio x_G is

$$\operatorname{var}(x_G^\top \rho) = \frac{1}{A}.$$

The usual graphical representation of the set of minimum-variance portfolios is in the so called *expected-return–variance plane* or in the *expected-return – standard-deviation plane*. The resulting plot is also known as *minimum-variance frontier*. From the expression for the variance of minimum-variance portfolios $\sigma^2(r)$ we immediately see that the dependence of the variance on any given expected return is expressed as a parabola while the dependence of the risk on any given expected return r is expressed as a hyperbola:

$$(13) \quad \sigma(r) = \sqrt{\frac{1}{\Delta}(A r^2 - 2 B r + C)}.$$

The focus of this hyperbola is at the point $x_G = B/A$ and $\sigma(r_G) = 1/\sqrt{A}$. The derivative of $\sigma(r)$ is

$$\sigma'(r) = \frac{A r - B}{\sqrt{\Delta(A r^2 - 2 B r + C)}}.$$

Taking the limits of this expression as $r \to \infty$, $r \to -\infty$, we get the slopes of the asymptotes of the hyperbola, $\sqrt{A/\Delta}$, $-\sqrt{A/\Delta}$, respectively. For historical reasons, the plot is in the form where the standard deviation (risk) is on the horizontal axis while the expected return is on the vertical one. The asymptotes expressed as functions of σ are $r(\sigma) = B/A \pm \sigma \sqrt{\Delta/A}$ in this case.

9.2.8 Remark

With the exception $\mu = B/A$, there are two minimum-variance portfolios with the same risk but two different expected returns. If we have the prescribed expected return $\mu < B/A$ then a simple calculation shows that the minimum-variance portfolio with the prescribed expected return $\tilde{\mu} = 2B/A - \mu$ has the same variance, $\sigma^2(\mu) = \sigma^2(\tilde{\mu})$. Thus, in accordance with the definition of an efficient portfolio, the *set of efficient portfolios* consists of all minimum-variance portfolios with expected returns $\mu \geq B/A$. In literature, the minimum-variance portfolios with expected returns less than B/A are often called *inefficient portfolios*.

9.2.9 Remark (Orthogonal Minimum-Variance Portfolios)

Let us seek the condition for expected returns of two minimum-variance portfolios a, b with uncorrelated returns. (Verify that this problem has no solution if either of these portfolios is the global minimum-variance portfolio.) From the formula in Remark 2.8.4 it follows that $Ar_ar_b - Br_a - Br_b + C = 0$ so that $r_a = (Br_b - C)/(Ar_b - B)$, $r_b \neq B/A$, or equivalently, $r_b = (Br_a - C)/(Ar_a - B)$, $r_a \neq B/A$. Note that portfolio a is efficient if and only if b is inefficient.

9.2.10 Remark

The global minimum-variance portfolio has a peculiar covariance property. We have cov$(\rho^\top x_G, \rho^\top x) = 1/A$ for every portfolio x. This is of course also valid for any single asset: cov$(\rho^\top x_G, \rho_n) = 1/A$, $n = 1, \ldots, N$.

9.2.11 Maximum of Sharpe's Ratio (Risky Assets, Short Sales Allowed)

There is no straightforward approach to the problem of direct maximizing the Sharpe's ratio defined in (4). Instead, we will solve the problem
maximize the square of the Sharpe's ratio

$$(14) \qquad \frac{(r^\top x)^2}{x^\top V x}$$

subject to $1^\top x = 1$, V positive definite.

It is important to emphasize that the two problems are not equivalent. The maximum of (14) may be reached for a portfolio giving negative expected return. Such a result is useless, of course. Despite the fact that such a case can be rarely met with on efficient markets, it is necessary to be careful when handling the emerging markets or in the cases where the investors are not risk averse or simply do not pay attention to the return to risk ratio.

To attack the problem first note that from the assumption of positive definiteness of the matrix V it follows that there exists a symmetric square root matrix $V^{1/2}$. From Schwarz inequality it follows that

$$(r^\top x)^2 = (r^\top V^{-1/2} V^{1/2} x)^2 \leq (r^\top V^{-1} r)(x^\top V x)$$

so that $r^\top V^{-1} r$ is the upper bound for the squared Sharpe's ratio, and the equality holds if and only if $x = \lambda V^{-1} r$ for some λ. Since x should be a portfolio, it follows that $\lambda = 1/\mathbf{1}^\top \mathbf{V}^{-1}\mathbf{r}$ provided the denominator is nonzero. If the denominator equals zero then there is no solution to the problem. With this exception, the optimal portfolio is

$$(15) \qquad \mathbf{x}^\star = \frac{\mathbf{V}^{-1}\mathbf{r}}{\mathbf{1}^\top \mathbf{V}^{-1}\mathbf{r}}.$$

9.2.12 General Solution (Riskless and Risky Assets, Short Sales Allowed)

In the presence of a *riskless asset* (also called *riskfree asset*) we can not fully adopt the above theory since the covariance matrix between returns becomes singular. The modification of the previous results is possible, however. The portfolio selection problem may now be formulated in the following way. Suppose we have N risky assets $1, \ldots, N$ with expected returns $\mathbf{r}$ as above, $\mathbf{r} \neq k\mathbf{1}$, and one riskless asset 0 with return r_0. Let $\widetilde{\rho} = (r_0, \boldsymbol{\rho}^\top)^\top$ denote the $(N+1) \times 1$ vector of the returns. It is economically plausible to suppose that on efficient markets the riskless return r_0 is less than the expected return on any risky efficient portfolio. Since the global minimum-variance portfolio possesses the expected return $r_G = B/A$, we will therefore assume

$$r_0 < \frac{B}{A}.$$

The covariance matrix of returns of the risky assets $\mathbf{V}$ is again assumed to be positive definite. The unit wealth is allocated among $N+1$ assets $0, 1, \ldots, N$ with weights $x_0, x_1, \ldots, x_N$, and we are seeking a portfolio p represented as $\widetilde{\mathbf{x}} = (x_0, \mathbf{x}^\top)^\top$ which minimizes the squared risk (independent of the portion of the riskless asset)

$$(16) \qquad \frac{1}{2}\mathbf{x}^\top \mathbf{V}\mathbf{x}$$

under the conditions

$$(17) \qquad \mathbf{1}^\top \widetilde{\mathbf{x}} = 1 \qquad x_0 r_0 + \mathbf{r}^\top \mathbf{x} = \mu$$

where μ is the prescribed expected return on the portfolio and symbol $\mathbf{1}$ now means the $(N+1) \times 1$ vector of 1's. Taking into account that $x_0 = 1 - \mathbf{1}^\top \mathbf{x}$, the second condition may be rewritten as

$$(18) \qquad (\mathbf{r} - r_0\mathbf{1})^\top \mathbf{x} = \mu - r_0 =: \mu_e$$

where μ_e is called *expected excess return*, that means the return over the return of the riskless asset. So we are forced to solve the problem of finding

$$(19) \qquad \min \frac{1}{2}\mathbf{x}^\top \mathbf{V}\mathbf{x} \quad \text{under condition} \quad (\mathbf{r} - r_0\mathbf{1})^\top \mathbf{x} = \mu_e.$$

Weight x_0 is not involved since afterwards it will be calculated using $x_0 = 1 - \mathbf{1}^\top \mathbf{x}$. The Lagrange function for this problem is

$$L(\mathbf{x}, \gamma) = \frac{1}{2}\mathbf{x}^\top \mathbf{V}\mathbf{x} + \gamma(\mu_e - (\mathbf{r} - r_0\mathbf{1})^\top \mathbf{x})$$

and from the equation

$$\frac{\partial L}{\partial \mathbf{x}} = \mathbf{V}\mathbf{x} + \gamma(r_0\mathbf{1} - \mathbf{r}) = \mathbf{0}$$

we obtain the optimal solution

(20)
$$\boxed{\mathbf{x}^\star = \gamma\mathbf{V}^{-1}(\mathbf{r} - r_0\mathbf{1}) \qquad x_0^\star = 1 - \mathbf{1}^\top\mathbf{x}^\star}$$

with γ satisfying the condition

$$(\mathbf{r} - r_0\mathbf{1})^\top\gamma\mathbf{V}^{-1}(\mathbf{r} - r_0\mathbf{1}) = \mu_e$$

or

$$\gamma = \frac{\mu_e}{Ar_0^2 - 2Br_0 + C},$$

where A, B, C are defined in 2.8. Such a portfolio is the portfolio with minimum risk with prescribed expected excess return μ_e and will be called *minimum-variance portfolio*.

9.2.13 Two Funds Separation Theorem with Riskless and Risky Assets

Define $\widetilde{\mathbf{x}}^1 := (1, 0, \ldots, 0)^\top$, the portfolio consisting of riskless asset only, and by

(21)
$$\mathbf{x}^t = \frac{\mathbf{V}^{-1}(\mathbf{r} - r_0\mathbf{1})}{B - Ar_0},$$

the so called *tangency portfolio*, and $\widetilde{\mathbf{x}}^2 := (0, \mathbf{x}^{t\top})^\top$.

9.2.14 Two Funds Separation Theorem. *Every minimum-variance portfolio can be expressed in the form*

$$\widetilde{\mathbf{x}}^\star = \delta\widetilde{\mathbf{x}}^1 + (1 - \delta)\widetilde{\mathbf{x}}^2$$

where

$$\delta = \delta(\mu_e) = 1 - \frac{\mu_e(B - Ar_0)}{Ar_0^2 - 2Br_0 + C}$$

Proof. The proof is obvious. $\square$

9.2.15 Corollary

Every portfolio consisting of minimum-variance portfolios is a minimum-variance portfolio.

9.2.16 Remark (Covariance Between the Returns of Two Minimum-Variance Portfolios)

Let $\tilde{\mathbf{x}}_a$, $\tilde{\mathbf{x}}_b$ be two minimum-variance portfolios with expected excess returns μ_{ea}, μ_{eb}, respectively. With weights δ's given by Theorem 2.10.1 we get

$$(22) \quad \text{cov}\,(\tilde{\mathbf{x}}_a^\top \tilde{\rho}, \tilde{\mathbf{x}}_b^\top \tilde{\rho}) = \text{cov}\,((\delta_a \tilde{\mathbf{x}}^1 + (1 - \delta_a)\tilde{\mathbf{x}}^2)^\top \tilde{\rho}, (\delta_b \tilde{\mathbf{x}}^1 + (1 - \delta_b)\tilde{\mathbf{x}}^2)^\top \tilde{\rho}) =$$

$$(1 - \delta_a)(1 - \delta_b)\text{cov}\,(\rho^\top \mathbf{x}^t, \rho^\top \mathbf{x}^t) = \frac{\mu_{ea}\mu_{eb}}{Ar_0^2 - 2Br_0 + C}.$$

Thus the variance of a minimum-variance portfolio in the presence of a riskless asset with expected return μ becomes

$$(23) \qquad \boxed{\sigma_0^2(\mu) = \frac{(\mu - r_0)^2}{Ar_0^2 - 2Br_0 + C}.}$$

9.2.17 Remark (Properties of Tangency Portfolio)

A simple calculation shows that the tangency portfolio has the expected return $r^t = (C - Br_0)/(B - Ar_0)$ so that the expected excess return is

$$(24) \qquad \mu_{et} = \frac{Ar_0^2 - 2Br_0 + C}{B - Ar_0}.$$

As a consequence, the variance of the tangency portfolio is

$$(25) \qquad \sigma_t^2 := \text{var}\,(\rho^\top \mathbf{x}^t) = \frac{Ar_0^2 - 2Br_0 + C}{(B - Ar_0)^2}.$$

Note that, since both the numerator and denominator in (24) are positive, also the expected excess return is positive.

9.2.18 Assertion. *The tangency portfolio belongs to the set of efficient portfolios of risky assets.*

Proof. Since the expected return on the tangency portfolio is $r^t = (C - Br_0)/(B - Ar_0)$, we just calculate the Lagrange multipliers:

$$\lambda_{1t} = \frac{r_0}{B - Ar_0}, \qquad \lambda_{2t} = \frac{1}{B - Ar_0}.$$

□

9.2.19 Remark

For expected excess return $\mu_e \in [0, \mu_{et}]$ we get $\delta \in [0, 1]$. This is the case of no short sales either of the riskless asset or of the tangency portfolio. A very unrealistic case is the case of borrowing the riskless asset, i.e., $\delta < 0$ which leads to higher expected returns than the tangency portfolio provides.

9.2.20 Remark (Geometry of the Minimum-Variance Portfolios with a Riskless Asset)

The dependence of the variance on the expected return in the expected-return – variance plane is again a parabola but in the expected-return – standard-deviation plane it becomes straight line

$$(26) \qquad \sigma_0(r) = \frac{r - r_0}{\sqrt{Ar_0^2 - 2Br_0 + C}}, \qquad r > r_0.$$

We have already mentioned that the tangency portfolio is a member of the set of efficient portfolios of risky assets. The point corresponding to this portfolio in the expected-return – standard-deviation plane is

$$\mathbf{P}_t = \{\frac{C - Br_0}{B - Ar_0}, \frac{\sqrt{Ar_0^2 - 2Br_0 + C}}{B - Ar_0}\}.$$

The line connecting the points $\mathbf{P}_0 = \{r_0, 0\}$ and $\mathbf{P}_t$ may be expressed as

$$y(r) = \frac{1}{\sqrt{Ar_0^2 - 2Br_0 + C}}(r - r_0).$$

For the derivative of the standard deviation of the return of the minimum-variance portfolio consisting of risky assets only we have

$$\frac{\partial \sigma(r)}{\partial r} = \frac{Ar - B}{\sqrt{\Delta(Ar^2 - 2Br + C)}}$$

and if we substitute the expected return on the tangency portfolio,

$$r \to (C - Br_0)/(B - Ar_0)$$

into the last expression, we get the tangency

$$\frac{1}{\sqrt{Ar_0^2 - 2Br_0 + C}}.$$

So $\mathbf{P}_t$ is the tangency point of the hyperbola and therefore line $y(r)$ is the tangency line to the hyperbola.

9.2.21 Remark (Short Sales not Allowed)

If short sales are not allowed, we are not able to give an explicit solution to the problem. The solution may be found by solving the *quadratic optimization problem*:

$$(27) \quad \min \tfrac{1}{2}\mathbf{x}^\top \mathbf{V}\mathbf{x} \quad \text{under the conditions} \quad x_0 = 1 - \mathbf{1}^\top \mathbf{x} \geq 0 \quad (\mathbf{r} - r_0\mathbf{1})^\top \mathbf{x} \geq \mu_e.$$

I.10   CAPITAL ASSET PRICING MODEL

market portfolio, Sharpe-Lintner model, security market line, capital market line

10.1   Sharpe-Lintner Model

In this Chapter we keep the notation of 9.2.12 and the assumptions of an efficient market. *Capital Asset Pricing Model*, shortly *CAPM*, expresses the expected excess returns of the individual assets in terms of the market expected excess return.

10.1.1   Alternative Form of the Expected Excess Return

Denote $\mu_t := \mathbf{r}^\top \mathbf{x}^t$ the expected return on the tangency portfolio and $\boldsymbol{\sigma}_t$ the vector of covariances between excess returns of the risky assets and the excess return on the tangency portfolio. We have

$$(1) \qquad \boldsymbol{\sigma}_t = \mathrm{cov}\,(\boldsymbol{\rho} - r_0\mathbf{1}, \boldsymbol{\rho}^\top \mathbf{x}^t) = \mathbf{V}\mathbf{x}^t = \frac{\mathbf{r} - r_0\mathbf{1}}{B - Ar_0}.$$

Hence the variance of the tangency portfolio is

$$(2) \qquad \sigma_t^2 = \mathbf{x}^{t\top}\mathbf{V}\mathbf{x}^t = \frac{\mathbf{r}^\top \mathbf{x}^t - r_0\mathbf{1}^\top \mathbf{x}^t}{B - Ar_0} = \frac{\mu_t - r_0}{B - Ar_0}$$

so that $B - Ar_0 = (\mu_t - r_0)/\sigma_t^2$. On the other hand, $\mathbf{r} - r_0\mathbf{1} = (B - Ar_0)\boldsymbol{\sigma}_t$ so that

$$(3) \qquad \mathbf{r} - r_0\mathbf{1} = \frac{\boldsymbol{\sigma}_t}{\sigma_t^2}(\mu_t - r_0).$$

10.1.2   Market Portfolio

Under the assumptions of an efficient market all investors on the market select their portfolios from the set of efficient portfolios. The investors differ only in their risk aversion. Mathematically it is expressed by weight δ in Theorem 9.2.14. Higher values of δ reflect higher risk aversion. Thus the weighted portfolio (according to the individual investors' wealth) consisting of the individual investors' portfolios also belongs to the set of efficient portfolios. The aggregate demand for risky assets is in the proportions of the tangency portfolio. In equilibrium demand and supply are equal and the proportions (both for riskless and risky assets) create the so called *market portfolio*. In other words, the market portfolio is a wealth-weighted average of the individual investors' optimal portfolios. If there is no supply of the riskless asset, the market portfolio coincides with the tangency portfolio. In practise, the market portfolio is often approximated by a composition of a stock exchange index. Let us denote such a *market portfolio* $\widetilde{\mathbf{x}}^M$. It may be expressed in the form

$$(4) \qquad \widetilde{\mathbf{x}}^M = \delta_M \widetilde{\mathbf{x}}^1 + (1 - \delta_M)\widetilde{\mathbf{x}}^2$$

for some $0 < \delta_M < 1$. Put $\mathbf{x}^M = (1 - \delta_M)\mathbf{x}^t$, the part of the market portfolio corresponding to risky assets only. The return on the market portfolio M is therefore

$\rho^M = \delta_M r_0 + (1 - \delta_M)\rho^\top \mathbf{x}^t$, the expected return on M is $\mu_M = \delta_M r_0 + (1 - \delta_M)\mu_t$, the variance of the return on M is $\sigma_M^2 = (1 - \delta_M)^2 \sigma_t^2$, the expected excess return on M is $\mu_{eM} = \mu_M - r_0 = (1 - \delta_M)(\mu_t - r_0)$, and the vector of covariances between excess returns of the risky assets and the excess return on the market portfolio reads

$$(5) \quad \boldsymbol{\sigma}_M = \operatorname{cov}(\rho - r_0 \mathbf{1}, \rho^\top \mathbf{x}^M) = (1 - \delta_M)\mathbf{V}\mathbf{x}^t = (1 - \delta_M)\frac{\mathbf{r} - r_0 \mathbf{1}}{B - Ar_0} = (1 - \delta_M)\boldsymbol{\sigma}_t.$$

Now we substitute into formula (3) and after cancelling the factor $(1 - \delta_M)$ we get

$$(6) \qquad \boxed{\mathbf{r} - r_0 \mathbf{1} = \frac{\boldsymbol{\sigma}_M}{\sigma_M^2}(\mu_M - r_0) = \boldsymbol{\beta}(\mu_M - r_0)}$$

where $\boldsymbol{\beta} = \boldsymbol{\sigma}_M / \sigma_M^2$. The last formula is known as *CAPM* also *Sharpe-Lintner model CAPM*. For individual assets the CAPM becomes

$$(7) \qquad \boxed{r_n - r_0 = \beta_n(\mu_M - r_0), \quad n = 1, \dots, N}$$

with $\beta_n = \sigma_{nM}/\sigma_M^2$ or

$$(8) \qquad \boxed{r_n - r_0 = \frac{\sigma_{nM}}{\sigma_M^2}(\mu_M - r_0),}$$

if we denote $\sigma_{nM} := \operatorname{cov}(\rho_n, \rho^M)$.

The concept of the market portfolio is a bit abstract. By definition, it is the wealth-weighted sum of the portfolio holdings of all investors. The weights can hardly be observed in practise however, so for calculation an observable indicator of the market performance is needed. Usually the market portfolio is approximated by some stock exchange index like DJIA, S&P 500, FTSI, etc. The stock exchange indexes serve as proxies for the market portfolio and the US Treasury bill rates proxy the riskless rate.

10.2   Security Market Line

The graphical representation of (7) and (8) is known as the *security market line*, *SML*. We see that (7) expresses the expected return on the nth asset as a function of β_n while (8) expresses the same quantity in dependence on the covariance σ_{nM}. We refer to (7) as to the β-version and to (8) as to the *covariance version* of the security market line. The quantity β for an asset or a portfolio is called *factor beta* and it plays an important role in equity (stock) valuation. For the market portfolio, the corresponding $\beta^M = 1$ and for the riskless asset $\beta_0 = 0$. Obviously, $\beta = 1$ also for any efficient portfolio. The factor beta may be considered as a risk factor. Assets with $\beta > 1$ are riskier than the market average and those with $\beta < 1$ are less risky than the market average.

For an arbitrary portfolio $\mathbf{x}$, the factor β_p of that portfolio is $\beta_p = \mathbf{x}^\top \boldsymbol{\beta}$, the weighted average of the respective β_n's, $\boldsymbol{\beta}$ defined in (6). While the variance is a risk measure of an efficient portfolio, beta may be considered as an indicator of the market risk of an individual security. For the asset n, we can express (7) in an alternative way

$$(9) \qquad \rho_n = r_0 + \beta_n(\rho^M - r_0) + \varepsilon_n, \quad n = 1, \ldots, N$$

where the ε_n's are disturbances, $E\,\varepsilon_n = 0$, $\mathrm{var}\,\varepsilon_n = \sigma_{n\varepsilon}^2$. From this equation we get the expression for the variance of ρ_n:

$$(10) \qquad \sigma_n^2 = \beta_n^2 \sigma_M^2 + \sigma_{n\varepsilon}^2 + 2\mathrm{cov}(\rho^M, \varepsilon_n).$$

Often it is supposed that ρ^M and ε_n are not correlated (questionable) so that (10) simplifies to

$$(11) \qquad \sigma_n^2 = \beta_n^2 \sigma_M^2 + \sigma_{n\varepsilon}^2$$

with the interpretation that the *total risk* σ_n^2 is decomposed into the *market risk* $\beta_n^2 \sigma_M^2$ and the *unique* or *specific risk* $\sigma_{n\varepsilon}^2$. Only the specific risk is diversifiable in the sense that by holding the asset n in a sufficiently large portfolio, the prevailing part of the risk of the whole portfolio is that of market risk. In practise, however, it is not necessary to hold a portfolio mirroring the whole market portfolio. A comparatively small portfolio of some tens of assets would eliminate most of the specific risk.

Betas have to be estimated. The most common approach is based on linear regression from historical data. If we have T observations of returns ρ_{nt} on the asset n and on the market return ρ_t^M (represented mainly by the actual value of a stock exchange index), $t = 1, \ldots, T$, we can rewrite (9) in the form of regression equations

$$(12) \qquad \rho_{nt} = r_0 + \beta_n(\rho_t^M - r_0) + \varepsilon_{nt}, \quad t = 1, \ldots T$$

for unknown parameter β_n if the riskless rate r_0 is supposed to be known or as

$$(13) \qquad \rho_{nt} = \alpha_n + \beta_n \rho_t^M + \varepsilon_{nt}, \quad t = 1, \ldots T$$

for unknown parameters α_n, β_n, if the excess return is not directly observable. The estimate of beta obtained by the least squares principle is

$$(14) \qquad \widehat{\beta}_n = \frac{\sum_{t=1}^{T}(\rho_{nt} - \overline{\rho}_n)(\rho_t^M - \overline{\rho^M})}{\sum_{t=1}^{T}(\rho_t^M - \overline{\rho^M})^2}$$

for model (12) where $\overline{\rho}_n$ and $\overline{\rho^M}$ denote the respective averages. Under (13), the estimate of β_n is (14) again, and for α_n the estimate is

$$(15) \qquad \widehat{\alpha}_n = \overline{\rho}_n - \widehat{\beta}_n \overline{\rho^M}.$$

In equilibrium the returns of all securities would lie along the security market line. If this is not the case, there is something wrong either with their risk parameter beta or with their pricing. If the beta on an asset is correct and the return is below SML, the asset is overpriced. If the return is above SML, the asset is underpriced. (Explain this phenomenon as an exercise. Note that with increasing price of a security the return decreases and vice versa.) The difference between the actual and expected (given by SML) return is called *Jensen measure*.

Betas are published in financial press and in the so called *Beta books* both for individual companies and for industries. For industry like essentials usually $\beta < 1$. This is typical for goods and services the demand for which is irrespective of the economic cycle. Thus food manufacturing may have $\beta = 0.9$ while car industry $\beta = 1.27$, and tourism $\beta = 1.66$.

10.3   Capital Market Line

Let us have a portfolio p with expected return μ_p and standard deviation σ_p. In the presence of a riskless asset we can modify the *Sharpe's measure of portfolio*:

$$(16) \qquad \frac{\mu_p - r_0}{\sigma_p}$$

and we will call it *modified Sharpe's measure of portfolio*.

10.3.1   Assertion. *All efficient portfolios have the same modified Sharpe's measure of portfolio.*

Proof. For an efficient portfolio p the expected excess return may be expressed as $\mu_p - r_0 = \delta_p r_0 + (1 - \delta_p)\mu_t - r_0 = (1 - \delta_p)(\mu_t - r_0)$ for some δ_p. Similarly, for the standard deviation we get $\sigma_p = (1 - \delta_p)\sigma_t$. Thus

$$\frac{\mu_p - r_0}{\sigma_p} = \frac{\mu_t - r_0}{\sigma_t}.$$

$\square$

Since we can take any of the efficient portfolios as a numeraire, we choose the market portfolio M. From the above assertion it follows that mean the μ_p and the standard deviation σ_p of any efficient portfolio fulfill the relationship

$$(17) \qquad \boxed{\mu_p = r_0 + \frac{\mu_M - r_0}{\sigma_M}\sigma_p.}$$

The dependence of the expected return on an efficient portfolio on its standard deviation is linear and its graphical representation is called *Capital Market Line, CML*.

The substantial difference between CML and SML is that CML expresses the excess return on the efficient portfolios while SML is valid for any security or portfolio.

I.11 ARBITRAGE PRICING THEORY

regression model, multifactor model, factor analysis, modified method of principal components

Arbitrage Pricing Theory, (APT), also known as *Arbitrage Pricing Model, APM*, serves as a generalization of the single factor CAPM to a multifactor model. The idea behind the APT is that the returns vary from their expected values due to unanticipated changes in production, inflation, term structure, and other economic factors. In the multifactor model it is supposed that the return on an asset is explained in terms of a linear combination of more factors or indexes. Note that in CAPM, the expected return on an asset is a linear function of the expected market return only. The development of APT is based on the assumptions of an efficient market, see I.9. A technical realization of APT uses two popular statistical methods; regression analysis and factor analysis.

11.1   Regression Model

We suppose that the return ρ_n on an asset n (n fixed in this Section) fulfills the usual model of linear regression

$$(1) \qquad \rho_n = \beta_{n1} F_1 + \cdots + \beta_{nm} F_m + \varepsilon_n$$

where $F_1, \ldots, F_m$ are explanatory variables independent of the asset return in question, ε_n is a zero mean random disturbance, and $\beta_{n1}, \ldots, \beta_{nm}$ are unknown parameters which are specific for the given asset. Usually an absolute term must be considered which can be simply done by setting $F_1 \equiv 1$. In the regression model we suppose that the values of F_i's are observable while the random deviate ε_n is not. If we have T observations of the vector $(\rho_n, F_1, \ldots, F_m)$ then (1) becomes

$$(2) \qquad \rho_{nt} = \beta_{n1} F_{1t} + \cdots + \beta_{nm} F_{mt} + \varepsilon_{nt}, \quad t = 1, \ldots, T.$$

Such observations are usually gathered historical data. It should be emphasized that $\beta_{n1}, \ldots, \beta_{nm}$ are characteristics of the underlying asset and $F_{1t}, \ldots, F_{mt}$ are independent of the asset but they take different values for different t's. In a simple regression model it is supposed that $E \varepsilon_{nt} = 0$ and $\mathrm{cov}(\varepsilon_{ns}, \varepsilon_{nt}) = \sigma_n^2 \delta_{st}$, $s, t = 1, \ldots, T$, where $\delta_{st} = 1$ for $s = t$ and $\delta_{st} = 0$ for $s \neq t$, σ_n^2 being also an unknown parameter. Put $\boldsymbol{\rho}_n := (\rho_{n1}, \ldots, \rho_{nT})^\top$, $\boldsymbol{\beta}_n := (\beta_{n1}, \ldots, \beta_{nm})^\top$, $\mathbf{F} := (F_{it})$, $i = 1, \ldots, m, t = 1, \ldots, T$ an $m \times T$ matrix, and $\boldsymbol{\epsilon}_n := (\varepsilon_{n1}, \ldots, \varepsilon_{nT})^\top$. Then we can express (2) in the matrix form

$$(3) \qquad \boxed{\boldsymbol{\rho}_n = \mathbf{F}^\top \boldsymbol{\beta}_n + \boldsymbol{\epsilon}_n,}$$

$E \boldsymbol{\epsilon}_n = \mathbf{0}$, $\mathrm{cov}\, \boldsymbol{\epsilon}_n = \sigma_n^2 \mathbf{I}_T$. Further let us suppose that $T > m$ and that $\mathbf{F}$ has the full rank, $r(\mathbf{F}) = m$ so that the inverse $(\mathbf{F}\mathbf{F}^\top)^{-1}$ exists. The ordinary least squares estimator of $\boldsymbol{\beta}_n$ is

$$(4) \qquad \widehat{\boldsymbol{\beta}}_n = (\mathbf{F}\mathbf{F}^\top)^{-1} \mathbf{F}^\top \boldsymbol{\rho}_n$$

with the covariance matrix $\mathrm{cov}\widehat{\beta}_n = \sigma_n^2 (\mathbf{FF}^\top)^{-1}$. An unbiased estimator of σ_n^2 is

$$(5) \qquad \widehat{\sigma}_n^2 = \frac{1}{T-m-1}(\rho_n - \mathbf{F}^\top\widehat{\beta}_n)^\top(\rho_n - \mathbf{F}^\top\widehat{\beta}_n).$$

The last statistic is used for the construction of the confidence intervals for β_n.

11.1.1 Remark

An empirical study of this type with $m = 7$ may be found in [143] together with further references. The variables, based on monthly observations are: $F_1 \equiv 1$, $F_2 =$ monthly growth in industrial production, $F_3 =$ change in expected inflation, $F_4 =$ unexpected inflation, $F_5 =$ risk premium as the difference in yields of corporate bonds and long-term Treasury bonds, $F_6 =$ change in the term structure as the difference in yields of long-term Treasury bonds and (short-term) Treasury bills, $F_7 =$ return on the market portfolio measured by the NYSE index.

11.1.2 Remark

In the preceding remark we have seen that one of the explanatory variables was the market return. Since we may always include this variable in APT consideration together with additional explanatory variables, we can not obtain worse fitting than that with CAPM. This is a well-known fact, the more parameters you have, the better fit you get. The number of explanatory variables has to be chosen with care, however, since including highly correlated variables brings the problems with multicollinearity etc. Refer to standard textbooks on regression analysis, like [180].

11.2 Factor Model

Instead of returns ρ_n we will now consider *standard scores* or *standardized returns*

$$(6) \qquad \rho_n := \frac{\rho_n - E\,\rho_n}{\sqrt{\mathrm{var}\,\rho_n}}, \quad n = 1,\ldots,N$$

at a given time instant. In the factor model we suppose that

$$(7) \qquad \rho_n = b_{n1}f_1 + \cdots + b_{nm}f_m + e_n, \quad n = 1,\ldots,N$$

where $f_1, \ldots, f_m$, e_n are random variables with zero means. $f_1, \ldots, f_m$ are called *common factors* or *sensitivities*, e_n is called *unique* or *specific factor*, and $b_{n1}, \ldots, b_{nm}$ are called *factor loadings* on the asset n. Note that in this context, e_n is also known as the *idiosyncratic risk*, the *asset-specific* or *firm-specific* component. The crucial assumption of the factor model is that neither the common nor the specific factors can be directly observed, i.e., they are *unobservable*. It is also supposed that all the factors are mutually uncorrelated. Denote, as usually, $\rho := (\rho_1,\ldots,\rho_N)^\top$, $\mathbf{B} := (b_{nj})$, $n = 1, \ldots, N$, $j = 1, \ldots, m$ the $N \times m$ matrix of factor loadings, $\mathbf{f} := (f_1,\ldots,f_m)^\top$ the vector of common factors, and $\mathbf{e} := (e_1,\ldots,e_N)^\top$ the vector of specific factors. The matrix form of (7) becomes

$$(8) \qquad \boxed{\rho = \mathbf{B}\mathbf{f} + \mathbf{e}.}$$

This is the *factor model* of returns. We summarize the above assumptions and make some additional ones:

$$E\mathbf{f} = \mathbf{0}, \quad E\mathbf{e} = \mathbf{0}, \quad \mathrm{cov}\,(\mathbf{f}, \mathbf{e}) = E\,\mathbf{f}\mathbf{e}^\top = \mathbf{0},$$

(9) $$\mathrm{cov}\,\mathbf{f} = \mathbf{I}_m, \quad \mathrm{cov}\,\mathbf{e} = \mathrm{diag}\,(\psi_1^2, \ldots, \psi_N^2) =: \boldsymbol{\Psi}.$$

The last assumption means that for different assets, the specific factors are uncorrelated and may have different variances. Under these assumptions, the covariance matrix of $\boldsymbol{\rho}$ is

(10) $$\boxed{\mathbf{R} = \mathrm{cov}\,\boldsymbol{\rho} = \mathbf{B}\mathbf{B}^\top + \boldsymbol{\Psi}.}$$

Since we have supposed that $\boldsymbol{\rho}$ is a standardized random vector, $\mathbf{R}$ coincides with the *correlation matrix of standardized returns*. Hence the nth element of the diagonal of $\mathbf{R}$ can be expressed in the form

(11) $$1 = 1 - \psi_n^2 + \psi_n^2 =: h_n^2 + \psi_n^2.$$

The quantity h_n^2 is called *communality* and ψ_n^2 is called *uniqueness, specificity*, or *specific variance* of the respective asset.

Note that the decomposition (10) is far from being unique. For example, if $\mathbf{U}$ is any $m \times m$ orthogonal matrix then

(12) $$\mathbf{R} = \mathbf{B}^*\mathbf{B}^{*\top} + \boldsymbol{\Psi}$$

where $\mathbf{B}^* = \mathbf{B}\mathbf{U}$ is called an *orthogonal rotation*.

The main *objective of the factor analysis* may be formulated in the following way: Given the correlation matrix $\mathbf{R}$, find the number m of common factors, a matrix of factor loadings $\mathbf{B}$ and a diagonal matrix $\boldsymbol{\Psi}$ with nonnegative elements such that (10) holds. The number of common factors should be small. This is a natural requirement since with a high number of common factors we loose the possibility of their proper interpretation.

There is a plenty of statistical methods aimed for solving the above problem. We just briefly mention one of the simplest but frequently used method with a clear motivation. The method is known as the *modified method of principal components*. The theoretical background of this method is based on the Lemma below. First recall that every symmetric $N \times N$ matrix allows a *spectral decomposition*

(13) $$\mathbf{A} = \lambda_1 \mathbf{x}_1 \mathbf{x}_1^\top + \cdots + \lambda_N \mathbf{x}_N \mathbf{x}_N^\top$$

where $\lambda_1 \geq \lambda_2 \geq \cdots \geq \lambda_N$ are the eigenvalues and $\mathbf{x}_1, \mathbf{x}_2, \ldots, \mathbf{x}_N$ the orthonormal eigenvectors corresponding to the eigenvalues $\lambda_1, \lambda_2, \ldots, \lambda_N$. By the Euclidean norm of a matrix $\mathbf{A}$ we mean $\|\mathbf{A}\| = \sqrt{\sum \sum a_{ij}^2}$.

11.2.1 **Lemma.** *Let* $\mathbf{A}$ *be an* $N \times N$ *symmetric positive definite matrix,* $r(\mathbf{A}) = r$, *and let* $m \leq r$. *Then the solution to the problem*

$$\text{minimize } \{ \, \|\mathbf{A} - \mathbf{B}\mathbf{B}^\top\|, \quad \mathbf{B} \text{ an } N \times m \text{ matrix} \, \},$$

i.e., the best approximation of $\mathbf{A}$ *by* $\mathbf{B}\mathbf{B}^\top$ *in the sense of Euclidean norm, is given by* $\widehat{\mathbf{B}} = (\sqrt{\lambda_1}\mathbf{x}_1, \ldots, \sqrt{\lambda_m}\mathbf{x}_m)$.

Proof. The proof can be found in textbooks on matrix calculus. $\square$

The estimation procedure starts with a guess of the number of common factors. A heuristic rule says that we take m equal to the number of the eigenvalues of $\mathbf{R}$ greater than or equal to one. Then we estimate the communalities. A good initial approximation is given by

$$(14) \qquad\qquad 1 - \widehat{\psi}_n^2 = \max_{i \neq n} |r_{in}|$$

or by the square of the multiple correlation coefficient $r_{n \cdot 1, \ldots, n-1, n+1, \ldots, N}^2$ in the regression of the nth variable on the remaining $N - 1$ variables. From (14) we form the estimate $\widehat{\mathbf{\Psi}}_0 = \text{diag}(\widehat{\psi}_1^2, \ldots, \widehat{\psi}_N^2)$. Now we define the *reduced correlation matrix* by

$$(15) \qquad\qquad \mathbf{R}_1 = \mathbf{R} - \widehat{\mathbf{\Psi}}_0.$$

Note that in the theoretical model (10) it is assumed that this matrix is positive semidefinite. It may not be the case for (15) since instead of $\mathbf{\Psi}$ we have used an estimate of it. Nevertheless, since we suppose $m \ll N$ we can expect that at least m eigenvalues of $\mathbf{R}_1$ are positive and we can therefore construct the best approximation of it based on Lemma 11.2.1:

$$(16) \qquad\qquad \widehat{\mathbf{R}}_1 = \widehat{\mathbf{B}}_1 \widehat{\mathbf{B}}_1^\top$$

where $\widehat{\mathbf{B}}_1 = (\sqrt{\lambda_1}\mathbf{x}_1, \ldots, \sqrt{\lambda_m}\mathbf{x}_m)$ from the spectral decomposition of the (surely symmetric) matrix $\mathbf{R}_1 = \sum_{n=1}^{N} \lambda_n \mathbf{x}_n \mathbf{x}_n^\top$. We then obtain a new estimate of specificities

$$(17) \qquad\qquad \widehat{\mathbf{\Psi}}_1 = \text{diag}\,(\mathbf{R} - \widehat{\mathbf{R}}_1).$$

We must take the diagonal only since $\mathbf{R} - \widehat{\mathbf{R}}_1$ may not be a diagonal matrix. We go back to (15), form the new reduced correlation matrix $\mathbf{R}_2 = \mathbf{R} - \widehat{\mathbf{\Psi}}_1$ and iteratively improve the estimates of $\mathbf{\Psi}$ and $\mathbf{B}$ until the differences in successive iterations are sufficiently small. Eventually we get the decomposition

$$(18) \qquad\qquad \mathbf{R} = \widehat{\mathbf{B}}\widehat{\mathbf{B}}^\top + \widehat{\mathbf{\Psi}}$$

or an analogy to the original model (8)

$$(19) \qquad\qquad \rho = \widehat{\mathbf{B}}\mathbf{f} + \mathbf{e}$$

with $\mathbf{f}$ still remaining an unknown vector of common factors. But with known matrix $\widehat{\mathbf{B}}$, we may look on (19) as on a linear regression model with unknown parameters $\mathbf{f}$. By ordinary least squares method we get an estimate of $\mathbf{f}$:

$$(20) \qquad\qquad \widehat{\mathbf{f}} = (\widehat{\mathbf{B}}^\top \widehat{\mathbf{B}})^{-1} \widehat{\mathbf{B}}^\top \rho.$$

11.2.2   Remark (Principal Components)

Factor analysis is a generalization of *principal components*. In the method of principal components we directly use the spectral decomposition (13) of the covariance matrix of returns $\mathbf{\Sigma}$:

$$\mathbf{\Sigma} = \sum_{n=1}^{N} \lambda_n \mathbf{x}_n \mathbf{x}_n^\top.$$

The random variable $Y_n := \mathbf{x}^\top \rho$ is called nth principal component, $n = 1, \ldots, N$. The principal components have some plausible properties: (i) they are uncorrelated, (ii) var $Y_n = \lambda_n$, (iii) the total dispersion of returns measured by $\sigma^2 = \mathrm{tr}\,\mathbf{\Sigma}$ is explained by all the principal components since $\mathrm{tr}\,\mathbf{\Sigma} = \sum_{n=1}^{N} \sigma_{nn}$. The eigenvalues are supposed to be ordered, hence the first principal component explains the greatest part of the total dispersion etc. In practise, often only a few components (3, say) explain most of the total dispersion (95 per cent, say). We see that the model of principal components coincides with that of the factor model if we put $m := N$ and $\mathbf{\Psi} := \mathbf{0}$, i.e., if no specific factors are considered.

11.2.3   Remark

The interpretation of the common factors represented by their factor loadings is a difficult and fairly controversial procedure. Roughly speaking, only the first two common factors may be usually identified with a more or less clear interpretation. The first factor represents an overall performance of economy giving higher loadings to the assets with greater importance. The second factor, often interpreted as a *bipolar factor*, usually divides the assets into industries which may act in opposite directions: oil – gas, nuclear power plants – heat power plants, etc.

The interpretation of the factor loadings is easier if each asset is highly loaded on at most one factor, and if all factor loadings are either large or close to zero. The assets are then grouped into disjoint sets, each of which is associated with one factor. The factor i has an influence on those assets for which b_{ni} is large. Since $|b_{ni}| \leq 1$, by term "b_{ni} large" we mean b_{ni} close to 1 or -1. We have seen that the decomposition (10) is also valid for any orthogonal rotation of $\mathbf{B}$. There is a lot of methods of rotation which, up to some extent, improve the interpretation of factors in the above sense. Generally, their principle is to find the orthogonal matrix $\mathbf{U}$ such that the rows of the transformed matrix $\widehat{\mathbf{B}}\mathbf{U}$ contain a few large loadings while the others are close to zero. The most popular orthogonal rotation method is *varimax*. There are also non-orthogonal methods (oblique rotations) like *quartimin*. Note that under oblique rotations the factors are no longer uncorrelated. All these methods are iterative and difficult from the computational point of view. See Rao's contribution in [109], pp. 489–505, for further discussion.

I.12 BIBLIOGRAPHICAL NOTES

Many of the topics treated in this Part are classical pieces of finance, financial mathematics, financial management, and partly of economics. Hence it is quite natural that there are hundreds of books on similar subjects but they differ in their viewpoint on the subjects. Also the material involved is treated on very different levels. Hence the following notes may cover only a small part of the vast existing literature on the related topics.

Money, capital, and securities. A thorough text on basic financial concepts and financial institutions is [138]. In [143] the reader will find both theory and many examples of investment management. [25] and [141] may serve as readable books on financial management together with accounting considerations which are not mentioned in this book, however. Only the most important securities (this applies particularly to derivatives) are mentioned. For more see [60], [88].

Interest rate. A simple arithmetics of interest rate is contained in [114], a deeper insight in [161]. The section on decomposition is based on various sources like [143], [25]. Inflation is also treated in I.7.3. Term structure is important in fixed-income securities' analysis. In continuous case, various models of the term structure are known as *Vasicek mean reversion*, *Cox-Ingersoll-Ross*, *Merton (Ho-Lee)*, *Hull-White*, *Heath-Jarrow-Morton*, and other models, see [43], [82], [105]. For modeling term structure in *Mathematica*® see [11] and [13].

Measures of cash flows. An elementary approach may be found in any book on financial arithmetics like [114] or on financial and investment management like [25], [141], [143]. A thorough discussion on the benefit to leasing is in [76]. The concept of duration has been ascribed to Frederick Macaulay. Yield curves are often treated in the context of term structure models.

Return, expected return, and risk. A comprehensive but still elementary treatise on return and random walk's hypothesis may be found in [26]. [159] is devoted to modeling returns as time series. Further recommended reading consists of [43], [85], [109]. The historical development of the log-normal model for a price development can be roughly traced as: Bachelier [4], Einstein [56], Merton [116, reprinted paper of 1973], Black-Scholes [23]. Concerning volatility, here we confine ourselves only to the case of a constant volatility. Stochastic models of volatility including popular GARCH are treated in [26], [105], [107], [109], e.g. VaR is ascribed to [118], despite in Statistics this statistic is known as the quantile for almost one hundred years. Some recent books on VaR and related topics are [42] and [89].

Valuation of securities. Valuation of coupon bonds is a simple application of the cash flows' measures but some literature do not handle the related cash flow properly. There is a vast literature on the derivatives' pricing, usually starting with the *Black-Scholes* model and covering a lot of generalizations. The pioneering works are [23], [81], [116]. Recommended reading with further references: [41], [82], [105]. An original approach based on so called *fundamental transform* taking into account stochastic volatility together with a *Mathematica*® code is presented by Lewis [107]. The valuation of stocks (value of the firm) is not explicitly covered here since it often depends on accounting principles which are beyond the scope of this book. We refer to [25], [36], and [143]. We should emphasize that the practical derivatives' valuation needs an extensive computing and some symbolic calculation is often necessary. See books [147], [162], [163], and papers on special related topics [9], [10], [12], all making use of *Mathematica*®.

Matching of assets and liabilities. Problem of matching of assets and liabilities likely originated in life insurance industry, see [114] for reference and description of *Redington's theory of immunization* of a life office. Further reading [143], pp. 638–658, [60], [178]. For a related actuarial model see [173].

Index numbers and inflation. Perhaps the first comprehensive study and theory of index numbers is the 1922 book *The Making of Index Numbers* by I. Fisher. Here we closely followed Bílý [18] who was one of the promoters of actuarial sciences and econometrics in former Czechoslovakia and before 1948 the chief official at the Ministry of Finance. Our notation has been adapted for the computational purposes.

There are hundreds of stock exchange indexes. If from related markets, they are usually highly correlated. See [61] and [143] for more information. An example of a relationship between stock prices and inflation is given in [3].

Basics of utility theory. The use of utility theory in modern financial decision making has origins in the von Neumann–Morgenstern theory. For a detailed analysis see [85]. Some particular observations are in [88], [116], and [178].

Markowitz mean-variance portfolio. The pioneering contribution to the modern portfolio theory is paper [112] of Markowitz. Many other authors elaborated his fundamental idea of portfolio diversification, let us mention [57], [85]. Basically, the Markowitz model is a one-period model. For multiperiod-selection models as well as for continuous-time models we refer to [116], [43], and [85]. Generalizations of portfolio separation theorems to more than two funds may be found in [85], [116], e.g. Useful nonlinear programming techniques suitable for portfolio selection algorithms via *Mathematica*® are in [17].

Capital asset pricing model. CAPM presented here is based on the mean-variance portfolio theory. For generalizations of the CAPM (consumption-based, continuous-time, intertemporal, and others) refer to [116].

Arbitrage pricing theory. Originally, the model has been developed as a multi-factor model (as a model of factor analysis) by S. A. Ross (Arbitrage theory of capital asset pricing, J. of Economic Theory 13 (1976), pp. 341–60). Due to its rather difficult tractability caused by a necessity of the interpretation of common factors, the regression form of APT with specified independent variables seems to be more convenient in practise, see [143], [109].

Part II

DISCRETE TIME STOCHASTIC
DECISION MODELS

Motto: "Investment is, in essence, present sacrifice for future benefit. But the present is relatively well known, whereas the future is always an enigma. Investment is also, therefore, certain sacrifice for uncertain benefit." [79]

II.1 INTRODUCTION AND PRELIMINARIES

problem of a private investor, stochastic dedicated bond portfolio, mathematical programs

Our motto clearly reflects the main features of investment problems: necessity to make *decisions under uncertainty* and *over more than one time period*. The uncertainties concern the future level of interest rates, returns, exchange rates, prepayments, external cash flows, inflation, technological innovations, future demand, etc. There exist various stochastic models describing or explaining these random parameters (cf. Chapters II.5 and III.3) that are used to build the input for decision models.

To build a decision model, one has to decide first about the purpose or goal; this includes identification of the uncertainties or risks one wants to hedge, of the hard and soft constraints, of the time horizon and its discretization, etc. The next is the data input and a subsequent algorithmic solution which concludes the first level of the procedure. Interpretation and evaluation of the results may lead to model changes and to a new solution or it may require a "what-if" analysis to get an information about robustness of the results. In the framework of the famous Markowitz model (cf. Chapter I.9), one may be forced to include further, e.g., regulatory constraints, one accepts a suitable model of the random returns, such as the factor model introduced in Section I.11.2, and uses it to get their expectations, variances and covariances. There are additional questions, for example, how sensitive is the investment strategy on the input values of moments of the random returns and on the chosen risk aversion level. Interpretation of the results should reflect the model assumptions (not necessarily fulfilled in real-life); for instance, the Markowitz model is a static model over a fixed period, it is based on the buy-and-hold strategy applied after the investment decision is made up to the horizon of the problem. Hence, investment decisions based on its repeated use over more that one period can be far from a good, suboptimal dynamic decision (cf. [27]).

In this Part we shall deal with *discrete time* stochastic decision models leaving the continuous time models to Part III. Let us illustrate the basic ideas on a slightly modified formulation of a simplified *problem of a private investor* introduced in [19]:

1.1   Problem of a Private Investor

The investor wishes to raise enough money for his or her child college education N years from now by investing w into some of I considered investments. Let the tuition goal be g; exceeding g after N years provides an additional income of $q\%$ of the excess while not meeting the goal would result in borrowing at the rate $r\%$, $r > q$. The investor plans to revise his investment at certain time instances prior to N using an additional information that will gradually become available in the future. These time instances (and the corresponding time periods) are indexed by $t = 1$ for the initial decision, by $t = 2, \ldots, T - 1$ for the revisions and by $t = T$ for the horizon N. The main uncertainty is the return $\rho_i(t) = \rho_i(t, \omega)$ on each investment i within each period t which depends on an underlying random element ω and is observable at the end of each period. The problem is that the investment decisions , say $x_i(t, \omega)$ made at the beginning of period t, can be only based on the already observed history, on the decisions made in the preceding periods and on the observed returns, i.e., they are *nonanticipative* of future outcomes. This means, inter alia, that the investment decisions $x_i(1, \omega) = x_i(1)$ at the beginning of the first period are fixed, independent on future realizations of returns.

Assume for a while that the future evolution of returns $\rho_i(1), \ldots, \rho_i(T - 1)$ for all considered investments is known, i.e., $\rho_i(t) = R_{it}$, the constant rate of return for asset i valid in period t, as defined in I.4.1. The investment problem can be formulated as

$$\text{maximize} \quad qy^+ - ry^-$$

subject to

$$\sum_i x_i(1) = w, \quad x_i(t) \geq 0 \quad \forall i, t$$

$$\sum_i (1 + \rho_i(t))x_i(t) - \sum_i x_i(t + 1) = 0, \quad t = 1, \ldots, T - 2$$

$$\sum_i (1 + \rho_i(T - 1))x_i(T - 1) - y^+ + y^- = g$$

with nonnegative variables y^+, y^- denoting the surplus and deficit, respectively; the middle line disappears for $T = 2$.

Let $T = 3$. Consider a discrete probability distribution of ω concentrated on a finite number of atoms (scenarios) $\omega^s, s = 1, \ldots, S$, with probabilities $p^s > 0$ $\forall s, \sum_s p^s = 1$. Denote $\rho_i(t, \omega^s)$ the corresponding returns, $x_i(2, \omega^s), y^+(\omega^s), y^-(\omega^s)$ the scenario dependent second- and third-stage variables, respectively, and solve the following *three-stage stochastic program*:

$$\text{maximize} \quad \sum_s p^s [qy^+(\omega^s) - ry^-(\omega^s)]$$

subject to

$$\sum_i x_i(1) = w, \quad x_i(1) \geq 0 \quad \forall i$$

$$\sum_i (1 + \rho_i(1, \omega^s)) x_i(1) - \sum_i x_i(2, \omega^s) = 0 \quad \forall s$$

$$\sum_i (1 + \rho_i(2, \omega^s)) x_i(2, \omega^s) - y^+(\omega^s) + y^-(\omega^s) = g \quad \forall s$$

$$x_i(2, \omega^s) \geq 0 \, \forall i, s, \quad y^+(\omega^s) \geq 0, \, y^-(\omega^s) \geq 0 \quad \forall s.$$

To solve this problem (now an ordinary linear program) one has to choose scenarios $\omega^s \, \forall s$, their probabilities p^s, to use them when evaluating the returns $\rho_i(1, \omega^s)$, $\rho_i(2, \omega^s)$ for all investments and all scenarios, and to fix the values of w, g, r, q.

The obtained first-stage decision does not depend on scenarios; using it the investor hedges (through optimally chosen second- and third-stage decisions $x_i(2, \omega^s)$ and $y^+(\omega^s)$, $y^-(\omega^s)$) against the *considered* future returns so that the expected value of the final outcome is maximal. Having formulated the problem, we tacitly assumed that y^+, y^- were the *minimal* values satisfying the listed requirements; otherwise, we could also borrow $y^- + c$, $c > 0$, having thus surplus of $y^+ + c$. In more sophisticated models, compensating the difference between the goal and the actual achievement is a nontrivial (last stage) decision problem. Hence we call also y^+, y^- *decision* variables.

Notice that the problem is always feasible; this is due to the assumed unlimited possibilities of borrowing. In addition to nonnegativity conditions, the investment strategies are explicitly constrained also by cash limitations and structural constrains; nonanticipativity enters implicitly.

1.2 Stochastic Dedicated Bond Portfolio

Assuming known short-term reinvestment interest rates i_t for period $(t, t + 1)$, the dynamic dedicated bond portfolio model was formulated in I.6.2.2. We rewrite it as

$$\text{minimize} \sum_{n=1}^{N} c_n x_n + y_0^+ \text{ subject to}$$

$$\sum_{n=1}^{N} f_{nt} x_n + (1 + i_{t-1}) y_{t-1}^+ - y_t^+ = l_t, \, t = 1, \ldots, T, \, \mathbf{x} \geq 0, \, \mathbf{y}^+ \geq 0$$

Here $\mathbf{x} = (x_1, \ldots, x_N)^\top$ is composition of the portfolio, $\mathbf{c} = (c_1, \ldots, c_N)^\top$ is the vector of acquisition prices and the T-vectors $\mathbf{CF}_n = \mathbf{f}_n$, $n = 1, \ldots, N$, l and $\mathbf{y}^+$ stand for the cash flows, liabilities and surpluses.

In reality, the future short-term reinvestment rates are hardly known. We assume instead that $\iota = (i_0, \ldots, i_{T-1})$ are random with a discrete probability distribution carried by a finite number of scenarios ι^s, $s = 1, \ldots, S$ with probabilities p^s. In addition, we allow for possibility of short-term shortfalls; this means that for some scenarios and time periods (except for the last one) nonzero discrepancies

$$y_t^{-s} = \left(l_t - \sum_n f_{nt} x_n - (1 + i_{t-1}^s) y_{t-1}^{+s} + y_t^{+s} \right)^+$$

may occur. In such case, the investor borrows this amount and is obliged to repay it including the interest rate (higher than i_t^s for a positive spread δ between the short-term reinvestment and borrowing rates) in the next period. For each s, t we consider now the cash flow constraints which include scenario dependent surpluses y_t^{+s} and shortfalls y_t^{-s}. In addition, there is a penalty $\sum_s p^s \mathbf{q}^{s\top} \mathbf{y}^{-s}$ for borrowing included into the objective function. The resulting problem reads

$$\text{minimize } \mathbf{c}^\top \mathbf{x} + y_0^+ + \sum_s p^s \mathbf{q}^{s\top} \mathbf{y}^{-s}$$

subject to

$$\sum_{n=1}^N f_{tn} x_n + (1 + i_{t-1}^s) y_{t-1}^{+s} - y_t^{+s} - (1 + i_{t-1}^s + \delta) y_{t-1}^{-s} + y_t^{-s} = l_t, \; \forall s, \; 1 \le t \le T - 1$$

$$\sum_{n=1}^N f_{Tn} x_n + (1 + i_{T-1}^s) y_{T-1}^{+s} - y_T^{+s} - (1 + i_{T-1}^s + \delta) y_{T-1}^{-s} = l_T \; \forall s$$

and nonnegativity of all variables $\mathbf{x}, \mathbf{y}^{+s}, \mathbf{y}^{-s}, s = 1, \ldots, S$.

This problem can be further generalized to accommodate not only random (scenario dependent) cash flows, liabilities and spread, but also for including trading possibilities. See the BONDS model 4.3 in Chapter II.4, Chapter II.6 and Section II.7.3.

In the general case of a T-stage problem a sequence of decisions is built along each of considered data trajectories in such a way that decisions based on the same partial trajectory, on the same history, are identical (nonanticipativity) and the expected outcome (e.g., the expected gain or cost) of the decision process at time T is the best possible. In the next Chapter the problem will be formulated as a minimization one but a parallel formulation of a maximization problem is straightforward. Hence, depending on the interpretation, the models are applied to maximization of the total expected gain, see Examples 2.2.1 and 2.4 in Chapter II.2, or to minimization of the total expected cost, etc. We shall use mainly the concepts and methods of stochastic programming.

As we shall illustrate by selected examples later on, see Chapters II.4 and II.6, the goal of the decision process can be specified in various ways; one of them is maximization of the expected utility of the final wealth. In other cases, one tries explicitly to find a decision which satisfies simultaneously several optimization criteria; recall the Markowitz model which aims at simultaneous maximization of expected return and minimization of a suitably defined risk, etc. Chapter II.3 contains a brief introduction to multi-objective programming and discusses selected applications and modeling issues.

1.3   Mathematical Programs

In this Part, we shall model and solve decision processes under uncertainty via stochastic programming. Prekopa [130] presents two definitions of stochastic programming:

• Stochastic programming deals with mathematical programming problems where some of parameters are random.

• Stochastic programming offers solutions for problems formulated in connection with stochastic systems, where the resulting problem to be solved is a mathematical program of a nontrivial dimension.

It is natural then to use terminology common in mathematical programming.

Mathematical program in $\mathbb{R}^n$ is a constrained optimization problem

$$\text{minimize } f(x_1, \ldots, x_n) \text{ on a set } \mathcal{X}$$

where the *set of feasible solutions* $\mathcal{X}$ is defined by *constraints* as follows:

$$\mathcal{X} = \{\mathbf{x} \,:\, h_j(\mathbf{x}) = 0, \, j = 1, \ldots, p, \, g_k(\mathbf{x}) \leq 0, \, k = 1, \ldots, m, \, \mathbf{x} \in \mathcal{X}_0\};$$

here f, $h_j \, \forall j$, $g_k \, \forall k$ are real functions and $\mathcal{X}_0$ is a set of specific (e.g., integrality) conditions. Function f is called the *objective function*. All functions involved may depend on parameters, which gives a rise to *parametric programs* or to *stochastic programs* if some of these parameters are random.

Linear programs correspond to $\mathcal{X}_0 = \mathbb{R}^n$ and all functions f, $h_j \, \forall j$, $g_k \, \forall k$ linear. Another important class are *convex programs* where $\mathcal{X}$ is a convex set and f a convex function.

In this Part, a basic knowledge of mathematical programming is assumed.

II.2 MULTISTAGE STOCHASTIC PROGRAMS

various formulations, nonanticipativity conditions, convexity properties, scenario-based stochastic linear programs, horizon and stages, the flower-girl problem, comparison with stochastic dynamic programming

2.1   Basic Formulations

The model that reflects the decision scheme described verbally at the beginning of the Introduction can be formulated in the following way:

In the general T-**stage stochastic program** we think of a stochastic data process

$$\omega = (\omega_1, \ldots, \omega_{T-1}) \quad \text{or} \quad \omega = (\omega_1, \ldots, \omega_T)$$

and of a decision process

$$\mathbf{x} = (\mathbf{x}_1, \ldots, \mathbf{x}_T).$$

The $\mathbf{x}_t$'s are real n_t-vectors, while the random elements ω_t may be of quite general nature; mostly, they are real random vectors as well. The realizations of ω are called also trajectories or scenarios. We denote by P the probability distribution of ω and by Ω its support.

The sequence of decisions and observations is

(1)
$$\mathbf{x}_1, \omega_1, \mathbf{x}_2, \omega_2, \ldots, \mathbf{x}_{T-1}, \omega_{T-1}, \mathbf{x}_T.$$

The decision process is **nonanticipative** in the sense that decisions taken at any stage of the process do not depend on future *realizations* of the data process or on future decisions whereas the past information as well as the knowledge of the probability distribution P of ω are exploited. We assume that the probability distribution P is *known* and *does not depend on* $\mathbf{x}$. We denote by $\omega^{t-1,\bullet} :=$ $(\omega_1, \ldots, \omega_{t-1})$ the part of the stochastic data process that precedes the stage t and, similarly, by $\mathbf{x}^{t-1,\bullet} := (\mathbf{x}_1, \ldots, \mathbf{x}_{t-1})$ the sequence of decisions at stages $1, \ldots, t-1$. Thus the decision at stage t is $\mathbf{x}_t = \mathbf{x}_t(\mathbf{x}^{t-1,\bullet}, \omega^{t-1,\bullet})$, or more precisely, $\mathbf{x}_t = \mathbf{x}_t(\mathbf{x}^{t-1,\bullet}, \omega^{t-1,\bullet}, P)$.

The outcome attributed to the sequence (1) is quantified by a function $f_0(\mathbf{x}, \omega)$. The aim is to minimize the expected value $Ef_0(\mathbf{x}, \omega)$ under both deterministic constraints $\mathbf{x}_t \in \mathcal{X}_t \, \forall t$ ($\mathcal{X}_t$ given sets in $\mathbb{R}^{n_t}$), $f_{1i}(\mathbf{x}_1) \leq 0, i = 1, \ldots, m_1$, and constraints

$$f_{ti}(\mathbf{x}^{t\bullet}, \omega^{t-1,\bullet}) \leq 0, i = 1, \ldots, m_t, t = 2, \ldots, T$$

that may depend on previous decisions and observations; here, $f_{ti} \, \forall t, i$ are real functions.

In the sequel we shall suppose that all functions are measurable with respect to ω and all expectations exist (this is certainly fulfilled if Ω is a finite set). Relations containing random elements are assumed to hold with probability 1. To simplify this exposition we shall assume in addition that all infima are attained; hence we shall write min instead of inf. This assumption implies that the sets defined by the tth-stage constraints, $t = 1, \ldots, T$,

$$(2) \qquad \mathbf{x}_t \in \mathcal{X}_t : f_{ti}(\mathbf{x}^{t-1,\bullet}, \mathbf{x}_t, \omega^{t-1,\bullet}) \leq 0, i = 1, \ldots, m_t$$

are nonempty for all histories $\mathbf{x}^{t-1,\bullet}, \omega^{t-1,\bullet}$. The first-stage constraints do not depend on the random element.

The form of (2) reflects the requirement that the choice of decisions $\mathbf{x}_t$ is not explicitly constrained by future decisions and observations. In general, however, this does not exclude the presence of *induced constraints* which must be fulfilled to guarantee the existence of a feasible nonanticipative decision process $\mathbf{x}$ and have to be detected within the algorithmic procedures; see the feasibility cuts in the Algorithms II.8.3.1 and II.8.4.1.

The corresponding T-stage stochastic program reads:

$$(3) \qquad \text{minimize } E f_0(\mathbf{x}_1, \mathbf{x}_2(\mathbf{x}_1, \omega_1), \ldots, \mathbf{x}_T(\mathbf{x}^{T-1,\bullet}, \omega^{T-1,\bullet}), \omega)$$

subject to $\mathbf{x}_t \in \mathcal{X}_t$, $t = 1, \ldots, T$ and

$$(4) \qquad f_{1i}(\mathbf{x}_1) \leq 0, i = 1, \ldots, m_1$$
$$f_{ti}(\mathbf{x}^{t\bullet}, \omega^{t-1,\bullet}) \leq 0, i = 1, \ldots, m_t, t = 2, \ldots, T.$$

Realizations of ω_T, i.e., those behind the horizon, do not affect the decision process, but they may contribute to the overall observed costs. Thus the decision process may be affected by the *probability distribution* of ω_T.

The special choice of the function f_0 in (3) as an indicator function of a certain interval leads to the **probability objective function** of the form

$$P[g_0(\mathbf{x}_1, \mathbf{x}_2, \ldots, \mathbf{x}_T, \omega) \notin \mathcal{I}]$$

where $\mathcal{I}$ is a given interval of desirable values of g_0. Similarly, the replacement of the condition that the tth-stage constraints are satisfied with probability 1, $t = 2, \ldots, T$, by the requirement that they hold true with a prescribed probability provides stochastic programs with **probabilistic** or **chance constraints**.

It is important to realize that the *stages do not necessarily refer to time periods*, they correspond to *steps in the decision process*. The main emphasis is on the first-stage decisions which consist of all decisions that have to be selected before a further information may be exploited whereas the second-stage decisions are allowed to adapt to this information, etc. In some applications the importance of the best first-stage decisions is evident: Examples are the decision about the capacity of a new water reservoir, the decision about an initial contract or allocation of funds or the initial charge decision for the metal melting process.

Various schemes were considered to reduce the T-stage stochastic program (3)–(4) to a sequence of similar t-stage programs, $t < T$. If ω_T is not considered, the objective functions are then defined recursively as

$$(5) \qquad \psi_T(\mathbf{x}^{T\bullet}, \omega^{T-1,\bullet}) \equiv f_0(\mathbf{x}, \omega)$$

$$\psi_t(\mathbf{x}^{t\bullet}, \omega^{t-1,\bullet}) = E_{\omega_t|\omega^{t-1,\bullet}} \left\{ \min_{\mathbf{x}_{t+1}} \psi_{t+1}(\mathbf{x}^{t+1,\bullet}, \omega^{t\bullet}) \right\}, t = 2, \ldots, T-1$$

$$\psi_1(\mathbf{x}_1) = E_{\omega_1} \min_{\mathbf{x}_2} \psi_2(\mathbf{x}^{2\bullet}, \omega_1).$$

The minimization is carried over the respective tth-stage constraints (2) and the symbol $E_{\omega|\omega'}$ denotes the expectation with respect to ω conditioned by ω'.

To relate an optimal solution of the T-stage problem to those minimizing the t-stage objective functions ψ_t, certain boundedness assumptions concerning sets defined by the t-stage constraints and convexity of f_0 as a function of $\mathbf{x}$, have to be fulfilled; see [133]. Then not only the canonical projections of the optimal solution $\hat{\mathbf{x}}^{T\bullet}$ of the T-stage problem are optimal solutions of the t-stage problems, $t < T$, but also the optimal solutions of the t-stage problems can be extended to an optimal solution of the T-stage problem. For instance, if $\hat{\mathbf{x}}_1 \in \arg\min \psi_1(\mathbf{x}_1)$ over the first-stage constraints $\mathbf{x}_1 \in \mathcal{X}_1$ and $f_{1i}(\mathbf{x}_1) \leq 0$, $i = 1, \ldots, m_1$, then the next component of the optimal solution, $\hat{\mathbf{x}}_2(\hat{\mathbf{x}}_1, \omega_1)$ is obtained by solving $\min_{\mathbf{x}_2} \psi_2(\hat{\mathbf{x}}_1, \mathbf{x}_2, \omega_1)$ over the second-stage constraints, etc. By introducing a fictious decision $\mathbf{x}_{T+1}$ which does not influence the value of the objective function f_0 these results may be extended also to problems which include ω_T.

Under additional assumptions, e.g., for

$$f_0(\mathbf{x}, \omega) = f_{10}(\mathbf{x}_1) + \sum_{t=2}^{T} f_{t0}(\mathbf{x}^{t-1,\bullet}, \mathbf{x}_t, \omega^{t-1,\bullet})$$

the scheme (5) can be written as a sequence of *nested two-stage stochastic programs* of the following type:

(6) $\text{minimize } E f_0(\mathbf{x}, \omega) := f_{10}(\mathbf{x}_1) + E_{\omega_1} \phi_1(\mathbf{x}_1, \omega_1)$

subject to

$$\mathbf{x}_1 \in \mathcal{X}_1 \text{ and } f_{1i}(\mathbf{x}_1) \leq 0, \, i = 1, \ldots, m_1,$$

where for $2 \leq t \leq T$, for given $\mathbf{x}_1, \ldots, \mathbf{x}_{t-1}$ and observed realizations of $\omega_1, \ldots, \omega_{t-1}$, $\phi_{t-1}(\mathbf{x}_1, \ldots, \mathbf{x}_{t-1}, \omega_1, \ldots, \omega_{t-1})$ denotes the optimal value of the stochastic program
(7)
$\text{minimize } f_{t0}(\mathbf{x}_t, \omega^{t-1,\bullet}) + E_{\omega_t|\omega^{t-1,\bullet}} \left\{ \phi_t(\mathbf{x}^{t-1,\bullet}, \mathbf{x}_t, \omega^{t-1,\bullet}, \omega_t) \right\}$ subject to (2).

Here, $\phi_T \equiv 0$ or it is an explicitly given function of $\mathbf{x}_1, \ldots, \mathbf{x}_T, \omega_1, \ldots, \omega_T$ if contribution of ω_T is considered. The two terms in the definition of functions ϕ_{t-1} may be interpreted as the costs attributed to the decision $\mathbf{x}_t$ at stage t augmented for the expected minimal future costs.

Formulation (6)–(7) resembles the backward recursion common in *stochastic dynamic programming*, see 2.5. In those models, the goal is to provide a sequence of *decision rules* that can be used in particular stages of the decision process and in any state of the system and that allow the decision maker to pass from observations to decisions in an optimal way, e.g., for minimal total expected costs. To solve the problem, i.e., to get the decision rules, one relies on the principle of optimality.

A special case of (6)–(7) is the following **multistage stochastic linear program with recourse** where all functions f (with arbitrary indices) in the above scheme are linear in the decision variables:

(8) $\text{minimize } \mathbf{c}_1^\top \mathbf{x}_1 + E_{\omega_1} \phi_1(\mathbf{x}_1, \omega_1)$

subject to

$$A_1 x_1 = b_1, \quad l_1 \leq x_1 \leq u_1,$$

where the functions $\phi_{t-1}, t = 2, \ldots, T$, are defined recursively by
(9)
$$\phi_{t-1}(x^{t-1,\bullet}, \omega^{t-1,\bullet}) = \min_{x_t} \left[c_t(\omega^{t-1,\bullet})^\top x_t + E_{\omega_t | \omega^{t-1,\bullet}} \left\{ \phi_t(x^{t-1,\bullet}, x_t, \omega^{t-1,\bullet}, \omega_t) \right\} \right]$$

subject to

$$\sum_{\tau=1}^{t-1} B_{t\tau}(\omega^{t-1,\bullet}) \, x_\tau + A_t(\omega^{t-1,\bullet}) \, x_t = b_t(\omega^{t-1,\bullet})$$

$$l_t(\omega^{t-1,\bullet}) \leq x_t \leq u_t(\omega^{t-1,\bullet})$$

and $\phi_T \equiv 0$ or a given function of x and ω.

Here, the A_t's are $m_t \times n_t$-matrices and the remaining vectors and matrices are of consistent dimensions. For the first stage, the values of all elements in b_1, c_1, A_1, l_1, u_1 are known. Again, the main decision variable is x_1 that corresponds to the first stage.

According to our assumption, an optimal solution of (9) exists for all t and all considered histories $x^{t-1,\bullet}, \omega^{t-1,\bullet}$ – the case of the *relatively complete recourse*. The term *recourse* was originally introduced in the context of two-stage stochastic linear programs, see Example II.3.3.2, for the cost of compensating constraints violation. Figuratively it has been used also for multistage stochastic programs even if the interpretation is not straightforward. We speak of a *fixed recourse* problem if the matrices A_t are known nonrandom for all t.

2.1.1   Exercise – Convexity Properties. Given the history, programs (9) are deterministic programs with linear constraints. For numerical solution of such nonlinear programs, convexity of the objective functions is important. To formulate the convexity results, let us start with a special version of problem (9):

Assume that $\omega_1, \ldots, \omega_T$ are *independent* real random vectors, $\Omega = \prod_{t=1}^{T-1} \Omega_t$, Ω_t convex supports of ω_t, $\phi_T \equiv 0$, $B_{t\tau} = 0$ for $\tau < t-1$, c_t, l_t, u_t, A_t and $B_{t,t-1} := B_t$ are known deterministic matrices $\forall t$ and the right-hand sides b_t are linear functions of $\omega_{t-1}, t = 2, \ldots, T$. Hence, (9) simplifies to

$$\phi_{t-1}(x_{t-1}, \omega_{t-1}) = \min_{x_t} \left[c_t^\top x_t + E_{\omega_t} \, \phi_t(x_t, \omega_t) \right]$$

subject to

$$B_t \, x_{t-1} + A_t \, x_t = b_t(\omega_{t-1}), \quad l_t \leq x_t \leq u_t$$

Using techniques of parametric programming [5] prove the following assertions:

  (i) The functions $\phi_{t-1}(x_{t-1}, \omega_{t-1})$ are convex in x_{t-1}. Hence, the first-stage objective function $c_1^\top x_1 + E_{\omega_1} \phi_1(x_1, \omega_1)$ is convex and the corresponding first-stage problem (8) is a convex program.

  (ii) The functions $\phi_{t-1}(x_{t-1}, \omega_{t-1})$ are convex in ω_{t-1} – an important property for constructing bounds (see 5.3.5, 5.4.9 and 7.3 in Chapters II.5 and II.7, respectively) and algorithms (see Chapter II.8).

(iii) Modify the assertion (i) for the general formulation of (9), i.e., for arbitrary random matrices of coefficients.

(iv) Under the independence assumption, the convexity result (ii) extends to problems with random matrices $\mathbf{B}_{t\tau}$ which are linear in ω_{t-1} and a parallel result holds true also for problems with random coefficients only in the objective functions. Is it possible to prove similar results also for mutually dependent components ω_{t-1} in the above problem specifications and/or for random matrices $\mathbf{A}_t$ (i.e., for the *random* recourse problems)?

2.2   Scenario-Based Stochastic Linear Programs

Assume now that the probability distribution of ω is concentrated on a finite number of scenarios. To simplify the exposition we shall work with problems (9) with the *staircase structure*, i.e. with $\mathbf{B}_{t\tau} \equiv \mathbf{0}$ for $\tau < t-1$ and we put $\mathbf{B}_{t,t-1} :\equiv \mathbf{B}_t$; an extension to the general case is straightforward. For disjoint sets of indices $\mathcal{K}_t$, $t = 2, \ldots, T$, let us list as $\tilde{\omega}_{k_t}$, $k_t \in \mathcal{K}_t$ all possible realizations of $\omega^{t-1,\bullet}$. Denote by the same subscripts k_t the corresponding values of the t-th stage coefficients. The total number of scenarios S equals the number of elements of $\mathcal{K}_T$. Each scenario $\omega^s = \{\omega_1^s, \ldots, \omega_{T-1}^s\}$ thus generates a sequence of coefficients $\{\mathbf{c}_{k_2}, \ldots, \mathbf{c}_{k_T}\}$, $\{\mathbf{A}_{k_2}, \ldots, \mathbf{A}_{k_T}\}$, $\{\mathbf{B}_{k_2}, \ldots, \mathbf{B}_{k_T}\}$, $\{\mathbf{b}_{k_2}, \ldots, \mathbf{b}_{k_T}\}$, $\{\mathbf{l}_{k_2}, \ldots, \mathbf{l}_{k_T}\}$, $\{\mathbf{u}_{k_2}, \ldots, \mathbf{u}_{k_T}\}$. Denote $\mathbf{x}(\omega^s) := \{\mathbf{x}_1, \mathbf{x}_{k_2}, \ldots, \mathbf{x}_{k_T}\}$ the vector of feasible solutions of the scenario ω^s subproblem, i.e., of the system

$$
\begin{aligned}
(10) \qquad \mathbf{A}_1 \mathbf{x}_1 &= \mathbf{b}_1 \\
\mathbf{B}_{k_2} \mathbf{x}_1 + \mathbf{A}_{k_2} \mathbf{x}_{k_2} &= \mathbf{b}_{k_2} \\
\mathbf{B}_{k_3} \mathbf{x}_{k_2} + \mathbf{A}_{k_3} \mathbf{x}_{k_3} &= \mathbf{b}_{k_3} \\
&\;\;\vdots \\
\mathbf{B}_{k_T} \mathbf{x}_{k_{T-1}} + \mathbf{A}_{k_T} \mathbf{x}_{k_T} &= \mathbf{b}_{k_T}
\end{aligned}
$$

$$
\mathbf{l}_1 \leq \mathbf{x}_1 \leq \mathbf{u}_1, \; \mathbf{l}_{k_2} \leq \mathbf{x}_{k_2} \leq \mathbf{u}_{k_2}, \; \ldots, \; \mathbf{l}_{k_T} \leq \mathbf{x}_{k_T} \leq \mathbf{u}_{k_T}.
$$

Denote further $\mathbf{A}(\omega^s)$ the matrix of system (10), $\mathbf{b}(\omega^s)$ the vector of the right-hand sides, $\mathbf{c}(\omega^s)$ the vector of the objective function coefficients and $\mathbf{l}(\omega^s)$, $\mathbf{u}(\omega^s)$ the vectors of the lower and upper bounds. Disregarding the nonanticipativity constraints we replace the multistage stochastic linear program by

$$
\text{minimize} \sum_{s=1}^{S} p^s \mathbf{c}(\omega^s)^\top \mathbf{x}(\omega^s)
$$

subject to

$$
\mathbf{A}(\omega^s) \mathbf{x}(\omega^s) = \mathbf{b}(\omega^s), \; s = 1, \ldots, S
$$

and the box constraints

$$
\mathbf{l}(\omega^s) \leq \mathbf{x}(\omega^s) \leq \mathbf{u}(\omega^s), \; s = 1, \ldots, S.
$$

This is already an ordinary (nonrandom) linear program. The components of the obtained optimal solution $\mathbf{x}^*(\omega^s)$, $1 \leq s \leq S$, depend on the underlying scenarios ω^s, they are not nonanticipative. To recover nonanticipativity, we must add constraints $\mathbf{x}_1(\omega^s) = \mathbf{x}_1(\omega^{s'}) \forall s, s'$ to get scenario independent first-stage decisions and, moreover, similar constraints to guarantee that the t-stage decisions based on the same history are equal. Such constraints can be expressed in the form $\mathbf{x} = \mathbf{U}\mathbf{x}$ where $\mathbf{x}$ consists of carefully grouped components of all decision vectors $\mathbf{x}(\omega^s)$ and $\mathbf{U}$ is a 0-1 matrix of coefficients of the nonanticipativity constraints, which enter now *explicitly* the problem constraints. This formulation of the scenario-based multistage stochastic program has been called the **split variable representation** or form.

Another scenario-based formulation of multistage stochastic linear programs is related to the implicit inclusion of nonanticipativity constraints (compare with (3)–(4)) and requires a specific organization of data in the form of a *scenario tree*. Each value $\tilde{\omega}_{k_{t+1}}$ of $\omega^{t\bullet}$ has a unique ancestor $\tilde{\omega}_{k_t}$ (the value of the corresponding $\omega^{t-1,\bullet}$); we denote it by subscript $a(k_{t+1})$. For instance, $a(k_2) = 1$ for all realizations $\tilde{\omega}_{k_2}$ of the component ω_1. Again, $\mathbf{x}_1$ denotes the first-stage decision and to each $\tilde{\omega}_{k_t}$ one assigns the tth stage decision vector $\mathbf{x}_{k_t} \in \mathbb{R}^{n_t}$. This allows to rewrite the T-stage stochastic linear program with recourse in the following **arborescent** form: minimize

$$(11) \qquad \mathbf{c}_1^\top \mathbf{x}_1 + \sum_{k_2 \in \mathcal{K}_2} p_{k_2} \mathbf{c}_{k_2}^\top \mathbf{x}_{k_2} + \sum_{k_3 \in \mathcal{K}_3} p_{k_3} \mathbf{c}_{k_3}^\top \mathbf{x}_{k_3} + \cdots + \sum_{k_T \in \mathcal{K}_T} p_{k_T} \mathbf{c}_{k_T}^\top \mathbf{x}_{k_T}$$

subject to

$$(12) \qquad \begin{aligned}
\mathbf{A}_1 \mathbf{x}_1 & & &= \mathbf{b}_1 \\
\mathbf{B}_{k_2} \mathbf{x}_1 + \; \mathbf{A}_{k_2} \mathbf{x}_{k_2} & & &= \mathbf{b}_{k_2}, & k_2 &\in \mathcal{K}_2 \\
\mathbf{B}_{k_3} \mathbf{x}_{a(k_3)} + \mathbf{A}_{k_3} \mathbf{x}_{k_3} & & &= \mathbf{b}_{k_3}, & k_3 &\in \mathcal{K}_3 \\
& \ddots \quad \ddots & \vdots & \\
\mathbf{B}_{k_T} \mathbf{x}_{a(k_T)} + \mathbf{A}_{k_T} \mathbf{x}_{k_T} &= \mathbf{b}_{k_T}, & k_T &\in \mathcal{K}_T
\end{aligned}$$

$$\mathbf{l}_1 \leq \mathbf{x}_1 \leq \mathbf{u}_1, \; \mathbf{l}_{k_t} \leq \mathbf{x}_{k_t} \leq \mathbf{u}_{k_t}, \; k_t \in \mathcal{K}_t, \; t = 2, \dots, T.$$

We adopt the natural choice $\mathcal{K}_t = \{K_{t-1}+1, \dots K_t\}$, $t = 2, \dots, T$, with $K_1 = 1$. The problem is thus based on the $S = K_T - K_{T-1}$ sequences $(\mathbf{c}_{k_t}, \mathbf{A}_{k_t}, \mathbf{B}_{k_t}, \mathbf{b}_{k_t}, \mathbf{l}_{k_t}, \mathbf{u}_{k_t})$, $t = 2, \dots, T$, of possible realizations of coefficients in the objective function (11), in recourse matrices $\mathbf{A}_*$, transition matrices $\mathbf{B}_*$, right-hand sides and bounds in the constraints for all stages, and on **path probabilities** $p_{k_t} > 0 \forall k_t$, $\sum_{k_t \in \mathcal{K}_t} p_{k_t} = 1, t = 2, \dots, T$, of *partial* sequences of these coefficients, hence, probabilities of realizations of $\omega^{t-1,\bullet} \forall t$. The path probabilities may be obtained by stepwise multiplication of the marginal probabilities p_{k_2} by the (conditional) **arc** or **transition** probabilities $\pi_{k_\tau, k_{\tau+1}}$ related to the corresponding partial sequences of realizations. Probabilities p^s of the individual scenarios ω^s are equal to the corresponding path probabilities p_{k_T}, $k_T \in \mathcal{K}_T$. On the other hand, given the structure of the scenario tree the path probabilities can be obtained from the scenario probabilities p^s. In some cases, it is expedient to use sets $\mathcal{D}(k_t)$ of descendants of k_t which consist of those indices $k_{t+1} \in \mathcal{K}_{t+1}$ for which the transition probability $\pi_{k_t, k_{t+1}} \neq 0$, see Examples 2.2.1 and 2.4.

The size of the linear program (11)–(12) can be very large; for instance, consider the two-stage problem with random right-hand sides $\mathbf{b}_{k_2}$, $k_2 = 2, \ldots, K_2$, each consisting of m_2 independent random components whose probability distributions have been approximated by alternative (zero-one) probability distributions: it gives $K_2 = 2^{m_2}$, hence, $m_1 + m_2 2^{m_2}$ equations in (12). The usefulness of special numerical techniques is obvious, see Chapter II.8. Both the implicit and explicit formulation of nonanticipativity constraints can be combined, for instance, the implicit form may relate only to the first-stage decision variables and the explicit one to the decision variables for stages $2, \ldots, T$; see Exercise 2.2.2.

2.2.1   Example – II.1.1 Revisited. Let $T = 3$ again. This is the case, e.g., if some of the considered investments (a term deposit or short term or medium term bonds) mature in $N_1 < N$ years and, consequently, the portfolio has to be restructured at time N_1. Hence, there is one more stage of the decision process. We put $\omega = (\omega_1, \omega_2)$, the trajectory up to N_1 and its continuation from N_1 to N. Following the scheme (11)–(12), we denote by $\tilde{\omega}_{k_2}$, $k_2 \in \mathcal{K}_2$ the considered realizations of ω_1, by p_{k_2} their probabilities and by $\tilde{\omega}_{k_3}$, $k_3 \in \mathcal{K}_3$ the possible realizations of $\omega^{2\bullet}$ grouped into sets $\mathcal{D}(k_2)$, $k_2 \in \mathcal{K}_2$ for which the conditional probabilities $\pi_{k_2, k_3} \neq 0$; notice that $\sum_{k_3 \in \mathcal{D}(k_2)} \pi_{k_2, k_3} = 1 \forall k_2$. This information about the discrete probability distribution may be represented by a scenario tree, see Figure 1.

The first-stage decisions $x_i(1)$ are scenario independent, the returns $\rho_i(1, \omega)$ and the decisions $x_i(2, \omega)$ at the second stage of the decision process depend only on the first part ω_1 of ω, the subsequent returns $\rho_i(2, \omega)$, the final decisions and compensations depend on the whole history, i.e., on scenarios ω^s which consist of $\tilde{\omega}_{k_2}$ and their "extension" to $\tilde{\omega}_{k_3}$ $k_3 \in \mathcal{D}(k_2)$, $k_2 \in \mathcal{K}_2$. Their probabilities p^s equal $p_{k_2} \pi_{k_2, k_3}$. Following the notation used in (11)–(12), we assign subscripts k_2, k_3 to random coefficients ρ and to the decision variables x_i and y^+, y^- which appear in the second and third stages. The problem – a *three-stage stochastic linear program* – reads

$$\text{maximize} \quad \sum_{k_2 \in \mathcal{K}_2} p_{k_2} \sum_{k_3 \in \mathcal{D}(k_2)} \pi_{k_2, k_3} \left[q y_{k_3}^+ - r y_{k_3}^- \right]$$

subject to

$$\sum_i x_{i1} = w, \quad x_{i1} \geq 0 \quad \forall i$$

$$\sum_i (1 + \rho_{ik_2}) x_{i1} - \sum_i x_{ik_2} = 0, \quad k_2 \in \mathcal{K}_2$$

$$\sum_i (1 + \rho_{ik_3}) x_{ik_2} - y_{k_3}^+ + y_{k_3}^- = g, \quad k_3 \in \mathcal{D}(k_2), \, k_2 \in \mathcal{K}_2$$

and nonnegativity of all variables.

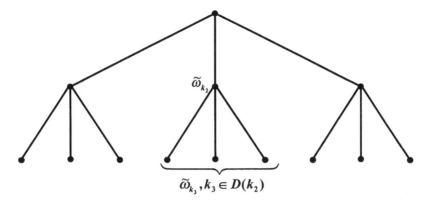

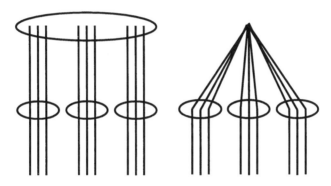

Figure 1: Scenario tree for 3-stage problem

Figure 2: Splitted 3-stage tree

2.2.2   Exercise. Let ω_t attain only two equiprobable values ω_t^u and ω_t^h, $t = 1, 2$. Rewrite the problem considered in Example 2.2.1 in the split variable form and also in the split variable form applied only to the second and third stage decision variables, i.e., with additional explicit nonanticipativity constraints for the second stage. See Figure 2 for illustration of this idea.

2.3   Horizon and Stages

In real-life applications, it is the modeling part of the problem which has become the most demanding task. Besides the formulation of goals and constraints and identification of the driving random process ω, building a scenario-based multistage stochastic program requires specification of the horizon, stages and generation of the input in the form of a scenario tree, see Chapter II.5.

In a majority of cases, the horizon and the stages are declared as given. In practise, various situations can be distinguished:
- Both the horizon and stages are determined ad hoc, often for purposes of testing numerical approaches and/or software.

- Both the horizon and stages are determined, e.g., by the nature of the real-life technological process, see application 4.8 in Chapter II.4; another example is the flower-girl problem which will be discussed in the next Section.
- The horizon is tied to a fixed date, e.g., to the end of the fiscal or hydrological year, to a date related with the annual Board of Directors' meeting, or to the end date of a screening study. Stages are sometimes dictated by the nature of the solved problem, e.g., by the dates of maturity of bonds, expiration dates of options or by periodic (quarterly, annual, etc.) management review meetings. In other cases, they are obtained by an application of heuristic rules and/or experience, taking into account limitations due to the numerical tractability. Rolling forward after the T-stage problem has been solved, the first-stage decision accepted and a new information exploited means to solve a subsequent $T - 1$-stage stochastic program with a reduced number of stages or possibly another T-stage problem with a different topology of stages.
- The horizon is connected with a time interval of a fixed (possibly even infinite) length, given for instance by the periodicity of the underlying random process, and the number of stages is chosen in dependence on the available computing facilities. Rolling forward means here the repeated solution of a T-stage problem of the same structure of stages with the initial state of the system determined by the applied first-stage decision and by observation of the value of ω_1, and using process ω shifted in time.

For example, the BONDS model, see application 4.2 in Chapter II.4, uses three one-year periods for the three-year planning horizon of the bank and rolling forward means that each year the bank is planning as if it wants to optimize its outcome at the end of the next three years. For production planning problems, a horizon of 12–18 months divided into three stages is used. Energy generation models, e.g., the unit commitment problem 4.7 in Chapter II.4, are usually built for a one-week horizon consisting of 168 hours arranged into $2 - 12$ stages. Short term hydro power system control may use a horizon of 3 hours subdivided more or less arbitrarily into stages, whereas the "long" term planning can be related to a weekend. In the last case, rolling forward will have the same meaning as for the fixed horizon problems (i.e., shortening the horizon).

There exist further specific features of multistage problems. For instance, the problem can be solved just once (to retire the debt by a given deadline as much as possible [40] or to raise money for the college education as in Section II.1.1 and in Example 2.2.1) or the problem and its solution persist in the future, with new horizons, taking always into account just the final state of the system at the previous termination date, i.e., at the previous horizon. To guarantee the possibility of such continuation, the models are usually extended for additional constraints and/or terms in the objective function to reduce the end effects, with or without reference to an additional, auxiliary stage.

For a chosen horizon, the crucial step is to relate the time instants and stages; this is a common problem both in applications of multistage stochastic programming models and in stochastic dynamic programming with discrete time. The main limitations of the number of stages are due to numerical tractability. Some recommendations are common for financial applications: Accept unequal lengths of time periods between subsequent stages, starting with a short first period. Together with repeated rolling of the model over time, this may replace well the full dynamics of the decision process even for problems with a few stages. Another, general suggestion is to break the problem with a long (possibly infinite) horizon into three phases: To use the scenario tree structure for $1 \leq t \leq T$, to design just one descendant from each node for $T + 1 \leq t \leq \tau$ (i.e., the horse-tail structure) and to aggregate the rest of the process into one additional stationary stage. Using these suggestions, one approximates the true probability distribution of ω by a simpler one for the sake of the numerical tractability. Moreover, in reality, the position of stages can be uncertain, random or scenario dependent — another interesting open problem.

2.4   The Flower-Girl Problem

The flower-girl sells roses at price c and has to buy them at cost p before she starts selling. Flowers left over at the end of the day can be stored and sold the next day, when she starts selling the old roses. The roses cannot be carried over more than one additional day at the end of which they are thrown away. The demand is random, ω_t denotes the demand on the t-th day. The flower-girl wants to maximize her total expected profit.

The horizon is related to the number of days for which the flower-girl continues selling roses without break (and also to the fact that our formulation treats only one-period carry-over). Assume first, that the flower-girl sells roses only during the weekend, orders the amount x_1 on Friday evening, registers the demand ω_1 on Saturday, stores the unsold roses (without any additional cost) and, possibly, buys $x_2(\omega_1)$ new roses. Denote $s_2(\omega_1)$ the stock left for the second day and $z(\omega_1, \omega_2)$ the amount of unsold roses at the end of the second day which is also affected by the demand ω_2 on Sunday.

All variables are nonnegative and subject to constraints

$$x_1 - s_2(\omega_1) \leq \omega_1, \quad x_2(\omega_1) + s_2(\omega_1) - z(\omega_1, \omega_2) \leq \omega_2;$$

the total profit is

$$(c - p)(x_1 + x_2(\omega_1)) - cz(\omega_1, \omega_2).$$

If the demand ω_1, ω_2 is known in advance, then one of the optimal solutions is to buy $x_1 = \omega_1$ and $x_2(\omega_1) = \omega_2$ roses which gives the maximal profit of $(c - p)(\omega_1 + \omega_2)$. Consider now a scenario-based version of this 3-stage problem. The scenario tree consists of S scenarios corresponding to the considered realizations $\omega^s = (\omega_1^s, \omega_2^s)$, $s = 1, \ldots, S$, of the demand on the first and second day, their probabilities are p^s, $s = 1, \ldots, S$. We denote again by $\tilde{\omega}_{k_2} = b_{k_2}$, $k_2 \in \mathcal{K}_2$ the possible realizations of ω_1, by p_{k_2} their probabilities, by ω_{k_3} the possible realizations of (ω_1, ω_2) conditional on ω_{k_2} and by π_{k_2, k_3} their (conditional) probabilities. The corresponding realizations of the demand on the second day will be denoted b_{k_3}. In the notation introduced in (11)–(12) the problem reads:

$$\text{maximize } (c - p)x_1 + \sum_{k_2 \in \mathcal{K}_2} p_{k_2} \left[(c - p)x_{2k_2} - c \sum_{k_3 \in \mathcal{D}(k_2)} \pi_{k_2, k_3} z_{k_3} \right]$$

subject to

$$x_1 - s_{2k_2} \leq b_{k_2}, \, k_2 \in \mathcal{K}_2$$
$$x_{2k_2} + s_{2k_2} - z_{k_3} \leq b_{k_3}, \, k_3 \in \mathcal{D}(k_2), \, k_2 \in \mathcal{K}_2$$

and nonnegativity constraints.

The generalization to the T-stage problem $(T > 3)$ is obvious. We index by t all variables related with the stage t, i.e., the amount of roses ordered (x), stored (s) and thrown away (z) at the end of the $(t-1)$st day; notice that z_T plays the role of the only decision variable at the last stage. We obtain:

$$\text{maximize } (c - p)x_1 + E\left\{ (c - p) \sum_{t=2}^{T-1} x_t(\omega^{t-1,\bullet}) - c \sum_{t=2}^{T} z_t(\omega^{t-1,\bullet}) \right\}$$

subject to

$$x_1 + s_1 - s_2(\omega_1) - z_2(\omega_1) \leq \omega_1$$
$$x_t(\omega^{t-1,\bullet}) + s_t(\omega^{t-1,\bullet}) - s_{t+1}(\omega^{t,\bullet}) - z_{t+1}(\omega^{t,\bullet}) \leq \omega_t, \, t = 2, \ldots, T - 1$$
$$s_t(\omega^{t-1,\bullet}) - z_{t+1}(\omega^{t,\bullet}) \leq \omega_t, \, t = 1, \ldots, T - 1$$

with $s_T(\omega) \equiv 0$ and nonnegativity of all variables. In case that the initial supply $s_1 = 0$, one gets $z_2(\omega_1) \equiv 0$. Recall that the number of stages equals one plus the number of days for which the flower-girl sells roses without break and that for the three stage problem considered above, i.e., for $T = 3$, the last inequalities are redundant.

The scenario-based formulation of the T-stage problem can be written in the arborescent form or in the split variable form with explicit nonanticipativity constraints. Notice that the flower-girl problem should be more realistically formulated as an *integer* stochastic program.

Imagine now that the flower-girl wants to earn as much as possible during the two months of her high-school vacations; such a 63 stage problem may be solvable thanks to its simple form. Still some other possibilities should be examined. The program may be rolled forward in time with an essentially shorter horizon, say, for $T = 8$ which covers a whole week. This means that the flower-girl decides as if she plans to maximize her profit over each one-week period and solves the problem every day with a known non-zero initial supply of roses and with a new scenario tree spanning over the horizon of the next 8 days. Another possibility is the aggregation of stages. With a long horizon and random parameters only on the right-hand side of the constraints, one may apply the idea of [71] designed for problems with an infinite horizon: One chooses a tractable horizon T and adds one stationary stage which takes into account the remaining stages $t \geq T$. Finally, one may reformulate the problem into the form of a stochastic dynamic program and solve it by the backward recursion; see Example 2.5.1.

2.4.1   Exercise.

(i) Modify the flower-girl problem to include unit carry-over costs q!

(ii) Rewrite the 3-stage flower-girl problem as a sequence of nested two-stage programs. Observe that the third stage program has an optimal solution $z^* = (x_2 + s_2 - \omega_2)^+$. This means that the problem can be alternatively formulated also as a two-stage problem in which $\phi_2(x_1, x_2, \omega_1, \omega_2) = -c(x_2 + s_2 - \omega_2)^+$ and the (conditional) probability distribution of ω_2 is used to evaluate its expectation.

2.5   Comparison with Stochastic Dynamic Programming

Multistage stochastic programs are in their nature similar to multistage stochastic control problems or to stochastic dynamic programming with discrete time; they deal essentially with the same types of problems – the dynamic and stochastic decision processes. The main distinction is in the solution concept. As we have observed, the main emphasis in multistage stochastic programs lies on the first-stage decisions. On the contrary, in stochastic dynamic programming it is the *decision rule* that is of the primary interest. Such decision rule should be available at any state of the system. An appropriate definition of the state is then the central point of the dynamic programming formulations whereas in the context of multistage stochastic programs states usually do not appear.

Given the state and control spaces $\mathcal{Z}_t$, $\mathcal{U}_t \, \forall t$, the initial state $\mathbf{z}_1$ and the initial control $\mathbf{u}_1$, the dynamics of the system is described by the system of equations

$$(13) \qquad\qquad \mathbf{z}_{t+1} = \mathbf{F}_t(\mathbf{z}_t, \mathbf{u}_t, \omega_t), \, t = 1, \ldots, T$$

with given vector-valued transition functions $\mathbf{F}_t$. (Equality means again equality with probability 1.) The goal is to control the evolution of the system by using an optimal decision rule $\mathbf{u}_t \, \forall t$ to get the best total expected outcome of the decision process.

Formally, one may define $\mathbf{x}_t = (\mathbf{z}_t, \mathbf{u}_t)$ and approach the above decision problem via multistage stochastic programming. Another possibility is to exploit the Markov character of the system dynamics: with *interstage independent* $\omega_t \, \forall t$ the dynamics as described by (13) is memoryless, and a further restriction of the problem structure allows to apply solution procedures akin to Bellman's principle of optimality. This restriction (besides of the already mentioned interstage independence) consists of a certain *separability property* of the random objective function f_0, e.g. the additive form $f_{10}(\mathbf{z}_1, \mathbf{u}_1) + \sum_t \overline{f}_t(\mathbf{z}_t, \mathbf{u}_t, \omega_t)$, further of *finiteness of state spaces* and of *compactness of control spaces*.

Assume for simplicity that $\mathcal{Z}_t \equiv \mathcal{Z} \, \forall t$, $\mathcal{U}_t \equiv \mathcal{U} \, \forall t$ and let $c_1(\mathbf{z}_1, \mathbf{u}_1) = f_{10}(\mathbf{z}_1, \mathbf{u}_1) + E_{\omega_1} \overline{f}_1(\mathbf{z}_1, \mathbf{u}_1, \omega_1)$, $c_t(\mathbf{z}_t, \mathbf{u}_t) = E_{\omega_t} \overline{f}_t(\mathbf{z}_t, \mathbf{u}_t, \omega_t)$ be the expected outcomes related with the respective stage only. Denote further for $\mathbf{z}, \mathbf{z}' \in \mathcal{Z}$, $\mathbf{u} \in \mathcal{U}$ by $\pi_t(\mathbf{z}'; \mathbf{z}, \mathbf{u})$ the transition probabilities $P[\mathbf{z}_{t+1} = \mathbf{z}' \,|\, \mathbf{z}_t = \mathbf{z}, \mathbf{u}_t = \mathbf{u}] = P[\mathbf{F}_t(\mathbf{z}, \mathbf{u}, \omega_t) = \mathbf{z}']$. Then the stochastic dynamic program can be reformulated as a **Markov decision problem**:

> Given the transition probabilities $\pi_t(\mathbf{z}'; \mathbf{z}, \mathbf{u})$, $\mathbf{z}, \mathbf{z}' \in \mathcal{Z}$, $\mathbf{u} \in \mathcal{U}$
> find $\mathbf{u}^{T\bullet} = (\mathbf{u}_1, \ldots, \mathbf{u}_T)$ that minimizes $E_\omega \sum_t c_t(\mathbf{z}_t, \mathbf{u}_t)$.

The collection of the transition probabilities along with the (possibly degenerated) probability distribution of the initial state $\mathbf{z}_1$ and the control $\mathbf{u}^{T\bullet}$ fully describe the stochastic evolution of the considered system without any reference to the transition relation (13). The usual approach is to define functions

$$V_t(\mathbf{z}) = \min_{\mathbf{u}^{T\bullet}} E_\omega \left[\sum_{\tau=t}^{T} c_\tau(\mathbf{z}_\tau, \mathbf{u}_\tau) \,\Big|\, \mathbf{z}_t = \mathbf{z} \right], \, t = 1, \ldots, T-1$$

and to evaluate them by the "backward recursion"

$$V_T(\mathbf{z}_T) = 0 \quad \text{and for } t = 1, \ldots, T-1$$

(14) $$V_t(\mathbf{z}) = \min_{\mathbf{u} \in \mathcal{U}} \left\{ c_t(\mathbf{z}, \mathbf{u}) + \sum_{\mathbf{z}' \in \mathcal{Z}} \pi_t(\mathbf{z}'; \mathbf{z}, \mathbf{u}) V_{t+1}(\mathbf{z}') \right\};$$

compare with (6)–(7). Let $\hat{\mathbf{u}}_t(\mathbf{z})$ be an optimal solution of (14). Hence, for a given initial state $\mathbf{z}_1 \in \mathcal{Z}$, $V_1(\mathbf{z}_1)$ is the optimal value of the objective function $\sum_t c_t(\mathbf{z}_t, \mathbf{u}_t)$ and $\mathbf{u}_1^*(\mathbf{z}_1) = \hat{\mathbf{u}}_1(\mathbf{z}_1)$ is the optimal control for the first stage. The optimal controls $\mathbf{u}_t^*$ for the subsequent stages depend on the state of the system, say, $\mathbf{z}_t^*$ which has been attained according to transition probabilities π_τ, $\tau = 1, \ldots, t-1$, corresponding to the already obtained optimal controls $\mathbf{u}_\tau^*$, $\tau = 1, \ldots t-1$. Hence, $\mathbf{u}_t^* = \hat{\mathbf{u}}_t(\mathbf{z}_t^*)$ – the optimal control at the first stage of the $(T-t)$-stages program (14) with the initial state $\mathbf{z}_t^*$.

Solution of the optimization problems (14) may be hard or easy depending on the application. However, implementation of the "backward recursion" means to solve the optimization problems (14) for all stages *in dependence on multidimensional parameters* $\mathbf{z}$ and to store the values of the functions V_t and of the optimal solutions of (14) for all states of the system. This puts considerable requirements on the memory and, therefore, in many applications the number of states turns out to be prohibitively large. On the other hand, problems with very distant or infinite horizon and problems with state and control dependent transition probabilities can be treated efficiently provided that the structure of the problem fits well the solution method (based on certain Markov properties, separability of the objective function, etc.), that the dimension of the state vector is not too big and that the number of constraints is limited. Notice that the considered structure of the problem is tied with the solution method – the backward recursion connected with the principle of optimality.

On the contrary, multistage stochastic programs do not use the notion of the state and their formulation is not connected with any prescribed solution technique. Therefore, there exists a variety of stochastic programs along with various solution procedures. The emphasis is on the first-stage decision, mostly with a continuum of possibilities. Even if it is often possible to characterize the optimal decision rules, it is not necessary to design a full backward recursion as in dynamic programming and, due to large dimensionality of stochastic programming problems, such procedure would be hardly tractable. Hence, the numerical methods are not based on the recursive form (6)–(7) or (8)–(9). It is possible to avoid special requirements on the Markov structure of the problem. Numerous constraints can be included, however, integrality of decision variables is a drawback, not an advantage from the computational point of view. The probability distribution of random parameters, which is assumed independent of the decisions, is approximated mostly by a discrete probability distribution before or in course of numerical procedures. The resulting problems are large mathematical programs, their dimensionality increases rapidly with increasing number of stages and scenarios.

Hence, the two discussed approaches used for modeling and solution of discrete time dynamic decision problems under uncertainty are not competitive, they are merely complementary having different favorable and unfavorable features. Certain type of the "curse of dimensionality" relates to each of them: In stochastic dynamic programs, it is connected with the number of states and the dimensionality of the state and control spaces whereas it is mostly the number of stages which puts limitations on tractability of multistage stochastic programs.

We conclude this discussion by a simple illustrative example and in the subsequent Sections we shall focus on applications of stochastic programming.

2.5.1 Example – The Flower-Girl Problem. Under simplifying assumption that the random demands ω_t are independent and identically distributed, the flower-girl problem 2.4 can be easily formulated as a stochastic dynamic program and can be also solved by the backward recursion of dynamic programming.

Assume that ω_t, $t = 1, 2, \ldots$ (demand on the tth day) are *independent* random variables such that for all t

$$P[\omega_t = k - 1] = p_k, \ k = 1, \ldots, K,$$

where $\sum_{k=1}^{K} p_k = 1$.

Since at most $K - 1$ roses can be sold every day, we may restrict on policies storing at most $K - 1$ roses on each day. Recalling that x_t (the control variable) denotes the number of fresh roses ordered for the t-th day ($t = 1, 2, \ldots$) and that s_t is reserved for the state variable (i.e. the stock of one day old roses left from the $(t-1)$st day and $s_1 \equiv 0$) we conclude immediately that for any $t \geq 1$

$$s_{t+1} \leq x_t, \quad s_t + x_t \leq K - 1.$$

Hence for any nonnegative integers i, k such that $i + k \leq K - 1$ and independently on t, the transition probabilities are

$$\pi(0; i, k) = P[s_t = 0 | s_{t-1} = i, x_{t-1} = k] = \sum_{h=i+k+1}^{K} p_h$$

$$\pi(k; i, k) = P[s_t = k | s_{t-1} = i, x_{t-1} = k] = \sum_{h=1}^{i+1} p_h, \ k \geq 1$$

$$\pi(j; i, k) = P[s_t = j | s_{t-1} = i, x_{t-1} = k] = p_{i+k+1-j}, \ j = 1, \ldots, k - 1$$

and are equal to zero otherwise.

Obviously, if the order is $x_t = k$ on the tth day when the stock left from the previous day is $s_t = i$ roses, the cost is pk and the expected amount of the money obtained by selling the roses is equal to $c\sum_{j=1}^{i+k+1}(j-1)p_j$, i.e., the expected profit of the t-th day in notation used for the additive objective function with $\mathbf{u}_t = x_t = k$, $\mathbf{z}_t = s_t = i$ equals

$$c_t(\mathbf{z}_t, \mathbf{u}_t) = c\sum_{j=1}^{i+k+1}(j-1)\,p_j - pk := c(i;k) \quad \forall t.$$

The flower-girl problem can be solved now as a standard stochastic dynamic program. The "backward recursion" (14) reads: $V_T(i) = 0$, $\forall i$ and

$$V_t(i) = \max_{k=0,\dots,K-1}[c(i;k) + \sum_{j=0}^{K-1}\pi(j;i,k)V_{t+1}(j)]$$

for $t = 1, 2, \dots, T-1$ and any $i = 0, 1, \dots, K-1$. It can be easily solved for the considered finite state space $\mathcal{Z} = \{0, 1, \dots, K-1\}$ and for a large horizon T; obviously, the integrality of states and of decision variables is exploited.

However, it is hard to apply the backward recursion in more complicated problems as to the dimensionality of the state vector and/or in presence of numerous state and control constraints. In the context of the flower-girl, think about a whole set of traded flowers with various carry-over constraints, inclusion of a limited store space or of a substitution effect in the (random) demand.

II.3 MULTIPLE CRITERIA

multi-objective programming, efficient solutions, ϵ-constrained problems, goal programming, Markowitz model, alternative definitions of risk, VaR, expected utility, tracking models, scenario-based stochastic programs, robust optimization

Very often, plausible economic decisions cannot be chosen only according to one criterion, such as the maximal profit, production efficiency or yield. In production planning, environmental criteria have to be taken into account, in macroeconomical problems regional aspects such as the local unemployment level play an essential role. Very different, even conflicting goals can be set for a short time horizon and for the long one, etc. Investment decisions should hedge against risks of various kinds, such as liquidity, volatility or currency risks. The mentioned disparate criteria will hardly be satisfied by a uniformly optimal decision. Problems of the above kind belong under *multi-objective programming* and we shall briefly introduce the main approaches for the case of continuous decision variables.

3.1 Theory

Without loss of generality we shall formulate the multi-objective programming problem for the case of *minimization* of two or more objective functions:
"minimize" $K \geq 2$ functions $f_1, \ldots, f_K$, $f_k : \mathbb{R}^n \to \mathbb{R}^1$ on a closed set $\mathcal{X}$ briefly

(1) $$\text{"min"} \, \mathbf{f}(\mathbf{x}) \text{ on } \mathcal{X}.$$

An *ideal solution* $\tilde{\mathbf{x}} \in \mathcal{X}$ of the multi-objective programming problem (1) is defined by the property

$$\tilde{\mathbf{x}} \in \bigcap_{k=1}^{K} \arg\min_{\mathbf{x} \in \mathcal{X}} f_k(\mathbf{x}).$$

Ideal solutions exist only rarely and the first task is to introduce another concept of solution $\mathbf{x} \in \mathcal{X}$ which is acceptable from the point of view of "minimization" of all considered objective functions.

3.1.1 Definition. A solution $\hat{\mathbf{x}} \in \mathcal{X}$ is an *efficient solution* of the multi-objective programming problem (1) if there is no element $\mathbf{x} \in \mathcal{X}$ for which $\mathbf{f}(\mathbf{x}) \leq \mathbf{f}(\hat{\mathbf{x}})$ and $\mathbf{f}(\mathbf{x}) \neq \mathbf{f}(\hat{\mathbf{x}})$.

The modification of Definition 3.1.1 to multicriterial *maximization* problems is straightforward and the interpretation of efficient decisions is clear: no other feasible decision is uniformly better with respect to all considered criteria. Multi-objective programming aims at location of efficient solutions and there are many ways how to achieve this goal; we shall introduce some of them.

3.1.2   Theorem. *Let $\mathcal{X}$ be compact, $f_k, k = 1, \ldots, K$, continuous on $\mathcal{X}$ and $h :$ $\mathbb{R}^K \to \mathbb{R}^1$ an arbitrary continuous function which is nondecreasing in its arguments. Then at least one solution belonging to*

$$(2) \qquad \mathcal{X}_h^* := \arg\min_{\mathbf{x} \in \mathcal{X}} h(f_1(\mathbf{x}), \ldots, f_K(\mathbf{x}))$$

is efficient for the multi-objective problem (1).

Proof. The set $\mathcal{X}_h^*$ defined by (2) is nonempty due to assumptions; let $\bar{\mathbf{x}} \in \mathcal{X}_h^*$. Similarly, there exists an optimal solution, say, $\mathbf{x}'$ of the auxiliary problem

$$\min \sum_{k=1}^{K} f_k(\mathbf{x})$$

subject to

$$\mathbf{x} \in \mathcal{X} \text{ and } f_k(\mathbf{x}) \le f_k(\bar{\mathbf{x}}), \ k = 1, \ldots, K.$$

According to the assumed monotonicity property of h, $\mathbf{x}' \in \mathcal{X}_h^*$.

Assume that $\mathbf{x}'$ is not an efficient solution of (1). Then there exists $\mathbf{x}^* \in \mathcal{X}$ such that $\mathbf{f}(\mathbf{x}^*) \le \mathbf{f}(\mathbf{x}')$ and $\mathbf{f}(\mathbf{x}^*) \ne \mathbf{f}(\mathbf{x}')$. It means that $\mathbf{x}^*$ is a feasible solution of the auxiliary problem and that for an index j, $f_j(\mathbf{x}^*) < f_j(\mathbf{x}')$ and $f_k(\mathbf{x}^*) \le f_k(\mathbf{x}')$ for all remaining indices. Hence, $\sum_{k=1}^{K} f_k(\mathbf{x}^*) < \sum_{k=1}^{K} f_k(\mathbf{x}')$ which contradicts the assumed optimality of $\mathbf{x}'$. $\square$

3.1.3   Comment

The optimal solution and the numerical tractability of the optimization problem (2) depend substantially on the choice of the function h. One of its properties is obvious: For h *increasing* in its arguments, all optimal solutions of (2) are efficient for (2). In general, an adequate choice of h is a difficult problem. It would not be necessary to deal with multi-objective programming if the choice of h was straightforward.

A special simple choice of the function h in Theorem 3.1.2 is

$$(3) \qquad h(\mathbf{z}) = \sum_{k=1}^{K} t_k z_k \text{ with a vector parameter } \mathbf{t} \in \mathbb{R}^{+K}, \mathbf{t} \ne 0.$$

We shall prove that under modest assumptions and using all possible choices of weights $\mathbf{t}$, (3) generates *all efficient solutions* of the solved multi-objective problem (1).

3.1.4   Theorem. *For a vector parameter $\mathbf{t} \in \mathbb{R}^{+K}$, let $\bar{\mathbf{x}}$ be an optimal solution of*

$$(4) \qquad \min_{\mathbf{x} \in \mathcal{X}} \sum_{k=1}^{K} t_k f_k(\mathbf{x}).$$

Assume further that the parameter vector $\mathbf{t}$ *in* (4) *is positive or that* $\bar{\mathbf{x}}$ *is a unique optimal solution of* (4). *Then* $\bar{\mathbf{x}}$ *is an efficient solution of the multi-objective programming problem* (1).

The proof follows directly from Definition 3.1.1 and is left to the reader.

3.1.5   Theorem. *Let* $\mathcal{X} \subset \mathbb{R}^n$ *be nonempty, convex, compact and* $f_k, k = 1, \ldots, K$ *be convex functions on* $\mathbb{R}^n$. *Then for an arbitrary efficient solution* $\hat{\mathbf{x}} \in \mathcal{X}$ *of the multi-objective programming problem* (1) *there exists a vector parameter* $\mathbf{t} \in \mathbb{R}^{+K}, \mathbf{t} \neq 0$ *such that* $\hat{\mathbf{x}}$ *is an optimal solution of* (4).

Proof. Denote $\mathcal{C}$ the convex hull of the set

$$\{\mathbf{y} \in \mathbb{R}^K : \mathbf{y} = \mathbf{f}(\hat{\mathbf{x}}) - \mathbf{f}(\mathbf{x}) \text{ for some } \mathbf{x} \in \mathcal{X}\}.$$

Efficiency of $\hat{\mathbf{x}}$ and convexity of functions f_k imply that $\mathcal{C}$ and $\mathbb{R}^{+K}$ do not contain common interior points. The two convex sets can be thus separated by a hyperplane with coefficients $\mathbf{t} \neq 0$:

$$\sum_k t_k y_k \leq \alpha \leq \sum_k t_k z_k \, \forall \mathbf{y} \in \mathcal{C}, \mathbf{z} \in \mathbb{R}^{+K}.$$

The special choice of $\mathbf{z} = 0$ implies $\alpha \leq 0$ and, moreover, $t_k \geq 0 \forall k$ (otherwise an appropriate choice of $\mathbf{z}$ in the unbounded set $\mathbb{R}^{+K}$ provides arbitrarily large negative values of $\sum_k t_k z_k$). This means that there is a $\mathbf{t} \in \mathbb{R}^{+K}, \mathbf{t} \neq 0$ such that $\sum_k t_k y_k \leq 0 \forall \mathbf{y} \in \mathcal{C}$. Substituting $\mathbf{y} = \mathbf{f}(\hat{\mathbf{x}}) - \mathbf{f}(\mathbf{x})$ with an arbitrary $\mathbf{x} \in \mathcal{X}$ we get the desired result:

$$\sum_k t_k f_k(\mathbf{x}) \geq \sum_k t_k f_k(\hat{\mathbf{x}}).$$

□

3.1.6   Comment

A stronger result holds true for *linear* functions $f_k, \forall k$ and a convex polyhedral set $\mathcal{X}$, namely, $\hat{\mathbf{x}}$ is an efficient solution of the multi-objective problem (1) if and only if it is an optimal solution of (4) for a *positive* parameter vector $\mathbf{t} \in \mathbb{R}^{+K}$. To prove this result it is enough to notice that the set $\mathcal{C}$ is polyhedral and to exploit this fact when constructing the separating hyperplane.

There are relatively many other approaches that provide efficient solutions:

3.1.7   Exercise - The ϵ-Constrained Approach.

Select one of the considered objective functions, say, f_1, choose a threshold vector $\epsilon \in \mathbb{R}^{K-1}$ and solve the classical optimization problem

(5)   minimize $f_1(\mathbf{x})$ subject to $\mathbf{x} \in \mathcal{X}$ and $f_k(\mathbf{x}) \leq \epsilon_k, k = 2, \ldots, K$.

Assume that for the choice of ϵ, the set $\mathcal{X}_\epsilon := \{\mathbf{x} \in \mathcal{X} | f_k(\mathbf{x}) \leq \epsilon_k, k = 2, \ldots, K\} \neq \emptyset$ and prove the following statements:

(i) Let $\bar{\mathbf{x}}$ be the unique optimal solution of (5). Then $\bar{\mathbf{x}}$ is an efficient solution of (1).

(ii) Let $\hat{\mathbf{x}}$ be an efficient solution of (1). Then there exists $\epsilon \in \mathbb{R}^{K-1}$ such that $\hat{\mathbf{x}}$ is an optimal solution of (5).

3.1.8   Exercise – The Mixed Approach. It is related to a different treatment of distinct objective functions: some of them are put into constraints as in the ϵ-constrained approach whereas weights $t_k > 0$ are assigned to the remaining objective functions. As a result, one solves a parametric programming problem with parameters in the objective function and on the right-hand sides, e.g.,

$$\min_{\mathbf{x} \in \mathcal{X}} \sum_{k=1}^{K_0} t_k f_k(\mathbf{x})$$

subject to

$$f_k(\mathbf{x}) \le \epsilon_k, \ k = K_0 + 1, \ldots, K$$

(see [72]). Explore the relationship between optimal solutions of this program and efficient solutions of (1).

3.1.9   Exercise – The Goal Programming Approach. The main idea is to try to get a solution from $\mathcal{X}$ for which the outcome measured by $\mathbf{f}(\mathbf{x}) = (f_1(\mathbf{x}), \ldots, f_K(\mathbf{x}))^\top$ is as close as possible to the K-vector of the best attainable outcomes $f_k^* = \min_{\mathbf{x} \in \mathcal{X}} f_k(\mathbf{x}), k = 1, \ldots, K$. The distances are defined in the space of function values, a subset of $\mathbb{R}^K$, and one is free to use any of (weighted) L_p-distances, $1 \le p \le \infty$. Hence, one solves a minimization problem

$$(6) \qquad\qquad \min_{\mathbf{x} \in \mathcal{X}} \| \mathbf{T}(\mathbf{f}^* - \mathbf{f}(\mathbf{x})) \|_p$$

with a diagonal matrix $\mathbf{T} = \mathrm{diag}\{t_1, \ldots, t_K\}$, $t_k > 0\, \forall k$. Using properties of the distances and Definition 3.1.1 prove the following statements:

(i) Let $\bar{\mathbf{x}}$ be an optimal solution of (6) with $1 \le p < \infty$. Then $\bar{\mathbf{x}}$ is an efficient solution of (1).

(ii) For $p = \infty$, at least one of optimal solutions of the minimax problem

$$\min_{\mathbf{x} \in \mathcal{X}} \max_k t_k |f_k^* - f_k(\mathbf{x})|$$

is efficient for (1).

3.1.10   Exercise – Modifications for Multi-objective Maximization Problems.

(i) Prove that with minimization replaced by maximization, Theorem 3.1.4 holds true and Theorem 3.1.5 is valid for concave functions f_k.

(ii) Modify Theorem 3.1.2 and the approaches described in Exercises 3.1.7 – 3.1.9 for maximization problems.

The techniques of multi-objective programming allow to exclude "bad", non-efficient solutions of the multi-objective problems (1) provided that the selected criteria can be quantified and that the considered objective functions are a priori given the same importance. In interactive numerical approaches, see [72], the user is allowed to change the values of parameters (weights and thresholds) to achieve an acceptable balance between the criteria. There are various other problems which have not been tackled here, e.g., an appropriate treatment of hierarchically ordered objective functions or the case of a finite list of feasible alternatives. We refer to [140].

3.2 Selected Applications to Portfolio Optimization

In the sequel we shall formulate several well-known problems of portfolio optimization, including the famous Markowitz model, which can be treated within the framework of multi-objective programming presented in Section 3.1.

3.2.1 The Markowitz Model

We have seen in Chapter I.9 that the Markowitz model was developed for investments in portfolio under various implicit simplifying assumptions. Among others it assumes a frictionless market and applies to small rational investors whose investments cannot influence the market prices and who prefer higher yields to lower ones and smaller risks to larger ones.

The composition of a portfolio consisting of J risky assets is identified with the vector $\mathbf{x}$ whose components are the fractions x_j, $j = 1, \ldots, J$, of the (disposable) unit wealth invested in the j-th asset, $\sum_j x_j = 1$. Depending on circumstances one may require $x_j \geq 0 \, \forall j$ or to drop the nonnegativity assumptions allowing for short sales. A unit investment in the j-th share gives a random return ρ_j over the considered unit period. The assumed probability distribution of the vector ρ of returns of all shares is characterized by a known vector of expected returns $E\rho = \mathbf{r}$ and by a covariance matrix $\mathbf{V} = [\mathrm{cov}(\rho_i, \rho_j), \quad i, j = 1, \ldots, J]$ whose main diagonal consists of variances σ_j^2 of individual returns.

The "yield from the investment" $\mathbf{x}$ is then quantified as the expectation

$$r(\mathbf{x}) = \sum_j r_j x_j = \mathbf{r}^\top \mathbf{x}$$

of the total portfolio return and the "risk of the investment" as the variance or standard deviation of the total portfolio return

$$\sigma^2(\mathbf{x}) = \sum_{i,j} \mathrm{cov}(\rho_i, \rho_j) x_i x_j = \mathbf{x}^\top \mathbf{V} \mathbf{x}$$

or $\sigma(\mathbf{x}) = \sqrt{\mathbf{x}^\top \mathbf{V} \mathbf{x}}$, respectively. According to the assumptions, the investors aim at maximal possible yields and, at the same time, at minimal possible risks – hence, a typical decision problem with two criteria, "max" $\{r(\mathbf{x}), -\sigma^2(\mathbf{x})\}$ or "min" $\{-r(\mathbf{x}), \sigma^2(\mathbf{x})\}$ and thus the mean-variance efficiency introduced by Markowitz is fully in line with general concepts of Section 3.1 and the approaches presented there can be exploited:

The mean-variance efficient portfolios can be obtained by solving various optimization problems, e.g.,

(7)
$$\max_{\mathbf{x} \in \mathcal{X}} \lambda \mathbf{r}^\top \mathbf{x} - \frac{1}{2} \mathbf{x}^\top \mathbf{V} \mathbf{x},$$

see (4), where the value of parameter $\lambda \geq 0$ is related to investor's risk aversion (small values of λ correspond to a risk averse investor whereas large values of λ are typical for risk seeking investors), or

$$(8) \qquad \min_{\mathbf{x} \in \mathcal{X}} \mathbf{x}^\top \mathbf{V} \mathbf{x}$$

subject to

$$\mathbf{r}^\top \mathbf{x} \geq r_p$$

where parameter r_p is the minimal acceptable return, cf. 3.1.7, or by maximization of a suitable aggregating objective function h such that $-h$ fulfills assumptions of Theorem 3.1.2. Provided that the covariance matrix $\mathbf{V}$ is positive definite, one possibility is

$$(9) \qquad h(r(\mathbf{x}), -\sigma^2(\mathbf{x})) := \frac{r(\mathbf{x})}{\sigma(\mathbf{x})},$$

hence, *Sharpe's ratio* introduced in I.9.2.

Notice that the same sets of efficient portfolios are obtained when the risk is quantified either by the variance or by the standard deviation of the portfolio return. To get this conclusion, consider (8) with the objective function $\mathbf{x}^\top \mathbf{V} \mathbf{x}$ and with its strictly increasing transform $\sqrt{\mathbf{x}^\top \mathbf{V} \mathbf{x}}$.

The set $\mathcal{X}$ is defined by $\sum_j x_j = 1$ and other conditions on the composition of the portfolio, e.g., the nonnegativity constraint and upper bounds.

3.2.2 Efficient Portfolios Based on Alternative Definitions of Risk

Objections against the symmetry of the variance of returns as a measure of risk has lead to various asymmetric risk definitions, such as the quadratic semivariance $E\{[\sum_j r_j x_j - \sum_j \rho_j x_j]^+\}^2$; the disadvantage are difficulties accompanying numerical solution of the resulting optimization problems.

Following the ideas of [145], Konno and Yamazaki [101] developed and applied efficient investment decisions based on "risk" defined as the mean absolute deviation

$$(10) \qquad m(\mathbf{x}) := E\left|\sum_j \rho_j x_j - \sum_j r_j x_j\right|.$$

To get a portfolio efficient with respect to $r(\mathbf{x})$ and $-m(\mathbf{x})$ means to solve, for instance, the program

$$(11) \qquad \min_{\mathbf{x} \in \mathcal{X}} E\left|\sum_j \rho_j x_j - \sum_j r_j x_j\right|$$

subject to

$$\sum_j r_j x_j \geq r_p$$

(compare with (8)).

Values of (10) are constant multiples of the standard deviations $\sigma(\mathbf{x})$ if the probability distribution of the random returns ρ is normal $N(\mathbf{r}, \mathbf{V})$.

The advantage of this approach is that the sample-based form of (11) with all expectations replaced by averages based on the sample or historical data can be transformed to solution of a linear program. It is not necessary to estimate the variance matrix, hence, it is easier to get the input data than for the Markowitz model. Also an extension to an asymmetric quantification of risk is straightforward. Indeed, using

$$
\mu E \left[\sum_j \rho_j x_j - \sum_j r_j x_j \right]^+ + (1-\mu) E \left[\sum_j \rho_j x_j - \sum_j r_j x_j \right]^-
$$

with $\mu \in (0,1)$ at the place of (10) reduces to maximization of (10) again. To see it, rewrite the positive and negative parts using known formulas $z = z^+ - z^-$ and $|z| = z^+ + z^-$.

Further approaches are related with the *safety-first* or probability criterion proposed by Roy [139] at the same time when Markowitz developed his mean-variance approach. The suggestion is to maximize on $\mathcal{X}$ the probability

$$
(12) \qquad\qquad P(\boldsymbol{\rho}^\top \mathbf{x} \geq r_p)
$$

with a given level r_p of the required return of the portfolio. Under normality assumption $\rho \sim N(\mathbf{r}, \mathbf{V})$, the deterministic equivalent is (compare with (9))

$$
(13) \qquad\qquad \max_{\mathbf{x} \in \mathcal{X}} \frac{r(\mathbf{x}) - r_p}{\sigma(\mathbf{x})}.
$$

Efficient portfolios with respect to the expected return $r(\mathbf{x})$ and the probability (12) can be found e.g. by application of the ϵ-constrained approach 3.1.7. This results in the *chance-constrained* criterion, cf. [160],

$$
(14) \qquad\qquad \max_{\mathbf{x} \in \mathcal{X}} \{ r(\mathbf{x}) \mid P(\boldsymbol{\rho}^\top \mathbf{x} \geq r_p) \geq 1 - \alpha \}
$$

with a prescribed level of the total return r_p and with a given probability $\alpha \in (0,1)$. The explicit form of (14) under normality assumption reads

$$
(15) \qquad\qquad \max_{\mathbf{x} \in \mathcal{X}} \{ r(\mathbf{x}) \mid r(\mathbf{x}) + \Phi^{-1}(\alpha) \sigma(\mathbf{x}) \geq r_p \};
$$

here, $\Phi^{-1}(\alpha)$ denotes the $100\alpha\%$ quantile of the standard normal distribution $N(0,1)$.

3.2.2.1 Exercise. Derive the form (13) and (15) of the safety-first and chance-constrained criteria under assumption of normally distributed returns.

3.2.3   Quantile Criterion and the Value at Risk (VaR)

The *quantile* criterion (cf. [97])

(16)   maximize r_p subject to $\mathbf{x} \in \mathcal{X}$ and $P(\boldsymbol{\rho}^\top \mathbf{x} \geq r_p) \geq 1 - \alpha$

for a prescribed probability $\alpha \in (0,1)$ can be evidently related with quantification of the risk of the investment by its *Value at Risk*, VaR introduced in I.4.2.3. Indeed, the probabilistic constraint in (16) can be written as $P(\boldsymbol{\rho}^\top \mathbf{x} < r_p) \leq \alpha$. If the distribution function F of $\boldsymbol{\rho}^\top \mathbf{x}$ is continuous and increasing,

$$r_p = -\operatorname{VaR}_\alpha(\mathbf{x}) = F^{-1}(\alpha),$$

the $100\alpha\%$ quantile of the probability distribution of $\boldsymbol{\rho}^\top \mathbf{x}$, satisfies the probabilistic constraint in (16) sharp. Naturally, small values of $\operatorname{VaR}_\alpha$ are desirable from the point of view of risk minimization.

To evaluate $\operatorname{VaR}_\alpha(\mathbf{x})$ one needs to get the probability distribution of $\boldsymbol{\rho}^\top \mathbf{x}$ which depends on the decision variables – the composition $\mathbf{x}$ of the portfolio. This is a demanding task which is mostly solved by simulations. An exception is the normal distribution $N(\mathbf{r}, \mathbf{V})$ of $\boldsymbol{\rho}$. In this case, the probability distribution of $\boldsymbol{\rho}^\top \mathbf{x}$ is $N(\mathbf{r}^\top \mathbf{x}, \mathbf{x}^\top \mathbf{V} \mathbf{x})$ whose $100\alpha\%$ quantile is

$$\mathbf{r}^\top \mathbf{x} + \Phi^{-1}(\alpha)\sqrt{\mathbf{x}^\top \mathbf{V} \mathbf{x}}.$$

Multiplying it by -1, we obtain the absolute value at risk, see (20) in I.4.2.3 for $u_\alpha = \Phi^{-1}(\alpha)$ – the $100\alpha\%$ quantile of the standard normal distribution $N(0,1)$.

3.2.3.1 Example. Assuming normal distribution of returns, derive an explicit form of an optimization problem which provides efficient solutions of

$$\text{``max''}\{r(\mathbf{x}), -\operatorname{VaR}_\alpha(\mathbf{x})\}$$

and compare it with (15). Derive also the explicit form of (16).

3.2.4   Expected Utility Criterion

The generally accepted decision criterion of maximal expected utility of the total final return offers further possibilities. As to utility functions $U : \mathbb{R} \to \mathbb{R}$, it is customary to assume that U is increasing and concave. The last assumption corresponds to preferences of a *risk averse investor* who never prefers to accept a fair gamble of the form – to move to wealth values w^1, w^2 with probabilities λ and $1-\lambda$, respectively, when he could remain at the wealth level $\lambda w^1 + (1 - \lambda)w^2$. Similarly, preferences of risk neutral or risk seeking investors are described by increasing linear or convex utility functions, respectively. Consult Chapter I.8.

Numerous early studies dealt with the relationship between various criteria mentioned in 3.2.1–3.2.4. It is apparent that for normal distribution of returns, the portfolios obtained by solving (13), (15) or by minimization of $\operatorname{VaR}_\alpha(\mathbf{x})$ are mean-variance efficient. Concerning the expected utility maximization, for normal distribution of returns and for concave utility functions, maximization of $EU(\boldsymbol{\rho}^\top \mathbf{x})$ provides mean-variance efficient portfolios as well.

3.2.4.1 Exercise. Assume that the returns ρ are normally distributed, $N(\mathbf{r}, \mathbf{V})$. Derive that for the exponencial utility function $U(w) = 1 - \exp(-aw)$ with $a \geq 0$ the optimal solutions of $\max_{\mathbf{x}} EU(\rho^\top \mathbf{x})$ can be found by solving $\max_{\mathbf{x}} \mathbf{r}^\top \mathbf{x} - \frac{a}{2}\mathbf{x}^\top \mathbf{V}\mathbf{x}$ and are thus mean-variance efficient.

3.3 Multi-objective Optimization and Stochastic Programming Models

We shall consider now various approaches to mathematical formulation of decision problems under stochastic uncertainty about the future values of the system parameters from the point of view of multi-objective programming.

3.3.1 Scenario-Based Stochastic Programs

Assume now that we are supposed to select the "best possible" decision which fulfills prescribed "hard" constraints, say, $\mathbf{x} \in \mathcal{X}$ where $\mathcal{X} \in \mathbb{R}^n$ is a closed nonempty set. We accept that the outcome of a decision $\mathbf{x}$ is influenced by a random element of a general nature whose realization is not known at the time of decision. The random outcome of a decision $\mathbf{x}$ is quantified by $f_0(\mathbf{x}, \omega)$ and different realizations of ω provide different optimal solutions, say, $\mathbf{x}^*(\omega) \in \arg\min f_0(\mathbf{x}, \omega)$. If the set of possible realizations of ω is finite, $\{\omega^s, s = 1, \ldots, S\}$, methods of multi-objective programming suggest to choose a solution efficient with respect to S objective functions $f_0(\mathbf{x}, \omega^s)$, $s = 1, \ldots, S$. According to Theorem 3.1.4 such efficient solutions can be obtained, e.g., by minimization (or maximization) of a weighted sum of $f_0(\mathbf{x}, \omega^s)$, $s = 1, \ldots, S$. In our case, it is natural to use probabilities p^s of scenarios ω^s at the place of weights t_s and the problem to be solved is

$$\min_{\mathbf{x} \in \mathcal{X}} \sum_{s=1}^{S} p^s f_0(\mathbf{x}, \omega^s).$$

The result is the widely used *expected value criterion*.

Notice that we get efficient solutions regardless the origin of probabilities p^s, e.g., for p^s – the true probabilites or subjective ones, for probabilities offered by experts, for equal probabilities obtained via simulation or coming from an empirical probability distribution.

Similarly, using the goal programming approach, see 3.1.9, we may get the tracking model

$$\min_{\mathbf{x} \in \mathcal{X}} \sum_{s=1}^{S} p^s |f_0(\mathbf{x}, \omega^s) - f_0(\mathbf{x}^*(\omega^s), \omega^s)|;$$

see [38] and 3.3.3.

To apply these ideas to multistage stochastic programs as formulated in II.2 one should first reformulate them so that the random elements enter only the objective function, see e.g. (6) or (8) of Chapter II.2.

3.3.2   Scenario-Based Two-Stage Stochastic Linear Programs

Using the general ideas of Section 2 for $T = 2$ let us consider problems of the form

(17) $$\text{minimize } \mathbf{c}^\top \mathbf{x} + \sum_{s=1}^{S} p^s {\mathbf{q}^s}^\top \mathbf{y}^s$$

subject to

(18)
$$
\begin{array}{llll}
\mathbf{Ax} & & & = \mathbf{b} \\
\mathbf{T}^1\mathbf{x}+\mathbf{W}^1\mathbf{y}^1 & & & = \mathbf{h}^1 \\
\mathbf{T}^2\mathbf{x}+ & \mathbf{W}^2\mathbf{y}^2 & & = \mathbf{h}^2 \\
\quad \vdots & & \ddots & \quad \vdots \\
\mathbf{T}^S\mathbf{x}+ & & \mathbf{W}^S\mathbf{y}^S & = \mathbf{h}^S
\end{array}
$$

$$\mathbf{x} \geq 0, \mathbf{y}^s \geq 0, s = 1, \ldots, S$$

where $(\mathbf{q}^s, \mathbf{T}^s, \mathbf{W}^s, \mathbf{h}^s)$ are coefficients determined by scenario ω^s, $s = 1, \ldots, S$ and $p^s \geq 0, s = 1, \ldots, S$, are their probabilities, $\sum_s p^s = 1$. The first-stage decision $\mathbf{x}$ is scenario independent. For each of considered scenarios, second-stage decisions $\mathbf{y}^s$ are introduced to maintain the fulfilment of constraints $\mathbf{T}^s\mathbf{x} = \mathbf{h}^s$ by compensating the discrepance $\mathbf{h}^s - \mathbf{T}^s\mathbf{x}$ for an additional cost ${\mathbf{q}^s}^\top \mathbf{y}^s$. To keep the linearity of the problem, various possibilities of compensation are modeled by the *recourse matrices* $\mathbf{W}^s \, \forall s$. The *recourse costs* $\phi(\mathbf{x}, \omega^s)$ are the minimal attainable costs of compensation. They depend on the first-stage decision and on the considered scenario and are defined as optimal values of an auxiliary, second-stage program

(19) $$\phi(\mathbf{x}, \omega^s) = \min_{\mathbf{y}^s} \left\{ {\mathbf{q}^s}^\top \mathbf{y}^s \; : \; \mathbf{W}^s\mathbf{y}^s = \mathbf{h}^s - \mathbf{T}^s\mathbf{x}, \quad \mathbf{y}^s \geq 0 \right\}.$$

The structure of the second-stage program (19) is influenced mainly by the recourse matrices $\mathbf{W}^s$ and special cases are distinguished. For instance, the second-stage program (19) with matrices $\mathbf{W}^s = (\mathbf{I} | - \mathbf{I}) \, \forall s$, where $\mathbf{I}$ is the unit matrix, distinguishes only between shortages $(\mathbf{T}^s\mathbf{x} - \mathbf{h}^s)^-$ penalized by $\mathbf{q}^{-s}$ and surpluses $(\mathbf{T}^s\mathbf{x} - \mathbf{h}^s)^+$ penalized by $\mathbf{q}^{+s}$. If $\mathbf{q}^{-s} + \mathbf{q}^{+s} \geq \mathbf{0}$ the recourse function

$$\phi(\mathbf{x}, \omega) = (\mathbf{T}^s\mathbf{x} - \mathbf{h}^s)^{+\top}\mathbf{q}^{+s} + (\mathbf{T}^s\mathbf{x} - \mathbf{h}^s)^{-\top}\mathbf{q}^{-s}.$$

This is called the *simple recourse* problem; see II.4.3 and II.8.2.

The assumption of the *relatively complete recourse* means that the set of feasible solutions (18) is nonempty, hence, it is possible to compensate discrepances for an arbitrary first-stage solution $\mathbf{x} \in \mathcal{X}$ and all considered scenarios. The *fixed recourse* problems refer to fixed matrices $\mathbf{W}^s \equiv \mathbf{W} \, \forall s$ whereas *random recourse* assumes scenario dependent matrices $\mathbf{W}^s$.

Using (19), the problem (17)–(18) can be rewritten as

minimize

$$(20) \qquad \mathbf{c}^\top \mathbf{x} + \sum_{s=1}^{S} p^s \phi(\mathbf{x}, \omega^s)$$

on the set

$$(21) \qquad \mathcal{X} = \{\mathbf{x} \ : \ \mathbf{A}\mathbf{x} = \mathbf{b}, \mathbf{x} \geq 0\} \ ;$$

compare with (8)–(9) in Chapter II.2. The optimal solution $\mathbf{x}^{HN}$ of (20)–(21) is the so-called *here-and-now* solution corresponding to the two-stage recourse problem.

Notice, that minimization of the total expected cost of the two-stage decision process (20) is in full agreement with general ideas of 3.3.1. The objective function (20) can be further modified; one possibility is parallel to expected utility maximization in 3.2.4:

$$(22) \qquad \text{minimize} \ \sum_{s=1}^{S} p^s \tilde{U}(\mathbf{c}^\top \mathbf{x} + \phi(\mathbf{x}, \omega^s))$$

subject to (21) with $\phi(\mathbf{x}, \omega)$ defined by (19), and $\tilde{U}$ a convex decreasing dis-utility or loss function.

A similar reformulation for scenario-based multistage stochastic programs is also possible and it corresponds to the first-stage decision problems (6) or (8) in Chapter II.2.

3.3.3 Tracking Models

In the two-stage stochastic program with recourse, such as (17)–(18), the second-stage variables $\mathbf{y}^s$ together with the recourse matrices $\mathbf{W}^s$ are used to compensate the possible discrepancies between $\mathbf{T}^s\mathbf{x}$ and $\mathbf{h}^s$ for different scenarios. In this way, relaxation of "soft" constraints $\mathbf{T}^s\mathbf{x} = \mathbf{h}^s, s = 1, \ldots, S$, is modeled. There are various other possibilities, for instance, a simple tracking model related to (17)–(18) can be formulated as follows: Let $v^s, s = 1, \ldots, S$, be the optimal values of the *individual* scenario problems

$$(23) \qquad \text{minimize} \ \mathbf{c}^\top \mathbf{x}$$

subject to

$$(24) \qquad \mathbf{A}\mathbf{x} = \mathbf{b}, \ \mathbf{T}^s\mathbf{x} = \mathbf{h}^s, \ \mathbf{x} \geq 0.$$

Assume that for each s the scenario problems (23)–(24) have an optimal solution. The basic compromising or *tracking model* follows the ideas of goal programming with respect to minimization of the objective functions (13) and of the possible differences between $\mathbf{T}^s\mathbf{x}$ and the right-hand sides $\mathbf{h}^s$ for all s:

$$(25) \qquad \text{minimize} \ \sum_{s=1}^{S} p^s \left(\|\mathbf{c}^\top \mathbf{x} - v^s\| + \|\mathbf{T}^s\mathbf{x} - \mathbf{h}^s\| \right)$$

subject to
$$\mathbf{A}\mathbf{x} = \mathbf{b}, \quad \mathbf{x} \geq 0.$$

The optimal solution obtained by solving this problem tracks the optimal solutions of the individual scenario problems (13)–(14) as closely as possible. Various types of the norm in (25) can be used, recall 3.1.9; the choice influences the optimal solution of (25) and the solution procedure.

A general form of the tracking model constructed for scenario problems

$$\min f_0(\mathbf{x}, \omega^s) \text{ subject to } \mathbf{x} \in \mathcal{X} \text{ and } g_k(\mathbf{x}, \omega^s) = 0, k = 1, \dots, K$$

reads

$$\min_{\mathbf{x} \in \mathcal{X}} \sum_s p^s (\alpha_0 \|f_0(\mathbf{x}, \omega^s) - v^s\| + \sum_k \alpha_k \|g_k(\mathbf{x}, \omega^s)\|)$$

where the optimal values v^s of all individual scenario problems are supposed to exist and the parameters $\alpha_k > 0 \, \forall k$.

Various types of tracking models have been suggested and applied in finance and also in water resources, cf. [38]. An example is the *portfolio immunization problem*, see I.5.1: The goal is to find the cheapest portfolio of fixed-income securities whose present value equals the present value of liabilities. Given scenario s of interest rates, the scenario subproblem is

$$\min \left\{ \sum_j c_j x_j : \sum_j PV_j^s x_j = PV_0^s, \mathbf{x} \in \mathcal{X} \right\}$$

where x_j is the amount of bond j in portfolio, c_j are the initial aquisition prices, PV_j^s, PV_0^s the present values of the bonds and of the liabilities and $\mathcal{X}$ consists of some scenario independent constraints. With probabilities p^s of individual scenarios and with optimal values v^s of scenario subproblems, the tracking model reads

$$\min_{\mathbf{x} \in \mathcal{X}} \sum_s p^s \left[\|\sum_j c_j x_j - v^s\| + \|\sum_j PV_j^s x_j - PV_0^s\| \right].$$

3.3.3.1 Exercise. Formulate a tracking model for the bond portfolio immunization problem in which the present values and dollar durations of the portfolio should be equal to those of liabilities. Consider also the possibility of assigning different priority weights to deviations from this goal for present values and for dollar durations.

3.3.4 Robust Optimization

Besides of the minimum risk and maximum yield criteria which appear frequently in finance, see 3.2.1 – 3.2.4, there are real-life problems which aim also at limited variability of the second-stage decisions and/or of their costs: It is not acceptable to apply an optimal compensation strategy which requires repeated changes of the labour allocation or suggests to spend funds or to repay the debts abruptly. The requirement of a limited variability is an additional criterion which can be quantified and included into the overall objective function or into constraints; also goal programming methods can be exploited.

An example is the *robust optimization model* [120] which modifies (17)–(18) to capture the required limited variability of costs. With the variability quantified by the *variance* of the costs, the problem is:

Minimize

(26)
$$\sum_{s=1}^{S} p^s \xi^s + \lambda \sum_{s=1}^{S} p^s \left[\xi^s - \sum_{k=1}^{S} p^k \xi^k \right]^2$$

subject to

(27)
$$\begin{aligned}
\mathbf{Ax} &= \mathbf{b} \\
\mathbf{T}^s \mathbf{x} + \mathbf{W}^s \mathbf{y}^s &= \mathbf{h}^s, \; s = 1, \ldots, S \\
\mathbf{c}^\top \mathbf{x} + \mathbf{q}^{s\top} \mathbf{y}^s - \xi^s &= 0, \; s = 1, \ldots, S
\end{aligned}$$

$$\mathbf{x} \geq 0, \mathbf{y}^s \geq 0, s = 1, \ldots, S.$$

The newly introduced variables ξ^s are equal to the cost of the decision $\mathbf{x}$ plus the corresponding cost of the compensation of discrepances by $\mathbf{y}^s$ if the scenario ω^s occurs. The additional term in the objective function equals the variance of these random costs ξ^s with its weight in the objective function expressed via a scalar parameter $\lambda \geq 0$. The efficient solutions which are obtained by solving (26)–(27) for different nonnegative values of λ can be also obtained via the ϵ-constrained approach 3.1.7:
Minimize

(28)
$$\sum_{s=1}^{S} p^s \left[\xi^s - \sum_{k=1}^{S} p^k \xi^k \right]^2$$

subject to (27) and

$$\sum_{s=1}^{S} p^s \xi^s \leq \epsilon$$

with an appropriately chosen parameter value ϵ, an upper bound on the acceptable expected costs.

A further possibility is to replace the objective function in the problem (26)–(27) by a general performance function of the decision variables $\mathbf{x}, \mathbf{y}^s, s = 1, \ldots, S$, and in addition, to relax the constraints. The resulting model can be formulated as follows:
Minimize

(29)
$$\sigma(\mathbf{x}, \mathbf{y}^1, \ldots, \mathbf{y}^S) + \mu\rho(\mathbf{u}^1, \ldots, \mathbf{u}^S)$$

subject to

(30)
$$\begin{aligned}
\mathbf{Ax} &= \mathbf{b}, \\
\mathbf{T}^s \mathbf{x} + \mathbf{W}^s \mathbf{y}^s + \mathbf{u}^s &= \mathbf{h}^s, \; s = 1, \ldots, S
\end{aligned}$$

$$\mathbf{x} \geq 0, \mathbf{y}^s \geq 0, s = 1, \ldots, S.$$

The first term in the objective function (29) corresponds for instance to (26), the second one, with a weight $\mu \geq 0$, penalizes possible violations of the original second-stage constraints $\mathbf{T}^s \mathbf{x} + \mathbf{W}^s \mathbf{y}^s = \mathbf{h}^s$ in (27).

Restricted recourse problems [167] aim at limitation of the dispersion of the second-stage *decisions.* According to the principles of multi-objective programming, the objective function (17) could be extended to

$$\mathbf{c}^\top \mathbf{x} + \sum_{s=1}^{S} p^s \mathbf{q}^{s\top} \mathbf{y}^s + \lambda \sum_{s=1}^{S} p^s \| \mathbf{y}^s - \sum_{k=1}^{S} p^k \mathbf{y}^k \|$$

with a nonnegative parameter λ, cf. (26), or the constraints (18) extended for an additional constraint

(31) $$\sum_{s=1}^{S} p^s \| \mathbf{y}^s - \sum_{k=1}^{S} p^k \mathbf{y}^k \| \leq \epsilon$$

where ϵ is a chosen tolerance level; see 3.1.7.

From the modeling point of view the choice among the introduced models depends on the nature of the solved real life problem and also on the properties of the resulting mathematical program, including its numerical tractability and sensitivity of its solution on the input data. Similar models can be developed without any reference to the two-stage stochastic program (17)–(18).

For multistage stochastic programs, the prevailing ways of including alternative criteria are based on the ϵ-constrained or mixed approaches.

II.4 SELECTED APPLICATIONS IN FINANCE AND ECONOMICS

stochastic programming in finance (portfolio revision, BONDS model, ALM model, general features), production planning, capacity expansion, unit commitment and economic power dispatch, and in melt control

Advances in theory and algorithms for stochastic programs have contributed to development of models which capture both the dynamic aspects and uncertainties of financial and economic problems. We shall introduce several models that have been, at least partly, applied in practise; to formulate them we shall use, with reference to Section II.2.2, the split variables form and the arborescent form of multistage stochastic programs as required.

We shall start with an example of the *two-stage stochastic programming problem*: A decision $\mathbf{x} \in \mathcal{X}$ should be selected *before* realizations of random parameters can be observed or their values revealed. After this information becomes available the decision process continues by the second-stage, i.e., by the choice of an auxiliary decision that depends on the first-stage decision and exploits the already obtained information. The second-stage decision is interpreted as an updating activity (portfolio revision, adjustment of the production plan, etc.) that brings about additional, *recourse* costs. The requirement that the first-stage decision $\mathbf{x}$ cannot depend on future observations of random parameters corresponds to the already mentioned more general *nonanticipativity* property of the multistage decision processes.

4.1 Portfolio Revision

Following [92], we assume that at the beginning of the first period, the investor chooses an initial portfolio of assets from a set of considered assets indexed by $i = 1, \ldots, I$. At the end of the period, he or she observes the realization of the random vector of security returns. At the start of the subsequent period, he can choose a revised portfolio conditional upon the initial portfolio and on the observed returns; the set of the considered assets remains unchanged and the investor pays transaction costs. Hence, contrary to the dedicated bond portfolio management problem II.1.2, the investor is allowed to trade. His goal is to satisfy the budget constraints in both periods and to maximize the expected utility of the final wealth, i.e., the wealth that results from the 1-st and 2-nd stage decisions and from the (random) returns in the both periods.

Denote

x_i the initial number of shares of security i, the main decision variables;

b_i/s_i numbers of shares of security i bought / sold at the beginning of period 2 for portfolio revision, the auxiliary decision variables;

h_i the resulting number of shares of security i held in period 2;

c_i the initial per-share cost of security i;

$w(1)$ $(w(2))$ nonrandom exogenous wealth allotments at the beginning of period 1 (2);

$\rho_i(1)$ $(\rho_i(2))$ random returns on security i in period 1 (2) which replace here the random element ω;

α the cost per unit transaction;

U the chosen utility function.

All decision variables are integers and constraints on the *initial holdings* are evident:

$$\sum_{i=1}^{I} c_i x_i = w(1), \ x_i \geq 0 \ \forall i.$$

The subsequent portfolio revision is limited by the total wealth at disposal (coming from the previous investment and from the additional external allotment), i.e., by

$$w(2) + \sum_{i=1}^{I}(1 + \rho_i(1))c_i x_i$$

and the corresponding *budget constraint* for the second period reads

$$\sum_{i=1}^{I}[((1 + \rho_i(1))c_i + \alpha)b_i - ((1 + \rho_i(1))c_i - \alpha)s_i] = w(2).$$

The second-stage variables b_i, s_i are limited by *inventory constraints* (recall that $h_i = x_i + b_i - s_i \ \forall i$)

$$0 \leq s_i \leq x_i, \ b_i \geq 0 \ \forall i.$$

The objective function

$$(1) \qquad E_{\boldsymbol{\rho}(1)} \left\{ \max_{\mathbf{b},\mathbf{s}} E_{\boldsymbol{\rho}(2)|\boldsymbol{\rho}(1)} U \left(\sum_{i=1}^{I}(1 + \rho_i(1))(1 + \rho_i(2))c_i(x_i + b_i - s_i) \right) \right\}$$

is maximized with respect to constraints on initial holdings. The optimal value of the second-stage problem

$$(2) \qquad \phi(\mathbf{x}, \boldsymbol{\rho}(1)) = \max_{\mathbf{b},\mathbf{s}} E_{\boldsymbol{\rho}(2)|\boldsymbol{\rho}(1)} U \left(\sum_{i=1}^{I}(1 + \rho_i(1))(1 + \rho_i(2))c_i(x_i + b_i - s_i) \right)$$

subject to budget and inventory constraints depends on the first-stage decision $\mathbf{x}$ and on a particular realization $\rho(1)$ of the first-period returns. In spite of a rather simple structure of the second-stage constraints, it is in general impossible to get the optimal value of (2) *explicitly* as a function of $\mathbf{x}$ and of $\rho(1)$ and substitute it into (1). The way out is to approximate the probability distribution of $\rho(1)$ by a discrete probability distribution concentrated on a finite number of scenarios, say, $\rho^s(1)$, $s = 1, \ldots, S$, with probabilities $p^s > 0$, $\forall s$, $\sum_{s=1}^{S} p^s = 1$. The considered portfolio revision problem reads now

$$\max \sum_{s=1}^{S} p^s E_{\rho(2)|\boldsymbol{\rho}^s(1)} U \left(\sum_{i=1}^{I}(1 + \rho_i^s(1))(1 + \rho_i(2))c_i(x_i + b_i^s - s_i^s) \right)$$

subject to

$$\sum_{i=1}^{I} c_i x_i = w(1),\ x_i \geq 0\ \forall i$$

$$\sum_{i=1}^{I} [((1 + \rho_i^s(1))c_i + \alpha)b_i^s - ((1 + \rho_i^s(1))c_i - \alpha)s_i^s] = w(2)\ \forall s$$

$$0 \leq s_i^s \leq x_i,\ b_i^s \geq 0,\ \text{integers } \forall i,\ s.$$

The necessary input consists of the initial selection of the considered securities, of the probability distribution of $\rho(1)$ and of its discretization, of the conditional probability distribution of $\rho(2)|\rho^s(1)\ \forall s$ that can be discretized as well, of deterministic parameters α and $c_i\ \forall i$ and of the utility function U. Some generalizations are possible, e.g., inclusion of tax on gross profit, possibility of borrowing and consumption.

4.2   The BONDS Model

The pioneering model [24] was designed as a decision support for multiperiod management of portfolios of fixed income securities, bonds, in commercial banks. At the beginning of each period, the portfolio manager has an inventory of bonds and cash and decides which bonds to hold, sell and buy. The possible composition of the portfolio depends on random cash inflows and outflows, on interest rates, etc., random variables which are supposed to have a *discrete* probability distribution. The decision variables at the beginning of the t-th period ($t = 1, \ldots, T$) depend on realizations of the random subsequences $\omega^{t-1,\bullet}$ and the stages of the decision process coincide with periods. The first-stage decision variables are scenario independent and the last stage decisions accepted at the beginning of the T-th period depend on $\omega^{T-1,\bullet}$; these nonanticipativity conditions will be spelled out implicitly. The decision process is affected by the probability distribution of ω_T, but not by its realizations.

The goal is to maximize the expected market value of the portfolio at the horizon T under constraints on cash flow, inventory balance, capital losses, initial holdings and under nonnegativity of all variables.

Denote

$b_i(t, \omega^{t-1,\bullet})$ the amount of bond i bought at the beginning of period t (in dollars of initial purchase price);

$s_i(\tau, t, \omega^{t-1,\bullet})$ the amount of bond i purchased at the beginning of period τ and sold at the beginning of period t (in dollars of initial purchase price);

$h_i(\tau, t, \omega^{t-1,\bullet})$ the amount of bond i purchased at the beginning of period τ and held (i.e., not sold) at the beginning of period t (in dollars of initial purchase price).

These decision variables have to be nonnegative, to fulfill constraints on *initial holdings* acquired before the beginning of the first period

$$h_i(0, 0) = h_i^0\ \forall i$$

and on *inventory balance*

$$(3)\quad -h_i(\tau, t-1, \omega^{t-2,\bullet}) + s_i(\tau, t, \omega^{t-1,\bullet}) + h_i(\tau, t, \omega^{t-1,\bullet}) = 0,\ \tau = 0, 1, \ldots, t-2,$$

$$-b_i(t-1, \omega^{t-2,\bullet}) + s_i(t-1, t, \omega^{t-1,\bullet}) + h_i(t-1, t, \omega^{t-1,\bullet}) = 0$$

for all $\omega^{t-1,\bullet}$, $t = 1, \ldots, T$, $i = 1, \ldots, I$.

Using (3) recursively, we obtain

$$(4) \qquad h_i(\tau, t, \omega^{t-1,\bullet}) = b_i(\tau, \omega^{\tau-1,\bullet}) - \sum_{\nu=\tau+1}^{t} s_i(\tau, \nu, \omega^{\nu-1,\bullet})$$

valid for all $\tau = 0, \ldots, t-1$, $t = 1, \ldots, T$, $i = 1, \ldots, I$ and for all scenarios.

Denote further

$g_i(\tau, t, \omega^{t-1,\bullet})$ the capital gain/loss on bond i purchased at the beginning of period τ and sold at the beginning of period t (per dollar of initial purchase price, after tax);

$r_i(t, \omega^{t-1,\bullet})$ the annual yield from coupons of bond i bought at the beginning of period t (per dollar of initial purchase price, after tax);

$w(t, \omega^{t-1,\bullet})$ exogenous incremental amount of funds at the beginning of period t.

The *cash flow constraints* read

$$(5) \quad \sum_{i=1}^{I} b_i(t, \omega^{t-1,\bullet}) - \sum_{i=1}^{I}\sum_{\tau=0}^{t-1}(1 + g_i(\tau, t, \omega^{t-1,\bullet}))s_i(\tau, t, \omega^{t-1,\bullet}) -$$

$$\sum_{i=1}^{I}\left(\sum_{\tau=0}^{t-2} r_i(\tau, \omega^{\tau-1,\bullet})h_i(\tau, t-1, \omega^{t-2,\bullet}) + r_i(t-1, \omega^{t-2,\bullet})b_i(t-1, \omega^{t-2,\bullet})\right) =$$

$$w(t, \omega^{t-1,\bullet}) \quad \forall \omega^{t-1,\bullet}, \ t = 1, \ldots, T.$$

(The transaction costs are taken into account by adjusting the gain coefficients for the broker's commission.)

The *constraints on capital losses* are

$$(6) \qquad -\sum_{i=1}^{I}\sum_{\tau=t'}^{t} g_i(\tau, t, \omega^{t-1,\bullet}))s_i(\tau, t, \omega^{t-1,\bullet}) \leq L(t, \omega^{t-1,\bullet}) \forall \omega^{t-1,\bullet}, \ t \in \mathcal{T}'$$

where $L(t, \omega^{t-1,\bullet})$ denotes the upper bound on the realized capital losses (after tax) from sales during a year, $\mathcal{T}'$ is the set of indices of periods which correspond to the end of fiscal years and t' is the index of the first period in the fiscal year indexed by $t \in \mathcal{T}'$.

Let finally $\bar{v}_i(\tau, T, \omega^{T-1,\bullet})$ denote the final expected cash value (i.e., expectation with respect to ω_T conditioned by $\omega^{T-1,\bullet}$) per dollar of the initial purchase price of the i-th bond purchased at the beginning of period τ and held at the beginning of period T or bought at the beginning of period T, and let $p(\omega^{T-1,\bullet})$ be the path probability of the partial sequence $\omega^{T-1,\bullet}$. The *objective function*

$$\sum_{\omega^{T-1,\bullet}} p(\omega^{T-1,\bullet}) \sum_{i=1}^{I}\left[\sum_{\tau=0}^{T-1}(r_i(\tau, \omega^{\tau-1,\bullet}) + \bar{v}_i(\tau, T, \omega^{T-1,\bullet}))h_i(\tau, T, \omega^{T-1,\bullet}) + \right.$$

$$\left. (r_i(T, \omega^{T-1,\bullet}) + \bar{v}_i(T, T, \omega^{T-1,\bullet}))b_i(T, \omega^{T-1,\bullet})\right]$$

should be maximized subject to constraints (3), (5), (6).

Again, it is a large scale linear program; the total number of constraints depends essentially on the number of possible partial sequences $\omega^{T-1,\bullet}$ for $t = 1, \ldots, T-1$, and it increases rapidly with the increasing number of stages. As the problem concerns bonds only, no conditions on liabilities or on liquidity are present.

The model was tested for $T = 3$ yearly periods. Scenarios were generated under the assumption that the conditional probability distributions of the random interest rates can be approximated in each period by discrete probability distributions concentrated at three points only. In the notation introduced in Section II.2.2 this gives 3 elements of $\mathcal{K}_2$, 9 elements of $\mathcal{K}_3$ and the total number of scenarios – elements of $\mathcal{K}_4$ – equals 27. See also Example II.5.4.4.

The reported results of backtesting the model on historical data (Salomon Brothers time series of good grade municipal bonds) for bonds characterized by their maturities (1, 2, 3, 4, 5, 10, 15, 20 and 30 years) indicate that the BONDS model is a true competition to the typical management tools of banks which use either the laddered portfolios (in which the amount invested in each of maturity classes is approximately equal for all maturities) or the barbell maturity structure (in which the maturities held are structured to the short and long ends with little investments into intermediate maturities). Moreover, BONDS is a truly dynamic model: The first-stage decision takes into account all considered future developments and its feasibility is guaranteed. At the end of the first period, the model is rolled forward, i.e., the horizon is moved out one more year and the model is solved again, with the initial holdings from the new portfolio obtained from the preceding application of the model and with new scenarios which take into account the already observed changes of the interest rates.

4.3   Bank Asset and Liability Management – Model ALM

The objective of the ALM model [104] is to maximize the discounted net value of bank profits minus the expected penalty costs for infeasibility with respect to soft constraints, subject to numerous constraints on budget, liquidity, structure of cash flows (deposits / withdrawals), including legal and policy restrictions. There are many sources of uncertainties such as rates of return, interest rates, deposits flows, etc. The authors focused on the random deposits flows assigning deterministic fixed values to the rates of return and to the interest rates. The problem is modeled as a two-stage multiperiod stochastic program over N periods, with discrete random variables ω_{dn} – the random balance sheet for deposit of the type d at the end of period n. Similarly as in Section 4.2 we denote $b_i(n)$, $s_i(m,n)$, $h_i(m,n)$ the decision variables on buying, selling and holding at period n the ith asset purchased in period m. In addition we introduce

$x_d(n)$ the planned deposit d inflow in period n;

$x_d(0)$ the initial holding of deposit d;

$x(n)$ the amount borrowed in period n;

γ_d the annual rate of withdrawals of deposits d;

$y_d^+(n)$, $y_d^-(n)$ the auxiliary variables that compensate the observed discrepancies of the deposit balance sheet;

$q_d^+(n)$, $q_d^-(n)$ the (nonnegative) proportional penalty costs associated with the auxiliary variables;

$r_d(n)$, $r(n)$ the interest rates paid on deposits of type d and on borrowing;

$\beta(n)$ the discount rate from period n to period 0.

The planned amount of the deposits of type d generated in period n

$$x_d(n) + \sum_{m=0}^{n-1} x_d(m)(1 - \gamma_d)^{n-m}$$

can differ from the observed value ω_{dn}; for each observed value ω_{dn}^s of ω_{dn}, the surplus

$$\omega_{dn}^s > x_d(n) + \sum_{m=0}^{n-1} x_d(m)(1 - \gamma_d)^{n-m}$$

is compensated by the second-stage variable $y_d^{+s}(n)$, the shortage

$$\omega_{dn}^s < x_d(n) + \sum_{m=0}^{n-1} x_d(m)(1 - \gamma_d)^{n-m}$$

is compensated by $y_d^{-s}(n)$. The resulting balance constraints on deposit flows read

$$(7) \qquad x_d(n) + \sum_{m=0}^{n-1} x_d(m)(1 - \gamma_d)^{n-m} + y_d^{+s}(n) - y_d^{-s}(n) = \omega_{dn}^s$$

$$y_d^{+s}(n) \geq 0, y_d^{-s}(n) \geq 0, \quad \forall s, d, n.$$

This form of constraints corresponds to the assumption that for all periods n, all decisions $b_i(n), s_i(m,n), h_i(m,n)$ on buying, selling, holding assets and on policy $x_d(n), x(n)$ on building deposits and borrowing are made all at once, for all periods, assets and deposits independently on the future realizations of the random sheet balances. The only second-stage variables are thus $y_d^{+s}(n)$ and $y_d^{-s}(n) \, \forall s, d, n$; the scenarios can be interpreted as N-dimensional vectors of various sheet balance forecasts up to the end of the planning horizon, say, $n = N$. In this case, the numerous inventory balance constraints (3) need not be spelled out separately for each period, they enter in their aggregated form (4) only and the scenarios consist of different outcomes $\omega_d \in \mathbb{R}^N$ of the balance sheet figures for all regarded deposits and all considered periods.

The remaining constraints are in form of linear equations and inequalities that do not include any random parameters; we shall list them without any details.

The *cash flow constraints* are of the same nature as (5) with transaction costs and taxation spelled out explicitly. The incremental funds $w(n)$ are generated by increments on deposits and borrowing. To approximate the continuous flow, one assumes that one half of period's net flows arrive at the beginning of the period and the other half at the end of the period (or at the beginning of the next period). Denote

$$\tilde{x}_d(n) = \sum_{m=0}^{n-1} x_d(m)(1 - \gamma_d)^{n-m-1}(1 - \gamma_d/2) + x_d(n)/2, \quad n = 1, \ldots, N$$

the amount of deposits d available for the n-th period computed according to the above rule. The increment of funds is

$$w(n) = x(n) - x(n-1)(1 + r(n-1)) + \sum_{d=1}^{D} [\tilde{x}_d(n) - (1 + r_d(n-1))\tilde{x}_d(n-1)] \, \forall n$$

with $\tilde{x}_d(0) = 0 \, \forall d$.

The *legal constraints* refer to the peculiar regulations of the case study; they express conditions like "the current assets (say, those with $i \in \mathcal{I}_0 \subset \{1,\ldots,I\}$) cannot be less that 10% of the total liabilities":

$$\sum_{i \in \mathcal{I}_0} \sum_{m=0}^{n} b_i(m) - 0.1 \left[\sum_{d=1}^{D} \tilde{x}_d(n) + x(n) \right] \geq 0, \quad n = 2,\ldots,N$$

$$\sum_{i \in \mathcal{I}_0} (b_i(0) + b_i(1)) - 0.1 \left[\sum_{d=1}^{D} x_d(0)(1 - \gamma_d/2) + 1/2x_d(1) + x(0) + x(1) \right] \geq 0.$$

Policy constraints are introduced to capture the internal policy of the bank; they put, e.g., limitations on personal loans and mortgages.

Liquidity constraints stem from the requirement that the market value of the bank's assets is adequate to meet depositors' withdrawal claims during adverse economic conditions. It means that the liquidity risk is taken into account.

The objective function consists of four terms: The discounted cost of direct borrowing, the net discounted cost of deposits and the expected minimal penalty cost for a deposit balance violation are subtracted from the discounted total returns and capital gains $g_i(m,n)$ on assets purchased in period m and sold in period n (after tax, with rates $a(n)$, $A(n)$, respectively):

(8)
$$\sum_{i=1}^{I} \left[\sum_{m=0}^{N-1} \sum_{n=m+1}^{N} s_i(m,n) \left(r_i(m) \sum_{l=m+1}^{n} (1 - a(l))\beta(l) + g_i(m,n)(1 - A(n))\beta(n) \right) + \right.$$
$$\left. \sum_{n=0}^{N} h_i(n,N)r_i(n) \sum_{l=n+1}^{N} (1 - a(l))\beta(l) \right] - \sum_{n=1}^{N} x(n)r(n)\beta(n+1) -$$
$$\sum_{d=1}^{D} \left[\sum_{n=1}^{N} \left(\sum_{m=1}^{n-1} x_d(m)(1 - \gamma_d/2)(1 - \gamma_d)^{n-m}r_d(n)\beta(n) + 1/2x_d(n)r_d(n)\beta(n) \right) \right] -$$
$$\sum_{s} p^s \left\{ \min_{y^+,y^-} \sum_{d=1}^{D} \sum_{n=1}^{N} (q_d^+(n)y_d^{+s}(n) + q_d^-(n)y_d^{-s}(n)) \right\}.$$

In the last term of (8), minimization is carried over nonnegative values of $y_d^{+s}(n)$ and $y_d^{-s}(n)$ that compensate the discrepancies of the balance sheet figures for scenario s and deposit d in the n-th period, see (7). This is an example of the *simple recourse* problem mentioned in II.3.3.2 and the *minimal* feasible discrepancies for a given value ω_{dn} equal

$$y_d^+(n) = \left(\omega_{dn} - x_d(n) - \sum_{m=0}^{n-1} x_d(m)(1 - \gamma_d)^{n-m} \right)^+$$

$$y_d^-(n) = \left(x_d(n) + \sum_{m=0}^{n-1} x_d(m)(1 - \gamma_d)^{n-m} - \omega_{dn} \right)^+ .$$

The last term of (8) can be then replaced by

$$\sum_{n=1}^{N} \sum_{d=1}^{D} \sum_{s \in S_{dn}} p_{dn}^s (q_d^+(n) y_d^{+s}(n) + q_d^-(n) y_d^{-s}(n))$$

and the resulting objective function is maximized with respect to all decision variables b_*, h_*, s_*, x_* and y_*^+, y_*^- subject to all introduced constraints including (7).

As a whole, ALM is a multiperiod two-stage stochastic linear program with simple recourse that can be solved by linear programming techniques if ω_{dn} are discrete random variables $\forall d, n$. It was developed for the Vancouver City Saving Credit Union for a 5-years planning horizon and it has been rolled forward continuously period after period.

4.4   General Features of Multiperiod Stochastic Programs in Portfolio Optimization

The most important requirements on realistic models supporting dynamic investment decisions can be summarized as follows:

• To reflect dynamic aspects including the intertemporal dependence of returns, to aggregate assets and liabilities in one model consistently with the accounting rules and to consider external cash flows;

• To include stochastic behavior of important parameters, such as external cash flows or returns;

• To express investor's risk attitudes in an adequate way;

• To include transaction costs, taxes, etc.

• To include legal, institutional and policy constraints;

• To respect trade-off among short, intermediate and long term goals;

• To keep the model understandable.

These general guiding lines will be discussed below. They can be traced to a certain extent already in the early models surveyed in 4.1–4.3 and are accented in recent applications.

A fundamental investment decision is the selection of asset *categories* and the wealth allocation over time. The allocation decision involves the proportion of major asset categories within a portfolio. The level of aggregation depends on investor's circumstances. For instance, a benchmark for US pension plans is the 60–40 mix, i.e., allocation of 60% of assets in a stock index and the remainder in a bond fund. The major assets (a large capitalized stock index, a government or corporate bond index and cash) are complemented by other assets depending upon investors requirements; these are foreign stock and bonds, loans, real-estate funds, commodities and precious metals, more risky assets, etc. (See for instance [27]). The portfolio can be rebalanced at the beginning of certain periods to cover the target ratio. Otherwise, one applies the *buy–and–hold* strategy which does not assume any transactions except reinvesting dividends and interest.

The planning horizon at which the outcome is evaluated is the endpoint T_0 of the interval $[0, T_0]$ which is further divided into nonoverlapping time intervals indexed by $t = 0, 1, \ldots, \tau$. The portfolio can be rebalanced at the beginning of each of these time intervals to cover the target ratio or to contribute to maximization of the final performance at T_0. In some cases, additional time instants $t > T_0$ are included at which some of economic variables are calculated; after T_0, no further active decisions are allowed. An example is the interest rates forecasts needed for pricing the long bonds included in portfolio.

A critical issue is the handling of uncertainty. This is done mainly via modeling probability distributions of random parameters by scenarios and their probabilities; see Chapter II.5.

The primary decision variables $h_i^s(t)$ represent the holding in asset category i at the beginning of time period t under scenario s *after the rebalancing decisions took place*; the initial holdings are $h_i(0)$. Holdings $h_i^s(t)$ can stand for the amounts of money invested in i at the beginning of time period t (e.g., in 4.1), or they can be expressed in dollars of the initial purchase prices (e.g., the BONDS model 4.2), in face values, in numbers of securities or in lots (e.g., Chapter II.6), etc. Accordingly, in the first case, the values of the holdings at the end of the period t may be affected by the returns on the market; the wealth accumulated at the end of the t-th period before the next rebalancing takes place is then

$$w_i^s(t) = (1 + \rho_i^s(t)) h_i^s(t) \; \forall i, t, s.$$

Purchases and sales of assets are represented by variables $b_i^s(t), s_i^s(t)$ with transaction costs defined via coefficients $\alpha_i(t)$ and assuming mostly the symmetry in the transaction costs; it means that purchasing one unit of i at the beginning of period t requires $1 + \alpha_i(t)$ units of cash and selling one unit of i results in $1 - \alpha_i(t)$ units of cash. The *inventory balance constraint* for each asset category (except for cash, the asset indexed by $i = 0$), scenario and time period is

$$h_i^s(t) = (1 + \rho_i^s(t-1)) h_i^s(t-1) + b_i^s(t) - s_i^s(t).$$

It restricts the cash flows at each period to be consistent.

The *flow balance equation for cash* for each time period and all scenarios is for instance

$$h_0^s(t) = h_0^s(t-1)(1 + \rho_0^s(t-1)) + c^s(t) + \sum_i s_i^s(t)(1 - \alpha_i(t)) - \sum_i b_i^s(t)(1 + \alpha_i(t))$$

$$+ \sum_i f_i^s(t) h_i^s(t) - y^{-s}(t-1)(1 + \delta^s(t-1)) - L^s(t) + y^{-s}(t)$$

with $f_i^s(t)$ the cash flow generated by holding one unit of the asset i during the period t (coupons, dividends, etc.) under scenario s, $c^s(t) = c^{+s}(t) - c^{-s}(t)$ the planned external cash flow at the period t under scenario s, $y^{-s}(t)$ borrowing in each period t under scenario s at the borrowing rate $\delta^s(t)$ and $L^s(t)$ the pay down of the liabilities at the beginning of the period t and under scenario s. (For simplicity we assume that all borrowing is done on a single period basis; a generalization is possible.)

For holdings, purchases and sales expressed in numbers or in face values, the cash balance equation contains purchasing and selling prices, $\xi_i^s(t) > \zeta_i^s(t)$:

$$h_0^s(t) = h_0^s(t-1)(1 + \rho_0^s(t-1)) + c^s(t) + \sum_i \zeta_i^s(t) s_i^s(t) - \sum_i \xi_i^s(t) b_i^s(t)$$

$$+ \sum_i f_i^s(t) h_i^s(t) - y^{-s}(t-1)(1 + \delta^s(t-1)) - L^s(t) + y^{-s}(t)$$

and the inventory balance constraints for all asset categories except for cash (i.e., for $i \neq 0$) assume a simpler form

$$h_i^s(t) = h_i^s(t-1) + b_i^s(t) - s_i^s(t)$$

as no wealth accumulation is considered. The network structure of the balance constraints for all i, t, s is an example of the *network recourse*.

The decision variables $h_i^s(t), b_i^s(t), s_i^s(t), y^{-s}(t)$ are nonnegative and the special form of the balance constraints displays a *network structure* which facilitates the algorithmic solution. The last stage constraints may be slightly different from those for intermediate stages, e.g., no revision and/or no borrowing is allowed.

Depending on the character of securities, amortization factors may be introduced in the inventory balance constraints. They express the fraction of the outstanding face value at the end of the interval which reflects the effects of call options, default, etc. It is easy to include further constraints which force a diversification, limit investments in risky or illiquid asset classes, limit borrowing, loan principal payments and turnovers, reflect legal and institutional constraints. It is also possible to force a specific decision policy, e.g., the *fixed-mix* policy which can be expressed as

$$h_i^s(t) = \lambda_i \sum_j h_j^s(t) \; \forall i$$

where λ_i is a fixed ratio of asset i in the portfolio.

Requirements of solvency are often formulated as probabilistic constraints on the level of total wealth with respect to total liabilities at the end of each period, e.g.,

$$P\left(\sum_i w_i(t) \geq \mu L(t)\right) \geq 1 - \alpha_t, \; t = 1, \ldots, T$$

with a given positive μ and given probabilities $\alpha_t \in (0, 1)$. At the same time, the expected costs due to insufficient solvency may be included into the objective function along with other types of shortfalls. Inclusion of probabilistic constraints, however, represents an increased complexity for numerical implementation of the model. Discrete probability distributions can be used but the related probabilistic constraints do not in general describe a convex set. The recommended approach is via auxiliary zero-one variables.

4.4.1   Exercise. Let $\tilde{\omega}_{k_t}$, $k_t \in \mathcal{K}_t$ be all different realizations of the partial sequences $\omega^{t-1,\bullet}$ which may arise at the beginning of stage t with probabilities p_{k_t}, and $D_{k_t} = \mu L_{k_t} - \sum_j w_{jk_t}$ denote the deficit or shortfall in case that scenario $\tilde{\omega}_{k_t}$ occurs. (For simplicity we do not consider here any interstage dependence.) Let M_{k_t}, $k_t \in \mathcal{K}_t$ be sufficiently large constants and Δ_{k_t}, $k_t \in \mathcal{K}_t$ be 0-1 variables. Show that the probabilistic constraint on solvency at the end of period t can be replaced by

$$D_{k_t} \leq \Delta_{k_t} M_{k_t}, \quad \Delta_{k_t} \in \{0,1\} \, \forall k_t \in \mathcal{K}_t, \quad \sum_{k_t \in \mathcal{K}_t} p_{k_t} \Delta_{k_t} \leq \alpha_t.$$

Whereas the random liabilities $L^s(t)$ are model parameters, various further *decisions* concerning liabilities can be included in the external cash flows: similarly as in the ALM model 4.3, one can distinguish decisions on accepting various types of deposits. Decisions concerning emission of additional debt instruments, on specific goal payments, on long term debt retirement, etc. can be included as well. Naturally, the cash balance equation has to take into account the cost of the debt service.

The objective function is related to the wealth at the end of the planning horizon T_0; this for each scenario consists of the amount of the total wealth $\sum_i w_i^s(T_0)$ reduced for the present value of future liabilities and loans outstanding at the horizon and for the costs attributed to shortfalls. At times, however, the investor's liabilities are not readily marketable and must be discounted and their market value estimated, for instance through actuarial models in case of pension plans. The risk factor can be included into constraints, or it enters the model through the choice of a suitable utility function. To include short term goals, cumulated penalties for shortfalls under scenario s (e.g., penalties for $y^{-s}(t) > 0$ or for $D^s(t) > 0$ in 4.4.1) are subtracted from the final wealth computed for the same scenario. An alternative way is to apply a utility function of several outcomes in certain time instants covered by the model.

Another serious problem is the *no-arbitrage property* which, whenever assumed to hold true in the reality, should be also captured by the models. Roughly speaking, no trading strategy which starts with zero or negative wealth should result in a nonnegative wealth for all scenarios and in a positive wealth for at least one scenario some time in the future. Such possibility is reduced by nonnegativity constraints on all variables (i.e., no short sales are permitted), by a restricted borrowing or other regulative constraints. Also inclusion of a nonlinear utility function can be recommended. In the linear case, e.g., for linear utility function of a risk neutral investor, no-arbitrage property of the model is connected with linear programming duality: If both the primal and the dual program are feasible, infinite values of the objective are excluded, see [85] for details.

To initiate the model, one uses scenarios $\rho_i^s(t), \delta^s(t), f_i^s(t), L^s(t)$ of the returns, interest rates and liabilities for all t and starts with the known, scenario independent initial holdings $h_i^s(0) \equiv h_i(0)$ of cash and all considered assets and with $y^{-s}(0) \equiv 0 \forall s$. If no ties in scenarios are considered we visualize them as a *fan* of individual scenarios which start from the common known values $\rho_i(0), \delta(0), f_i(0), L(0)$ valid for $t = 0$. In this case, no additional information will be released later on and all decisions $h_i^s(t), b_i^s(t), s_i^s(t), y^{-s}(t), c^s(t) \,\forall i, t, s$ can be computed all at once thanks to the full foresight of the evolution of random coefficients for each scenario up to the horizon. In this case, only one additional requirement must be met: the initial decision $h_i^s(1), b_i^s(1), s_i^s(1), y^{-s}(1), c^s(1)$ must be *scenario independent*. This is a simple form of the *nonanticipativity constraints* and the resulting problem is a *multiperiod two-stage stochastic program*. For an example see the ALM model 4.3 or Chapter II.6.

For *multistage stochastic programs*, the input is mostly in the form of a scenario tree, the nodes of the tree are related with revealing of an additional information which makes a basis for the subsequent decision. To build the scenario tree is a complicated task. The nonanticipativity constraints can enter in an explicit way by forcing the decisions based on an identical part of several scenarios to be equal (as it was in the case of the two-stage model) or implicitly by using a decision tree which follows the structure of the already given scenario tree.

The main interest lies in the first-stage decisions which consist of all decisions that have to be selected before an additional information is revealed, just on the basis of the given probability distribution P, i.e., on the basis of the already designed scenario tree. The second-stage decisions are allowed to adapt to the additional information, etc.

With the explicit inclusion of the nonanticipativity constraints, the scenario based multiperiod and multistage stochastic programs can be written as large scale deterministic programs decomposed along scenarios, see the split variable form discussed in Section II.2.2. An example is

$$\max_{\mathcal{X} \cap \mathcal{N}} \left\{ \sum_s p^s f_0(\mathbf{x}(\omega^s), \omega^s) \,:\, \mathbf{A}(\omega^s)\mathbf{x}(\omega^s) = \mathbf{b}(\omega^s),\ s = 1, \ldots, S \right\}$$

where $\mathcal{X}$ is a set of "hard" constraints on the decision vectors $\mathbf{x}(\omega^s)$, such as non-negativity or scenario independent bounds, $\mathcal{N}$ represents the nonanticipativity constraints and $f_0(\bullet, \omega^s)$ is the performance measure (random objective function) in case of scenario ω^s. To solve this program, one can apply general purpose software for nonlinear programs, to implement special decomposition algorithms which make use of the structure of the problem, etc.; see Chapter II.8. Nonlinear or integer constraints can be included but for the cost of an increased numerical complexity. On the other hand, if the resulting problem is a large *linear* program, there are at disposal special decision support systems which are able to manage efficiently large scenario based stochastic programming portfolio problems; see Chapter II.8.

4.5   Production Planning

Production planning and utilization of production capacity is an important task for manufacturing managers. To complete it, they have to face the demand uncertainty and they are asked to provide production plans over several periods to avoid abrupt changes and problems connected with the continuation of the production. The problem is usually designed for several monthly periods and the plans are revised on a rolling horizon basis. In the problem formulation, see [59], several simplifications are introduced:

• Production is always completed in the period in which it begins;
• At any given period, whatever cannot be produced in-house can be produced at a vendor;
• All products can be manufactured on the same set of machines and the raw materials are available in the required quantities.

The objective is to minimize the total cost (discounted to its net present value) of inventory holding and vendor production. An alternative objective can be to maximize the expected difference between total revenue and cost.

Let the considered products be indexed by $j = 1, \ldots, J$, the periods by $t = 1, \ldots, T$, the manufacturing machines by $r = 1, \ldots, R$, and the scenarios (demand outlooks up to the horizon T) by $s = 1, \ldots, S$.

The *deterministic data* consist of

$h_j(t)$ the inventory holding cost per unit of product j in period t;
$q_j(t)$ the unit cost of product j obtained from the vendor in period t;
a_{jr} the amount of capacity of machine r needed for producing one unit of product j;
$K_r(t)$ the available capacity of machine r in period t;
I_{j0} the initial inventory of product j.

The *stochastic data* consist of

$d_j^s(t)$ the demand for product j in period t under scenario s.

The *decision variables* are

$I_j^s(t)$ the inventory volume of product j at the end of period t under scenario s;
$x_j^s(t)$ the production volume of product j in period t under scenario s;
$y_j^s(t)$ the amount of product j obtained from the vendor in period t under scenario s.

Again, the decisions can be made sequentially using the past information only; this is the nonanticipativity property called sometimes also the *implementability constraint*. In the problem formulation, the nonanticipativity constraints will be briefly summarized as $\mathbf{x} \in \mathcal{N}, \mathbf{y} \in \mathcal{N}, \mathbf{I} \in \mathcal{N}$. The problem reads

$$\min \sum_{s=1}^{S} p^s \left\{ \sum_{j=1}^{J} \sum_{t=1}^{T} h_j(t) I_j^s(t) + \sum_{j=1}^{J} \sum_{t=1}^{T} q_j(t) y_j^s(t) \right\}$$

subject to

$$I_j^s(t-1) - I_j^s(t) + x_j^s(t) + y_j^s(t) = d_j^s(t) \quad \forall j, t, s$$

$$\sum_{j=1}^{J} a_{rj} x_j^s(t) \leq K_r(t) \quad \forall r, t$$

$$x_j^s(t), y_j^s(t), I_j^s(t) \text{ nonnegative} \quad \forall j, t, s$$

$$\mathbf{x} \in \mathcal{N}, \mathbf{y} \in \mathcal{N}, \mathbf{I} \in \mathcal{N}$$

where p^s is the probability (or the weight) of the scenario s.

An interesting modification is the model where no alternative source of production exists. In this case, decision variables $y_j^s(t)$ are replaced by $b_j^s(t)$, the lost demand of product j in period t under scenario s, and $q_j(t)$ replaced by $l_j(t)$, the per unit revenue for product j in period t; a further possibility is to introduce into the objective function another measure of risk or a penalty function related with the unserved demand.

4.5.1   Exercise. Extend the problem formulation to the case when there is a choice of vendors from a given set of vendors with prescribed capacities and costs for delivery of product j in period t.

4.6   Capacity Expansion of Electric Power Generation Systems – CEP

Electric power generation systems are supposed to serve the uncertain time varying demand for electricity, called *load*. The statistical analysis of load can be based on extensive data sets and historical time series. The required information is the total length t of subintervals belonging to a given time interval $[0, T]$ at which the load exceeds a given level L: this can be described by a nonincreasing function $t = h(L)$. The inverse function $l(t) := h^{-1}(t)$ is the *load duration curve*. It is again a nonincreasing function $l : [0, T] \to [0, L_{\max}]$ whose values $l(t)$ equal the load L which is surpassed during time periods of a total length $t \in [0, T]$, $l(T) = 0$, see Figure 3.

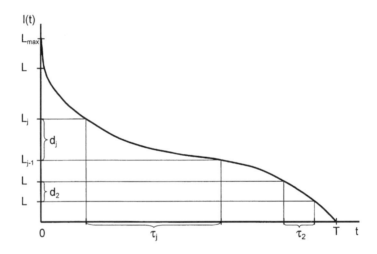

Figure 3: The load duration curve and its approximation

The load can be discretized by assuming that $L^1 < L^2 < \cdots < L^S (\leq L_{\max})$ are its possible values and $p^1, \ldots, p^S$ the corresponding probabilities.

Consider first a simple problem: The aim is to determine the numbers or capacities of R different equipments/ technologies characterized by unit investment costs $c_1, \ldots, c_R$ and unit operating costs $q_1 < \cdots < q_R$ to serve the demand in such a way that the expected total cost for the capacity expansion is minimal. Denote x_r the total capacity of equipments of the type r and

z_r^s the capacity of equipments of the type r used to generate power if load equals L^s, i.e., for the s-th operation mode.

The resulting problem

$$(9) \qquad \text{minimize } \sum_{r=1}^{R} c_r x_r + \sum_{s=1}^{S} p^s \sum_{r=1}^{R} q_r z_r^s$$

subject to

$$x_r - z_r^s \geq 0 \; \forall r, s \quad \sum_{r=1}^{R} z_r^s = L^s \; \forall s, \quad x_r, z_r^s \geq 0 \; \forall r, s$$

is a two-stage stochastic program whose second-stage constraints resemble those of the well known transportation problem.

A more realistic formulation tries to incorporate dynamic features of the capacity expansion problems which should capture the time evolution of costs and of the load duration curve, the possible appearance of new technologies, the construction delays, etc. The vector of random parameters ω includes not only the random demand in the considered periods, but also the costs; in more general cases, one may also consider random life-time of equipments, random construction delays Δ_r, random date of appearance of new technologies, etc.

An example is a model over T periods with the decision variables and model parameters dependent on the index of the period and on the considered random factors, demands and costs; for simplicity, the possibly limited life-time of equipments will not be taken into account. Let us denote $x_r(t, \omega)$ the new capacity decided at time t for equipment r which becomes available at time $t + \Delta_r$; $x_r(t, \omega) = 0$ for $t \leq 0$. The *existing* capacity of r-th equipment at time t that was decided before $t = 1$ is denoted $\tilde{x}_r(t)$; $w_r(t, \omega)$ the cumulated new capacity of r-th equipment which is available or already ordered at time t; $w_r(t, \omega) = 0$ for $t \leq 0$.

Assume now that the load duration curve has been approximated by piecewise constant function whose different (load) values $L_1 < L_2 < \ldots L_J$ correspond to different operation modes, the differences $d_j = L_j - L_{j-1}, j = 2, \ldots, J, d_1 = L_1$ represent the additional power demanded in the mode j for a duration τ_j; see Figure 3. We assume here that both d_j and $\tau_j \; \forall j$ and also the investment costs c_r and operation costs $q_r \; \forall r$ depend on time and on random parameters ω.

The stochastic model is then

$$(10) \qquad \min E \sum_{t=1}^{T} \left(\sum_{r=1}^{R} c_r(t, \omega) w_r(t, \omega) + \sum_{r=1}^{R} \sum_{j=1}^{J} q_r(t, \omega) \tau_j(t, \omega) y_{rj}(t, \omega) \right)$$

subject to

$$w_r(t, \omega) = w_r(t-1, \omega) + x_r(t, \omega) \quad \forall r, t$$

$$\sum_{r=1}^{R} y_{rj}(t, \omega) = d_j(t, \omega) \quad \forall j, t$$

$$\sum_{j=1}^{J} y_{rj}(t,\omega) \le a_r(t)(\tilde{x}_r(t) + w_r(t - \Delta_r, \omega)) \quad \forall r, t$$

$$w_r(t,\omega),\ x_r(t,\omega),\ y_{rj}(t,\omega) \ge 0 \quad \forall r, j, t$$

where $y_{rj}(t,\omega)$ is the capacity of equipment r operating in mode j at time t allocated to meet the random demand and $a_r(t)$ is the availability factor of equipment r in period t. Notice that one distinguishes between the capacity which is available at time t, e.g., $x_r(t) + w_r(t - \Delta_r, \omega)$, the extension of capacity $x_r(t,\omega)$ decided at t and the cumulated existing and ordered capacity $w_r(t,\omega)$ up to time t.

Both continuous and discrete probability distributions of the random element ω can be considered and, similarly as in Chapter II.2, constraints involving random coefficients are fulfilled with probability 1. The comparison of the model with the general T-stage stochastic program (3)–(4) introduced in Chapter II.2 leads to the observation that the constraints have to be extended for nonanticipativity conditions on all decision variables. It is also expedient to consider an additional emergency backstop technology which is always available but for a high cost.

4.6.1   Exercise. Following the methodology explained in 4.2 reformulate the capacity expansion problem for the case of a discrete probability distribution of ω.

4.6.2   Comment

For $T = 2, \tilde{x}_r(t) = 0 \, \forall r, t, \Delta_r = 1, a_r = 1 \, \forall r$ we have $w_r(1) = x_r(1), y_{rj}(1) = 0 \, \forall r, j$ and besides the nonnegativity constraints, the constraints of (10) reduce to

$$\sum_r y_{rj}(2,\omega) = d_j(2,\omega) \, \forall j, \quad \sum_j y_{rj}(2,\omega) \le x_r(1) \, \forall r$$

Hence, for fixed costs $c_r, q_r \, \forall r$, the difference between (9) and (10) lies only in a different way of approximating the load curve.

4.6.3   Loss of Load Probability

An important issue for electricity generation problems is the *reliability* of the system. The relevant characteristic is called LOLP – *Loss of Load Probability* and it is defined by the property that the demand can be satisfied with a probability 1-LOLP, on which a lower bound, say, 1-LOLP $\ge 1 - \varepsilon, \varepsilon > 0$ small, is imposed. It is often limited by the authorities. Using the introduced notation we may formulate this requirement via probabilistic constraint

$$(11) \qquad P\left\{ \sum_{j=1}^{J} d_j(t,\omega) \le \sum_{r=1}^{R} a_r(t)(\tilde{x}_r(t) + w_r(t - \Delta_r, \omega)) \right\} \ge 1 - \varepsilon \quad \forall t.$$

4.6.3.1 Exercise. Assume that $T = 2, \Delta_r = 1 \forall r$. The probabilistic constraint applies for $t = 2$ and the capacity $w_r(2 - \Delta_r, \omega) = w_r(1)$ is nonrandom for all equipments. Let F be the probability distribution function of $\sum_{j=1}^{J} d_j(2, \omega)$. The deterministic equivalent of the probabilistic constraint (11) is

$$\sum_{r=1}^{R} a_r(2)(\tilde{x}_r(2) + w_r(1)) \geq F^{-1}(1 - \varepsilon)$$

where $F^{-1}(1 - \varepsilon)$ is the $(1 - \varepsilon)$-quantile of F. Elaborate in detail.

4.7 Unit Commitment and Economic Power Dispatch

The problem is an optimal scheduling of the generating capacity among generating units of an existing power system. Such schedule attempts to minimize the generation costs while meeting the demand and other constraints imposed by the physical characteristic of the system. We are supposed to decide which units to commit at each time period and (if committed) at what generating capacity.

Disregarding randomness, considering R generating units and T time periods and using the notation introduced in 4.6 as much as possible we can write a deterministic model

$$(12) \qquad \min_{y,z} \sum_{r=1}^{R} \sum_{t=1}^{T} \left[g_r^t(y_r(t)) z_r(t) + h_r^t(z_r(t-1), z_r(t)) \right]$$

subject to

$$(13) \ \sum_{r=1}^{R} y_r(t) \geq d(t), \quad \sum_{r=1}^{R} x_r z_r(t) \geq d(t) + r(t), \quad l_r z_r(t) \leq y_r(t) \leq u_r z_r(t) \ \forall r, t.$$

Here, functions g_r^t evaluate the cost of generating $y_r(t)$ provided that the generating unit r was committed. The status of units is modeled via 0-1 decision variables $z_r(t)$, $z_r(t) = 1$ if unit r is committed at period t and $z_r(t) = 0$ otherwise. Functions h_r^t provide the costs due to changing the status of generating units in subsequent time periods. There are lower and upper bounds on the capacity of the units and a prescribed safety volume $r(t) \geq 0$ which expresses the fact that the total capacity at each period must exceed the predicted demand to avoid disturbances of the system. There are further constraints on the status of the units, e.g., the minimum up-time or switch-off requirements; we shall aggregate them as $z_r \in Z_r$. Further constraints, say, $y_r \in \mathcal{Y}_r(z_r)$ relate the generation capacity and the status of the unit; an example are the upper and lower bounds in (13).

Besides thermal- and hydro-generating units, which differ by generating costs and whose physical character influences essentially the construction of functions h, there is also a possibility to sell or buy electricity on an open electricity spot market. We may include these options into constraints (13) as additional generating units indexed by $R + 1$ and $R + 2$, with capacities y_{R+1}, y_{R+2}, with a predetermined status and assign them unit costs $c_{R+1}(t) = (1 + \delta_1)e(t)$ and $c_{R+2} = (1 - \delta_2)e(t)$ in the objective function (12). We denote here by $e(t)$ the quoted spot market prices, the parameters $\delta_1 \geq 0$, $0 \leq \delta_2 \leq 1$ are spreads for buying and selling, respectively.

The size of the resulting mixed-integer nonlinear program depends on the number of generating units and on the number of periods, usually 168 hourly periods for a horizon of one week. However, neither the demands $d(t)$ nor the spot market prices $e(t)$ are known in advance. To hedge these uncertainties we consider S scenarios – T-dimensional vectors of couples $(d^s(t), e^s(t))$, $s = 1, \ldots, S$, with probabilities $p^s > 0$, $\sum_s p^s = 1$, scenario dependent vectors $\mathbf{y}^s_r, r = 1, \ldots, R + 2$, and $\mathbf{z}^s_r, r = 1, \ldots, R$, and minimize the expected costs

$$\sum_{s=1}^{S} p^s \sum_{t=1}^{T} \{\sum_{r=1}^{R} [g_r^{t,s}(y_r^s(t))z_r^s(t) + h_r^{t,s}(z_r^s(t-1), z_r^s(t))] + (1 + \delta_1)e^s(t) + (1 - \delta_2)e^s(t)\}$$

with respect to all decision variables and subject to

$$\sum_{r=1}^{R+2} y_r^s(t) \geq d^s(t), \ \mathbf{z}_r^s \in \mathcal{Z}_r, \ \mathbf{y}_r^s \in \mathcal{Y}_r(\mathbf{z}_r^s) \ \forall r, s, t.$$

This problem – a *mixed-integer stochastic program* – has to be further elaborated to take into account stages of the decision process (the nonanticipativity constraints will be then included into $\mathcal{Z}_r, \mathcal{Y}_r$), to include the specific character of thermal- and hydro-generating units, availability and costs of the fuel, environmental restrictions, etc. End effects have to be treated, namely for hydro-generating units. In addition to the spot operations on the electricity market, the existing term documents may be included. In the special case that the commitment decisions $\mathbf{z}_*$ have been already accepted the problem reduces to the *economic power dispatch problem*.

Besides the power generation capacity, it is also the *capacity and reliability of the transmission network* which influences the performance of the system. We refer to [129] for a possible approach based on the stochastic programming methodology.

In spite of its complexity the unit commitment problem belongs presently to the most popular applications of stochastic programming. It is a mixed-integer stochastic program and for its efficient solution, both the stochastic programming methods and stochastic dynamic programming may be combined; cf. [158]. Problems of this type are solved within large deregulated electricity markets and they combine various features of stochastic production problems, water resources management and financial decision problems. The peculiarity is that electricity is a nonstorable commodity.

4.8   Melt Control: Charge Optimization

Melt control problems belong to the broad field of production control applications. They are studied as one of the production steps in iron and steel works. Melt control problems may be fully separated from other foundry optimization problems, which simplifies the model building and its solution. Their importance stems from the fact that foundries usually have high overheads, and hence, even small percentual savings may recover a significant amount of money. In addition, material inputs represent the biggest part of the total melting costs.

The produced alloys and input materials are composed of certain basic elements (iron, carbon, etc.). The production process consists of several steps (e.g., charge, alloying). In each of them, the hot melt in the furnace is enriched with certain input materials (return materials, scrap, ferroalloys, etc.) and a new mixture is melted again. Hence, the problem has a natural multistage decision structure. In each step of the process the melt composition changes and particularly, random losses of elements in the melt must be considered. During heating of the melt the amounts of elements change randomly, e.g., due to the rise in slag and oxidation. These losses are influenced by the composition of the melted material. The remaining amount of an element is expressed as a linear function in the input quantities of all considered elements, the coefficients are called *utilizations* of the considered element related to the amount of other elements in the melt. In the following simplified examples, these losses, and hence the related utilizations of elements are taken as the only random variables. Historical melt reports are available and may be used to construct scenarios or scenario trees of utilizations for the melt control problems.

The goal is to find amounts of the input materials in the cheapest way under the main requirement that the prescribed output alloy composition is achieved. We use scenario-based two- and three-stage stochastic linear programs to illustrate basic modeling ideas for charge optimization of induced and electric-arc furnaces. For a general approach to melt control, developed for any alloy, furnace, and technology see [126].

4.8.1   Example: Two-Stage Induced Furnace Charge Optimization. We begin with a simple model for charge optimization of iron production in an induced furnace – a model with a common two-stage structure: Through the initial charge decision the final cost of the melt is minimized taking into account also the consequences of possible random losses and the requirements on the final composition of the melt. The problem written in the arborescent form (11)–(12) of Section II.2.2 reads

$$\text{minimize} \sum_{j \in J_1} c_j x_{j1} + \sum_{k_2 \in \mathcal{K}_2} p_{k_2} \sum_{j \in J_2} c_j x_{jk_2}$$

subject to

$$l_{i1} \leq \sum_{l=1}^{m_1} \tau_{il}^E \sum_{j \in J_1} a_{lj} x_{j1} \leq u_{i1}, i = 1, \ldots, m_1$$

$$l_{i2} \leq \sum_{l=1}^{m_1} \tau_{il}^{k_2} \sum_{j \in J_1} a_{lj} x_{j1} + \sum_{j \in J_2} a_{ij} x_{jk_2} \leq u_{i2}, i = 1, \ldots, m_2, k_2 \in \mathcal{K}_2$$

$$x_{j1} \geq 0, j \in J_1, \quad x_{jk_2} \geq 0, j \in J_2, k_2 \in \mathcal{K}_2$$

where $t = 1, 2$ are stages, J_t is the set of indices of input materials available at stage t,
m_t is the number of considered elements at stage t and indices i and l specify them,
$c_j \geq 0$ are known unit costs of jth input material,

$l_{it}, u_{it} \geq 0$ are prescribed lower and upper goal bounds for the amount of the ith element in melt composition at stage t,

$a_{ij} \geq 0, \sum_i a_{ij} \leq 1 \forall j$ denote the amount of ith element in the unit amount of jth input material,

$x_{j1} \geq 0$ denote the first-stage decision variables, the amount of jth input material at the beginning of the melt process,

$x_{jk_2}, k_2 \in \mathcal{K}_2$ denote the second stage decision variables, which stay for the additional amount of jth input material assigned under scenario k_2.

The only random elements are utilizations and we denote

$\tau_{il}^{k_2}$ the utilization of ith element related to the amount of lth element in the melt when scenario k_2 occurs, $0 \leq \tau_{il}^{k_2} \leq 1$; k_2 are indices of scenarios and $p_{k_2} \geq 0$ are their probabilities. The frequently considered case $\tau_{il}^{k_2} = 0$ when $i \neq l$ means that interactions of random losses are ignored.

In the first-stage constraints, τ_{il}^E stands for an expert-designed 'standard' utilization which applies to the first stage. Usually, $\tau_{il}^E = \sum_{k_2 \in \mathcal{K}_2} p_{k_2} \tau_{il}^{k_2} \forall i, l$ is chosen. These constraints reflect the metallurgical rules which aim at the process control stability and in general, they cannot be neglected. On the other hand, possible losses of the materials added in the second stage of the melting process are not considered.

4.8.2   Example – Three-Stage Electric-Arc Furnace Charge Optimization.

The situation becomes more complicated with a steel production in an electric-arc furnace. Because of two alloying phases, the whole process must be modeled as a three-stage one. To simplify the model description, we mostly utilize the notation of Example 4.8.1 and the arborescent form (11)–(12) from II.2.2.

To get the Markovian structure of the model constraints, the melt composition is described explicitly by additional auxiliary variables h_{ik_t} describing the state of the decision process – the amount of i-th melt element at node k_t of stage t before a subsequent decision was taken. We assume an empty furnace at the beginning of the process, hence, $h_{i1} = 0, \forall i$ at stage 1, then $h_{ik_2} = \sum_{l=1}^{m_1} \tau_{il}^{k_2} \sum_{j \in J_1} a_{lj} x_{j1}$, etc.

The resulting form of the model is

$$\text{minimize} \sum_{j \in J_1} c_j x_{j1} + \sum_{t=2}^{3} \sum_{k_t \in \mathcal{K}_t} p_{k_t} \sum_{j \in J_t} c_j x_{jk_t}$$

subject to

$$\sum_{l=1}^{m_{t-1}} \tau_{il}^{k_t} \left(h_{la(k_t)} + \sum_{j \in J_{t-1}} a_{lj} x_{ja(k_t)} \right) - h_{ik_t} = 0 \; \forall i, k_t \in \mathcal{K}_t, t = 2, 3$$

$$l_{it-1} \leq \sum_{k_t \in \mathcal{K}_t} p_{k_t} h_{ik_t} \leq u_{it-1}, i = 1, \dots, m_t, t = 2, 3$$

$$l_{i3} \leq h_{ik_3} + \sum_{j \in J_3} a_{ij} x_{jk_3} \leq u_{i3}, i = 1, \dots, m_3$$

$$x_{j1} \geq 0,\, j \in J_1,\, x_{jk_t} \geq 0,\, k_t \in \mathcal{K}_t,\, j \in J_t,\, t = 2, 3.$$

Again, the expected cost of melt is minimized subject to constraints requiring that during the whole melt process the average melt composition satisfies the given bounds and that the final product satisfies these boundes for all considered scenarios. In the last stage, full utilization of added materials is assumed again.

Let us point out that both models introduced in this Section need further extensions (e.g., involving technological constraints, inventory constraints, and uncertain scrap composition or for more than three stages) to be a more realistic tool in the melt control.

At the end, two specific problem and model properties have to be underlined:
• Stages are not defined by modeler's choice because they are given by the modeled production process.
• Because the filled furnace cannot be enlarged or emptied during the process (contrary to the assumed unlimited borrowing and lending possibilities in financial applications, to the simplifying assumptions of the production planning problem 4.5 or to trading possibilities on an open electricity market in 4.7), the related hard constraints imply the fact that relatively complete recourse cannot be assumed. Hence, feasibility of the first-stage solution must be analyzed.

II.5 APPROXIMATION VIA SCENARIOS

scenarios and their generation, various levels of the available information, examples of specific models used for scenario generation (vector autoregressive model, Black-Derman-Toy model, Vašíček model), sample information, postoptimality and output analysis (consistency, contamination and minimax bounds), scenario tree generation (clustering, sampling, fitting moments)

5.1   Introduction

A typical, even though not quite realistic assumption in the general formulation of the stochastic programming models is that the probability distribution P of the random element ω is known. With P known, the main stumbling block for algorithmic solution of stochastic programs, see problem (3)–(4) in Chapter II.2 for instance, is the necessity to compute repeatedly at least the values of the multidimensional integrals of the involved functions, which themselves need not be defined explicitly. To overcome this problem, various approximation schemes, both stochastic and deterministic ones, were designed: One can replace the function f by a simpler one or approximate the true assumed probability distribution P or both. The goal is to get a numerically tractable optimization problem, or a sequence of such problems, whose solution would be acceptable as an approximation of the solution of the true underlying decision problem.

When approximating the true probability distribution P, one should exploit the structure of the problem and also the available information about P that comes from theory, historical data and experience. There are relatively many prospects if the approximation of P reduces to the approximation of (possibly many) one-dimensional probability distributions: Besides approximation by a discrete probability distribution, one can use piecewise uniform distributions (histograms) or approximate the density by kernel estimates, etc. In the multi-dimensional case, approximation by discrete probability distributions is the prevailing approach. It means that the true probability distribution P is replaced by a *discrete* probability distribution concentrated on a finite number of points, say, $\omega^1, \ldots, \omega^S$ with probabilities $p^1, \ldots, p^S$. The atoms ω^s of this discrete probability distribution are the *scenarios* which enter the scenario-based formulations of the stochastic programming problems.

The origin of scenarios can be very diverse; they can be atoms of a known genuine discrete probability distribution, can be obtained in the course of a discretization / approximation scheme, by simulation or by a limited sample information, they can result from recognized regulations or from a preliminary analysis of the problem with probabilities of their occurrence that may reflect an ad hoc belief or a subjective opinion of an expert, etc.

The use of scenarios as a representation of uncertainty is not limited to applications of multiperiod and multistage stochastic programs; several other possibilities were introduced in Chapter II.3. In addition, scenarios are built to evaluate the average performance of an already proposed sequence of decisions, to get the covariance structure of intersecurity returns required, e.g., for the Markowitz model, or to compute durations for stochastic versions of duration matching and immunization models for asset/liability management problems.

Naturally, one is interested in results for the *true* underlying problem. Various techniques of *output analysis* for the obtained solution of the approximate problem help to draw inference about the solution of the true problem. They must be tailored to the structure of the solved problem and to the approximation technique which is supposed to reflect the nature of the input data. The next item is the robustness of the obtained approximate optimal solution and the optimal value: The procedure should be robust in the sense that small perturbances of the input, i.e., of the chosen scenarios and of their probabilities, should impair the outcome only slightly so that the obtained results remain close to the unperturbed ones and that somewhat larger perturbations do not cause a catastroph. The importance of robust procedures increases with the complexity of the model and with its dimensionality.

5.2   Scenarios and their Generation

We have introduced scenarios as atoms of the true discrete probability distribution P or of that discrete probability distribution which approximates the true one. However, the primary aim of scenario generation is to build a manageable problem which will provide good decisions for the true underlying real-life problem. This is an ambitious task in which compromise is needed between precision of the approximation of the probability distribution P and the size and goal of the approximate problem, which, moreover, often requires a specific form of the input (e.g., a scenario tree for multistage problems). Generation of scenarios is problem specific and it should reflect both the problem structure and the available information on the underlying probability distribution. It is natural to use historical data (if any) in conjunction with an assumed background model, to apply suitable estimation, simulation and sampling procedures and also to reflect the opinion of the experts based on their experience, heuristics, etc. For all these reasons the generation of scenarios cannot be reduced solely to forecasting the future development of the complex system under consideration.

Concerning the level of the available information, four basic types of problems may be distinguished:

5.2.1   Full Knowledge of the Probability Distribution

The probability distribution P is fully specified, hence, scenarios can be obtained by sampling from this probability distribution or by application of a discretization or simulation scheme. Related to the chosen approximation technique, there are various possibilities how to draw conclusions about the optimal solution of the original problem. In its pure form this situation appears mostly in the context of testing the designed models and/or the performance of newly developed solvers. The assumed fully known probability distribution (mostly normal, uniform or discrete one) may stem equally from a theoretical model, from historical data or from an experience of an expert. An example is generation of scenarios for the ALM model [104] discussed in Section II.4.3.

5.2.2 Known Parametric Family

In this case, only a parametric family of probability distributions based on a theoretical model is specified while the parameters of the probability distribution P are estimated from the available data. The choice of the parametric form of the probability distribution or of the stochastic process corresponds to the choice of the model, the estimation of parameters to the calibration of the model and a subsequent simulation, sampling or discretization procedure follow similarly as in 5.2.1.

This type of information appears frequently in stochastic programming problems in finance and also in water resources management and planning. This is partly due to the fact that the relevant stochastic models of interest rates and assets prices or those of water inflows came to the attention relatively early and both discrete and continuous time models have been well developed and supported by time series of historical data. The most common situation is that the random element ω is in fact a stochastic process, $\omega = (\omega_1, \ldots, \omega_T)$; it explains the use of the word "scenario".

5.2.3 Vector Autoregressive Models

As an example of the discrete time stochastic models we introduce the vector autoregressive model of the first order

$$(1) \qquad \omega_t = \mu + \mathbf{H}(\omega_{t-1} - \mu) + \varepsilon_t, \quad \varepsilon_t \sim N(0, \Sigma)$$

where the eigenvalues of $\mathbf{H}$ fulfill $|\lambda(\mathbf{H})| < 1$ and ε_t's are jointly independent. The parameters $\mu, \mathbf{H}, \Sigma$ are estimated from historical data and possibly further adapted to distinct sources of information (e.g., experts' forecasts or values of related global parameters). Let $\hat{\mu}, \hat{\mathbf{H}}, \hat{\Sigma}$ be these estimates.

Starting with a known vector ω_0 and using the calibrated model (1), scenarios are constructed step by step as

$$\omega_t^s = \hat{\mu} + \hat{\mathbf{H}}(\omega_{t-1}^s - \hat{\mu}) + \hat{\varepsilon}_t^s,$$

where $\hat{\varepsilon}_t^s$ is obtained as an observation from $N(0, \hat{\Sigma})$ by a suitable discretization or simulation technique.

Similarly, also autoregressive and ARMA models of higher orders or econometric models with lagged variables may be applied. Factor analysis can be used to get a small number, say M, of one-dimensional independent factors $\phi_t^1, \ldots, \phi_t^M$ such that the covariance matrix of $\tilde{\varepsilon}_t := \mathbf{C}\phi_t$ is approximately equal to $\hat{\Sigma}$; the elements of matrix $\mathbf{C}$ are time-independent factor loadings.

5.2.4 The Black-Derman-Toy Model

This model introduced in [22] is one of frequently used discrete time models aimed at generation of interest rate scenarios. Its primary purpose was to provide a tool for pricing bond options. The model assumptions can be summarized as follows:

• The short interest rates are lognormally distributed, the volatility of their logarithms depending only on time.

• Binomial approximation of the normal distribution of $\ln r(t)$ with time dependent drift and volatility is used at all selected time instants t.

• In the binomial lattice, the short interest rate can move up or down with equal probability over the next time period; the pairs of "up - down" and "down - up" moves from any fixed state at a time point t result into the same value of short interest rate at the time point $t + 2$ (the path independence property).

As a result, at each time point t, there are $t + 1$ possible states and for the given horizon T there are 2^{T-1} equiprobable scenarios. Each of them can be represented by a vector with $T - 1$ zero-one digits, say

$$\boldsymbol{\omega}^s = (\omega_1^s, \omega_2^s, \ldots, \omega_{T-1}^s)^\top$$

and the probability of each scenario is $p^s = 2^{-(T-1)}$ $\forall s$. The digit 1 at the tth position corresponds to the "up" move, the digit 0 corresponds to the "down" move of the one-period short term interest rate in the step t. The corresponding one-period short term interest rates for scenario s and for the time interval $(t, t+1]$ are then given as

(2) $$\rho_t^s = r_{t0} k_t^{i_t(s)}, \; i_t(s) = \sum_{\tau=1}^{t} \omega_\tau^s$$

That is, $i_t(s)$ is a realization of a binomial random variable, its value equals the number of the "up" moves for the given scenario s which occur at time points $1, \ldots, t$. See Figure 4 for the lattice with the scenario corresponding to vector $(1, 0, 0, 1, 0, 1)^\top$ marked out.

The base rates r_{t0} and the volatility factors k_t for all t are computed numerically so that the obtained binomial lattice fits the yield curve and yield volatilities for zero-coupon government bonds at a given date; this input should be available for all maturities.

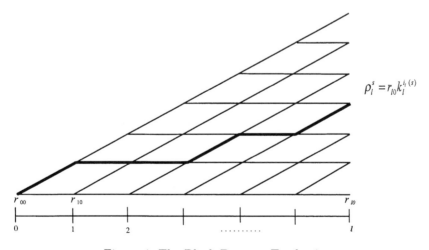

$$\rho_l^s = r_{l0} k_l^{i_l(s)}$$

Figure 4: The Black-Derman-Toy lattice

The Black-Derman-Toy model is currently close to the industry standard. For its application see Chapter II.6.

Continuous time stochastic models are mostly represented by stochastic differential equations of the type

$$(3) \qquad d\omega(t) = b(t, \omega(t))dt + \sigma(t, \omega(t))dW(t), \ \omega(0) = \omega_0, \ t \geq 0,$$

where W is the Wiener process and coefficients $b(t, \omega(t)), \sigma(t, \omega(t))$ fulfill some assumptions. For instance, solutions of (3) with b, σ independent of t and satisfying certain integrability conditions are known as diffusion processes; see III.2.2.

5.2.5 Vašíček's Model for Spot Rates

Consider the following form of the model introduced in [164]:

$$d\omega(t) = b(k - \omega(t))dt + \sigma dW(t), \ t \geq 0$$

with b a positive constant. It corresponds to the choice $b(t, \omega(t)) = b(k - \omega(t))$ and $\sigma(t, \omega(t)) = \sigma$ in (3). It is the so called Ornstein-Uhlenbeck process, a Markov process with normally distributed increments. In contrast to the Wiener process, it has a stationary distribution. The instantaneous drift $b(k - \omega(t))$ forces the process towards its long-term mean k, the so called *mean reversion property*, and the constant instantaneous variance σ^2 makes it to fluctuate around k in a continuous but erratic way. This also means that, contrary to the Black-Derman-Toy model, negative values of $\omega(t)$ cannot be excluded.

To apply the model for generation of scenarios of interest rates means to estimate the parameters and to choose a suitable time discretization $\Delta < 1/b$. Then

$$\omega_{(n+1)\Delta} = bk\Delta + \omega_{n\Delta}(1 - b\Delta) + \sigma\sqrt{\Delta}\,\varepsilon, \quad n = 0, 1, \ldots$$

with an arbitrarily given initial value ω_0 and with ε independent, $N(0, 1)$.

There are various extensions and modifications of the initial Vašíček model; alternative choices of coefficients $b(t, \omega(t))$, $\sigma(t, \omega(t))$ provide other one-factor models. More dimensional continuous time models result by considering multiple sources of uncertainties, e.g., default possibilities of corporate bonds, prepayment of mortgages, influence of inflation on indexed bonds, random behavior of volatilities, etc. The stochastic differential equations may include more that one factor, i.e., differentials of several independent or correlated Wiener processes instead of one. This helps for instance to distinguish differences in behavior of the short term and long term rates. Also the Wiener process can be replaced by the Poisson process to reflect jumps of the interest rates process, etc.

5.2.6 Multidimensional, Multifactor Models

An example is the model of evolution of the vector $\boldsymbol{\omega}(t) = (\omega_1(t), \ldots, \omega_N(t))^\top$ of key zero coupon interest rates $\boldsymbol{\omega} = (\omega_1, \ldots, \omega_N)^\top$:

$$d\omega_i(t) = b_i(t)\omega_i(t)dt + \sigma_i\omega_i(t)dW_i(t), \ i = 1, \ldots, N$$

with N factors - the intercorrelated Wiener processes W_i, see [87]. Principal components method is used to reduce the dimension of the problem. As a result, the initial N-dimensional Wiener process with correlation matrix $\mathbf{R}$ is represented by $n < N$ non-correlated components and for $i = 1, \ldots, N$,

$$dW_i(t) \approx \beta_{i1} d\hat{W}_1(t) + \cdots + \beta_{in} d\hat{W}_n(t)$$

with time independent loadings $\beta_{i1}, \ldots, \beta_{in}$. The advantage is that the N- dimensional scenario can be now obtained by generating repeatedly n *independent* normal variables $\varepsilon_1, \ldots, \varepsilon_n$:

$$\omega_{i,t+1} = (b_{it} + 1)\omega_{it} + \sigma_i \omega_{it}[\beta_{i1}\varepsilon_{1t} + \cdots + \beta_{in}\varepsilon_{nt}], \, i = 1, \ldots, N.$$

5.2.7 Sample Information

The available sample information about the true probability distribution is based mostly on observed past data. If the data are homogeneous enough, if they can be viewed as independent, identically distributed random variables (vectors), the use of the empirical distribution is straightforward. Otherwise, one could think of a preprocessing procedure to treat the missing data, smoothing, etc., or of an adjustment to fit specific values of (sample) moments.

The simplest idea is to use as scenarios the past observations obtained under comparable circumstances and assign them equal probabilities; see for instance [157] for scenarios of future electricity demand in a given period of year. Similarly, [121] suggests to construct scenarios of joint assets returns for a T-period model as $n - T$ distinct T-tuples of their subsequent observations from n previous periods and to assign them equal probabilities, $(n - T)^{-1}$.

5.2.8 Low Information Level

The above mentioned procedures fail if there are no reliable data. Under such circumstances, scenarios and their probabilities are mostly based on experts' forecasts or even on governmental regulations. For instance, to test the surplus adequacy of an insurer, New York State Regulation 126 suggests seven interest rate scenarios to simulate the performance of the surplus. Postoptimality (are the seven scenarios enough?) and stability analysis of the obtained solutions is crucial even though it is hardly possible to draw conclusions about the optimal solution of the true underlying problem.

Under the heading of "low information level" we can also include the cases when the true probability distribution is described only by several moment values or/and some simple qualitative properties.

5.2.9 Miscelaneous Sources of Uncertainties

In most applications one can trace interactions of various information levels and the best thing to do is to use all available information. Past information is often combined with experts' opinions and this is probably one possible origin of the postulated form of the true probability distribution. Different information levels and different time-scales of collecting and recording data may apply to distinct parameters of the model separately.

In portfolio management, different classes of securities require different treatment, deposits and liabilities can be driven by conceptually different external factors such as mortality rates. For management of a new pension plan, for instance, the random factors which run the bond market can be well described by a relevant interest rates model, basic requirements on premium and pension payments are partly known thanks to extensive demographic data whereas the uncertainty relates to the future preferences of clients. The scenarios and their probabilities are then based on experts' forecasts, they may even reflect only certain extremal cases.

If there are *independent* sources of uncertainties the number of scenarios needed to represent their mutual influence on the results is the product of the number of scenarios used to represent the impact of each source separately.

5.3   How to Draw Inference about the True Problem?

Possibilities of drawing conclusions about the optimal solutions and the optimal value of the true stochastic program using the results of the approximate scenario based program depend essentially on the structure of the solved problem as well as on the origin of scenarios. Generally speaking, the output can hardly be more precise than the input and it is easier to answer questions concerning precision of the obtained optimal values than those concerning the sets of optimal solutions.

We shall discuss briefly the general methodological devices which can be applied to the analysis of the scenario based problems that arise as an approximation of the true problem. The main tools are selected methods of probability theory, asymptotic and robust statistics, simulation methods and parametric optimization, the main sources of errors come from simulation, sampling, estimation and also from incomplete or unprecise input information. To simplify the exposition, let us concentrate on stochastic programs written in the form (compare with (6) in Chapter II.2 and with (20)–(21) or (25) in Chapter II.3)

$$(4) \qquad\qquad \text{minimize } F(\mathbf{x}, P) := E\, f_0(\mathbf{x}, \omega)$$

on a closed nonempty set $\mathcal{X} \subset \mathbb{R}^n$ which does not depend on P. This notation is used to underline the dependence of the problem on the chosen probability distribution P of ω. We denote
- $\varphi(P)$ the optimal value of (4),
- $\mathcal{X}(P) = \arg\min_{\mathbf{x}\in\mathcal{X}} f(\mathbf{x}, P)$ the set of optimal solutions of (4), not necessarily a singleton,
- $\mathbf{x}(P)$ the unique optimal solution of (4) in case that $\mathcal{X}(P)$ is a singleton.

We accept that the true probability distribution P has been replaced by another probability distribution $\hat{P}$ obtained by parametric or nonparametric methods and by sampling, discretization and simulation techniques. The precision of the approximation can be quantified by means of a suitable distance of the two probability distributions. The question is how the chosen/evaluated precision of the approximation influences the differences of the optimal values $\varphi(P) - \varphi(\hat{P})$ and the distances of the sets of optimal solutions $\mathcal{X}(P), \mathcal{X}(\hat{P})$ for the true problem and for the approximate one, respectively. This problem setting falls into the frame of quantitative stability of parametric programs and presentation of the relevant results is beyond the scope of this book. Instead, let us start with the notion of *weak convergence* to get qualitative stability results.

5.3.1 Definition. Let P, $P_\nu, \nu = 1, \ldots$, be probability measures on Borel sets of the same Euclidean space Ω. Then P_ν is said to converge weakly to P as $\nu \to \infty$ if for any bounded continuous function $h : \Omega \to \mathbb{R}^1$,

$$\int_\Omega h(\omega) P_\nu(d\omega) \to \int_\Omega h(\omega) P(d\omega).$$

To obtain continuity of expectation functionals $F(\mathbf{x}, P) = \int_\Omega f_0(\mathbf{x}, \omega) P(d\omega)$ means to restrict the class of the considered functions $f_0(\mathbf{x}, \bullet)$ to bounded continuous functions of ω or to restrict the set of probability measures to a subset with respect to which the functions $f_0(\mathbf{x}, \bullet)$ are uniformly integrable. In addition, we need convergence results also for the optimal values $\varphi(P_\nu) \to \varphi(P)$ and for the optimal solutions.

5.3.2   Classical Consistency Results

Under assumptions that $f_0(\mathbf{x}, \omega)$ is a continuous bounded function of ω for every $\mathbf{x} \in X$ and $P_\nu \to P$ weakly, the pointwise convergence of the objective functions in (4) follows from Definition 5.3.1:

$$F(\mathbf{x}, P_\nu) \to F(\mathbf{x}, P) \quad \forall \mathbf{x} \in X.$$

If X is compact and the convergence is uniform on X we get immediately

$$\varphi(P_\nu) \to \varphi(P).$$

If, moreover, X is convex and $f_0(\bullet, \omega)$ is strictly convex on X for all ω it is easy to get in addition the convergence of the (unique) optimal solutions $\mathbf{x}(P_\nu)$ of $\min_{\mathbf{x} \in X} F(\mathbf{x}, P_\nu)$ to the unique optimal solution $\mathbf{x}(P)$ of the initial problem (4).

5.3.3   Example – One-Sided Bias.

Consider the case where P_ν are empirical probability distributions obtained by simple random sampling from P. It means that $P_\nu(A)$ is the relative frequency of the event $\{\omega^i \in A\}$ among independent copies $\omega^1, \ldots, \omega^\nu$ of ω and, accordingly, it is a random variable. However, the weak convergence of P_ν to P still holds true with probability 1 (that is, for almost all sequences of realizations $\omega^1, \omega^2, \ldots$).

In this case, it is possible to prove that the obtained optimal values $\varphi(P_\nu)$ have a *one-directional bias* in the sense that

$$E\varphi(P_\nu) \le \varphi(P).$$

Indeed, for any fixed $\mathbf{x} \in X$, the function values $f_0(\mathbf{x}, \omega^i)$ are independent, identically distributed and

$$E\varphi(P_\nu) = E \min_{\mathbf{x}} \frac{1}{\nu} \sum_{i=1}^{\nu} f_0(\mathbf{x}, \omega^i) \le \min_{\mathbf{x}} \frac{1}{\nu} E \sum_{i=1}^{\nu} f_0(\mathbf{x}, \omega^i) = \min_{\mathbf{x}} E f_0(\mathbf{x}, \omega) = \varphi(P).$$

5.3.3.1 Exercise. An empirical point estimate of $E\varphi(P_\nu)$ follows from the Law of Large Numbers and an asymptotic confidence interval can be obtained from the Central Limit Theorem.

Prove that $E\varphi(P_\nu) \leq E\varphi(P_{\nu+1})$. Construct the asymptotic confidence interval for $E\varphi(P_\nu)$.

An upper bound for $\varphi(P)$ is the expected value $Ef_0(\hat{\mathbf{x}}, \omega)$ evaluated for an arbitrary $\hat{\mathbf{x}} \in \mathcal{X}$. Construct a point estimate of $Ef_0(\hat{\mathbf{x}}, \omega)$ and an asymptotic confidence interval in a similar way as before. The resulting bounds are important for designing stopping rules and for tests of quality of a "candidate" solution $\hat{\mathbf{x}}$.

To get asymptotic results under less stringent assumptions concerning the approximating probability measures and the functions f_0 requires a different methodology which also extends to properties of the sets of optimal solutions. A simplification is possible whenever the general stability properties with respect to the probability distribution can be reduced to a finite dimensional parameter case. An example are probability distributions of a given parametric form (cf. 5.2.2) and the results concern differences between the optimal values $\varphi(\boldsymbol{\theta}_0)$ and $\varphi(\boldsymbol{\theta}_\nu)$ obtained for the true parameter value $\boldsymbol{\theta}_0$ and for its estimate, respectively, etc.

5.3.4   Asymptotic Results for a Parametric Family

Assume that the true probability distribution P is known to belong to a parametric family $\mathcal{P} = \{P_{\boldsymbol{\theta}}, \boldsymbol{\theta} \in \Theta\}$ of probability distributions indexed by a parameter vector $\boldsymbol{\theta}$ belonging to an open set $\Theta \subset \mathbb{R}^q$. The objective function now depends on $\boldsymbol{\theta}$, $F(\mathbf{x}, P_{\boldsymbol{\theta}}) := F(\mathbf{x}, \boldsymbol{\theta})$ and (4) is a standard parametric program $\min_{\mathbf{x} \in \mathcal{X}} F(\mathbf{x}, \boldsymbol{\theta})$. Let us assume that the optimal value $\varphi(\boldsymbol{\theta})$ exists for all $\boldsymbol{\theta} \in \Theta$ and is a continuous function of $\boldsymbol{\theta}$ on a neighborhood of the true parameter value, say, $\boldsymbol{\theta}_0$. Having a statistical estimate $\boldsymbol{\theta}_\nu$ of $\boldsymbol{\theta}_0$ and knowing its asymptotic properties we can obtain parallel asymptotic properties of the optimal value $\varphi(\boldsymbol{\theta}_\nu)$:

Whenever $\boldsymbol{\theta}_\nu \to \boldsymbol{\theta}_0$ with probability 1 or in probability, then $\varphi(\boldsymbol{\theta}_\nu) \to \varphi(\boldsymbol{\theta}_0)$ with probability 1 or in probability, respectively.

This assertion can be complemented by the *rate of convergence* based on the δ-theorem:

Let $\boldsymbol{\theta}_\nu$ be an asymptotically normal estimate of $\boldsymbol{\theta}$, i.e., $\sqrt{\nu}(\boldsymbol{\theta}_\nu - \boldsymbol{\theta}) \sim N(\mathbf{0}, \boldsymbol{\Sigma})$ and φ be continuously differentiable at $\boldsymbol{\theta}$ with $\nabla\varphi(\boldsymbol{\theta}) \neq 0$. Then $\varphi(\boldsymbol{\theta}_\nu)$ is asymptotically normal,

$$(5) \qquad \sqrt{\nu}(\varphi(\boldsymbol{\theta}_\nu) - \varphi(\boldsymbol{\theta})) \sim N(\mathbf{0}, \nabla\varphi(\boldsymbol{\theta})^\top \boldsymbol{\Sigma} \nabla\varphi(\boldsymbol{\theta})).$$

We shall apply this result in Chapter II.7. Similar assertions can be obtained for optimal solutions provided that these solutions are unique, continuous, differentiable in $\boldsymbol{\theta}$. However, even uniqueness of optimal solutions requires special assumptions which are not always realistic and their verification is not straightforward.

Low information level corresponds to the case when the true probability distribution is known to belong to a *nonparametric family* $\mathcal{P}$ of probability distributions defined, e.g., by fixed values of some moments, by a fixed support or by qualitative attributes such as unimodality. Also the case of *scenarios designed by experts* without any obvious relation to the true probability distribution will be included into this category.

5.3.5   Moment Bounds

In the first case, one may use the results concerning the moment problem: *Required consistent moment conditions can be attained by discrete probability distributions carried by a relatively small number of atoms.* These probability distributions qualify as natural discrete approximations of the true probability distribution and given the available information, they provide a sensible choice of scenarios and probabilities. Moreover, one can try to construct *minimin and minimax bounds* on the optimal value of the true program:

$$(6) \qquad \min_{\mathbf{x} \in \mathcal{X}} \inf_{P \in \mathcal{P}} F(\mathbf{x}, P) \leq \varphi(P) \leq \min_{\mathbf{x} \in \mathcal{X}} \sup_{P \in \mathcal{P}} F(\mathbf{x}, P) \, \forall P \in \mathcal{P}.$$

Under special assumptions about the family $\mathcal{P}$ and about the structure of the underlying stochastic program (4), e.g., for the random objective function $f_0(\mathbf{x}, \omega)$ convex or concave, and/or separable with respect to ω , this approach provides the best case and the worst case discrete probability distributions carried by fully specified scenarios.

A well-known simple example is related with the *Jensen inequality*. Assume that $f_0(\mathbf{x}, \bullet)$ is convex for all $\mathbf{x} \in \mathcal{X}$ and that the expectation $E_P \omega$ exists. Then the Jensen inequality applies, $E_P f_0(\mathbf{x}, \omega) \geq f_0(\mathbf{x}, E_P \omega)$. Consider now family $\mathcal{P}$ of probability distributions with a fixed expectation, say $E_P \omega = \bar{\omega} \, \forall P \in \mathcal{P}$. Jensen's inequality provides a lower bound

$$F(\mathbf{x}, P) = E \, f_0(\mathbf{x}, \omega) \geq f_0(\mathbf{x}, \bar{\omega})$$

valid for all $\mathbf{x} \in \mathcal{X}$ and for all $P \in \mathcal{P}$. This bound is attained for the degenerated probability distribution Q concentrated on $\bar{\omega}$. Notice that this *worst case* probability distribution does not depend on $\mathbf{x}$, a property which cannot be in general expected. The above discussion implies that, in the considered case, the lower bound in (6) is the optimal value of the deterministic convex program $\min_{\mathbf{x} \in \mathcal{X}} F(\mathbf{x}, Q) = \min_{\mathbf{x} \in \mathcal{X}} f_0(\mathbf{x}, \bar{\omega})$.

5.3.6   The Contamination Method

This method applies again to stochastic programs rewritten into the form (4) and it does not require any specific properties of the probability distribution P. It is suitable for the *postoptimality analysis* as it may be used to support conclusions about resistance of the *already obtained optimal output* to changes of scenarios and their probabilities and to check possible influence of out-of-sample scenarios.

Inclusion of additional scenarios or branches of the scenario tree means to pass from the initial probability distribution P to

$$(7) \qquad P_\lambda = (1 - \lambda)P + \lambda Q, \quad 0 \leq \lambda \leq 1,$$

the probability distribution P *contaminated* by the probability distribution Q which is carried by the additional scenarios or branches of the scenario tree. For fixed probability distributions P, Q, the expected value in (4) computed for the contaminated probability distribution P_λ is *linear* in the parameter λ and under mild assumptions, its optimal value

$$\varphi(\lambda) := \min_{\mathbf{x} \in \mathcal{X}} F(\mathbf{x}, P_\lambda)$$

is a finite concave function on $[0, 1]$ with a derivative $\varphi'(0^+)$ at $\lambda = 0^+$.

5.3.6.1 Exercise. Let $\mathcal{X}$ be convex compact and assume that $F(\mathbf{x}, \lambda) := F(\mathbf{x}, P_\lambda)$ is continuous in $\mathbf{x} \in \mathcal{X}$ for all $\lambda \in [0, 1]$ (a simplifying assumption). Prove that $\varphi(\lambda)$ is concave on $[0, 1]$.

Bounds on the optimal value $\varphi(\lambda)$ for an arbitrary $\lambda \in [0, 1]$ follow by properties of concave functions:

(8) $\qquad (1 - \lambda)\varphi(0) + \lambda\varphi(1) \leq \varphi(\lambda) \leq \varphi(0) + \lambda\varphi'(0^+) \quad \forall \lambda \in [0, 1].$

An upper bound for the derivative $\varphi'(0^+)$ equals $F(\mathbf{x}(0), Q) - \varphi(0)$ where $\mathbf{x}(0)$ is an arbitrary optimal solution of the initial problem (4) obtained for the probability distribution P; in case of the unique optimal solution, this upper bound is attained. Hence, the evaluation of bounds in (8) requires the solution of another stochastic program of the type (4) for the new probability distribution Q to get $\varphi(1)$ and evaluation of the expectation $F(\mathbf{x}(0), Q)$ at an already known optimal solution of the initial problem (4) but for the contaminating probability distribution Q.

Contamination technique can be useful also in analysis of results obtained under a fully specified probability distribution P. It is another example of reduction of the stability problem to a finite dimensional parameter case. Small values of the contamination parameter λ are typical for various stability studies. The choice of λ may reflect the degree of confidence in expert opinions represented as the contaminating probability distribution Q, and so on. By a suitable choice of the contaminating probability distribution Q, one can study the influence of including additional "out-of-sample" scenarios and emphasize the importance of a scenario by increasing its probability.

5.3.6.2 Example. Consider the problem of investment decisions in the debt and equity markets in the US, Germany and Japan. Historical data enable to construct many scenarios concerning returns of investments in the considered assets categories. We denote these (presumably equiprobable) scenarios by ω^s, $s = 1, \ldots, S$, and let P be the corresponding uniform discrete probability distribution. Assume that for each of these scenarios, an outcome of a feasible investment strategy, say, $\mathbf{x} \in \mathcal{X}$ can be evaluated as $f(\mathbf{x}, \omega^s)$. Maximization of the expected outcome

$$F(\mathbf{x}, P) = \frac{1}{S} \sum_{s=1}^{S} f(\mathbf{x}, \omega^s) \text{ with respect to } \mathbf{x} \in \mathcal{X}$$

provides the optimal value $\varphi(P)$ and an optimal investment strategy $\mathbf{x}(P)$.

The historical data do not cover all possible extremal situations on the market. However, experts in the investment committee may foresee such events. Assume that they agreed on one additional scenario ω^* capturing this extremal event. This scenario is the only atom of the degenerated probability distribution Q for which the best investment strategy is $\mathbf{x}(Q)$ - an optimal solution of $\max_{\mathbf{x} \in \mathcal{X}} f(\mathbf{x}, \omega^*)$. The contamination method explained above is based on the probability distribution P_λ, carried by the scenarios ω^s, $s = 1, \ldots, S$, and on the experts scenario ω^* with probabilities $\frac{1-\lambda}{S}$ for $s = 1, \ldots, S$, and $p^* = \lambda$. The probability λ assigns a weight to the view of the investment committee and the bounds (8) (multiplied by -1) are valid for all $0 \leq \lambda \leq 1$. They clearly indicate how much the weight λ, interpreted as the degree of confidence to the investor's view, affects the outcome of the portfolio allocation.

The impact of a modification of every single scenario according to the investor's views on the performance of each asset class can be studied in a similar way. We use the initial probability distribution P contaminated by Q, which is carried now by equiprobable scenarios $\hat{\omega}^s = \omega^s + \delta^s$, $s = 1, \ldots, S$. The contamination parameter λ relates again to the degree of confidence to the expert's view.

Also the *structure of the problem* influences essentially the possibilities of an adequate scenario generation. The most complicated problems – the multistage stochastic programs with interstage dependent coefficients – will be discussed in the next Section.

5.4 Scenario Trees for Multistage Stochastic Programs

For scenario-based multistage stochastic programs one assumes that the probability distribution P of ω is concentrated on a finite number of scenarios $\omega^s = (\omega^1, \ldots, \omega^S)$ having probabilities p^s, $s = 1, \ldots, S$. Partial sequences $(\omega_1^s, \ldots, \omega_{t-1}^s)$, *scenarios at stage $t - 1$, $t = 2, \ldots, T$* are listed as $\tilde{\omega}_{k_t}$, $k_t \in \mathcal{K}_t$, as already introduced in Section II.2.2, together with their path probabilities p_{k_t} and arc (transition) probabilities $\pi_{k_t, k_{t+1}}$. The set of all scenarios ω^s consists of $\tilde{\omega}_{k_T}$, $k_T \in \mathcal{K}_T$, their probabilities p^s equal the path probabilities p_{k_T} and are obtained as products of the corresponding arc probabilities.

The described structure of the input data is represented as a *scenario tree*. We can think of it as of an oriented graph which starts from a root (the only node at level 0) and branches into nodes at level 1, each corresponding to one of the possible realizations of ω_1, and the branching continues up to nodes at level T. This arrangement is based on the one-to-one correspondence between the partial sequences $\omega^{t-1,\bullet}$ and the nodes of the tree at stage t for $t = 2, \ldots, T$. This means that for any node at level t, each of the new observations ω_t must have only one immediate predecessor $\omega^{t-1,\bullet}$, i.e., a node at level $t - 1$, and a (finite) number of descendants ω_{t+1} which result in nodes at level $t+1$, $t < T$; compare with (11)–(12) in Section II.2.2. The number of descendants of all nodes at a given level $0 \leq t < T$ of the scenario tree may be equal. If this occurs for all stages the structure of such *balanced* tree can be coded as a product of numbers of descendants of the root and of nodes at levels $1, \ldots, T-1$. For instance $3^1 2^2 1^1$ describes the structure of a scenario tree with 3 branches from the root, 2 branches from nodes at the first and second levels and no branching at the third level. The total number of scenarios equals the numerical value of this product, $S = 12$; see Figure 5. Similarly, Figure 1 depicts a balanced tree of the type 3^2.

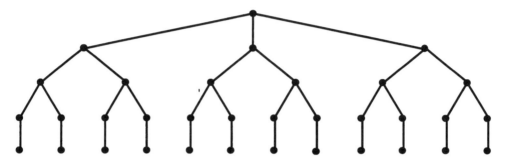

Figure 5: $3^1 2^2 1^1$ balanced tree

Two special cases of the scenario tree are to be pointed out:

- For all stages $t = 2, \ldots, T$, the conditional probability distributions $P_{\omega_t | \omega^{t-1, \bullet}}$ are equal to the marginal probability distributions P_{ω_t} – the *interstage independence*; in this case, the scenario generation methods apply to each stage separately.

- For all stages $t = 2, \ldots, T$, the supports of conditional probability distributions of ω_t conditioned by realizations $\omega^{t-1, \bullet}$ are singletons. This means that the scenario tree is nothing else but a "fan" of individual scenarios $\omega^s = (\omega_1^s, \ldots, \omega_T^s)$ which occur with probabilities $p^s = P(\omega_1^s) \, \forall s$ and, independently of the number of periods, the multiperiod stochastic program reduces to the *two-stage* one; see Figure 6.

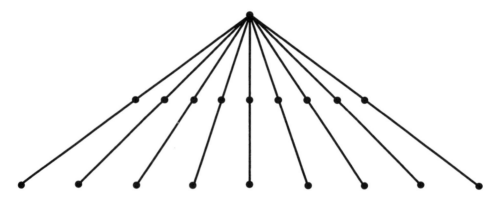

Figure 6: Fan of scenarios for multiperiod two-stage problem

Except for the two special cases mentioned above, to build a representative scenario tree seems to be presently the crucial problem for applications. It can be approached from the point of view of a suitable data manipulation, it should reflect both the underlying probability assumptions and the existing data, and be linked with the purpose of the application. It often asks for compromises between a manageable problem size and the desired precision of the results. In this context, the important problems are possibilities of trimming the tree, designing strategies for refining the tree, or tests of the influence of including additional scenarios and stages; one of the tools is the contamination technique introduced briefly in 5.3.6.

Number of nodes of a scenario tree grows exponencially with the number of stages. Therefore, an alternative data arrangement may be considered, obtained for instance by relaxation of the requirement of unique predecessors at the previous stages. For example, the Black-Derman-Toy binomial lattice, see Figure 4, assumes a special *recombining* property of the data paths and consequently, the number of nodes grows linearly in the number of time periods.

5.4.1 From Data Paths to a Scenario Tree

We shall focus now on generation of scenario trees for the cases when the main random factors have been detected and enough data paths of their realizations can be generated in accordance with a parametric or nonparametric model, see 5.2.1 – 5.2.7. In general, these data paths do not display any nonanticipativity property and do not necessarily follow the time partition imposed by the stages of the stochastic programming formulation. They are used to generate the coefficients of the solved problem, e.g. the matrices $(\mathbf{c}_t, \mathbf{A}_t, \mathbf{B}_{t\tau}, \mathbf{b}_t, \mathbf{l}_t, \mathbf{u}_t)$, $t = 1, \ldots, T$, in stochastic linear program (8)–(9) in Chapter II.2. For example, industrial project evaluation, as well as asset and liability management will often include an interest rate process as one of random elements. Long term pension fund and insurance models also have to take into account the inflation process. Whereas the data process can be modeled solely according to theoretical and empirical assumptions, the coefficient process for scenario-based multistage stochastic programs is by construction a discrete time path-dependent process defined mostly as a deterministic transformation of the data process.

Hence, the first important step is to delineate the initial structure of the scenario tree, i.e., the horizon, the number and the character of stages and the branching scheme. The stages are characterized by the possibility to take additional decisions based on a *newly released information*. Such information can be obtained at a specific date (expiration of an option), every day, week, month, quarter, year, etc. The horizon of the model can be relatively large when measured in these time units whereas for computations, it is impossible to use a scenario tree with too many stages; the majority of contemporary real-life models use 4 – 10 stages. The stages do not necessarily correspond to time periods of an equal length. Typically, the first stage relates to a relatively short time period whereas the last one may cover several years.

For initial screening studies the degree of aggregation of possible future outcomes, which results in number of branches from the individual nodes, is quite high. It can be verbally described as distinction of "high" and "low" or "up" and "down" for branching into two descending nodes, "high", "medium", "low" or "dry", "medium", "wet" for branching into three descending nodes. Another strategy is to use an extensive branching from the root leading to relatively many, say 10, nodes on level 1, use a modest branching from the nodes at the middle of the tree and a relatively poor branching, e.g., into two descendants for the nodes at the last levels of the tree. There are some hints concerning the minimal number of descendants which come from *problem specific requirements*, such as necessity to build a model without arbitrage opportunities or to fit some moments of the probability distribution; see 5.4.7 and 5.4.8.

5.4.2   Clustering

Generating scenarios by procedures which do not account for the tree structure of input data requires additional steps to build a scenario tree of a prescribed structure. This has been done often by ad hoc crude methods, by cutting and pasting the data paths in a more or less intuitive way. Another possibility is to apply *cluster analysis*. It is easy to cluster according to the first component (or subvector) ω_1 of ω and to continue by clustering according to the second components (subvectors) ω_2 of the objects included into the created clusters, etc. To treat properly the interstage dependences, consider instead a *multi-level clustering* scheme which exploits the whole sequences of observed/simulated data $(\omega_1, \ldots, \omega_T)$:

- For each pair of scenarios $\omega^s, \omega^{s'}$ evaluate a suitable dissimilarity measure, e.g.,

$$d(\omega^s, \omega^{s'}) = \sum_{t=1}^{T} w_t \|\omega_t^s - \omega_t^{s'}\|$$

where $w_t \geq 0$ are suitably chosen nonincreasing positive weights. This allows us to give emphasis on differences at the beginning of the sequence.

- Measures of dissimilarity among the compared objects are used in definitions of the standard measures of dissimilarity of clusters and used subsequently in the cluster analysis approaches; see e.g. [74]. The result is $K_2 - 1$ clusters, $C_1^2, \ldots, C_1^{K_2}$ represented by $\tilde{\omega}_{k_2}, k = 2, \ldots, K_2$; these can be the mean values or modal values of the first components ω_1^s of the scenarios ω^s included into the relevant cluster. Probabilities of $\tilde{\omega}_{k_2}, k = 2, \ldots, K_2$, equal the sum of probabilities of the individual ω^s belonging to the respective cluster.

- The clustering procedure continues for each cluster C_1^k separately, starting with the second component ω_2 of the observations included into C_1^k, or equivalently, with the first component ω_1 replaced by $\tilde{\omega}_{k_2}$ and so on.

Assume now that a scenario tree of a *prescribed structure* has been created. We may think of the scenarios as of representatives of certain regions which cover Ω. A natural question is: Could the approximation of the initial probability distribution by the discrete probability distribution corresponding to this scenario tree be improved, i.e., is there a better representation given the prescribed scenario tree structure? Such iterative procedure related with the ideas of cluster analysis is suggested in [124].

5.4.3   Sampling Methods

Development of special sampling methods for creating scenario trees instead of sampling individual scenarios first is another possibility. *Markov structure of data,* cf. the vector autoregressive model of the first order (1), can be exploited for conditional sampling of scenarios in a way which takes into account the already created structure of the tree. In this case, ω_t depends only on the preceding component ω_{t-1} and on an additional random vector ε_t which is independent of the history $\omega^{t-1,\bullet}$:

$$(9) \qquad\qquad \omega_t = \mathbf{P}\omega_{t-1} + \varepsilon_t.$$

Contrary to (1), the transition matrix $\mathbf{P}$ may depend on t. Interstage independence can be regarded as a special form of the Markov structure (9) with $\mathbf{P}$ the zero matrix.

The Markov property (9) allows for a direct sampling from the probability distribution of ε_t at each node which corresponds to an already obtained realization of ω_{t-1}. Another possibility is to discretize the probability distribution of ε_t at a given number of points and add the obtained realizations to the already known past values of $\mathbf{P}\omega_{t-1}$. The arc probabilities are fixed according to the used discretization method.

5.4.4   Example. In the BONDS model 4.2 in Chapter II.4, data on one-year yields $u(\tau)$ on good grade municipal bonds of maturities $\tau = 1, 2, 5, 10, 20$ and 30 years were available for a period of the past seventeen years. The yield curves of the specified parametric form

$$(10) \qquad\qquad u(\tau) = \theta\tau^\beta e^{\gamma\tau},$$

where τ is the time to maturity, were considered. According to the analysis of the results, the authors decided to determine the three parameters of the yield curve (10) using only the one-year, 20-year and 30-year yields and to approximate the remaining maturities by points on the fitted curve.

Monthly time series were used to determine the probability distribution of the one-year changes in the one-year rates and the changes in the 20- and 30-year rates were forecast conditionally on the one-year rate using two separate regression models. The two regression equations were then used as deterministic functions to obtain the one-year changes in the long rates from the estimated one-year changes of the one-year rates.

Scenarios correspond to a discrete approximation of the obtained (normal) distribution of the one-year changes of the one-year rates concentrated at three points – the expected value and the expected value ± standard deviation. The parameters of the future yield curves are fitted for each branch of the obtained scenario tree using the three maturities. The approximated future yield curve was used to get the input coefficients of the designed model of the type (11)–(12) from Section II.2.2.

Comparing this procedure with the general hints, the three stages correspond to the planning horizon of three years, the chosen structure of the scenario tree, 3^3, was partly given by the limitations of the computational technology twenty years ago. Another reported structure used four stages covering two years by periods of

3 months, 3 months, 6 months and one year which reflects the increasing uncertainty about events more distant in the future. For the same reason, a five point approximation of the probability distribution was applied for the first two levels and a three point approximation for the remaining two levels of the scenario tree: hence, the total $2^5 2^3$ scenarios.

As the main random factor, the one-year *changes* of the one-year rates were selected, their probability distribution fitted from historical data, discretized and used to generate the input coefficients. For testing the performance of the model on historical data, only the information from the past seven years was used. The first stage decision was accepted, and the model rolled over for one year forward, i.e., the earliest year of the used data was dropped and the most recent year of data was added.

In spite of many simplifications, the optimal BONDS portfolio outperformed (for the same history of interest rates) the classical investment strategies including the barbell portfolios based on immunization and the laddered portfolios which use a forced diversification.

5.4.5   Sequential Importance Sampling-Based Scenario Generation

The recent sequential importance sampling method [35] elaborates further techniques based on the Markov structure of data. It takes into account a given suitably labeled tree structure already in the course of simulation. Moreover, the sequential procedure can be adopted for an iterative refinement of the discrete representation of the underlying continuous data process.

The framework is based on the definition of a *scenario tree nodal partition matrix* $\mathbf{M} = \{m(s,t)\}$ that uniquely identifies the structure of the associated scenario tree. The matrix, with the number of rows equal to the number of scenarios and the number of columns equal to the number of stages, provides the necessary labeling scheme for the specification of the index sets $\mathcal{K}_t$, $t = 2,\ldots T$, introduced in Section II.2.2. In the iterative procedure, the matrix is an input to the *conditional* scenario generator, and an output from the sampling algorithm. A balanced $3^1 2^1 1^1$ tree structure would be described in the matrix as

$$\mathbf{M} = \begin{pmatrix} 1 & 2 & 5 & 11 \\ 1 & 2 & 6 & 12 \\ 1 & 3 & 7 & 13 \\ 1 & 3 & 8 & 14 \\ 1 & 4 & 9 & 15 \\ 1 & 4 & 10 & 16 \end{pmatrix}$$

where, row-wise, every scenario is explicitly labeled. The matrix can clearly accommodate any tree configuration in a straightforward way.

The sequential sampling algorithm is designed as an iterative procedure. It requires initially the specification of the number of stages, T, the maximum number of possible iterations, a stopping criterion and the initial scenario tree structure described by the associated nodal partition matrix $\mathbf{M}$. At every iteration, some version of the stochastic program, such as (11)–(12) in Section II.2.2, is specified and solved, the nodal values of the importance sampling criterion are evaluated along the tree and a new tree structure is defined through an update of the nodal partition matrix.

The scenario generation procedure is general, and independent of both the mathematical characterization of the random data process and the adopted sampling criterion. The most frequently used models for the data process cover the vector autoregressive models, see 5.2.3, the random walk models - with or without drift - with fine time discretization for short term decision problems and Gaussian noise, see (3), and binomial or trinomial models, e.g., the Black-Derman-Toy model 5.2.4.

5.4.6 Problem Oriented Requirements

When building the scenario tree one tries to avoid as much as possible any distortion of the available input information. Moreover, the goal of this procedure does not reduce to an approximation of the probability distribution P but rather to creating an input which provides applicable solutions of the real-life problem. This means, inter alia, that problem oriented requirements should be respected. The motivation comes from various problem areas.

• Scenarios based solely on past observations may ignore possible time trends or exogenous knowledge or expectations of the user; see Example 5.3.6.2. Moreover, scenarios coming from historical data need not be directly applicable. See [126] for a scenario tuning procedure which helps to discover the information hidden in indirect measurement results contained in past records on specific metal melting processes.

• In financial applications, one prefers that scenario-based estimates of future asset prices in a portfolio optimization model do not allow arbitrage opportunities; this may put additional requirements on scenario selection.

Explicitly formulated additional requirements concerning properties of the probability distribution may help. They can be made concrete through a suitable massaging of the data to obtain the *prescribed moments values*, given a fixed tree structure. This idea has appeared for instance in [27] where at the given stage of a multistage stochastic program, the observed data were grouped and scaled to retain the prescribed values of expectations and variances. One of the reasons was the sought possibility of comparisons with the Markowitz mean-variance model.

We shall follow [80] who suggest to build the scenario tree in such a way that some of statistical properties of the data process are retained, for instance, there are specified expectations, correlation matrices and skewness of the marginal probability distributions of ω_t.

5.4.7 Why Matching Moments?

The question of a possible representation of probability distributions by (infinite) sequences of moments and approximating them using only a few moments goes back to Chebyshev and is connected with the moment problem. Moreover, it is possible to prove that given m admissible values of moments, there exists a *discrete* probability distribution with these moments and its support has at most $m + 2$ points. We refer to Chapter 5 of [130] for a brief introduction and selected results in this direction. For our purposes it means that given values of certain moments or expectations of continuous functions, say, $\mu_k = \int_\Omega g_k(\omega)P(d\omega), k = 1, \ldots, m,$

there exists a modest number of scenarios $w^s, s = 1, \ldots, S$, and their probabilities $p^s, s = 1, \ldots, S, \sum_s p^s = 1$ so that the moment values are retained, i.e.,

$$(11) \qquad \sum_s p^s g_k(w^s) = \mu_k, \ k = 1, \ldots, m.$$

To get the scenarios and their probabilities means to find a solution w^s and $p^s, s = 1, \ldots, S$, of the system (11) extended for nonnegativity conditions on probabilities and for the additional constraint $\sum_s p^s = 1$. This is a highly nonlinear numerical problem.

The system of equations (11) can be further extended for other constraints on selection of scenarios to represent certain strata, to cover extremal cases, etc.

5.4.8 Fitting Moment Values

For simplicity assume that $w = (w_1, w_2)$ is a two-dimensional random vector with the first three moments $\mu_k(1), \mu_k(2), k = 1, 2, 3$ of the marginal probability distributions and with the covariance ρ of their joint probability distribution. To cover an important extremal case, we require in addition that for at least one scenario, $w_1^s \geq l_1, w_2^s \geq l_2$ holds true. Let the discrete two-dimensional probability distribution which matches the true one be carried by S atoms $w^s = (w_1^s, w_2^s), s = 1, \ldots, S$, with probabilities $p^s \geq 0 \forall s, \sum_s p^s = 1$. Hence, we search values of pairs $(w_1^s, w_2^s), s = 1, \ldots, S$, and scalars $p^s, s = 1, \ldots, S$, such that

$$\sum_{s=1}^{S} p^s (w_1^s)^k = \mu_k(1) \text{ for } k = 1, 2, 3$$

$$\sum_{s=1}^{S} p^s (w_2^s)^k = \mu_k(2) \text{ for } k = 1, 2, 3$$

$$\sum_{s=1}^{S} p^s (w_1^s - \mu_1(1))(w_2^s - \mu_1(2)) = \rho$$

$$w_1^1 \geq l_1, w_2^1 \geq l_2$$

$$p^s \geq 0, s = 1, \ldots, S, \ \sum_s p^s = 1.$$

For S large enough and for consistent moments' values, this nonlinear system has a solution. For a small number of scenarios or for inconsistent moment values the existence of solution is not guaranteed. Still an almost feasible solution can be found by the goal programming technique, see II.3.1.10. This means that scenarios w^s and probabilities p^s can be obtained for instance by solving a weighted least squares minimization problem

minimize

$$(12) \qquad \sum_{k=1}^{3} \alpha_k \left(\sum_{s=1}^{S} p^s (w_1^s)^k - \mu_k(1) \right)^2 + \sum_{k=1}^{3} \beta_k \left(\sum_{s=1}^{S} p^s (w_2^s)^k - \mu_k(2) \right)^2$$

$$+ \gamma \left(\sum_{s=1}^{S} p^s (w_1^s - \mu_1(1))(w_2^s - \mu_1(2)) - \rho \right)^2$$

subject to

$$\omega_1^1 \geq l_1, \omega_2^1 \geq l_2$$

$$p^s \geq 0, s = 1, \ldots, S, \sum_s p^s = 1.$$

From the optimization point of view, problem (12) is non-convex and may have many local minima. Nevertheless, the advantage of this formulation is that the optimal value is zero if the data is consistent and S is large enough, but that the optimal solution is also a good representation of data in the case of inconsistency; recall the goal programming approach introduced in II.3.1.10. The parameters α, β and γ can be used to reflect importance and quality of data.

Inconsistency can appear if the information about moments comes from different sources, if implicit specifications are inconsistent with explicit ones, etc. Consider for instance a problem which covers two periods. Let us specify the variance of ω_1 and the variance of the sum $\omega_1 + \omega_2$. But specifying these two variances, we have said something about the correlation over time. If we now explicitly specify correlations over time, we are likely to end up with two inconsistent specifications of the same entity.

There is a numerical evidence in favor of performance of stochastic programs based on scenario trees with moment values fitted at each node over those based only on a few randomly sampled realizations. Moreover, taking into account the wish to approximate well the expectation of function $f_0(\mathbf{x}, \omega)$ which appears in the objective function of the stochastic program (4), it is possible to search for extremal scenarios, the atoms of the worst or best discrete probability distributions which fulfill the moment conditions (11). These probability distributions appear in the lower or upper bounds for the expected value of $f_0(\mathbf{x}, \omega)$ at a point $\mathbf{x}$ or in the bounds for the optimal value (6). They are valid for all probability distributions with the given moment values, hence also for the true probability distribution. Their construction requires that certain convexity properties of $f_0(\mathbf{x}, \bullet)$ hold true.

5.4.9 Example – The Edmundson-Madansky Bound.

Let Ω be a bounded convex polyhedron in $\mathbb{R}^q$, $\Omega = \mathrm{conv}\{\omega^1, \ldots, \omega^K\}$, $\omega^k \in \mathbb{R}^q \; \forall k$, and assume that for all $\mathbf{x} \in \mathcal{X}$, $f_0(\mathbf{x}, \bullet)$ is a convex function of ω on Ω. Let the probability distribution P on Ω be described only by the condition $E\omega = \mu$, with $\mu \in \Omega$ (consistence condition); denote the family of such probability distributions by $\mathcal{P}$. For an arbitrary fixed $\mathbf{x} \in \mathcal{X}$ consider a linear function $h(\omega) = a + \mathbf{b}^\top \omega$ such that

$$h(\omega) \geq f_0(\mathbf{x}, \omega) \; \forall \omega \in \Omega.$$

Evidently, $Eh(\omega) = a + \mathbf{b}^\top E\omega \geq Ef_0(\mathbf{x}, \omega)$ for all probability distributions $P \in \mathcal{P}$. Moreover, thanks to the assumed convexity property, the above infinite dimensional system of inequalities can be reduced to the finite dimensional system of constraints on coefficients $a, \mathbf{b}$:

(13) $$a + \mathbf{b}^\top \omega^k \geq f_0(\mathbf{x}, \omega^k), k = 1, \ldots, K$$

Hence, the sharpest upper bound for $Ef_0(\mathbf{x}, \omega)$, the Edmundson-Madansky bound (cf. [110]), can be obtained by solving the *linear program*

(14)
$$\min_{a, \mathbf{b}} a + \mathbf{b}^{\top} \boldsymbol{\mu}$$

subject to (13). Notice also that the optimal solutions of the dual linear program to (13), (14)

$$\max_{p^1, \ldots, p^K} \sum_{k=1}^{K} p^k f_0(\mathbf{x}, \omega^k)$$

subject to

$$\sum_{k=1}^{K} p^k = 1, \ \sum_{k=1}^{K} p^k \omega^k = \boldsymbol{\mu}, \ p^k \geq 0 \, \forall k$$

identify fully the probability distribution for which the expected value $Ef_0(\mathbf{x}, \omega)$ attains its maximum with respect to all probability distributions $P \in \mathcal{P}$: by duality, there is a worst case probability distribution carried by at most $q+1$ extremal points ω^k with positive probabilities p^k which solve the dual problem.

In the special case of Ω a nondegenerated closed interval $\Omega = [\omega^1, \omega^2] \subset \mathbb{R}^1$ and $E\omega = \mu \in (\omega^1, \omega^2)$, there is a unique feasible solution of the dual problem,

$$p^{1*} = \frac{\omega^2 - \mu}{\omega^2 - \omega^1}, \quad p^{2*} = 1 - p^{1*}$$

and $p^{1*} f_0(\mathbf{x}, \omega^1) + p^{2*} f_0(\mathbf{x}, \omega^2)$ is the tight upper bound for $F(\mathbf{x}, P) = Ef_0(\mathbf{x}, \omega)$ for all probability distributions $P \in \mathcal{P}$. In this case the corresponding worst case probability distribution $P^* \in \mathcal{P}$ does not depend on $f_0(\mathbf{x}, \bullet)$. For evaluation of the minimax bounds (6), this is an attractive property:

$$\min_{\mathbf{x} \in \mathcal{X}} F(\mathbf{x}, P^*) \geq \varphi(P) \geq \min_{\mathbf{x} \in \mathcal{X}} f_0(\mathbf{x}, \mu) \, \forall P \in \mathcal{P}$$

where the upper bound is the optimal value of a scenario based stochastic program and the lower bound is the optimal value of the deterministic *expected value* problem.

This bounding technique can be modified to Ω – a Cartesian product of simplices, i.e., of bounded convex polyhedra whose points have uniquely determined barycentric coordinates q^k in their representation via extremal points. If $\Omega = \text{conv}\{\omega^1, \ldots, \omega^K\}$ is a simplex, then there is a unique solution $q^k, k = 1, \ldots, K$, of $\omega = \sum_{k=1}^{K} q^k \omega^k, \ \sum_{k=1}^{K} q^k = 1, \ q^k \geq 0 \, \forall k$ for an arbitrary $\omega \in \Omega$. An example is Ω a closed interval in $\mathbb{R}^1$ considered above.

For all of these results, *convexity of the random objective function* $f_0(\mathbf{x}, \omega)$ *with respect to* ω plays an essential role; consider Exercise II.2.1.1 once more.

5.4.10 Conclusions

We have presented several techniques for generating scenario trees for multistage stochastic programs of a general structure.

The common starting point for all approaches is the selected or prescribed number of stages and their allocation. The branching scheme may be developed sequentially. Otherwise, the approaches are tailored to the available input information:

• In case of an external scenario generator, cluster analysis can help to build the scenario tree.

• When using a well calibrated stochastic model or its time discretization, the tree structure can be built within an importance sampling procedure.

• In cases of a low information level and for specific requirements we suggest to build the tree in such a way that the known relevant information (e.g., some prescribed moments of the marginal and conditional probability distributions) is recovered.

Still, it is impossible to provide a general recipe for generating scenario trees. All introduced methods leave a space for the method, problem, data, solver, and computer specific considerations. Within the steps of each method, there are choices to be made. For example, we must choose a clustering technique, an importance sampling criterion, or weights and distances used in various goal programming versions of (12).

Particular requirements of real-life applications are for instance the no-arbitrage property, the consistency with past records and reflection of an exogenous knowledge of experts or a foresight. The last of these specific requirements can be treated by the contamination method.

II.6 CASE STUDY: BOND PORTFOLIO MANAGEMENT PROBLEM

scenario-based multiperiod two-stage stochastic program, stability properties, scenario generation, data, numerical results, out-of-sample scenarios, errors due to estimation

6.1 The Problem and the Input Data

The problem considered here is preserving the value of a bond portfolio of a risk averse or risk neutral institutional investor over time. This is a problem of allocation and management of resources, not of trading. It may include additional features, e.g., presence of fixed or uncertain external inflows or outflows in the future or a required balance between assets and liabilities. There are various options concerning the choice of an appropriate model, starting with deterministic duration based immunization models or dedicated bond portfolio management models formulated in Chapter I.6, through their simple stochastic analogs, e.g. Example II.1.2, or tracking models introduced in Chapter II.3, up to multistage stochastic programming models discussed in Chapter II.4.

Why not to rely on the duration based immunization models? A good answer is the following quotation, cf. [90]:

> "Many years ago, bonds were boring. Returns were small and steady. Fixed income risk monitoring consisted in watching duration and avoiding low qualities. But as interest-rate volatility has increased and the variety of fixed income instruments has grown, both opportunities and dangers have flourished..."

Yield curves are not flat, do not move in a parallel way, the interest rates are not constant and an investment in long maturity bonds requires an active trading strategy. These are the reasons that has led us to an exploitation of multiperiod stochastic programs with the main random element to be included - the evolution of the short interest rate over time.

Given a sequence of equilibrium future short term interest rates r_t valid for the time interval $(t, t+1], t = 0, \ldots, T-1$, the fair price of the j-th bond at time t just after the coupon was paid equals the total cash flow $f_{j\tau}, \tau = t+1, \ldots, T$, generated by this bond in subsequent time instances discounted to t:

$$(1) \qquad P_{jt}(\mathbf{r}) = \sum_{\tau=t+1}^{T} f_{j\tau} \prod_{h=t}^{\tau-1} (1 + r_h)^{-1}$$

where T is greater or equal to the time to maturity.

However, the considered time points need not coincide with the dates of coupon payments. Also the sequence of the future short term rates r_h that determines the prices (1) is not known precisely, but prescribed ad hoc or modeled in a probabilistic way. The cash flows $f_{j\tau}$ need not be known with certainty; this is for instance the case of indexed bonds, bonds with options or default. Hence, the formula (1) should be extended for the accrued interest A_{jt} and revised to take into account the effect of options and other risks related with the j-th bond. The resulting selling and purchasing prices do reflect also the transaction costs and the bid/ask spread.

Assume for simplicity that the only random factor which influences the fair prices is the evolution of short term interest rates. This is then the *data process* assumed to be governed by a discrete probability distribution P of T-dimensional vector ρ of the short rates $\rho_t, t = 0, \ldots, T - 1$, where r_0 (the rate valid in the first period) is supposed to be known. The possible trajectories of ρ (scenarios) are indexed as $\rho^s, s = 1, \ldots, S$, with probabilities $p^s > 0, s = 1, \ldots, S, \sum_s p^s = 1$.

The bond portfolio management problem concerns the Italian bond market which, prior to the European Monetary Union, was the fourth largest fixed-income market in the world. The government fixed-income securities represented more than 85% of this market and they included zero-coupon bonds of maturities up to 2 years, coupon bonds without option (the so called BTPs), with different maturities (3, 5, 10 and 30 years) issued two times per month through a marginal auction without minimal price, floater bonds, there used to be puttable bonds (CTOs) etc. There are futures and options on some of BTPs and also bonds with maturities between 3 and 30 years issued by corporations. See Table 1 for BTPs traded on September 1, 1994, their maturities and yields. The yields are obtained from the quoted prices and the accrued interests. The size and liquidity of the market provides a sound basis for application of various models of interest rates and for their calibration; its liquidity can be taken for granted when designing models which admit rebalancing strategies.

Table 1. Government bonds traded on Sept. 1, 1994

Bond	Maturity	Yield	Bond	Maturity	Yield
BTP12669	16/06/1997	.105156	BTP12673	01/11/1997	.105118
BTP12674	01/01/1996	.097577	BTP12675	01/01/1998	.105953
BTP12676	01/03/1996	.099117	BTP12677	01/03/2001	.106883
BTP12678	19/03/1998	.106756	BTP12679	01/06/2001	.109037
BTP12680	01/06/1996	.104325	BTP12681	20/06/1998	.107101
BTP12682	01/09/1996	.102685	BTP12683	01/09/2001	.108558
BTP12684	18/09/1998	.107772	BTP12685	01/11/1996	.104167
BTP12686	01/01/1997	.104281	BTP12687	01/01/2002	.108846
BTP12688	17/01/1999	.108441	BTP36605	01/05/2002	.109130
BTP36606	01/05/1997	.106852	BTP36607	18/05/1999	.108290
BTP36613	01/09/1997	.102081	BTP36614	01/09/2002	.107825
BTP36615	01/10/1995	.091232	BTP36621	01/01/1996	.098457
BTP36622	01/01/1998	.106039	BTP36623	01/01/2003	.107671
BTP36630	01/03/1996	.100311	BTP36631	01/03/1998	.105644
BTP36632	01/03/2003	.108137	BTP36634	01/05/1996	.103588
BTP36635	01/05/1998	.105841	BTP36640	01/06/1996	.101091
BTP36641	01/06/1998	.106830	BTP36642	01/06/2003	.107617
BTP36649	01/08/1996	.099297	BTP36650	01/08/1998	.103564
BTP36651	01/08/2003	.105266	BTP36658	01/10/1996	.101362
BTP36659	01/10/1998	.104957	BTP36660	01/10/2003	.104684
BTP36665	01/11/2023	.108862	BTP36674	01/01/1997	.104849
BTP36675	01/01/1999	.107896	BTP36676	01/01/2004	.106509
BTP36682	01/04/1997	.105092	BTP36683	01/04/1999	.108172
BTP36684	01/04/2004	.106617	BTP36691	01/08/1997	.105096
BTP36692	01/08/1999	.107886	BTP36693	01/08/2004	.106280

6.2   The Model and the Structure of the Program

The model designed below is a two-stage multiperiod stochastic program, its formulation allows for the possibility of intertemporal *rebalancing* the portfolio. Let

$j = 1, \ldots, J$ be indices of the bonds and T_j the dates of their maturities, $T = \max_j T_j$;

$t = 0, \ldots, T_0$ the discretization of the planning horizon;

$b_j \geq 0$ the initial holdings of bond j (in lots);

b_0 the initial holding in riskless asset (in cash);

f_{jt}^s cash flow generated under scenario s from the unit quantity of bond j at time t;

ξ_{jt}^s and ζ_{jt}^s are the selling and purchasing prices of bond j at time t for scenario s obtained from the corresponding fair prices (1) adding the accrued interest A_{jt}^s and subtracting or adding scenario independent transaction costs and spread; the initial prices ξ_{j0} and ζ_{j0} are known constants, i.e., scenario independent;

L_t is an external cash flow at time t, $L_t > 0$ corresponds to an outflow due to liabilities, $L_t < 0$ stays for external inflows;

x_j / y_j are quantities of bond j purchased / sold at the beginning of the planning period, i.e., at $t = 0$;

z_{j0} is the quantity of bond j held in portfolio after the initial decisions x_j, y_j have been made.

All first-stage decision variables x_j, y_j, z_{j0} are nonnegative and subject to conservation of holdings,

$$(2) \qquad y_j + z_{j0} = b_j + x_j \quad j = 1, \ldots, J$$

and

$$(3) \qquad y_0^+ + \sum_j \zeta_{j0} x_j = b_0 + \sum_j \xi_{j0} y_j$$

where the nonnegative variable y_0^+ denotes the surplus. Notice that in the first stage, no borrowing is permitted. This limitation together with the assumed *positive market value of the initial portfolio*, $b_0 + \sum_j \xi_{j0} b_j > 0$ implies that *the set of the feasible first-stage solutions is nonempty and bounded.* The same property holds true also in case that a *restricted borrowing* possibility in the first stage is permitted.

The second-stage decisions on rebalancing the portfolio, borrowing or reinvestment of the surplus depend on individual scenarios. They have to fulfill constraints on conservation of holdings in each bond at each time period and for each of scenarios

$$(4) \qquad z_{jt}^s + y_{jt}^s = z_{j,t-1}^s + x_{jt}^s \quad \forall j, s, \text{ and } t = 1, \ldots, T_0$$

where $x_{jt}^s, y_{jt}^s, z_{jt}^s$ denote the quantities of bond j purchased, sold, held in the portfolio at time $t, t = 1, \ldots, T_0$, under scenario s, and constraints on rebalancing the portfolio at each time period $t = 1, \ldots, T_0$

(5) $\quad \sum_j \xi_{jt}^s y_{jt}^s + \sum_j f_{jt}^s z_{j,t-1}^s + (1 - \delta_1 + \rho_{t-1}^s) y_{t-1}^{+s} + y_t^{-s} =$

$$L_t + \sum_j \zeta_{jt}^s x_{jt}^s + (1 + \delta_2 + \rho_{t-1}^s) y_{t-1}^{-s} + y_t^{+s} \quad \forall s, t$$

with $y_0^{-s} = 0$, $y_0^{+s} = y_0^+$, $z_{j0}^+ = z_{j0} \; \forall j$. The variables y_t^{+s}, y_t^{-s} describe the (unlimited) lending/borrowing possibilities for period t under scenario s and the spreads δ_1, δ_2 are model parameters to be fixed. Non-zero values of δ_1 account for the difference between the returns for bonds and for cash. Assume that $\delta_2 > 0$, i.e., there is a positive cost of borrowing.

The optimization problem is maximizing the expected utility of the final wealth at the horizon T_0

(6) $$\sum_s p^s U(w_{T_0}^s)$$

subject to constraints (2)–(5) and nonnegativity constraints on all variables, with

(7) $$w_{T_0}^s = \sum_j \xi_{jT_0}^s z_{jT_0}^s + y_{T_0}^{+s} - \alpha y_{T_0}^{-s} \quad \forall s.$$

The multiplier $\alpha \geq 1$ is fixed according to the problem area. For instance, α may be scenario dependent and values $\alpha^s > 1$ take into account the debt service in the future. In case of preservation of portfolio value (with no liabilities considered) or for an investment project terminating at time T_0, an arbitrarily large value of α plays a role of a penalty for borrowing at the end of the accounting or planning period.

Because of the possibility of reinvestment and of unlimited borrowing, the problem has always a feasible solution. It is a *multiperiod two-stage stochastic programming model with random relatively complete recourse* and with nonlinearities in the objective function. The existence of optimal solutions is guaranteed for a large class of utility functions that are *increasing and concave* which will be assumed henceforth. Moreover, for strict inequalities $\xi_{j0} < \zeta_{j0} \; \forall j$, $\xi_{jt}^s < \zeta_{jt}^s \; \forall j, t, s$ and $\delta_1 \geq 0, \delta_2 > 0$, the *optimal* solutions satisfy

$$y_j x_j = 0 \;\; \forall j, \quad y_{jt}^s x_{jt}^s = 0 \;\; \forall s, j, t, \quad y_t^{+s} y_t^{-s} = 0 \;\; \forall s, t.$$

It means that *at optimality there is no unnecessary trading and borrowing*, which is a natural property.

We obtain a large scale deterministic program with a concave objective function and numerous linear constraints. The size and the numerical values of the coefficients of the program result from the application and the available data: the choice of bonds, their characteristics (initial prices and future cash flows) and initial holdings, from the scheduled stream of liabilities, transaction costs and spread and from *the way how scenarios of future interest rates are generated and sampled*. The main outcome is the optimal value of the objective function (the maximal expected utility of the final wealth) and the optimal values of the first-stage variables x_j, y_j (and y_0^+, z_{j0}) for all j. In a dynamic setting, this decision is applied and, at the end of the first period, the model is solved again for the changed input information on holdings and on scenarios of interest rates; this is the rolling forward technique.

Assume that the portfolio consists of *default free, liquid bonds with maturities* $T_j > T_0 \forall j$, *all cash flows are after tax, the transaction costs and bid/ask spreads are constant.*

The following reformulation of the problem is useful for stability and postoptimality analysis. Assume that an initial trading strategy determined by scenario independent first-stage decision variables x_j, y_j, y_0^+ (and z_{j0}) for all j has been accepted, then the subsequent scenario dependent decisions have to be made in an optimal way regarding the goal of the model. It means that given the values of the first-stage variables y_0^+ and $\mathbf{x}, \mathbf{y}, \mathbf{z_0}$ with components $x_j, y_j, z_{j0} \forall j$, the required maximal contribution of the portfolio management under the s-th scenario to the value of the objective function is obtained as the value of the utility function computed for the maximal value of the wealth $w_{T_0}^s$ attainable for the s-th scenario under the constraints of the model, i.e., the utility of the optimal value $w_{T_0}^s$ of the linear program

$$\text{maximize } w_{T_0}^s$$

subject to

(8) $$z_{jt}^s + y_{jt}^s = z_{j,t-1}^s + x_{jt}^s \quad \forall j, t$$

(9) $$\sum_j \xi_{jt}^s y_{jt}^s + \sum_j f_{jt}^s z_{j,t-1}^s + (1 - \delta_1 + \rho_{t-1}^s)y_{t-1}^{+s} + y_t^{-s} =$$

$$L_t + \sum_j \zeta_{jt}^s x_{jt}^s + (1 + \delta_2 + \rho_{t-1}^s)y_{t-1}^{-s} + y_t^{+s}, \quad t = 1, \ldots, T_0,$$

(10) $$x_{jt}^s \geq 0, \ y_{jt}^s \geq 0, \ z_{j,t}^s \geq 0, \ y_t^{-s} \geq 0, \ y_t^{+s} \geq 0 \quad \forall j, t,$$

$$\text{with } y_0^{-s} = 0, \ y_0^{+s} = y_0^+, \ z_{j0}^s = z_{j0} \ \forall j \text{ and with}$$

(11) $$w_{T_0}^s = \sum_j \xi_{jT_0}^s z_{jT_0}^s + y_{T_0}^{+s} - \alpha y_{T_0}^{-s}.$$

Denote the corresponding *maximal value* by $w_{T_0}^*(\boldsymbol{\rho}^s; \mathbf{x}, \mathbf{y}, \mathbf{z_0}, y_0^+)$ and rewrite the program (2)–(7) as

(12) $$\text{maximize } \sum_{s=1}^S p^s U(w_{T_0}^*(\boldsymbol{\rho}^s; \mathbf{x}, \mathbf{y}, \mathbf{z_0}, y_0^+))$$

subject to nonnegativity constraints and subject to (2)–(3). Scenarios enter now only the objective function (12).

The remaining part of this section reviews results on the stability of the problem (2)–(7), or (2), (3), (12), which are *independent of the way how the scenarios of interest rates were generated and selected.*

6.2.1 Out-of-Sample Scenarios

Assume that the stochastic program (2)–(7) has been solved for a fixed set of scenarios $\rho^s, s = 1, \ldots, S$, and that the influence of including other out-of-sample scenarios should be considered. Such problem can be related to the "what-if" analysis, to various stability and sensitivity studies, to incorporating investors' views, etc. One could rewrite the program (2)–(7) for the extended set of scenarios, with additional variables and additional constraints of the type (4), (5), (7) and solve it. Another possibility is to use the form (2), (3), (12) whose set of feasible solutions is not influenced by inclusion of additional scenarios. The additional scenarios appear only in the objective function which is an expected value of the utility of the final wealth under a discrete probability distribution carried by a finite number of scenarios. Being an expected value, the objective function (12) is linear in the probability distribution.

Denote by P the initial probability distribution carried by S interest rates scenarios indexed as $s = 1, \ldots, S$, with probabilities $p^s > 0 \, \forall s, \sum_s p^s = 1$. Let $\varphi(P)$ be the optimal value of (7) and $\mathbf{x}(P), \mathbf{y}(P), \mathbf{z}_0(P), y_0^+(P)$ be an optimal first-stage solution. For simplicity, assume that the optimal first-stage solution is unique.

Inclusion of other out-of-sample scenarios means to consider another discrete probability distribution which is carried by the extended set of scenarios. Such distributions can be modeled as a convex mixture of two discrete probability distributions: P that is carried by the initial scenarios indexed by $s = 1, \ldots, S$, with probabilities $p^s > 0, \sum_s p^s = 1$ and Q carried by the out-of-sample scenarios indexed by $\sigma = 1, \ldots, S'$, with probabilities $\pi^\sigma > 0, \sum_\sigma \pi^\sigma = 1$. The weights of the two probability distributions are given by the contamination parameter λ and the contaminated distribution

$$P_\lambda = (1 - \lambda)P + \lambda Q, \quad 0 \leq \lambda \leq 1$$

is carried by the *pooled sample* of $S + S'$ scenarios that occur with probabilities $(1 - \lambda)p^1, \ldots, (1 - \lambda)p^S, \lambda\pi^1, \ldots, \lambda\pi^{S'}$. For instance, if both P and Q are carried by *equiprobable* scenarios and this is required also for the pooled sample, we get $\lambda = \frac{S'}{S+S'}$. Notice that for fixed probability distributions P and Q, the objective function (12) which corresponds to the contaminated distribution P_λ is a *linear* function of λ.

The bounds for the optimal value $\varphi(P_\lambda)$ of the problem based on the pooled sample of $S + S'$ scenarios follow according to (8) in II.5.3.6:

$$(13) \quad (1 - \lambda)\varphi(P) + \lambda \sum_\sigma \pi^\sigma U\left(w_{T_0}^*(\rho^\sigma; \mathbf{x}(P), \mathbf{y}(P), \mathbf{z}_0(P), y_0^+(P))\right)$$

$$\leq \varphi(P_\lambda) \leq (1 - \lambda)\varphi(P) + \lambda\varphi(Q), \, 0 \leq \lambda \leq 1.$$

The additional numerical effort consists in solving the stochastic program

$$(14) \qquad \text{maximize} \quad \sum_\sigma \pi^\sigma U(w_{T_0}^*(\rho^\sigma; \mathbf{x}, \mathbf{y}, \mathbf{z}_0, y_0^+))$$

subject to (2)–(3) and nonnegativity constraints for the distribution Q carried by S' out-of-sample scenarios to obtain $\varphi(Q)$ and in evaluation and averaging the S'

function values $U(w_{T_0}^*(\boldsymbol{\rho}^\sigma; \mathbf{x}(P), \mathbf{y}(P), \mathbf{z}_0(P), y_0^+(P)))$ for the new scenarios at the already obtained optimal first-stage solution; these are in fact the main numerical indicators which appear in various simulation studies of the portfolio performance under out-of-sample scenarios.

Similarly, one can approximate the optimal value $\varphi(P_\lambda)$ using an optimal solution $\mathbf{x}(Q), \mathbf{y}(Q), \mathbf{z}_0(Q), y_0^+(Q)$ and the optimal value $\varphi(Q)$ of the stochastic program with the objective function (14) based on the alternative probability distribution Q:

$$\lambda\varphi(Q) + (1-\lambda)\sum_s p^s U\left(w_{T_0}^*(\boldsymbol{\rho}^s; \mathbf{x}(Q), \mathbf{y}(Q), \mathbf{z}_0(Q), y_0^+(Q))\right)$$

$$\le \varphi(P_\lambda) \le (1-\lambda)\varphi(P) + \lambda\varphi(Q), \, 0 \le \lambda \le 1$$

so that

$$(15) \quad \max\left\{(1-\lambda)\varphi(P) + \lambda\sum_\sigma \pi^\sigma U\left(w_{T_0}^*(\boldsymbol{\rho}^\sigma; \mathbf{x}(P), \mathbf{y}(P), \mathbf{z}_0(P), y_0^+(P))\right);\right.$$

$$\left.\lambda\varphi(Q) + (1-\lambda)\sum_s p^s U\left(w_{T_0}^*(\boldsymbol{\rho}^s; \mathbf{x}(Q), \mathbf{y}(Q), \mathbf{z}_0(Q), y_0^+(Q))\right)\right\}$$

$$\le \varphi(P_\lambda) \le (1-\lambda)\varphi(P) + \lambda\varphi(Q), \, 0 \le \lambda \le 1;$$

see Figure 7.

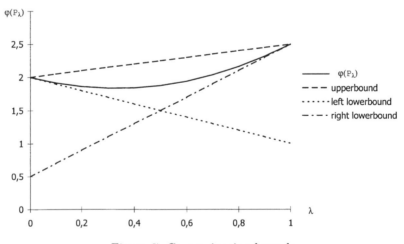

Figure 7. Contamination bounds

6.2.2   Stability Results

We shall discuss now the stability properties of the stochastic program (2)–(7) with respect to changes in the numerical values of its coefficients. These changes are consequences of changes in numerical values of the components ρ_{jt}^s of the selected S scenarios of interest rates.

It is easy to prove that for an arbitrary fixed scenario ρ^s, the optimal value of
(8)–(11), $w^*_{T_0}(\rho^s; \mathbf{x}, \mathbf{y}, \mathbf{z}_0, y_0^+)$ is concave, piecewise linear in the first-stage decision
variables. (They appear only on the right hand sides of the linear program (8)–(11).)
Moreover, it is possible to prove, that for an arbitrary fixed scenario ρ^s and for an
arbitrary feasible first-stage decision $\mathbf{x}, \mathbf{y}, \mathbf{z}_0, y_0^+$ the scenario subproblems (8)–(11)
are *stable linear programs* in the sense of [132] provided that $0 < \xi^s_{jt} < \zeta^s_{jt} \, \forall j, t$ and
$\alpha > 1, \delta_2 > 0$. This implies, inter alia, that the sets of optimal solutions of the pairs
of the dual scenario subproblems are nonempty and bounded and that the optimal
value functions $w^*_{T_0}(\rho^s; \mathbf{x}, \mathbf{y}, \mathbf{z}_0, y_0^+)$ are *jointly* continuous in $\rho^s, \mathbf{x}, \mathbf{y}, \mathbf{z}_0, y_0^+$.

Continuity properties with respect to scenarios, their probabilities and with respect to the first-stage decision variables apply also to the objective function (2) and
the expectations $\sum_{s=1}^{S} p^s w^*_{T_0}(\rho^s; \mathbf{x}, \mathbf{y}, \mathbf{z}_0, y_0^+)$ and $\sum_{s=1}^{S} p^s U(w^*_{T_0}(\rho^s; \mathbf{x}, \mathbf{y}, \mathbf{z}_0, y_0^+))$
are concave with respect to the first-stage decision variables for an arbitrary nonde-
creasing concave utility function U. This means (recall that the set of the feasible
first-stage decisions is nonempty and bounded) that also the optimal value function
of the full problem (12), (2), (3) is continuous with respect to the input parameters
$\rho^s, p^s, s = 1, \ldots, S$ and a certain continuity property (upper semicontinuity in the
sense of Berge) holds true also for the sets of the first-stage optimal solutions; see
e.g. [5].

These results imply that *small* errors in evaluation of scenarios of interest rates,
of their probabilities and consequently of prices ξ^s_{jt}, ζ^s_{jt} cause only small changes
to the best available scenario-based market values $w^*_{T_0}(\rho^s; \mathbf{x}, \mathbf{y}, \mathbf{z}_0, y_0^+)$ and also to
the optimal value of the overall performance function (12). However, (contrary to
the postoptimality with respect to additional scenarios) there does not yet exist
any general numerically tractable method to *quantify* these errors. This suggests
to turn the attention to simulation studies. To provide well interpretable results,
these simulation studies have to be tailored to the way the scenarios have been
generated and selected.

6.3   Generation of Scenarios

Our primal goal is to get a sensible investment strategy for the considered bond
portfolio management problem. We expect that this investment strategy depends
on the way the underlying data process has been approximated. In our analysis of
sensitivity of the optimal value of (6) and of the optimal first-stage solution with
respect to the selected scenarios of interest rates we can rely on the theoretical
stability results presented in 6.2.2.

In general, scenarios can be obtained by discretization of a true continuous probability distribution, from a model calibrated by market data, from historical observations and in principle, they can be also fixed ad hoc using experts' forecasts, see
Chapter II.5. The chosen model of the short rates has to be calibrated so that it fits
the market data reasonably well; this can be formulated as a requirement to price
precisely some of traded financial documents, e.g., the fixed coupon government
bonds.

In this study, we use interest rate scenarios sampled from the binomial lattice constructed according to the Black-Derman-Toy model, see II.5.2.4. The data
from the Italian bond market considered here gave a solid base for its applications,

whereas interest rate scenarios for thin emerging markets can hardly be based on historical data or on the estimation techniques discussed below.

The obtained one-period short term rates valid for scenario s and for the time interval $(l, l+1]$ are

$$(16) \qquad \rho_l^s = r_{l0} k_l^{i_l(s)}, \quad i_l(s) = \sum_{\tau=1}^{l} \omega_\tau^s.$$

They are determined by the base rates r_{l0}, lattice volatilities k_l and by the chosen sampling strategy which determines the exponent $\sum_{\tau=1}^{l} \omega_\tau^s$, i.e., the number of "up" moves in the time discretization points $\tau = 1, \ldots, l$. The theoretical binomial lattice consists of 2^{T-1} different paths or vectors of interest rates; their components are given by (16).

To *calibrate the Black-Derman-Toy model*, i.e., to get the base rates r_{l0} and the volatility factors or lattice volatilities $k_l \, \forall l$, means to use the yield and volatility curve related to yields to maturity of zero coupon government bonds of all maturities corresponding to the chosen time steps of the lattice. Such bonds are rare in the market and have to be replaced by synthetic zero coupon bonds whose yields correspond to yields of fixed coupon government bonds that do not contain any special provision such as call or put options.

Various numerical and statistical methods have been used to *fit or estimate the yield curve* from the existing market data on yields of fixed coupon government bonds at the given day. We have applied regression analysis to estimate and test the analytical form of the yield curve. Instead of yields one could use the corresponding prices of these bonds as the input. Regarding the assumption of homoskedasticity commonly present in regression models we decided to use yields.

Having tried different parametric nonlinear models as well as nonparametric ones, we chose to use a simple form of the yield curve applied in [24] – see I.3.8 and Example II.5.4.4:

$$g(t; \theta, \beta, \gamma) = \theta t^\beta e^{\gamma t}$$

We applied its linearized form to the logarithms of yields: For the market information consisting of yields $u_i, i = 1, \ldots, N$ of various fixed coupon government bonds (without option) characterized by their maturities T_i, the postulated model is

$$(17) \qquad \ln u_i = \ln \theta + \beta \ln T_i + \gamma T_i + \varepsilon_i, \quad i = 1, \ldots, N$$

where the random errors $\varepsilon_i, i = 1, \ldots, N$ are independent, normal $\mathcal{N}(0, \sigma^2)$. There is a good reason to accept the hypothesis of approximately normal errors in (17) which is in line with the assumed log-normal process of short rates approximated by the Black-Derman-Toy binomial lattice.

The yield curve for September 1, 1994 estimated according to the linearized Bradley and Crane model (12) is plotted on Figure 8. Naturally, the fit is sensitive to data. Compare the fit with and without the long bond BTP36665 maturing in 2023. (Maturities of all remaining bonds are less than 10 years.)

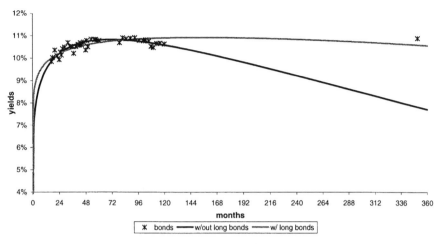

Figure 8. Yield curve on Sept. 1, 1994

The least squares estimates $\widehat{\ln\theta}, \hat{\beta}, \hat{\gamma}$ of parameters $\ln\theta, \beta, \gamma$ are approximately normal, with the mean values equal the true parameter values and the covariance matrix $\sigma^2\Sigma^{-1}$, $\Sigma = \mathbf{G}^\top\mathbf{G}$ where σ^2 is estimated by

$$(18) \qquad s^2 = \frac{1}{N-3} \min_{\ln\theta,\beta,\gamma} \sum_{i=1}^{N} (\ln u_i - \ln\theta - \beta\ln T_i - \gamma T_i)^2$$

and the $N \times 3$ matrix $\mathbf{G}$ consists of rows $(1, \ln T_i, T_i)$, $i = 1, \ldots, N$.

To estimate the yields of zero coupon bonds of all required maturities which are not directly observable, $\tilde{t} \neq T_i$, we replace the unobservable logarithm of yield by the corresponding value on the already estimated log-yield curve. Such estimates are subject to an additional error.

Assume that the logarithm of the yield $\tilde{u} = u(\tilde{t})$ for maturity $\tilde{t}$ is

$$\ln\tilde{u} = \ln\theta^* + \beta^*\ln\tilde{t} + \gamma^*\tilde{t} + \tilde{\varepsilon}$$

with $\tilde{\varepsilon}$ normal, independent of ε_i, $i = 1, \ldots, N$, $E\tilde{\varepsilon} = 0$, var $\tilde{\varepsilon} = \sigma^2$ and with the true parameter values denoted by asterisks. Then $\ln\tilde{u}$ is approximately normal,

$$(19) \qquad \ln\tilde{u} - \widehat{\ln\theta} - \hat{\beta}\ln\tilde{t} - \hat{\gamma}\tilde{t} \sim \mathcal{N}(0, \sigma^2(1 + Q^2(\tilde{t})))$$

where

$$(20) \qquad Q^2(t) := [1, \ln t, t]\Sigma^{-1}[1, \ln t, t]^\top.$$

The corresponding approximate $100(1-\alpha)\%$ confidence interval for the logarithm of yield $\ln\tilde{u}$ for a *fixed* maturity $\tilde{t} \neq T_i, i = 1, \ldots, N$ is

$$\ln g(\tilde{t}; \hat{\theta}, \hat{\beta}, \hat{\gamma}) \pm s(1 + Q^2(\tilde{t}))^{1/2}t_{N-3}(1 - \alpha/2)$$

and $t_{N-3}(1 - \alpha/2)$ is the corresponding quantile of the t-distribution with $N - 3$ degrees of freedom.

The techniques for obtaining *volatilities of yields* or *log-yields* are less obvious. There is not enough data for fitting the volatility curve by a regression model. Most of the authors work with an ad hoc fixed constant volatility; the volatility curve may be estimated from historical data or based on the Risk Metrics datasets which provide historical volatilities computed daily for several main maturities, 1 year, 2, 3, 4, 5, 7, 9, 10, 15, 20 and 30 years. It is suggested to estimate the missing yields by linear interpolation and to use the volatilities and correlations of the reported yields to compute the approximate values of yield volatilities for these nonincluded maturities.

In contrast to the volatility curves obtained independently on the yield curve model one could get *approximate* standard deviations of $\ln u(t)$ from the chosen parametric model of the yield curve provided that the errors in the applied regression model are normally distributed. For the linearized Bradley and Crane model (17) one can use directly the standard deviation which comes from (19)–(20).

Based on the obtained yield and volatility curves, the *calibration of the binomial lattice* in agreement with no-arbitrage valuation principles, recall I.5.2.1, follows by a numerical procedure suitable for solving the large system of nonlinear equations for the base rates r_{l0} and lattice volatilities k_l, $l = 1, \ldots, T$.

According to (16), the *fitted binomial lattice* provides different 2^{T-1} scenarios of interest rates identified by the binary fractions of $T - 1$ ones or zeroes, see II.5.2.4. A smaller, manageable number of scenarios has to be *selected* or *sampled* from this large set.

The *nonrandom sampling strategy* [177] is based on a uniform approximation of the expected utility of final wealth, computed with respect to the uniform distribution over the *full* set of the 2^{T-1} scenarios, by an expected value over a subset of these scenarios. Its simplified version can be described as follows: We fix $L, 1 < L < T$ and assign the index $s, s = 1, \ldots, 2^L$ to each possible binary fraction of length L. The sample point ω^s from $(0, 1)$ is determined by one of these binary fractions and by an *arbitrary continuation* up to binary fraction of length $T - 1$.

The components of selected S scenarios ρ^s, are then computed according to (16) using the *scenario independent* base rates r_{t0}, $t = 1, \ldots, T - 1$, volatilities k_t, $t = 1, \ldots, T - 1$, and the *scenario dependent* position on the lattice given by the exponent $i_t(s)$ in (16) which equals the number of "up" moves needed to reach the position on the lattice within t periods. The prices ξ_{jt}^s, ζ_{jt}^s and cash flows f_{jt}^s are evaluated along each of these scenarios and the problem (2)–(7) is solved. The main output is the optimal value - the maximal attainable expected utility of final wealth at time T_0 and an optimal first-stage solution, say $\mathbf{x}^*, \mathbf{y}^*, \mathbf{z}_0^*, y_0^{+*}$.

6.4   Selected Numerical Results

To simulate the behavior of a value preserving portfolio of fixed income securities on the Italian bond market we use the model described by (2)–(7) with monthly steps and for the time horizon of one year ($T_0 = 12$).

The initial term structure and the portfolio are related to September 1, 1994. It is composed of cash (500 mil. Liras) and of typical government bonds, paying semi-annual coupons and covering two year forward till 29 years maturities (BTPs) as well as puttable bonds (CTOs), paying semi-annual coupons with the maturity of 8 years and a possible exercise of the option in the 4th year or with the maturity of 6 years and an exercise at the 3rd year; see Table 2. The quantities (Qt) of bonds included into the initial portfolio are expressed in lots (hundreds) of million items so that the nominal value of the portfolio is 10500 mil. Liras, its initial value in *market prices* of September 1st, 1994 is $w = 10465.86$ mil. Liras. The coupon yields and the redemption prices are after tax.

Table 2. Initial portfolio composition, Sept. 1, 1994

Bonds	Qt	Coupon	payment dates	exercise	redemp.	maturity
BTP36658	10	3.9375	01Apr & 01Oct		100.187	01Oct96
BTP36631	20	5.0312	01Mar & 01Sep		99.531	01Mar98
BTP12687	15	5.2500	01Jan & 01Jul		99.231	01Jan02
BTP36693	10	3.7187	01Aug & 01Feb		99.387	01Aug04
BTP36665	5	3.9375	01May & 01Nov		99.218	01Nov23
CTO13212	20	5.2500	20Jan & 20Jul	20Jan95	100.000	20Jan98
CTO36608	20	5.2500	19May & 19Nov	19May95	99.950	19May98

In this application, liabilities are not considered, liquidity may be obtained from the interbank market at a rate greater than that one at which the surplus can be always reinvested. The additive transaction costs are fixed at $\pm .01$, $\delta_2 = .0016$, $\delta_1 = .0005$ and $\alpha = 1$. We shall report here only results obtained for the *linear utility function*.

To estimate the term structure of interest rates we used the linearized model (17) applied to the yields obtained by the market quotation of the BTPs on the relevant day, see Figure 8. Volatilities of log-yields were set equal to the standard deviations of the normal distribution in (19). The parameters r_{t0} and k_t of the binomial lattice, see (16), were computed with a monthly discretization along 5 years. To evaluate the prices of bonds with longer maturities, the computed interest rates have been kept constant for each scenario after the 5th year. This means that 2^{60} scenarios were at disposal. Alternative experiments were based on the full binomial lattice with 2^{350} scenarios.

The first sampling strategy was based on the Zenios and Shtilman approach [177] with different choices of $L = 3, 4, 5, 6$ in covering fully the beginning of the lattice, and proceeding with alternating up-down movements, which resulted accordingly into 8, 16, 32 and 64 scenarios; the acronyms are ZS(# of scenarios). It gives identical first-stage optimal solutions with slightly different optimal values for 8 and 32 scenarios and also for 16 and 64 scenarios.

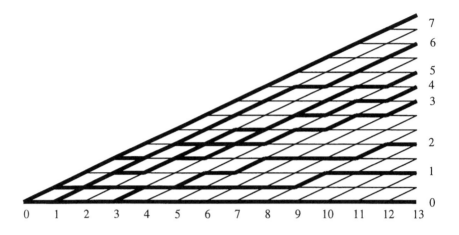

Figure 9: Particular sampling strategy

A better coverage of the lattice along the investment time horizon $(t = 1, ..., T_0)$ was achieved for a particular, ad hoc choice of 8 scenarios, as reported in Figure 9; the acronym is Part(8). A different first-stage optimal solution was obtained, the optimal value was close to the previous ones. Table 3 lists the results. They illustrate the influence of the sampling strategy on the first-stage optimal decisions and give a motivation for further sensitivity, postoptimality and simulation studies.

Table 3. First-stage optimal solutions

Portfolio	Initial	Part(8)	ZS(8)	ZS(16)	ZS(32)	ZS(64)
Cash	5	0	0	79.5	0	79.5
BTP36658	10	0	0	10	0	10
BTP36631	20	0	0	0	0	0
BTP12687	15	15	103.4	15	103.4	15
BTP36693	10	0	0	0	0	0
BTP36665	5	114.4	0	0	0	0
CTO13212	20	0	0	0	0	0
CTO36608	20	0	0	0	0	0
Optimal value		11499	11560	11472	11559	11470

6.5   "What if" Analysis

Assume that the models applied on the input side of the bond portfolio management problem have been fixed according to our past experience. In the context of the Black-Derman-Toy model of interest rates it means that the yield curve (17) has been accepted to get the term structure and a sampling strategy has been used to get a modest number of scenarios out of the fitted binomial lattice. Even in this case there are numerous sources of errors that influence the input of the large scale mathematical program (2)–(7):

• The market data of the given day are used to fit the yield curve, i.e., to estimate the coefficients in the chosen nonlinear regression model and to estimate the yields or prices of zero coupon government bonds of all required maturities $t = 1, \ldots, T$. In addition, a plausible hypothesis about volatility of these yields (i.e., about standard deviations of log-yields) is needed. The estimated prices or yields of zero coupon government bonds of all maturities together with their volatilities are called the initial *term structure*. Evidently, both statistical and numerical errors enter the initial term structure.

• The Newton-Raphson method is used to fit the base rates and lattice volatilities of the Black-Derman-Toy model in accordance with the term structure. It requires the solution of a system of $2T$ nonlinear equations which can be done in several ways. Additional errors which stem from the chosen numerical procedure seem to be of minor importance than errors due to estimation of the yield curve and, namely, due to more or less ad hoc assessment of the volatility curve.

• There is an indeterminacy as to the choice and number of scenarios to be used.

The final task is the solution of the large mathematical program (2)–(7) whose coefficients are burdened by errors of various kinds. The question is the sensitivity of the optimal first-stage decision (the first-period trading strategy) and of the optimal value of the objective function on the above mentioned errors.

The form of the fitted interest rates, see (16), allows us to separate the influence of the input data and of methods used for the lattice calibration from the impact of the chosen sampling strategy. Hence, we may concentrate separately on the *influence of the out-of-sample scenarios* and on an *analysis of errors in the estimated term structure and their effect on the results*.

6.5.1 Influence of Out-of-Sample Scenarios

Instead of the 2^{350} scenarios which may be theoretically obtained from the already fitted *full* binomial lattice we select 2^L scenarios, $L \ll 350$, following the Zenios and Shtilman sampling strategy: Fix the number of periods L for which all possibilities (choices of zeros and ones on the first L positions) are fully covered. The remaining digits necessary to complete the full length paths may be selected for example according to Table 4. For $l > T_0 + 1$, the components w_l^s alternate regularly up and down (1 or 0) starting with the indicated value of $w_{T_0+1}^s$.

Table 4. The sampling strategies

case	L	$l = L + 1$	$l = L + 2, \ldots, T_0$	$l = T_0 + 1$
B1	3	$w_{L+1}^s = 0$	$w_l^s = 0$	$w_{T_0+1}^s = 1$
B2	3	$w_{L+1}^s = 0$	$w_l^s = 0$	$w_{T_0+1}^s = 0$
B3	3	$w_{L+1}^s = 0$	$w_l^s = 1$	$w_{T_0+1}^s = 1$
B4	3	$w_{L+1}^s = 1$	$w_l^s = 1$	$w_{T_0+1}^s = 1$
C4	4	$w_{L+1}^s = 1$	$w_l^s = 1$	$w_{T_0+1}^s = 1$

The numerical result reported in Figure 10 records the optimal values and optimal initial strategies for two alternative cases based on distributions P and Q
carried by scenarios selected according to B3 and B4, respectively.

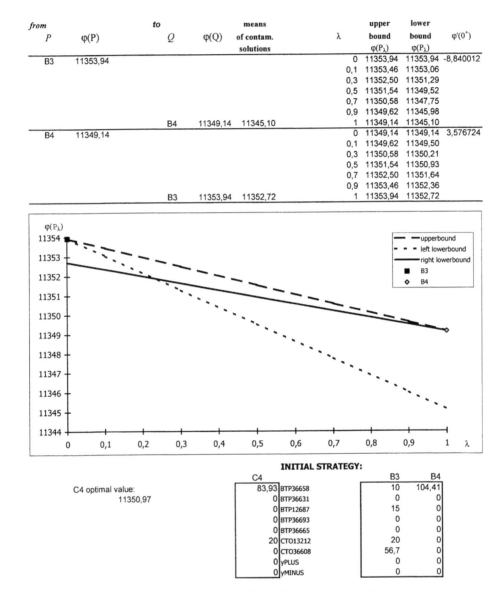

from P	$\varphi(P)$	*to* Q	$\varphi(Q)$	means of contam. solutions	λ	upper bound $\varphi(P_\lambda)$	lower bound $\varphi(P_\lambda)$	$\varphi'(0^+)$
B3	11353,94				0	11353,94	11353,94	-8,840012
					0,1	11353,46	11353,06	
					0,3	11352,50	11351,29	
					0,5	11351,54	11349,52	
					0,7	11350,58	11347,75	
					0,9	11349,62	11345,98	
		B4	11349,14	11345,10	1	11349,14	11345,10	
B4	11349,14				0	11349,14	11349,14	3,576724
					0,1	11349,62	11349,50	
					0,3	11350,58	11350,21	
					0,5	11351,54	11350,93	
					0,7	11352,50	11351,64	
					0,9	11353,46	11352,36	
		B3	11353,94	11352,72	1	11353,94	11352,72	

INITIAL STRATEGY:

C4 optimal value: 11350,97	C4		B3	B4
	83,93	BTP36658	10	104,41
	0	BTP36631	0	0
	0	BTP12687	15	0
	0	BTP36693	0	0
	0	BTP36665	0	0
	20	CTO13212	20	0
	0	CTO36608	56,7	0
	0	yPLUS	0	0
	0	yMINUS	0	0

Figure 10: Contamination bounds and optimal decisions

Figure 10 contains also values of directional derivatives and of the lower and
upper bounds computed according to (13) for distinct values of λ and a graphical
representations of these bounds. Expectations

$$\sum_\sigma \pi^\sigma U(w_{T_0}^*(\rho^\sigma; \mathbf{x}(P), \mathbf{y}(P), \mathbf{z}(P), y_0^+(P))), \quad \sum_s p^s U(w_{T_0}^*(\rho^s; \mathbf{x}(Q), \mathbf{y}(Q), \mathbf{z}(Q), y_0^+(Q)))$$

are listed under headings "means of contam. solutions".

These results illustrate a possible application of the contamination bounds for supporting decisions concerning the required number of scenarios (i.e., concerning the value of L in Table 4): For $\lambda = .5$, this example gives the interval [11350.93, 11351.54] for the optimal value of C4 based on 2^4 scenarios – a pooled sample of scenarios from the beds B3 and B4. Exploitation of the more complicated bound (15) helped to increase the original lower bound 11349.52 obtained according to (13). (Indeed, the optimal value for C4 computed directly is 11350.97.) It means that using the double number of scenarios does not increase essentially the precision of the obtained approximate of the true optimal value for the full hypothetical problem which would be based on all 2^{350} possible scenarios of interest rates. However, the listed optimal first-stage investment strategies are quite different.

A similar experiment can be run for the pooled sample of two randomly chosen beds of scenarios. Bounds for the optimal value based on a pooled sample carried by $S + S'$ scenarios, are obtained for $\lambda = \frac{S'}{S+S'}$.

We considered also the the case B2-shifted with the rates based on the case B2 perturbed by the fixed additive shift of $-.000355$ (which corresponds to the shift of 5% of the current B2 rates). Such choice of the couple of scenario beds allows to analyze the influence of the fixed additive shift of rates on the optimal value. For the shift of -.000355, both the left lower bound and the upper bound obtained for contamination of B2-shifted by B2 are very tight, hence, the optimal final wealth is untouched by this perturbation of the input interest rates.

6.5.2   Errors due to Estimation

The results summarized in the context of estimating the yield curve by parametric regression, cf. (17)–(20), provide a basis for simulation of log-yields at individual points t which are needed for fitting the binomial lattice provided that the *volatility curve is not subject to any perturbations*:

(i) At each point t of the discretization of the time horizon generate the random error e by sampling from the normal distribution $\mathcal{N}(0, \sigma^2(1 + Q^2(t)))$; the corresponding simulated log-yield at the given time instant t is $\ln u = \ln g(t; \hat{\theta}, \hat{\beta}, \hat{\gamma}) + e$. Let $\mathbf{e}$ be the vector of the independent normally distributed components e obtained in the described way.

(ii) For each vector of log-yields obtained according to (i) get the vector of simulated yields $\mathbf{u}$, fit the lattice and evaluate the interest rates ρ_t^s, prices P_{jt}^s, ξ_{jt}^s, ζ_{jt}^s and cash flows f_{jt}^s.

To guarantee that the numerical method used to fit the Black-Derman-Toy lattice gave reasonable interest rates when applied to the perturbed yield curves, the range of perturbations of the initial yield curve was reduced: standard deviations equal $h10^{-2}\sigma[1 + Q^2(t)]^{1/2}$ with $0 < h \le 1$ were used instead of those obtained from (19)–(20).

By repeated soluving of the scenario based programs (2)–(7) for various sets of coefficients obtained by the simulation procedure (i), (ii) one gets repeated "observations" of the optimal value and of the optimal initial trading strategy which allows to construct empirical distribution of the maximal expected utility of the final wealth, a useful information for subsequent, sample-based statistical inference, and to classify the considered bonds.

Notice that the specific parametric form of the yield curve, such as (17), is not essential for the simulation experiments delineated above. They can be applied whenever there is a sound basis for assuming *random errors in the model input*; the examples are other regression models and/or other assumed distribution of errors and also random sampling procedures for selection of scenarios.

The simulation experiment was run for the problem based on the 8 particular scenarios selected from the binomial lattice fitted up to the horizon of 5 years and with $K = 100$. The obtained different optimal first-stage strategies together with the respective optimal values φ^k are listed in Table 5 (the first column lists the result for the unperturbed yield curve), a survey on how the considered bonds are distributed with respect to the strategies of selling, holding and buying, is displayed in Figure 11. Hence, there seem to be only a few typical optimal strategies which can be verbally described for instance as "cash only", or "long bond only", and which are far from being similar. The reason of such differences is that the short rates obtained for the slightly perturbed yield curves differ essentially; for interest rates we found differences up to 80% of values obtained for the unperturbed yield curve. These results indicate that, under the fitted yield curve, *the longest bond is a dominant investment* as it appears in the most of perturbed cases. This is in agreement with our earlier observations; in comparison with the assumed behavior of short term interest rates, the long bond is underpriced by the market.

Table 5. Selected results of the simulation experiment

	Unperturbed								
Cash	0	104.6	0	94.6	0	0	89.4	79.5	0
BTP36658	0	0	0	10	105	0	0	10	10
BTP36631	0	0	0	0	0	0	0	0	0
BTP12687	15	0	0	0	0	100	15	15	89.7
BTP36693	0	0	0	0	0	0	0	0	0
BTP36665	114.4	0	133.8	0	0	5	0	5	5
CTO13212	0	0	0	0	0	0	0	0	0
CTO36608	0	0	0	0	0	0	0	0	0
Opt. value	11499	11478	11910	11481	11473	11485	11487	11490	11498
B&H value	11344	11171	11383	11124	11029	11339	11343	11350	11346

For 100 simulations, the mean of the (empirical) optimal values φ^k is 13 041 with the standard deviation of 2 479, for 1500 simulations this changes to the mean value 12 911 with the standard deviation of 2 461. The resulting 2σ confidence intervals around the true optimal value are relatively large. The distribution of these empirical optimal values is skewed and nonnormal.

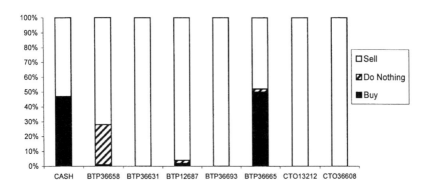

	CASH	BTP36658	BTP36631	BTP12687	BTP36693	BTP36665	CTO13212	CTO36608
Buy	47	1	0	2	0	50	0	0
Do Nothing	0	27	0	2	0	2	0	0
Sell	53	72	100	96	100	48	100	100

Figure 11. Trading strategies

Table 5 provides also a comparison of the performance of expected values for the Buy & Hold (B&H) strategy at the horizon T_0. The expected values are computed again for the same 100 simulations of the particular 8 scenarios selected according to Figure 9 which are used in the simulated stochastic programs, and their evaluation takes into account expiration of the CTOs and the cash flows due in the considered time period of September 1, 1994 – August 31, 1995. Comparing the statistics for the expected values of the Buy & Hold strategies with that for optimal values for the 100 simulated stochastic programs indicates that the distribution of the optimal stochastic programming values is shifted to higher function values, is nonsymmetrical and provides possibilities of rather large values.

In addition, sensitivity of the obtained "candidate" solutions listed in Table 2 on perturbations of the yield curve was tested using bounds constructed along the lines of Exercise II.5.3.3.1. The optimal solution obtained for the Part(8) sampling strategy proved to be the most robust one.

6.6   Discussion

We have applied a very simple model, based on one factor only. Similarly as in the continuous time models of bond prices with one factor, cf. [81] or [164], the bonds of the same risk class are equivalent to each other in the terms of their fair prices; some deviations appear due to transaction costs and due to differences of the initial market prices and the theoretical ones. This exactly was the case of the long bond which was underpriced by the market.

For a more realistic model, one should use two or more random factors, for instance, a two-dimensional data process of the short and long rates or short rates and inflation. Inclusion of foreign bonds will add the exchange rate process, etc. Also the presence of liabilities or external cash flows add new sources of uncertainties; see II.7.3.

It is possible to consider further constraints, such as *integrality of some decision variables, limited borrowing, restrictions on certain investments* or *duration matching*, to use a nonlinear utility function, etc. Inclusion of other types of assets, e.g., stock or real estate, means a passage to an asset allocation or an asset and liability management within aggregated asset classes and to models based on the value of the investments; see discussions in II.4.4.

An essential generalization is in direction to a *multistage* version of the considered bond portfolio management problem. The theoretical results of 6.2.1 and 6.2.2 can be extended to this case and the generalizations mentioned above can be in principle built in. The problem is the generation of the scenario tree, see Section II.5.4, which includes fixing meaningful stages as the first step.

II.7 INCOMPLETE INPUT INFORMATION

sensitivity to estimated parameters (volatility in Black-Scholes formula, expected returns in Markowitz model), incomplete information about liabilities

We shall present now selected approaches suitable for analysis of results of financial decision models solved under *uncertainty about the stochastic input*. Even when these approaches seem to be cast for a specific model and/or specific assumptions, e.g., for sensitivity of the Black-Scholes formula for pricing European call options to volatilities in 7.1, or sensitivity to the assumed expected returns for the mean-variance Markowitz model in 7.2, or for incorporation of incomplete knowledge about distribution of future liabilities in 7.3, there is an open possibility to exploit them under other quite disparate circumstances. Recall that the contamination technique explained in II.5.3.6 and applied in II.6.5.1 can be also regarded as one of suitable methods for postoptimality analysis with respect to incomplete input information.

7.1 Sensitivity for the Black-Scholes Formula

Consider a European call option on a unit quantity of a given stock for the striking price K at the expiration date T. Assume that the stock price follows the geometric Brownian process

$$dS_t = aS_t + \sigma S_t dW_t, \, 0 < t \leq T$$

with given constants S_0, drift a, instantaneous annualized volatility σ and with the Wiener process W_t. No-arbitrage reasoning and the Itô formula are used to get the European call price c_t at time t as

$$c_t = S_t\Phi(d_1) - K\exp\{-r(T-t)\}\Phi(d_2),$$

where r is the riskless instantaneous annualized interest rate,

$$d_1 = \frac{1}{\sigma\sqrt{T-t}}\left[\ln\frac{S_t}{K} + (r + \frac{\sigma^2}{2})(T-t)\right],$$

$d_2 = d_1 - \sigma\sqrt{T-t}$ and Φ denotes the distribution function of the standard normal distribution; see I.5.2.5.3 and III.3.3.

The parameter values are hardly known precisely and various sensitivity measures have been introduced and applied to hedge against their changes, cf. [81]. We shall focus on sensitivity of c_t to volatility σ; to indicate it we include σ as a parameter in c_t. The value of σ can be calculated by comparing the formula for a similar traded option with its market price, which gives so called *implied volatility*, recall I.5.2.5.9, or it can be estimated. The sensitivity measure *Vega* introduced in I.5.2.5.7 as

$$\mathcal{V}_c(\sigma) = \frac{\partial c_t(\sigma)}{\partial\sigma} = S_t\varphi(d_1)\sqrt{T-t},$$

where φ denotes the density of the standard normal distribution, is a continuous function of σ.

Assume now that there is at disposal an asymptotically normal estimate s_ν of σ, so that $s_\nu \sim N(\sigma, v_\nu^2)$ with $v_\nu \to 0$ for $\nu \to \infty$. The results of [144], Chapter 3, briefly presented in II.5.3.4 imply that $c_t(s_\nu)$ is asymptotically normal

$$c_t(s_\nu) \sim N(c_t(\sigma), \mathcal{V}_c^2(\sigma)v_\nu^2);$$

compare with (5) in Chapter II.5. Moreover, the true unknown value σ in $\mathcal{V}_c(\sigma)$ can be replaced by its sample counterpart s_ν. As a result

$$\frac{c_t(s_\nu) - c_t(\sigma)}{\mathcal{V}_c(s_\nu)v_\nu} \sim N(0, 1),$$

hence, asymptotic confidence intervals for the option price $c_t(\sigma)$ can be easily computed.

7.2   Markowitz Mean-Variance Model

From the point of view of optimization, an application of Markowitz mean-variance model in selection of optimal portfolio of risky assets can be reduced to solution of the following parametric quadratic program, see (7) in II.3.2.1:

(1)   maximize $\lambda \mathbf{r}^\top \mathbf{x} - \mathbf{x}^\top \mathbf{V} \mathbf{x}$

on a given convex polyhedral set

(2) $\mathcal{X} = \left\{ \mathbf{x} \in \mathbb{R}^{+n} : \mathbf{A}\mathbf{x} \leq \mathbf{b} \right\},$

and with λ a nonnegative scalar parameter.

One assumes that the vector $\mathbf{r}$ of expected returns and the positive definite matrix $\mathbf{V}$ of their variances and covariances are known whereas λ can be chosen according to the investor's attitude towards risk.

The optimal solution $\mathbf{x}(\mathbf{r}, \mathbf{V}; \lambda)$ and the optimal value $\varphi(\mathbf{r}, \mathbf{V}; \lambda)$ of (1)–(2) depend on $\mathbf{r}, \mathbf{V}$ (and on the chosen value of λ, of course) and at the same time, one can hardly assume full knowledge of these input values. The impact of errors in expected returns, variances and covariances on the optimal return φ of the obtained portfolio was investigated, e.g., in [16] and [31]. Results of simulation studies indicate that errors in expected values are more important than those in the second order moments and that these errors influence essentially the composition of the resulting portfolio whereas the portfolio returns are less sensitive.

Inspired by the cited results we shall deal with sensitivity analysis of the optimal composition of the portfolio and of the optimal value of (1)–(2) on the input values of the expected returns $\mathbf{r}$ of the risky assets; we shall complement results based on parametric programming by *stochastic sensitivity analysis* delineated in II.5.3.4. The variance matrix $\mathbf{V}$ and the parameter λ will be kept fixed and they will not be indicated in our denotation of the optimal value and of the optimal solution of (1)–(2).

The set of feasible solutions $\mathcal{X}$ of the quadratic program (1)–(2) can be decomposed into finitely many relatively open facets (i.e., open with respect to the linear

manifold of the smallest dimension which contains them) that are identified by in-
dices of the active constraints; interior of $\mathcal{X}$ and vertices of $\mathcal{X}$ are special cases of
these facets. The parametric space $\mathbb{R}^n$ of vectors $\mathbf{p} := \lambda\mathbf{r}$ can be also decomposed
into finitely many disjoint stability sets linked with the facets by the requirement
that for all $\mathbf{p}$ belonging to a stability set, the optimal solutions $\mathbf{x}(\mathbf{p})$ of the quadratic program (1)–(2) lie in the same facet. It is possible to prove (see [5]) that
$\mathbf{x}(\mathbf{p})$ is continuous on $\mathbb{R}^n$, it is linear on each stability set and differentiable on its
interior. If, however, $\mathbf{p}$ belongs to the boundary of a stability set, $\mathbf{x}(\mathbf{p})$ looses the
differentiability property and is only directionally differentiable. The optimal value
function $\varphi(\mathbf{p})$ is a piecewise linear–quadratic convex function of $\mathbf{p}$. Thanks to the
assumption that $\mathbf{V}$ is positive definite, the optimal value function φ is differentiable
on the whole parameter space provided that the vectors of coefficients of the active
constraints of (2) are linearly independent; see e.g. [62], Theorem 2.4.5. These
results explain the observed cases of a stability of the optimal value and, at the
same time, of an extremal sensitivity of optimal solutions to small changes of the
vector $\mathbf{r}$ of expected returns: Whenever the initial value of $\mathbf{p} = \lambda\mathbf{r}$ belongs to the
boundary of a stability set, arbitrarily small changes in $\mathbf{r}$ can cause transition to
one of the neighboring stability sets. Hence, for each type of transition different
assets are included into portfolio and the composition of the optimal portfolio is
regarded unstable. At the same time, the change of the maximal value of (1) is
small for small changes of $\mathbf{r}$. Notice that similar situations can be observed also
in case of changes of the parameter λ (i.e., when tracing the mean-variance efficient frontier) but they are more easy to take in as the changes concern only a
scalar parameter. There exist some generalizations of the cited results to the case
of $\mathbf{V}$ positive semidefinite and bounded $\mathcal{X}$, however, the fact that from the point of
view of quadratic programming there might be multiple optimal solutions indicates
clearly the limitations.

7.2.1 Simple Clarifying Example.

Consider the quadratic program maximize
$p_1 x_1 + p_2 x_2 - \frac{1}{2}x_1^2 - x_1 x_2 - x_2^2$ on the set $\mathcal{X} = \{x_1, x_2 : x_1 \geq 0, x_2 \geq 0, x_1 + x_2 \leq 1\}$.

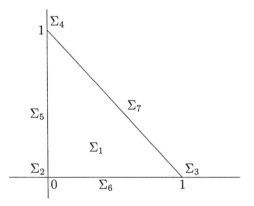

Figure 12: Set of feasible solutions $\mathcal{X}$

Set $\mathcal{X}$ can be decomposed into relatively open facets $\Sigma_1, \ldots, \Sigma_7$, see Figure 12.

The corresponding stability sets $\sigma(\Sigma_k)$, $k = 1, \ldots, 7$, for parameters p_1, p_2 are drawn on Figure 13.

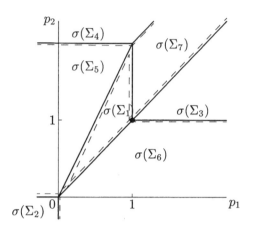

Figure 13: Stability sets

Consider $p_1 = p_2 = 1$. For this parameter value, the optimal solution is the vertex Σ_3, however, a small change of parameter values causes moving the optimal solution into the adjacent facets Σ_6 or Σ_7 or into the interior Σ_1 of $\mathcal{X}$. The corresponding changes of the optimal value and of the first component of the optimal solution are illustrated for fixed $p_2 = 1$ and $p_1 \geq 0$ on Figure 14.

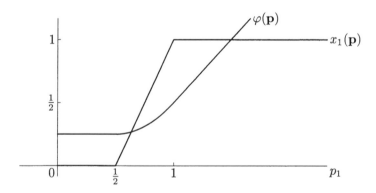

Figure 14: $\varphi(\mathbf{p})$ and $x_1(\mathbf{p})$ for $p_2 = 1$

Differentiability of φ is important for obtaining the form of the approximate probability distribution of the optimal returns of the portfolio, which are based on estimates of the true expected returns $\mathbf{r}$ obtained by a known appropriate statistical method. Such results are useful for constructing approximate confidence intervals for the true optimal value of (1)–(2).

7.2.2 Theorem. *Assume that* $\mathbf{V}$ *is positive definite and that the linear independence condition is fulfilled at the point of the true optimal solution* $\mathbf{x}(\mathbf{r})$ *of* (1)–(2). *Let* $\mathbf{r}_\nu$ *be an asymptotically normal estimate of the true expectation* $\mathbf{r}$,

$$(3) \qquad\qquad \sqrt{\nu}\,(\mathbf{r}_\nu - \mathbf{r}) \sim N\,(\mathbf{0}, \mathbf{\Sigma})\,.$$

Then the optimal values $\varphi(\mathbf{r}_\nu)$ *are asymptotically normal*

$$(4) \qquad\qquad \sqrt{\nu}\,(\varphi(\mathbf{r}_\nu) - \varphi(\mathbf{r})) \sim N\,\left(\mathbf{0}, \nabla\varphi(\mathbf{r})^\top \mathbf{\Sigma} \nabla\varphi(\mathbf{r})\right)\,.$$

This type of result was presented in II.5.3.4, formula (5), and the statistical technique behind is nothing else but asymptotic normality of differentiable functions of asymptotically normal vectors, so called δ-theorem; see e.g. [144], Chapter 3. The variance matrix of the asymptotic distribution (4) depends on the unknown expectation $\mathbf{r}$. However for ν large enough the gradients in the variance matrix of the distribution (4) can be evaluated at the estimated expectations $\mathbf{r}_\nu$ and also variance matrix $\mathbf{\Sigma}$ can be replaced by its sample counterpart $\mathbf{\Sigma}_\nu$. Well known results from parametric programming (e.g. [62]) imply that

$$\nabla\varphi(\mathbf{r}_\nu) = \nabla_r\,\left(\lambda\mathbf{r}_\nu^\top\mathbf{x} - \mathbf{x}^\top\mathbf{V}\mathbf{x}\right)$$

evaluated at the optimal solution $\mathbf{x}(\mathbf{r}_\nu)$, i.e.,

$$\nabla\varphi(\mathbf{r}_\nu) = \lambda\mathbf{x}(\mathbf{r}_\nu).$$

The final applicable result is

$$\sqrt{\nu}\,(\varphi(\mathbf{r}_\nu) - \varphi(\mathbf{r})) \sim N\,\left(\mathbf{0}, \lambda^2\mathbf{x}(\mathbf{r}_\nu)^\top\mathbf{\Sigma}_\nu\mathbf{x}(\mathbf{r}_\nu)\right)$$

where $\mathbf{\Sigma}_\nu$ is the sample counterpart of the variance matrix $\mathbf{\Sigma}$.

The precision of the point estimate $\varphi(\mathbf{r}_\nu)$ of $\varphi(\mathbf{r})$ based on the average returns $\mathbf{r}_\nu$ depends on the sample size ν and also on the value of λ.

7.2.3 Exercise. Assume that λ and $\mathbf{V}$ are fixed, that $\mathbf{r}_\nu$ is the arithmetic mean of ν observed independent vectors of assets returns and that the assumptions of the Markowitz model hold true (i.e., the observed returns come from populations characterized by fixed but unknown expected values $\mathbf{r}$ and a fixed known variance matrix $\mathbf{V}$). Apply Theorem 7.2.2 to get the asymptotic 95% confidence interval for the true optimal value $\varphi(\mathbf{r})$ of (1)–(2).

The lack of differentiability of the optimal solutions $\mathbf{x}(\mathbf{r})$ implies that a parallel asymptotic result for $\mathbf{x}(\mathbf{r}_\nu)$ cannot be in general expected: The asymptotic distribution of optimal solutions $\mathbf{x}(\mathbf{r}_\nu)$ is normal only if the true optimal solution is differentiable at the true parameter vector $\mathbf{r}$, i.e., if $\lambda\mathbf{r}$ belongs into the interior of a stability set.

7.3   Incomplete Information about Liabilities

As the next task we shall consider the bond portfolio management problem (2) – (7) treated in Chapter II.6. We assume that the interest rate scenarios have been already fixed and we shall turn our attention to liabilities. In II.6, the liabilities L_t were modeled as known amounts to be paid at the time t. Their full knowledge, however, need not be realistic for instance in management of pension funds, portfolios of insurance companies, etc. We shall assume now that liabilities are random, $\mathbf{L}(\eta)$, with η *independent of random interest rates*. The objective function (the expected utility of the final wealth) assumes the form

$$(5) \qquad E_\eta \sum_s p^s U(w^*_{T_0}(\boldsymbol{\rho}^s; \mathbf{x}, \mathbf{y}, \mathbf{z}, y_0^+; \mathbf{L}(\eta)))$$

where $w^*_{T_0}(\boldsymbol{\rho}^s; \mathbf{x}, \mathbf{y}, \mathbf{z}, y_0^+; \mathbf{L}(\eta)))$ denotes, similarly to (12) in II.6, the maximal contribution of portfolio management for already fixed first-stage decisions, for a given scenario $\boldsymbol{\rho}^s$ of interest rates and for a realization $\mathbf{L}(\eta)$ of liabilities, and is obtained as the optimal value of the corresponding linear program.

We shall further assume that the probability distribution of $\mathbf{L}(\eta)$ is not known completely. Using the available information, we shall try to get bounds on the optimal value of (5) subject to the first-stage constraints (2)–(3) from II.6 and to nonnegativity of all variables.

As the first step, it is easy to realize that $w^*_{T_0}(\boldsymbol{\rho}^s; \mathbf{x}, \mathbf{y}, \mathbf{z}, y_0^+; \mathbf{L}(\eta))$ in the objective function (5) are concave in the right-hand sides $L_t(\eta)$ when taken as parameters in evaluating the maximal final wealth under scenario $\boldsymbol{\rho}^s$ for fixed feasible first-stage variables $\mathbf{x}, \mathbf{y}, \mathbf{z}, y_0^+$. (This familiar property of the optimal value of a linear program in dependence on right-hand sides was exploited already in Exercise II.2.1.1 for a minimization problem.) For a concave increasing utility function U, also the individual terms $U(w^*_{T_0}(\boldsymbol{\rho}^s; \mathbf{x}, \mathbf{y}, \mathbf{z}, y_0^+; \mathbf{L}(\eta)))$ in expectation (5) are concave in the right-hand sides $L_t(\eta)$. It means that the Jensen inequality provides an upper bound for the objective function (5)

$$E_\eta \sum_s p^s U(w^*_{T_0}(\boldsymbol{\rho}^s; \mathbf{x}, \mathbf{y}, \mathbf{z}, y_0^+; \mathbf{L}(\eta))) \le \sum_s p^s U(w^*_{T_0}(\boldsymbol{\rho}^s; \mathbf{x}, \mathbf{y}, \mathbf{z}, y_0^+; E_\eta \mathbf{L}(\eta))).$$

The corresponding upper bound for the *optimal* value of (5) subject to constraints on the first-stage variables equals

$$\max \sum_s p^s U(w^*_{T_0}(\boldsymbol{\rho}^s; \mathbf{x}, \mathbf{y}, \mathbf{z}, y_0^+; E_\eta \mathbf{L}(\eta))).$$

Hence, *replacing the random liabilities by their expectations in the bond management problem leads to overestimating the maximal expected gain.*

The lower bound can be based on the Edmundson-Madansky inequality (see II.5.4.9) if, in addition, for each t the random liabilities $L_t(\eta)$ are known to belong to finite intervals, say $[L'_t, L''_t]$. The general bound, however, is computationally expensive unless the objective function is separable in individual liabilities, which is not our case. A trivial lower bound can be obtained by replacing all liabilities by their upper bounds L_t''; this bound will be rather loose. Another possibility is to assume a special structure of liabilities (their independence, a Markov property, etc.) in which case the lower bound can be simplified provided that the objective function remains concave with respect to the random variables used to model the liabilities. Hence, assume that

$$\mathbf{L}(\eta) = \mathbf{Ga}(\eta)$$

with a given matrix $\mathbf{G}$ of the size $T_0 \times I$ and $a_i(\eta)$, $i = 1, \ldots, I$, mutually independent random variables with known expectations $E_\eta a_i$ and known supports $[a_i', a_i'']$ $\forall i$. Accordingly, the individual objective functions

$$U(w_{T_0}^*(\rho^s; \mathbf{x}, \mathbf{y}, \mathbf{z}, y_0^+; \mathbf{L}(\eta))) = U(w_{T_0}^*(\rho^s; \mathbf{x}, \mathbf{y}, \mathbf{z}, y_0^+; \mathbf{Ga}(\eta)))$$
$$:= U^{s*}(\mathbf{x}, \mathbf{y}, \mathbf{z}, y_0^+; \mathbf{a}(\eta)))$$

are concave in $\mathbf{a}(\eta)$.

For I small enough, the following string of inequalities valid for each of scenarios ρ^s and for all feasible first-stage solutions may be useful:

$$(6) \quad E_\eta U^{s*}(\mathbf{x}, \mathbf{y}, \mathbf{z}, y_0^+; \mathbf{a}(\eta)) \geq \lambda_1 E_\eta U^{s*}(\mathbf{x}, \mathbf{y}, \mathbf{z}, y_0^+; a_1', a_2(\eta), \ldots, a_I(\eta)) \quad +$$
$$(1 - \lambda_1) E_\eta U^{s*}(\mathbf{x}, \mathbf{y}, \mathbf{z}, y_0^+; a_1'', a_2(\eta), \ldots, a_I(\eta)) \geq$$
$$\sum_{\mathcal{I} \subset \{1, \ldots, I\}} \prod_{i \in \mathcal{I}} \lambda_i \prod_{i \notin \mathcal{I}} (1 - \lambda_i) U^{s*}(\mathbf{x}, \mathbf{y}, \mathbf{z}, y_0^+; \mathbf{a}_\mathcal{I})$$

where the components of $\mathbf{a}_\mathcal{I}$ equal a_i' for $i \in \mathcal{I}$ and a_i'' for $i \notin \mathcal{I}$ and

$$\lambda_i = \frac{a_i'' - E_\eta a_i}{a_i'' - a_i'} \quad \forall i.$$

The lower bound for the maximal value of the objective function (5) based on inequalities (6) can be thus obtained by maximization of the weighted sum of the last terms in (6) for all scenarios subject to the first-stage constraints, i.e., by solving the corresponding stochastic program with $2^I S$ scenarios.

For instance for *pension funds* it is natural to assume that $\mathbf{G}$ is a lower triangular matrix: The liabilities L_1 to be paid at the end of the first period are known with certainty and their portion, say, δL_1 corresponds to unrepeated payments (e.g., final settlements or premiums) whereas the remaining main part of L_1 will be paid also in the subsequent period (continuing pensions). The liabilities $L_2(\eta)$ to be paid at the end of the period 2 can be modeled as

$$L_2(\eta) = (1 - \delta)L_1 + a_2(\eta),$$

etc. Moreover, it is possible to assume that $a_i(\eta)$ are mutually independent so that (6) is a valid and tight lower bound that applies in the case that the intervals $[a_i', a_i'']$ $\forall i$ and the expectations $E_\eta \mathbf{a}$ are known.

II.8 NUMERICAL TECHNIQUES AND AVAILABLE SOFTWARE*

illustrative examples, stochastic knapsack problem, EVPI, solution techniques for two-stage stochastic programs (L-shaped algorithm, progressive hedging algorithm) and their multistage versions, approximations, model management

8.1 Motivation

In this Chapter, we deal with solution techniques including algorithms and software. This is necessary since only very elementary educational stochastic programs can be solved without any computer support. In spite of this fact, we still begin with a simple example to present various modeling and solution principles in an instructive way.

8.1.1 Example. Consider a deterministic investment problem formulated as the knapsack problem:

$$\max\{\mathbf{c}^\top\mathbf{x} \ : \ \mathbf{a}^\top\mathbf{x} \leq b, \mathbf{x} \in \mathbb{R}^n, x_j = 0 \text{ or } 1 \ \forall j\}$$

where $\mathbf{c} \in \mathbb{R}^n$ contains yields on the unit investments, $\mathbf{a} \in \mathbb{R}^n$ the costs of realization of the investments, and b is the available budget. If the *constraints are relaxed* to $\mathbf{x} \in [0,1]^n$ the optimal policy is implied by the nonincreasing sequence of fractions $\frac{c_j}{a_j}$. Assume now that $\mathbf{a}, \mathbf{c}$ are fixed but the budget is not known at the moment of decision making and equals $b_0 + \omega$ where ω is a nonnegative random variable. Hence, the total yield $\mathbf{c}^\top\mathbf{x}$ is maximized over a set which depends on the random right-hand side. Several deterministic equivalents can be formulated and solved. We shall illustrate them on a simple numerical example

$$\max\{10x_1 + 15x_2 + 20x_3 \ : \ 5x_1 + 10x_2 + 20x_3 \leq 3 + \omega, 0 \leq x_j \leq 1, j = 1,2,3\}$$

with the probability distribution of ω specified as follows:

$$P(\omega = 0) = .2 = p^1, \ P(\omega = 9) = .3 = p^2, \ P(\omega = 22) = .5 = p^3.$$

Scenarios $\omega^1 = 0$, $\omega^2 = 9$, $\omega^3 = 22$ can be interpreted as the pessimistic one (no additional external cash flow), the standard one and the optimistic one, respectively. Notice that the intuitive decision rule does not depend on the budget and it means to include the first investment at the maximal possible level, then the second one and at the last place the third one.

The first possibility is to postpone the decision until the budget is known, the *wait-and-see* (WS) approach. It provides three different optimal decisions, one for each of the three possible realizations of ω,

$$\mathbf{x}^{\mathrm{WS}}(\omega^1) = (.6,0,0)^\top, \ \mathbf{x}^{\mathrm{WS}}(\omega^2) = (1,.7,0)^\top, \ \mathbf{x}^{\mathrm{WS}}(\omega^3) = (1,1,.5)^\top$$

---

*by Pavel POPELA

and three corresponding maximal values of the yield

$$z^{WS}(\omega^1) = 6,\ z^{WS}(\omega^2) = 20.5,\ z^{WS}(\omega^3) = 35.$$

The expected optimal yield is then $z^{WS} := E z^{WS}(\omega) = \sum_{i=1}^{3} p^i z^{WS}(\omega^i) = 24.85$.

The *pessimistic approach* exploits only the guaranteed budget $b_0 = 3$, i.e., provides the defensive optimal solution $\mathbf{x}^{MM}$ and optimal value z^{MM} obtained already for scenario ω^1. Notice that z^{MM} is the maximin bound for the optimal value with minimization over all probability distributions carried by the three considered scenarios; compare with II.5.3.5, inequality (6). Using the *expected value* of the budget means to replace the right-hand side by $b_0 + E\omega = 3 + 2.7 + 11 = 16.7$ and to solve the linear program

$$\max\{10x_1 + 15x_2 + 20x_3 : 5x_1 + 10x_2 + 20x_3 \leq 16.7,\ 0 \leq x_j \leq 1,\ j = 1, 2, 3\}$$

with results $\mathbf{x}^{EV} = (1, 1, .085)^\top$ and $z^{EV} = 26.7$; in our case, it is an optimistic solution (recall results of II.7.3).

As the next step, we relax the budget constraint allowing for a *recourse activity*, if necessary. For each scenario, we consider a two-stage decision model

$$\max\{\mathbf{c}^\top\mathbf{x} - q(\omega)y^-(\omega) : \mathbf{a}^\top\mathbf{x} + y^+(\omega) - y^-(\omega) = b_0 + \omega,\ \mathbf{x} \in [0, 1]^n\ y^+(\omega), y^-(\omega) \geq 0\}$$

with penalty coefficient $q(\omega)$ interpreted as the cost of additional borrowing or penalty for delayed payments, etc. The penalty term can be incorporated into the already discussed models. For illustration, we set up $q(\omega^1) = 2$, $q(\omega^2) = 3$, $q(\omega^3) = 6$. In this case, we get $\mathbf{x}$-components of the optimal solutions as those for the original example (which does not include any recourse activity). This outcome, of course, depends on the proportions between the yields $\mathbf{c}$ and penalty costs $q(\omega)$.

The two-stage counterpart of the expected value problem

$$\max\{\mathbf{c}^\top\mathbf{x} - [Eq(\omega)]y^- : \mathbf{a}^\top\mathbf{x} + y^+ - y^- = b_0 + E\omega,\ \mathbf{x} \in [0, 1]^n, y^+, y^- \geq 0\}$$

does not reflect the possible dependence of the recourse variables y^+, y^- on ω. Reflecting dependence leads to the *here-and-now* model formulation

(1) $$\text{maximize }\ \mathbf{c}^\top\mathbf{x} - \sum_i p^i q(\omega^i) y^-(\omega^i)$$

subject to

$$\mathbf{a}^\top\mathbf{x} + y^+(\omega^i) - y^-(\omega^i) = b_0 + \omega^i,\ y^+(\omega^i), y^-(\omega^i) \geq 0\ \forall i,\ \mathbf{x} \in [0, 1]^n.$$

In our example, this gives the optimal, scenario independent first-stage solution $\mathbf{x}^{HN} = (1, 1, 0)^\top$, the non-zero scenario dependent recourse variables $y^-(\omega^1) = 12$, $y^-(\omega^2) = 3$, $y^+(\omega^3) = 6$ and the optimal value of the objective function $z^{HN} = 17.5$. To get these results, it was necessary to solve a linear program of a larger size, without possibility to apply the intuitive rule valid for the relaxed knapsack problem.

At the end of the example, let us compare the obtained optimal values. First, we compute the expectation EEV of the expected value solution as follows. We replace $\mathbf{x}$ by the solution $\mathbf{x}^{EV}$ in (1). To achieve feasibility of (1) in the cheapest way, the nonzero recourse actions need to have the following values $y^-(\omega^1) = 13.7$, $y^-(\omega^2) = 4.7$, and $y^+(\omega^3) = 8.3$. Then, we compute the objective value of (1), and hence, EEV $= 10(1) + 15(1) + 20(.085) - .2(2)(13.7) - .3(3)(4.7) - .5(6)(0) = 26.7 - 9.71 = 16.99$. Notice that the following inequalities are valid in our case:

$$z^{MM} < \text{EEV} < z^{HN} < z^{WS} < z^{EV},$$

with the numerical values $6 < 16.99 < 17.5 < 24.85 < 26.7$. We see that in our example the choice of stochastic programming approach (see z^{HN}) is advantageous in comparison with common sense approaches often used in practise which result in z^{MM} and EEV; moreover, the lower bound EEV may be replaced by the function value of (1) evaluated for an arbitrary feasible solution.

A similar chain of inequalities may be constructed also for general stochastic programs with the lower and upper bounds depending substantially on the structure of the problem. The interested reader will find a detailed general discussion about relations of objective function values for various deterministic equivalents in Chapter 4 of [19]. There are also introduced the following important quantities together with their bounds (we rewrite them for maximization case with numerical values taken from Example 8.1.1):

$$\text{EVPI} = z^{WS} - z^{HN} = 7.35, \qquad 100\% \cdot \text{EVPI}/z^{HN} = 42\%,$$
$$\text{VSS} = z^{HN} - \text{EEV} = .51, \qquad 100\% \cdot \text{VSS}/z^{HN} = 2.91\%.$$

The **EVPI** denotes the **Expected Value of Perfect Information** and compares here-and-now and wait-and-see approaches. It measures how much is reasonable to pay to obtain perfect information about the future. A small value of the EVPI informs about a low additional profit when we reach perfect information; the large EVPI says that the information about the future is valuable. The **VSS** represents the **Value of the Stochastic Solution** and compares here-and-now and expected value approaches. It is an important characteristic that measures how much can be saved when the true here-and-now approach is used instead of the expected-value approach. A small value of the VSS means that the approximation of the stochastic program by the program with expected values instead of random variables is a good one.

8.1.2 Exercise. Prove that both EVPI and VSS are always nonnegative, whereas z^{EV} is an upper bound for z^{WS} only under special assumptions, e.g., if the objective function $c(\mathbf{x}, \omega)$ in II.3.2.4 is concave in ω.

8.2   Common Optimization Techniques

We see from Example 8.1.1 that there are various ways how to reduce stochastic programs to deterministic ones whose objective and constraints are usually in the form of expectations having integral representations. For certain cases those integrals can be evaluated by direct integration methods and standard mathematical programming algorithms may be used to solve the problem. We choose a simple recourse program as an example.

8.2.1   Example. *Simple recourse* refers to the situation when the recourse function ϕ in

$$\min_{\mathbf{x}}\{E_\omega\{\mathbf{c}^\top\mathbf{x} + \phi(\mathbf{x};\omega)\} : \mathbf{Ax} = \mathbf{b}, \mathbf{x} \geq \mathbf{0}\}$$

is determined for each $\mathbf{x}$ and ω as follows (recall II.3.3.2 and the ALM model 4.3 in Chapter II.4):

(2)
$$\phi(\mathbf{x};\omega) = \inf_{\mathbf{y}^+(\omega),\mathbf{y}^-(\omega)} \{\mathbf{q}^+(\omega)^\top\mathbf{y}^+(\omega) + \mathbf{q}^-(\omega)^\top\mathbf{y}^-(\omega) :$$
$$\mathbf{T}(\omega)\mathbf{x} + \mathbf{y}^+(\omega) - \mathbf{y}^-(\omega) = \mathbf{h}(\omega), \mathbf{y}^+(\omega), \mathbf{y}^-(\omega) \geq \mathbf{0}\}.$$

Provided that $q_i^+(\omega) + q_i^-(\omega) \geq 0$ $\forall i$, it is easy to solve the second-stage program (2) (a special form of (19) in II.3.3):

$$\phi(\mathbf{x};\omega) = \sum_{i=1}^{m_2}\left[q_i^+(\omega)(h_i(\omega) - \sum_{j=1}^{n_1}t_{ij}(\omega)x_j)^+ + q_i^-(\omega)(h_i(\omega) - \sum_{j=1}^{n_1}t_{ij}(\omega)x_j)^-\right],$$

where $(u)^+$ is the positive part of u and $(u)^-$ is the negative part of u. For simple recourse programs having only $\mathbf{h}(\omega) = \omega$ random, we denote $\chi_i = \sum_{j=1}^{n_1}t_{ij}x_j$, $\forall i$, and assuming absolutely continuous probability distribution of ω, we can express the recourse function as the sum of one-dimensional integrals (separability):

(3)
$$E_\omega\phi(\mathbf{x};\omega) = \sum_{i=1}^{m_2}E_{\omega_i}\{q_i^+(\omega_i - \chi_i)^+ + q_i^-(\omega_i - \chi_i)^-\} =$$
$$\sum_{i=1}^{m_2}\left(q_i^+E\omega_i - (q_i^+ - q_iF_i(\chi_i))\chi_i - q_i\int_{-\infty}^{\chi_i}zdF_i(z)\right)$$

where $F_i(z)$ denotes the marginal distribution function of ω_i. The separable terms in (3) offer computational advantages for both discrete and absolutely continuous probability distributions of ω. Explicit form of the integrals can be obtained for various distribution functions, e.g., for uniform or normal ones. This allows for a direct exploitation of mathematical programming methods. Notice, that the simple recourse models do not use any information about the joint probability distribution of ω and only marginal probability distributions are utilized.

The software code called SPORT (Stochastic Programming Optimizer with Recourse and Tenders) was written for simple recourse linear programs with discrete probability distribution, see Chapter 14 in [58] for details.

Also stochastic programs with individual probabilistic constraints and random right hand sides may be transformed into deterministic programs which exploit only the quantiles of the marginal probability distributions; cf. Chapter 8 of [130].

The two mentioned cases concern problems of a special structure which allows for a possible splitting of the probability distribution to one-dimensional marginal probability distributions, with the subsequent use of one-dimensional numerical integration routines (see (3)) and of general mathematical programming software. Another manageable case occurs when the probability distribution of ω is discrete and concentrated on not too many atoms (scenarios). In the sequel, we shall focus on the latter case.

For instance, Example 8.1.1 may be treated as a linear program. Mostly, stochastic linear programs with discrete probability distributions discussed in Part II are large-scale linear programs. Therefore, the well-known revised simplex and interior point methods (e.g., see Chapters 5 and 8 of [7]) are useful. Also nonlinear scenario-based stochastic programs or other models formulated in Chapter II.3 are frequently solvable as deterministic nonlinear programs (e.g., see algorithms described in Part 3 of [8]). For an overview of classical mathematical programming algorithms for the solution of stochastic programs with expectations see Chapter 4 in [58].

At first, there are *libraries of numerical subroutines*, which have extensive optimization capabilities (e.g., IMSL and NAG Fortran and C libraries) and their recent evaluations are contained in lp.faq and nlp.faq files available on Internet.

As the next step in the optimization software development, the optimization algorithms were implemented as 'black boxes' called *solvers*, using partially or fully precompiled code. Their main advantages for business applications are commercial availability, professional vendor support, and only basic input/output format understanding required from beginning users. At this moment, we present briefly several frequently used linear and nonlinear programming solvers; see [117] for details:

OSL (Optimization Subroutine Library), distributed by IBM corporation, contains primal and dual versions of simplex method, three versions of interior-point methods for large programs, and network simplex method for specially structured problems discussed, e.g., in [121]. It uses successive linear approximations for quadratic programmming problems, e.g., applicable for the Markowitz model II.3.2.1. CPLEX and XA solvers implement fast and reliable branch-and-bound methods for mixed-integer programs useful for modified problems with integer variables, cf. applications mentioned in Chapter II.4. MINOS and CONOPT are nonlinear programming solvers, e.g., useful for expected utility maximization problems. MINOS is useful for cases with majority of linear or near-linear constraints and CONOPT gives better results for programs with 'highly nonlinear' constraints. To unify input/output format for optimization solvers, the *MPS standard format* for linear programs has been developed in the early seventies, see [117].

8.2.2   Example. The MPS input file is specified for the following instance of investment program of Example II.1.1 based on expected values:

$$\text{maximize} \quad 1.1y^+ - 1.4y^-$$

subject to

$$x_1(1) + x_2(1) + x_3(1) = 1.8$$
$$1.067x_1(1) + 1.039x_2(1) + 1.05x_3(1) - x_1(2) - x_2(2) - x_3(2) = 0$$
$$1.073x_1(2) + 1.048x_2(2) + 1.055x_3(2) - y^+ + y^- = 2.0$$

$$x_1(1), x_2(1), x_3(1), x_1(2), x_2(2), x_3(2), y^+, y^- \geq 0.$$

Using the notation of Example II.1.1, we assign rate $r = 1.4$, additional income $q = 1.1$, goal $g = 2.0$, wealth $w = 1.8$, investment indices $i \in \{1, 2, 3\}$, stages $t \in \{1, 2\}$, and expected returns $\rho_1(1) = 0.067$, $\rho_2(1) = 0.039$, $\rho_3(1) = 0.05$, $\rho_1(2) = 0.073$, $\rho_2(2) = 0.048$, $\rho_3(2) = 0.055$.

```
NAME            Investment
ROWS
 N  OBJ
 E  R0000001
 E  R0000002
 E  R0000003
COLUMNS
    C0000001    R0000001    1.00000000    R0000002    1.06700000
    C0000002    R0000001    1.00000000    R0000002    1.03900000
    C0000003    R0000001    1.00000000    R0000002    1.05000000
    C0000004    R0000002   -1.00000000    R0000003    1.07300000
    C0000005    R0000002   -1.00000000    R0000003    1.04800000
    C0000006    R0000002   -1.00000000    R0000003    1.05500000
    C0000007    R0000003   -1.00000000    OBJ        -1.20000000
    C0000008    R0000003    1.00000000    OBJ         1.40000000
RHS
    RHS         R0000001    1.80000000
    RHS         R0000003    2.00000000
ENDATA
```

The presented INVEST.MPS input file has several data sections separated by keywords, and the end of the file is identified by the keyword ENDATA. The NAME section gives the name to the contained data set. The ROWS section contains the identifier of a matrix row together with constraint classifying letters E, L, G, N in each row. The letter N is used for the row of an objective function that is minimized. The COLUMNS section presents a matrix in column order. Each row of the file contains a column identifier followed by row identifiers (in the matrix) and nonzero coefficients. The RHS section contains the row identifiers and nonzero right-hand-side coefficients. The RANGES and BOUNDS sections are optional and define additional limits for constraints and another then nonnegative bounds for variables.

As the example shows, the MPS standard does not represent a user-friendly interface. For instance, it has unpleasant archaic features such as the limited size of identifiers (8 characters) and the fixed numerical field format. Therefore, we may simplify writing of the MPS file using popular *spreadsheet programs* (e.g., MS EXCEL utilized for business calculations) as the MPS generators. They can also be used as optimization tools because they include some of the previously mentioned solvers as optimization engines. However, the size of solved programs is limited to several hundreds of variables and the tabular form of the input data is convenient only for small educational programs with dense matrices, which are not typical in practise.

General purpose mathematical tools seem to be more promising than spreadsheets as they also involve optimization capabilities. MATLAB as a *matrix-oriented programming language* is a suitable tool, particularly, for algorithm development and testing. *Packages including symbolic rules* for differentiation such as MATHE-MATICA and MAPLE also have efficient optimization procedures and specifically may help with preprocessing and postprocessing tasks. However, these systems have not been originally developed for the solution of large-scale programs, and hence, they cannot successfully compete with special purpose systems discussed below.

The MPS format is a basic interface for test computations but the interfaces between solvers and data sources become important for the applications (see, e.g., preface of [66]). As a further step in the optimization system development, *algebraic modeling languages* describe a program by its set of constraints, obvious in mathematics, and they generate the MPS or similar files for solvers. They are based on the declarative kernel and are extended with some properties of procedural languages such as statements that control operations flow. They have Windows-like environments and have libraries of sample programs for learning by example. GAMS (General Algebraic Modeling System) [66] is widely used with different computer platforms. It is designed for linear, nonlinear, and mixed integer programming, and supports many different solvers (e.g., CPLEX, OSL, MINOS, and CONOPT). Other widely used modeling languages are AMPL [65] and AIMMS [21].

8.2.3   Example. We continue with the program from Example 8.1.1 and we introduce its GAMS source code that ends with solution and display statements. It is contained in file INVEST.GMS:

```
$TITLE     Investment example
$OFFSYMXREF OFFSYMLIST OFFUELLIST OFFUELXREF
OPTION     LIMROW = 0, LIMCOL = 0, SYSOUT = OFF, SOLPRINT = OFF

SET        I considered investments, T time, S scenarios;
PARAMETER RHO(I,T,S) random returns, P(S) scenario probabilities;
SCALAR     W wealth, G goal, Q additional income, R rate;
$INCLUDE   "DATA.GMS"

VARIABLE Z objective function value, X(I,T,S) investment decision;
VARIABLE YPLUS(S) surplus variable, YMINUS(S) deficit variable;
  X.LO(I,T,S) = 0; YPLUS.LO(S) = 0; YMINUS.LO(S) = 0;

EQUATION NONANTICIP(I,S), INITIAL(S), INTERMED(T,S), FINAL(T,S), OBJFUNC;
NONANTICIP(I,S)$(ORD(S) NE 1)..
  X(I,"1","1")              =E= X(I,"1",S);
INITIAL(S)..
  SUM(I, X(I,"1",S))        =E= W;
INTERMED(T,S)$(ORD(T) LT CARD(T))..
  SUM(I,(1+RHO(I,T,S))*X(I,T,S)=E= SUM(I,X(I,T+1,S));
FINAL(T,S)$(ORD(T) EQ CARD(T))..
  SUM(I,(1+RHO(I,T,S))*X(I,T,S)=E= G + YPLUS(S)-YMINUS(S);
```

```
OBJFUNC..
   Z                         =E= SUM(S, P(S)*(Q*YPLUS(S) - R*YMINUS(S)));

MODEL    INVESTMENT / NONANTICIP, INITIAL, INTERMED, FINAL, OBJFUNC/;
SOLVE    INVESTMENT MAXIMIZING Z USING LP;
DISPLAY  Z.L, X.L;
```

In some sense, modeling languages follow the design steps of a modeler. At first, compiler directives and options are set up. Then the SET keyword allows the declaration of indices of a program giving a name to each index. The PARAMETER keyword declares real parameters, giving them a name and joining them with indices. Their values are assigned in the included file DATA.GMS. The VARIABLE keyword starts a block that declares variables of a program in a manner similar to parameter declaration. Variable identifiers allocate a data record in memory, and these identifiers will be used with different suffices for computational or displaying purposes. Suffixes .LO, .UP, .L, and .M allow separate storing of lower and upper bounds, solutions, and marginal values respectively. The EQUATION block declares constraints, which are defined separately. Conditioned generation of individual constraints is specified with $ symbol. The objective is identified by variable Z on the left hand side of one constraint. MODEL is a named list of constraints. Its suffixes often serve to obtain compilation and computation results in the form of model and solver status. The SOLVE statement joins the model name INVESTMENT, the target of optimization for the objective (MINIMIZING Z), the solver type (USING LP), and then starts a computational process. DISPLAY is the original GAMS output statement.

In our case, data definitions are contained in the included file DATA.GMS (see the compiler directive $INCLUDE "DATA.GMS" in the source code of INVEST.GMS):

```
SET      I / 1 * 3 /, T / 1 * 2 /, S / 1 * 3 /;
SCALAR   W / 1.8 /, G / 2.0 /, Q / 1.1 /, R / 1.4 /;
TABLE    RHO(I,T,S)
            1      2      3
   1.1     0.10   0.05   0.07
   2.1     0.12   0.03   0.03
   3.1     0.08   0.04   0.05
   1.2     0.04   0.07   0.08
   2.2     0.06   0.06   0.04
   3.2     0.01   0.08   0.05;
PARAMETER P(S) / 1 0.1, 2 0.3, 3 0.6 /;
```

The members of index domains are specified within the SET statements. The values of parameters may be defined as a part of the PARAMETER declaration using rules for sparse data structures or with SCALAR or TABLE declarations either directly or through assignment = (see forthcoming Example 8.6.1). Used notation and parameter values are inherited from Example 8.2.2. In addition, instead of expected values $\rho_i(t)$, we consider random returns $\rho_i(t, \omega)$ based on scenarios ω^s, $s \in \{1, 2, 3\}$ with probabilities $p^1 = .1$, $p^2 = .3$, and $p^3 = .6$.

The GAMS user may find some difficulties during its use in financial and industrial applications. They stem from the GAMS original field of use, which was the support of macroeconomic studies. These studies have often needed software codes 'for one use', in contrast to the financial and industrial systems that require codes 'for permanent use'. For instance, the GAMS does not support database management directly and it must be added separately when it is required. Our discussion continues in Section 8.6.

8.3   Solution Techniques for Two-Stage Stochastic Programs

The main advantage of the aforementioned common optimization algorithms and software tools is that they are easy to use and well-tested for mid size problems. However, the main disadvantage is that they are inefficient for large-scale stochastic programs as they do not utilize their special structure.

As the next step, we discuss linear programming algorithms modified for two-stage scenario-based stochastic linear programs (see II.3.3.2):

$$(4)\quad \min_{\mathbf{x}, \mathbf{y}^s \forall s} \{\mathbf{c}^\top \mathbf{x} + \sum_{s=1}^{S} p^s \mathbf{q}^{s\top} \mathbf{y}^s : \mathbf{A}\mathbf{x} = \mathbf{b}, \mathbf{x} \geq 0, \mathbf{T}^s \mathbf{x} + \mathbf{W}^s \mathbf{y}^s = \mathbf{h}^s, \mathbf{y}^s \geq 0, s \in \mathcal{S}\}$$

where $\mathcal{S}$ identifies a finite set of scenarios $\omega^s = (\mathbf{h}^s, \mathbf{q}^s, \mathbf{T}^s, \mathbf{W}^s), s \in \mathcal{S}$ with probabilities p^s, $\mathbf{x}$ and $\mathbf{y}^s$ are decision variables, $\mathbf{A}$, $\mathbf{b}$, and $\mathbf{c}$ contain only deterministic coefficients.

First, we denote as $\mathbf{H}$ the LHS matrix composed of all zero and nonzero technological submatrices $\mathbf{A}$, $\mathbf{T}^s$, and $\mathbf{W}^s, \forall s \in \mathcal{S}$. For fixed recourse programs (i.e., $\mathbf{W}^s = \mathbf{W} \ \forall s \in \mathcal{S}$) *a special basis factorization* was suggested. It utilizes separate submatrix factorizations corresponding to diagonal blocks with $\mathbf{W}$ components and an additional working basis, see Section 12.3 in [130] and references therein.

Interior-point algorithms have become popular, because they proved to be efficient, especially, with parallel implementations. The most time consuming step in the interior-point methods is usually related to computations with a matrix $\mathbf{M} = \mathbf{HDH}^\top$, where $\mathbf{D}$ denotes a suitable diagonal matrix. The obtained $\mathbf{M}$ is much denser (see [19]) and [7] for explanatory references) than the original constraint matrix $\mathbf{H}$, and hence, a direct use of the interior point method need not be efficient enough. Therefore, a factorization scheme that significantly reduces the number of necessary arithmetic operations has been developed, see [19]. An alternative is a split variable representation, which uses an explicit treatment of nonanticipativity constraints (recall II.2.2). This means that the nonanticipativity constraint $\mathbf{x}^r = \mathbf{x}^u$ is included only if $u = r + 1, r = 1, \dots, S - 1$. The new $\mathbf{M}$ is sparser and larger than the original one, hence we have to decide whether we prefer a small size or a reduced density. Another idea [19] is to form the dual program of (5), and this dual is then solved with the interior-point method. In this case, the $\mathbf{M}$ matrix is again large, however, it has a special so called sparse block-row-and-column-bordered structure. We may summarize that due to factorization enhancements, interior-point methods are reported as more efficient than direct simplex algorithms for large-scale stochastic programs.

Decomposition algorithms advantageously use a dual block-angular program structure of (17)–(18) in Section II.3.3 to save memory requirements and speed up computations. It was already noticed in the fifties that the program dual to (5) has a structure suitable for the Dantzig-Wolfe decomposition. However, the dual program solution often requires more computations than the primal, because a larger program has to be solved. In addition, extra calculations are needed to recover the primal program solution. For these reasons, the Benders (dual) decomposition is applied to the primal program instead of employing the Dantzig-Wolfe (primal) decomposition for the dual program, although both decompositions are computationally similar. To detail these ideas, we reformulate the two-stage stochastic linear program as follows:

(5)
$$\min_{\mathbf{x}, \theta}\{\mathbf{c}^\top \mathbf{x} + \theta : \theta \geq \mathcal{Q}(\mathbf{x}), \mathbf{x} \in \mathcal{X}\},$$

where $\mathcal{Q}(\mathbf{x})$ is the expectation of the recourse function $\phi(\mathbf{x}; \omega)$ and $\mathcal{X}$ denotes the first-stage feasible set. The L-shaped algorithm (L refers to the shape of matrix nonzero blocks) discussed further is based on the application of the Benders decomposition to two-stage stochastic programs. Additional linear constraints called *feasibility cuts* approximate the domain K of $\mathcal{Q}(\mathbf{x})$, and another linear constraints called *optimality cuts* approximate the shape of the function $\mathcal{Q}(\mathbf{x})$. The stochastic linear program is decomposed into the first-stage (master) program and a series of second-stage programs (subprograms). Each subprogram corresponds to one scenario. The modified master program is solved, the solution is sent to the subprograms to identify more precisely their feasible sets. The information about all solved subprograms is returned to the master program in the form of cuts, and the algorithm continues with the next iteration (see [91] for further explanatory and detailed comments). BDECOM (Benders DECOMposition) is a Benders decomposition-based solver included in SLP-IOR model management system, see [113] and Section 8.6. To show the principles of the L-shaped algorithm, we assume that (5) has a finite optimal value. The general case is discussed, e.g., in [19] and [130].

8.3.1 Algorithm *(LSHAPED)*

1. Initialization of parameters: the number of iteration $n := 0$, the number of feasibility cuts $r := 0$, the number of optimality cuts $u := 0$, and hence, vectors $\mathbf{d}, \mathbf{f}$ and matrices $\mathbf{D}, \mathbf{F}$ used for the description of cuts are initialized as empty.

2. Set $n := n + 1$ and solve the following master program

(6)
$$\min_{\mathbf{x}, \theta}\{\mathbf{c}^\top \mathbf{x} + \theta : \mathbf{A}\mathbf{x} = \mathbf{b}, \mathbf{x} \geq \mathbf{0},$$

(7)
$$\mathbf{F}\mathbf{x} \geq \mathbf{f}, \mathbf{D}\mathbf{x} + 1\theta \geq \mathbf{d}\}.$$

Let $(\mathbf{x}[n]^\top, \theta[n])^\top$ be an optimal solution of this program (If $r = u = 0$, then cuts (7) are not included, θ is not considered in the objective and constraints, and $\theta[n] := -\infty$).

3. Solve the following linear programs $\forall s \in \mathcal{S}$:

(8)
$$z^s[n] = \min_{\mathbf{y}, \mathbf{v}}\{\mathbf{1}^\top \mathbf{v}^+ + \mathbf{1}^\top \mathbf{v}^- :$$

$$\mathbf{W}^s \mathbf{y} + \mathbf{v}^+ - \mathbf{v}^- = \mathbf{h}^s - \mathbf{T}^s \mathbf{x}[n], \mathbf{y} \geq \mathbf{0}, \mathbf{v}^+ \geq \mathbf{0}, \mathbf{v}^- \geq \mathbf{0}\}.$$

If $\forall s \in S$ the optimal value $z^s[n] = 0$, then $\boxed{\text{GOTO 4}}$. Otherwise, for given $s \in S$ with $z^s[n] > 0$ use the associated dual multipliers $\sigma[n]$ (representing unboundedness of the dual to (6)) to generate a new feasibility cut. Increase the number of feasibility cuts $r := r+1$, and define the new rth row of matrix $\mathbf{F}$ and the related rth component of the RHS vector $\mathbf{f}$:

$$(9) \qquad \mathbf{F}_{r\cdot} := \sigma[n]^\top \mathbf{T}^s, f_r := \sigma[n]^\top \mathbf{h}^s,$$

and $\boxed{\text{GOTO 2}}$.

$\boxed{4.}$ Solve the following subprograms $\forall s \in S$:

$$(10) \qquad \min_{\mathbf{y}^s} \{ \mathbf{q}^{s\top} \mathbf{y}^s \ : \ \mathbf{W}^s \mathbf{y}^s = \mathbf{h}^s - \mathbf{T}^s \mathbf{x}[n], \mathbf{y}^s \geq 0 \}.$$

Let $\pi^s[n]$ be the optimal multipliers associated with the optimal solution of the sth subprogram. Increase the number of optimality cuts $u := u+1$ and define the new uth row of matrix $\mathbf{D}$ and the related uth component of the RHS vector $\mathbf{d}$:

$$(11) \qquad \mathbf{D}_{u\cdot} := \sum_{s=1}^{S} p^s \pi^s[n]^\top \mathbf{T}^s, \quad d_u := \sum_{s=1}^{S} p^s \pi^s[n]^\top \mathbf{h}^s.$$

If $\theta[n] \geq d_u - \mathbf{D}_{u\cdot}\mathbf{x}[n]$ then $\boxed{\text{STOP}}$, and $\mathbf{x}[n]$ is an optimal solution. Otherwise, add this constraint as a new optimality cut to (8), and $\boxed{\text{GOTO 2}}$.

Since its discovery in the late sixties, the L-shaped algorithm has been discussed and improved in many details (see [19], [58], [91], and [130] for bibliographical references). Notice that with *relatively complete recourse* (see II.3.3.2) no feasibility cuts are necessary, so step $\boxed{3.}$ can be omitted. Having *fixed recourse* $\mathbf{W}$ in step $\boxed{3.}$, we may try to construct $\mathbf{h}^{s*} - \mathbf{T}^{s*}\mathbf{x}[n]$ a lower bound of possible values of the random vector $\mathbf{h}(\omega) - \mathbf{T}(\omega)\mathbf{x}[n]$ and to solve only one program of type (5). Then, we may utilize a bunching procedure improving the efficiency of computations in step $\boxed{4.}$. The basic idea lies in the fact that the basis $\mathbf{B}$ optimal for some scenario $\omega^s = (\mathbf{q}^s, \mathbf{h}^s, \mathbf{T}^s)$ is also optimal for any other scenario $l \in S$ satisfying $\mathbf{B}^{-1}(\mathbf{h}^l - \mathbf{T}^l\mathbf{x}[n]) \geq 0$ and $\mathbf{q}^l - \mathbf{q}_B^l \mathbf{B}^{-1}\mathbf{W} \geq 0$ where $\mathbf{q}_B^l$ is the subvector of $\mathbf{q}^l$, whose elements correspond to the columns of $\mathbf{W}$, contained in $\mathbf{B}$. Then solving linear *programs that differ only in the right-hand side*, we may apply the idea of building a dual feasible basis tree, i.e. to create a search tree, where each node is related to a dual feasible basis for the linear program and each arc to a dual pivot step. Nodes are added to the search tree as they are needed. Then the different right-hand sides are 'trickled down' in the tree until primal feasibility is achieved. The alternative method called sifting uses linear programming parametric analysis applied to suitably ordered vectors $\mathbf{q}^s$ and $\mathbf{h}^s - \mathbf{T}^s\mathbf{x}[n]$, $s \in S$, see 5.4 of [19] for further references.

Another L-shaped algorithm variant introduces *multiple cuts* and deals with two-stage stochastic quadratic programs, see [19]. In step $\boxed{4.}$, $\pi^s[n]$ are not aggregated to generate piecewise linear approximation of $\mathcal{Q}(\mathbf{x})$ but $\pi^s[n]$ are used to construct a piecewise linear approximations of $\phi(\mathbf{x}; \omega^s)$. Therefore, for sth scenario-related optimality cut $\mathbf{D}^s$ denotes a left-hand-side matrix and $\mathbf{d}^s$ is a right-hand-side vector. Then, uth row of $\mathbf{D}^s$ is $\mathbf{D}^s_{u\bullet} = \pi^s[n]^\top \mathbf{T}^s$ and uth component of $\mathbf{d}^s$ is $d^s_u = \pi^s[n]^\top \mathbf{h}^s$.

If we add a regularizing term then we get the quadratic objective function (see [19]) and in step $\boxed{2.}$, the master program is replaced by the following program:

$$(12) \qquad \min_{\mathbf{x}, \theta^s} \{ \mathbf{c}^\top \mathbf{x} + \sum_{s=1}^S p^s \theta^s + \frac{1}{2} \| \mathbf{x} - \mathbf{x}[n-1] \|^2 :$$

$$\mathbf{A}\mathbf{x} = \mathbf{b}, \mathbf{x} \geq \mathbf{0}, \mathbf{F}\mathbf{x} \geq \mathbf{f}, \mathbf{D}^s \mathbf{x} + \mathbf{1}\theta^s \geq \mathbf{d}^s, s \in \mathcal{S} \}.$$

The idea to speed up computations with adding a *regularizing quadratic term* in the objective as in (12) has resulted in QDECOM (Quadratic DECOMposition) procedure used in [113]. This method is advantageous when initial cuts force the approximating solutions $\mathbf{x}[n]$ to oscillate widely after initialization.

Financial problems may lead to models with a specific structure. If matrix $\mathbf{W}$ is derived from network flow problems (see [121]), then cuts are generated more easily, see discussion in [91]. In contrast, the presence of integer variables in financial models causes big solution difficulties, see [99]. If a modeler cannot avoid this situation, either SIRD2CR developed for SLP-IOR [113] or a modified implementation of the L-shaped method can be used.

Although the regularizing term introduces the possibility to use nonlinear terms in the objective, we suggest to employ a *Lagrangian-based algorithm* for the solution of scenario-based two-stage stochastic *nonlinear* programs, having the following form:

$$(13) \qquad \min_{\mathbf{x}, \mathbf{y}^s: s \in \mathcal{S}} \left\{ c(\mathbf{x}) + \sum_{s=1}^S p^s \phi(\mathbf{x}; \mathbf{y}^s; \omega^s) : \mathbf{x} \in \mathcal{X}, \mathbf{y}^s \in \mathcal{Y}(\mathbf{x}; \omega^s), s \in \mathcal{S} \right\},$$

where $\mathcal{X}$ is the first-stage feasible set and $\mathcal{Y}(\mathbf{x}; \omega^s), s \in \mathcal{S}$ are the second-stage feasible sets. Program (13) may be rewritten with an explicit description of nonanticipativity constraints (recall II.2.2) in the following form:

$$(14)$$

$$\min_{\mathbf{x}^s, \mathbf{y}^s \forall s} \left\{ \sum_{s=1}^S p^s (c(\mathbf{x}^s) + \phi(\mathbf{x}; \mathbf{y}^s; \omega^s)) : \mathbf{x}^s \in \mathcal{X}, \mathbf{y}^s \in \mathcal{Y}(\mathbf{x}^s; \omega^s), \mathbf{x}^s = \sum_{u=1}^S \mathbf{x}^u, \forall s \right\}$$

If nonanticipativity constraints are relaxed (e.g., with penalty terms involved in the objective), then the resulting program is separable with respect to scenario-related variables, and hence, the solution algorithm is reduced to the repeated solution of separate updated programs for each scenario. Practitioners using scenario analysis often utilize the average value of these first stage decisions obtained for individual scenarios as their decision, which is not the best idea.

The *progressive hedging algorithm* (PHA) [134] is based on the scenario aggregation principle. During one PHA iteration, all scenario programs are solved, and the obtained solutions are averaged to get an approximation of the searched solution. The objective of each scenario is augmented by terms based on nonanticipativity constraints and updated with use of iterations' results. The whole progressive hedging algorithm has the following form:

8.3.2 Algorithm *(PHA)*

> 1. Initialization: Set the iteration counter $n := 1$, $\forall s \in \mathcal{S}$: assign weights $\mathbf{w}^s[n] := \mathbf{0}$, and choose $\hat{\mathbf{x}}[n]$, $\rho > 0$, and $\varepsilon > 0$.
>
> 2. $\forall s \in \mathcal{S}$ find the optimal solution $\mathbf{x}^s[n], \mathbf{y}^s[n]$ of the program:

$$(15) \quad \min_{\mathbf{x}^s, \mathbf{y}^s} \{ c(\mathbf{x}^s) + q(\mathbf{x}^s, \mathbf{y}^s; \omega^s) + \mathbf{w}^s[n-1]^\top \mathbf{x}^s + \frac{\rho}{2} \| \mathbf{x}^s - \hat{\mathbf{x}}[n-1] \|^2 :$$

$$\mathbf{x}^s \in \mathcal{X}, \mathbf{y}^s \in \mathcal{Y}(\mathbf{x}^s; \omega^s) \}.$$

> 3. Compute the new average solution: $\hat{\mathbf{x}}[n] := \sum_{s \in \mathcal{S}} p^s \mathbf{x}^s[n]$. Update perturbation terms $\forall s \in \mathcal{S}$: $\mathbf{w}^s[n] := \mathbf{w}^s[n-1] + \rho \cdot (\mathbf{x}^s[n] - \hat{\mathbf{x}}[n])$. If $\sum_{s \in \mathcal{S}} p^s \mathbf{w}^s[n]$ is near to $\mathbf{0}$ (measured by ε), then $\boxed{\text{STOP}}$, otherwise $n := n+1$ and $\boxed{\text{GOTO2}}$.

The main advantage from the implementation point of view is that one may use any locally convergent nonlinear programming algorithm discussed in 8.2 to solve (15). The updating step 3. is then quite simple to be implemented, even in parallel. The sequence of $\hat{\mathbf{x}}[n]$ converges to the optimal solution for convex programs (13), and the algorithm's convergence is also reported for certain nonconvex cases, see [19]. The use of solution averages in the PHA guarantees robustness of the computational process, but the algorithm convergence is slow. In the linear case (5), this algorithm is not as efficient and fast as the L-shaped decomposition. Parallel implementation of PHA was successfully realized in [122] for problems with network structure (recall II.4.4).

Another variant of the augmented Lagrangian method called Diagonal Quadratic Approximation (DQA) [19] is based on writing the nonanticipativity constraints equivalently in a specific order. For any scenario $s \in \mathcal{S}$ we choose a unique successor $\sigma(s)$. So, that the bijection mapping $\sigma : \mathcal{S} \longrightarrow \mathcal{S}$ defines a cyclic permutation of all scenarios, and hence, specifies the set of used nonanticipativity constraints $\mathbf{x}^s = \mathbf{x}^{\sigma(s)}$, $\forall s \in \mathcal{S}$. Then, the augmented terms in the objective $\| \mathbf{x}^s - \mathbf{x}^{\sigma(s)} \|^2$ are approximated by replacing $\mathbf{x}^{\sigma(s)}$ by the known value $\mathbf{x}^{\sigma(s)}[n-1]$. The original augmented Lagrangian program is decomposed into S subprograms:

$$(16) \quad \min_{\mathbf{x}^s, \mathbf{y}^s} \{ c(\mathbf{x}^s) + Q(\mathbf{x}^s, \mathbf{y}^s, \omega^s) + (\mathbf{w}^s[n-1] - \mathbf{w}^{\sigma^{-1}(s)}[n-1])^\top \mathbf{x}^s +$$

$$\frac{\rho}{2} (\| \mathbf{x}^s - \mathbf{x}^{\sigma(s)}[n-1] \|^2 + \| \mathbf{x}^s - \mathbf{x}^{\sigma^{-1}(s)}[n-1] \|^2) : \mathbf{x}^s \in \mathcal{X}, \mathbf{y}^s \in \mathcal{Y}(\mathbf{x}^s, \omega^s) \},$$

where ρ is a penalty coefficient, $\mathbf{w}^s$, $s \in \mathcal{S}$ represent updated multipliers with values known from previous algorithm iteration, and σ^{-1} denotes the mapping inverse to σ.

8.4   Solution Techniques for Multistage Stochastic Programs

We focus on numerical techniques for multistage stochastic programs with finite discrete probability distributions. For this purpose, we may employ and generalize our knowledge about two-stage stochastic programming algorithms from 8.3.

However, new difficulties arise, because the number of possible realizations exponentially increases with the number of stages and scenarios (so called 'curse of dimensionality'). Therefore, several input data manipulation ideas how to reach a *manageable size of the problem* have been developed. For instance, we can waive the stochastic character of chosen data and replace these random parameters by some fixed values; we can ignore the requirement of adapting the decisions according to the past information; we can aggregate scenarios or stages; we can select only a few 'important' scenarios using statistical techniques or expert's opinion; we can decompose the problem into manageable ones to use parallel procedures, etc.

Then, the resulting large-scale linear program can be solved either by general purpose algorithms or by techniques taking advantages from the special multistage structure of the problem. So, for multistage programs, there are versions of basis factorization (see [130] for details) and interior-point methods based on the efficient rewriting of the program description (see **BL). We shall focus on decomposition-based algorithms; we refer to [30] for parallel implementations of interior-point-based methods.

At first, a Scenario-Based Multistage Stochastic Linear Program (SBMSLP) that is written in the arborescent form (11)–(12) in Section II.2.2 can be solved with the use of nested Benders decomposition based on generalization of Algorithm 8.3.1 (see [19] for historical references). To simplify the notation in the forthcoming algorithm description, we assume that $\mathbf{x}_{k_t} \geq \mathbf{0}$ (instead of $\mathbf{l}_{k_t} \leq \mathbf{x}_{k_t} \leq \mathbf{u}_{k_t}$ introduced in (11)–(12)) in Section II.2.2 for all $k_t \in \cup_t \mathcal{K}_t = \mathcal{K}$ where $\mathcal{K}_t = \{K_{t-1}+1, \ldots, K_t\}, t = 1, \ldots, T$ and $K_0 = 0$. In addition, we suppose that the considered program has a finite optimal solution or is infeasible. (This is indeed valid for (11)–(12) in Section II.2.2.) The unbounded case for our changed description may be treated, e.g., as in [19].

The nested Benders decomposition is based on the idea of arranging subprograms in a tree-like structure. For problems with the staircase structure, the subprogram assigned to tree node $k_t \in \mathcal{K}$ has the form:

$$(17) \qquad \min\{\mathbf{c}_{k_t}^\top \mathbf{x}_{k_t} + \theta_{k_t} : \mathbf{A}_{k_t}\mathbf{x}_{k_t} = \bar{\mathbf{b}}_{k_t}, \mathbf{x}_{k_t} \geq \mathbf{0},$$

$$(18) \qquad \mathbf{F}_{k_t}\mathbf{x}_{k_t} \geq \mathbf{f}_{k_t},$$

$$(19) \qquad \mathbf{D}_{k_t}\mathbf{x}_{k_t} + \mathbf{1} \cdot \theta_{k_t} \geq \mathbf{d}_{k_t}\},$$

where $\bar{\mathbf{b}}_{k_t} = \mathbf{b}_{k_t} - \mathbf{B}_{k_t}\mathbf{x}_{a(k_t)}$ and $a(k_t)$ denotes the ancestor of k_t. We may see that the program (17)–(19) extracts its structure from (11)–(12) of Section II.2.2 using fixed k_t, however the recourse term including objective values of subprograms situated in the tree below is replaced by the approximating variable θ_{k_t} in (17). At the tree's topmost level (we assign $t = 1$, $k_t = 1$), the master program does not change the right-hand side (cf. (17)), at the tree bottom level ($t = T$), subprograms are neither augmented by θ_{k_T} nor by cutting constraints (18)–(19). In general, in k_tth node, the lth feasibility cut (18) is formed as follows:

$$(20) \qquad l: \alpha_{kl}^\top \mathbf{B}_k \mathbf{x}_{k_t} \geq \alpha_{kl}^\top \mathbf{b}_k + \beta_{kl}^\top \mathbf{f}_k + \gamma_{kl}^\top \mathbf{d}_k,$$

where k is the index of a subsequent infeasible subprogram. The lth optimality cut (19) reads:

$$(21) \qquad l: \sum_{k \in \mathcal{D}(k_t)} p^k \pi_{kl}^\top \mathbf{B}_k \mathbf{x}_{k_t} + \theta_{k_t} \geq \sum_{k \in \mathcal{D}(k_t)} p^k (\pi_{kl}^\top \mathbf{b}_k + \lambda_{kl}^\top \mathbf{f}_k + \mu_{kl}^\top \mathbf{d}_k).$$

We denote by $\mathcal{D}(k_t)$ the set of immediate descendants of k_t, see II.2.2. During the solution process, primal information about the optimal solution $\mathbf{x}_{k_t}$ is sent down through the tree and bounds subsequent programs modifying $\bar{\mathbf{b}}_{k_t}$. Dual information (solutions π_{kl}, λ_{kl}, μ_{kl}, and rays α_{kl}, β_{kl}, γ_{kl}) is returned back from nodes $k \in \mathcal{D}(k_t)$ in the form of additional constraints (18)–(19). If the program is solvable, the computation process stops and the optimum is reached when no new information about the optimal solution at the top-most level is obtained. So, the multistage version of the L-shaped algorithm follows:

8.4.1   Algorithm *(MSLSHAP)*

1. Initialization of parameters: Set the stage $t := 1$, the subprogram $k_t := 1$, the number of iteration $n := 0$, and $\forall k \in \mathcal{K}$ assign the number of feasibility cuts $r_k := 0$, the number of optimality cuts $u_k := 0$, and the value $\theta_k := -\infty$. Assign the variable that controls the sequencing protocol: DIRECTION:=FORWARD. Continue with the step 2.

2. Set $n := n + 1$ and solve the current k_tth subprogram (18)–(20), $k_t \in \mathcal{K}_t$. If $r_{k_t} = u_{k_t} = 0$, then cuts are not included. When $\theta_{k_t} := -\infty$, then θ_{k_t} is not considered in the objective.

3. If solved subprogram (18)–(20) is infeasible and $t = 1$, then STOP and the original program is infeasible.
   If solved subprogram is infeasible and $t > 1$, then $r_{a(k_t)} := r_{a(k_t)} + 1$ and generate a new $r_{a(k_t)}$th feasibility cut for $a(k_t)$ subprogram by (21) using the dual ascent extremal direction given by $\alpha_{k_t}[n]$, $\beta_{k_t}[n]$, and $\gamma_{k_t}[n]$. Select the next subprogram with DIRECTION:=BACKWARD, $k_t := a(k_t)$, $t := t-1$, and GOTO 2.

4. If solved subprogram (18)–(20) is feasible, then $\mathbf{x}_{k_t}[n]$ and $\theta_{k_t}[n]$) denote its optimal solution. This solution is stored together with the dual solution denoted as $\pi_{k_t}[n]$, $\lambda_{k_t}[n]$, and $\mu_{k_t}[n]$ for further generation of the optimality cut by (22). If $k_t < K_t$, then $k_t := k_t + 1$ and GOTO 2.
   Otherwise, $k_t = K_t$, and continue depending on the value of DIRECTION:
   If DIRECTION=FORWARD then for $t < T$ assign $t := t + 1$ and $k_t := k_t + 1$ and GOTO 2.
   If DIRECTION=FORWARD and $t = T$, then set DIRECTION:=BACKWARD and continue.
   If DIRECTION=BACKWARD then $\forall k \in \mathcal{K}_t$ repeatedly assign $u_{a(k)} := u_{a(k)} + 1$ and generate new $u_{a(k)}$th optimality cut for $a(k)$ subprogram by (22) using stored information about dual solution.

5. Redundancy check of the additional optimality cuts: The inequality (22) is tested for the $a(k)$ subprograms using stored values $\theta_{a(k)}[n]$, and $\mathbf{x}_{a(k)}[n]$. If this inequality is satisfied for certain $a(k)$, then no optimality cut is added to the related subprogram. If for an $a(k)$ the inequality (22) does not hold true the related optimality cut is added.
   If $t = 2$ and no optimality cuts are added to the master program (identified by $t = 1$, $k_t = 1$), then STOP, and the optimal solution is found.

Otherwise ($t > 2$ or cuts are added), select the next subprogram assigning $t := t - 1$ and $k_t := K_{t-1} + 1$. Afterwards, if $t = 1$ then DIRECTION:=FORWARD else DIRECTION:=BACKWARD and $\boxed{\text{GOTO 2}}$.

Algorithm 8.4.1 is implemented as the MSLiP software [68]. It works with random transition matrices $\mathbf{B}_{k_t}$ and with more than three stages. It involves efficient bunching strategies inspired by those discussed in 8.3. An interested reader may obtain this freeware directly from its author. It was shown that the used fast-forward-fast-back sequencing protocol is advantageous in comparison with other tree traversing strategies. MSLiP was adapted to solve each node program using OSL, see [34]. Further developments mainly deal with parallel computations and incorporation of sampling (see 8.5) into the nested decomposition. The key problem is how to avoid the occurrence of computational bottlenecks caused by the need of synchronization among one master program and many subprograms.

Bender's-based decomposition for stochastic programs was also implemented within the SP/OSL library. The first flexible commercial code called IBM Stochastic Solutions designed for SBMSLP is supplemented by a set of callable modules and particularly the user has node-by-node access to data and solutions for postoptimality analysis. As the aforementioned MSLiP, it may read a stochastic extension of the MPS input format, called the SMPS (Stochastic MPS) format.

The SMPS format [20] is based on splitting the program description into three files:

a) CORE file: This is a standard MPS file, which describes an underlying deterministic program. Arbitrarily chosen fixed values are used instead of random parameters.

b) TIME file: splits the CORE file into stages using the north-west corner coordinates of each stage.

c) STOCH file: describes the actual probability distribution for each random parameter. The idea is that values from the probability distribution support replace deterministic values in the CORE file.

8.4.2 Example.

Let us return to Example 8.2.2. The CORE file INVEST.COR is the same as INVEST.MPS. The following TIME file INVEST.TIM specifies the program's dynamic structure:

```
TIME         Example
PERIODS      LP
   C1        R1           PERIOD1
   C4        R2           PERIOD2
ENDATA
```

Rows and columns in the CORE file have to respect the ordering given by successive periods in the TIME file (i.e. C1, C2, C3 must appear in CORE file before C4, and C5, ... must come after). The random parameters definition is contained in STOCH file INVEST.STO:

```
STOCH        Example
BLOCKS       DISCRETE
 BL BLOCK1   PERIOD2 0.10
```

```
         C1       R2       1.10    R3      1.04
         C2       R2       1.12    R3      1.06
         C3       R2       1.08    R3      1.01
   BL BLOCK2   PERIOD2  0.30
         C1       R2       1.05    R3      1.07
         C2       R2       1.03    R3      1.06
         C3       R2       1.04    R3      1.08
   BL BLOCK3   PERIOD2  0.60
         C1       R2       1.07    R3      1.08
         C2       R2       1.03    R3      1.04
         C3       R2       1.05    R3      1.05
   ENDATA
```

The BLOCKS keyword used above allows to introduce a dependence between different random parameters *of the same stage*. For the compact description of independent random variables the SMPS format uses the INDEP keyword. It does not require the explicit definition of all parameters for all scenarios as the MPS format presented in 8.2. The SCENARIOS keyword is designed for the SBMSLPs with interstage dependencies, and hence, it also serves in the case of two-stage multiperiod structure, see II.5.4. Then, sections headed by different keywords may be combined in one STOCH file to build a complex program structure.

8.4.3 Exercise. Study the SMPS format specification from [20], change Example 8.2.2 input data, and write the corresponding CORE, TIME and STOCH files.

The SMPS format allows unified storing of previously introduced real-life problems and their data. Therefore, *test batteries* have been completed and are available via Internet (see, e.g., WATSON pension fund management test problems and POSTS collection including ALM, CEP1 – related to Chapter II.4 examples).

Several limitations have been found during the SMPS format use. Particularly, a manipulation with the dependence structure is cumbersome. For these reasons, several SMPS format extensions have been proposed and tested by Gassmann and Schweitzer. In addition, the idea to support decomposition algorithms by modeling languages has been developed for GAMS. SETSTOCH is a new tool that may serve as an example of how to link modeling languages and multistage stochastic programming solvers without the necessity of modifying either language or solver. Since 1999, a similar GAMS connection is available for the aforementioned IBM Stochastic Solutions.

After this software intermezzo, we turn back to decomposition algorithms. We may decompose the large SBMSLP to many relatively small problems related to individual scenarios. For instance, the problem corresponding to scenario $\omega^s = \{(\mathbf{A}_t^s, \mathbf{B}_t^s, \mathbf{b}_t^s, \mathbf{c}_t^s),\ t = 2, \dots, T\}$ (see II.2.2), is the following linear program:

$$(22) \qquad \min \mathbf{c}_1^\top \mathbf{x}_1 + \mathbf{c}_{k_2}^\top \mathbf{x}_{k_2} + \mathbf{c}_{k_3}^\top \mathbf{x}_{k_3} + \cdots + \mathbf{c}_{k_T}^\top \mathbf{x}_{k_T}$$

subject to constraints (10) in II.2.2. The $\mathbf{x}_1$ solution is a part of the optimal solution for the used scenario ω^s. However, this solution need not be either optimal or feasible for other scenarios. Hence, for decomposition methods with respect to scenarios, nonanticipativity constraints should be included in the form of a large system of simple binding linear equations for the second stage that spell out explicitly the corresponding requirements $\mathbf{x}_{k_t} = \mathbf{x}_{k'_t}$ for those k_t, k'_t that have the same data up to t, see II.2.2. A scenario tree can be described by the sequence of partitions $\mathcal{P}_t, t = 1, \ldots, T$ of the set $\mathcal{S}$. Given t, scenarios $s \in \mathcal{S}$ belong to the same set A in partition $\mathcal{P}_t$ when the related decision variables $\mathbf{x}_{k_t}$ belong to the same node of the scenario tree.

As done already in II.2.2, for a given scenario ω^s we denote by $\mathbf{x}(\omega^s)$ the vector of all decision variables $\mathbf{x}_{k_t}$, $t = 1, \ldots, T$ in the individual scenario problems such as (22) and (10) in II.2.2. Then, the system of nonanticipativity constraints may be written as $\mathbf{x} = \mathbf{U}\mathbf{x}$, where $\mathbf{x}$ contains grouped decision vectors $\mathbf{x}(\omega^s) \forall s \in \mathcal{S}$ and $\mathbf{U}$ is the 0-1 matrix of coefficients of the nonanticipativity constraints.

8.4.4   Exercise. Write down various matrices $\mathbf{U}$ for different two-stage and multistage problems (advice: employ permutation or projection matrices). Omit redundant constraints and analyze properties of the related penalty terms $\| \mathbf{x} - \mathbf{U}\mathbf{x} \|$ that may be used in the forthcoming MSPHA algorithm. Focus on the algorithm stopping criteria.

The *multistage progressive hedging algorithm* (MSPHA) (see Algorithm 8.3.2 for its two-stage version) is not limited to linear problems, so we further denote the program's objective as $\sum_{s \in \mathcal{S}} p^s c(\mathbf{x}^s; \omega^s)$ and scenario-related constraints as $\mathbf{x}^s \in \mathcal{X}(\omega^s)$. Nonanticipativity constraints are treated explicitly by $\mathbf{x} = \mathbf{U}\mathbf{x}$ where $\mathbf{U}$ is a projection matrix. For instance, given t, $A \in \mathcal{P}_t$, and $s \in A$, the nonanticipativity constraints may be expressed as $\mathbf{x}_t^s = \sum_{u \in A}(p^u \mathbf{x}_t^u)/(\sum_{u \in A} p^u)$. When these constraints are omitted, the corresponding penalty terms are included in the scenario-related programs objectives. Then, the scenario-based multistage stochastic programs may be solved with the following algorithm:

8.4.5   Algorithm *(MSPHA)*

> **1.** Initialization: Set the iteration counter $n := 1$, $\forall s$, t assign weights $\mathbf{w}_t^s[n] := \mathbf{0}$, and choose $\hat{\mathbf{x}}^s[n]$, $\rho > 0$, and $\varepsilon > 0$.

> **2.** $\forall s \in \mathcal{S}$ find $\mathbf{x}^s[n]$ solving $\min\{c(\mathbf{x}^s; \omega^s)[n] : \mathbf{x}^s \in \mathcal{X}(\omega^s)\}$, where
> $$c(\mathbf{x}^s; \omega^s)[n] := c(\mathbf{x}^s; \omega^s) + \sum_{t \in \mathcal{T}} \left((\mathbf{w}_t^s[n-1])^\top \mathbf{x}_t^s + \tfrac{1}{2}\rho \| \mathbf{x}_t^s - \hat{\mathbf{x}}_t^s[n-1] \|^2\right).$$

> **3.** $\forall t \in \mathcal{T} = \{1, \ldots, T\}$, $\forall A \in \mathcal{P}_t$, and $\forall s \in A$ compute new average solutions:
> $$\hat{\mathbf{x}}_t^s[n] := \left(\sum_{u \in A} p^u \mathbf{x}_t^u[n]\right)/\left(\sum_{u \in A} p^u\right).$$
> Update perturbation terms: $\mathbf{w}_t^s[n] := \mathbf{w}_t^s[n-1] + \rho(\mathbf{x}_t^s[n] - \hat{\mathbf{x}}_t^s[n])$.
> If $\hat{\mathbf{x}}_t^s[n-1]$ is close to $\hat{\mathbf{x}}_t^s[n]$ and $\mathbf{w}_t^s[n-1]$ is close to $\mathbf{w}_t^s[n]$
> (e.g., $\max\{\max_{t,s} \| \mathbf{w}_t^s[n-1] - \mathbf{w}_t^s[n] \|, \max_{t,s} \| \hat{\mathbf{x}}_t^s[n-1] - \hat{\mathbf{x}}_t^s[n] \|\} < \varepsilon$)
> then $\boxed{\text{STOP}}$, otherwise $n := n + 1$ and $\boxed{\text{GOTO 2}}$.

The implementation of MSPHA is quite straightforward. Scenario related problems are solvable, e.g., in GAMS and the external program may repeatedly update augmented terms in the objectives.

8.5 Approximation Techniques

The algorithms presented till now are designed for scenario-based stochastic programs and for programs with explicitly computed expectations. If the probability distribution of ω is continuous or discrete with a very large support, then some random sampling or deterministic approximation technique must be employed.

Random sampling may be utilized either internally or externally with respect to the optimization algorithm. For example, the *stochastic decomposition algorithm* (SDECOMP) [77] is based on random sampling within the LSHAPED algorithm 8.3.1. It does not use all scenarios but only a single incrementally increasing sample to build new and update previous cuts. In this way, the outer linearization of $q(\mathbf{x}; \omega)$ is asymptotically created for two-stage stochastic linear programs with relatively complete recourse (see [91] for an explanatory discussion and references). *Stochastic quasigradient methods* (SQG) [58] are another class of internal-sampling-based algorithms useful for nonlinear programs.

Another possibility is to approximate the original program by *external random sampling*. The main idea is to replace the expected value of the objective function by the realization of the related sample mean. Randomly generated observations of ω serve to computation of the point estimate of the objective function. Then, theoretical results about consistency, sensitivity, and asymptotic normality may be applied (see Chapter II.5). The considered stochastic programs may also be approximated by methods that are not necessarily based on random sampling.

For instance, one may *relax certain constraints* (e.g., nonanticipativity constraints in MSPHA algorithm 8.4.5), and then solve the program more easily. This provides a lower bound of the original objective. In the opposite way, one may *add several constraints*, and the optimal value of the modified program gives an upper bound.

The approximation of the objective is often based on the replacement of the objective $c(\mathbf{x}; \omega)$ with a function $\sum_i c_i(\mathbf{x}; \omega_i)$ separable in ω, where c_i are suitable approximating functions (cf. simple recourse in Example 8.2.1). Then, through this continuous approximation one replaces the original high-dimensional expectation with a combination of low-dimensional expectations to obtain an upper bound for the convex recourse program. The presence of significant *interactions between various components* ω_i of ω restricts the usefulness of this technique.

The majority of approximation methods involve *discretization* of an underlying probability measure, see Chapter II.5. The discretization may be chosen to provide upper or lower bounds on the objective function value. For example, let $\{\Omega_l, l \in \mathcal{L} = \{1, \dots, L\}\}$ be a partition of Ω ($\cup_{l \in \mathcal{L}} \Omega_l = \Omega$ and $\forall l, j \in \mathcal{L}, l \neq j : \Omega_l \cap \Omega_j = \emptyset$). Then:

$$\mathcal{Q}(\mathbf{x}) = E\, c(\mathbf{x}; \omega) = \int_\Omega c(\mathbf{x}; \omega) P(d\omega) = \sum_{l=1}^L \int_{\Omega_l} c(\mathbf{x}; \omega) P(d\omega) \approx$$

$$\sum_{l=1}^L c(\mathbf{x}; \omega^l) \int_{\Omega_l} P(d\omega) = \sum_{l=1}^L p^l c(\mathbf{x}; \omega^l) = \mathcal{Q}_L(\mathbf{x}),$$

where $\approx$ denotes 'approximately equal', $p^l = P(\omega \in \Omega_l)$, and $\omega^l \in \Omega_l$ are representative scenarios of Ω_l subsets. These scenarios are often defined as conditional expectations $\omega^l = E\{\omega \mid \omega \in \Omega_l\}$. In addition, if $c(\mathbf{x}; \omega)$ is a *convex function with respect to* ω, then $\geq$ instead of $\approx$ is valid (see Jensen's inequality in II.5.3.5), and hence, an objective function lower bound is obtained. When $\mathcal{Q}_{L'}(\mathbf{x})$ is obtained using a refinement of $\{\Omega_l : l \in \mathcal{L}\}$, then the aforementioned bound can be made tighter: $\mathcal{Q}_L(\mathbf{x}) \leq \mathcal{Q}_{L'}(\mathbf{x}) \leq \mathcal{Q}(\mathbf{x})$. This approximation may be either considered as the approximation of the objective by a step function in ω, which is constant on each set Ω_l, or it may be interpreted as an approximation of ω by a discrete random element ω_L attaining only values ω^l with probabilities p^l.

The introduced approximation techniques can be *iteratively used* for solving stochastic programs by so called approximation schemes. Firstly, for absolutely continuous probability distributions, the appropriate solver starts its work with a discretization, which is based either on random sampling or a deterministic approximation. Secondly, the relation between the original program and the approximate program is analyzed using error bounds. Thirdly, a technique for improving the accuracy, usually based on a refinement of the approximation, is applied.

We illustrate these ideas by using the L-shaped decomposition algorithm for two-stage programs (see Algorithm 8.3.1) within an approximation scheme (see [19] for details and [84] for additional references). Notice that fixed recourse is needed for the construction of valid bounds and that, in general, applicability of this bounding approach for multistage problems is restricted due to the required convexity of the random objective function with respect to ω.

8.5.1 Algorithm *(LSHAPPROX)*

1. The same initialization as in Algorithm 8.3.1 (LSHAPED) step 1. is used.
2. The same master program involving cuts is solved as in Algorithm 8.3.1 step 2. .
3. Similar programs as in the L-shaped algorithm (see Algorithm 8.3.1)step 3. are solved to obtain a new feasibility cut. The main difference in comparison with the original L-shaped algorithm is that instead of all realizations $\omega^s \in \Omega$, a lower bounding approximation is utilized and resulting ω_L and $\mathcal{Q}_L(\mathbf{x})$ are stored. If a new feasibility cut is generated, then GOTO 2 , otherwise continue with GOTO 4 .
4. Similar programs as in the L-shaped algorithm (see Algorithm 8.3.1 step 4. are solved to obtain a new optimality cut. Again, the existing lower bounding approximation is utilized. If $\theta[n] \geq d_u - \mathbf{D}_u.\mathbf{x}[n] = \mathcal{Q}_L(\mathbf{x}[n])$, then $\mathbf{x}[n]$ is optimal relative to the lower bound and GOTO 5 . Otherwise, add a new optimality cut and GOTO 2 .
5. For an upper bounding approximation ω_U find $\mathcal{Q}_U(\mathbf{x}[n])$ by a suitable procedure, see, e.g., [19]. If $\theta[n] \geq \mathcal{Q}_U(\mathbf{x}[n])$, then STOP , $\mathbf{x}[n]$ is optimal. Otherwise, refine the lower and upper bounding approximations, $n := n+1$, and GOTO 4 .

8.6   Model Management

All discussed numerical techniques for solving multistage stochastic programs are demanding as to the input data (including scenarios) and to the programming efforts. In addition, if a multistage stochastic program is to be applied repeatedly as a part of a decision support system, additional routines for data processing, scenario generation, and evaluation of results should be developed.

We can see that data of programs must be transformed in six steps to reach a solution: a formulation (schemes, texts), an algebraic form (used in modeling language), an algorithm input (MPS input files), a solver output (MPS output files), a solution description (produced by modeling language), an analysis form (tables and graphs). So, model management software should support this data flow and must coordinate it for different programs and models.

At first, we have already noted (cf. Section 8.2) that optimization techniques coded as solvers on computers do not stand alone. We know that they are packaged by modeling tools such as modeling languages. In addition, the communication with the user is established by the user interface. Such an interface may be a tool of an operating system, a separate program, or it can be a procedure joint with an optimization system. Therefore, a modeling tool is again packaged by a user interface. The user interface helps to realize the well-known idea that the principal benefit of modeling should be 'insight, not numbers'. The insight relates to understanding, and it is reached by a user with the help of various forms of user interface. A modeling language often uses a standalone editor as the input interface, and it has built-in procedures as the output interface for the presentation of results in the form defined by the user.

8.6.1   Example. The source code in INVEST.GMS file from Example 8.2.3 may continue as follows:

```
PARAMETER CPLUS(S), CMINUS(S);
  CPLUS(S) = Q*YPLUS.L(S); CMINUS(S) = R*YMINUS.L(S);
FILE    OUTPUT / RESULTS.TXT /; PUT OUTPUT; PUT 'PROFIT:'/;
PUT     'Initial W',@14,'| Goal G',@31,'| Objective Z',
PUT     @48,'| Income Q',@65,'| Rate R'/;
PUT     W:4:2,@14,'| ',G:4:2,@31,'| ',Z.L:4:2,@48,'| ',Q:4:2, @65,'| ',R:4:2/;
PUT     '------------------------------------'/;
PUT     '------------------------------------'/;
PUT     'DECISIONS X(I,T,S):'/;
PUT     'Time period ';
LOOP    (T, PUT ' | ', T.TL:14);
PUT     @48,'|',@65,'|'/; PUT 'Investment ';
LOOP    (T, PUT ' |';
  LOOP (I, PUT ' ',I.TL:4));
PUT     @48,'| Surplus YPLUS',@65,'| Defic. YMINUS'; PUT /;
LOOP    (S, PUT 'Scenario ',S.TL:3;
  LOOP (T, PUT ' |';
    LOOP(I, PUT X.L(I,T,S):5:2));
  PUT @48,'| ',YPLUS.L(S):4:2,' ',CPLUS(S):6:2;
  PUT @65,'| ',YMINUS.L(S):4:2,' ',CMINUS(S):6:2/;);
```

First, assignments and algebraic operations help to prepare report variables.

Then, PUT statements generate printout pages, see Addendum of [66] for syntax details. A tabular form of the output is formed with LOOP statements representing a cyclic flow of operations. Then, the output file RESULTS.TXT has the following content:

```
PROFIT:
Initial W   | Goal G       | Objective Z   | Income Q      | Rate R
1.80        | 2.00         | 0.08          | 1.10          | 1.40
--------------------------------------------------------------------------
DECISIONS X(I,T,S):
Time period | 1                | 2                |               |
Investment  | 1    2    3      | 1    2    3      | Surplus YPLUS | Defic.  YMINUS
Scenario 1  | 1.80 0.00  0.00  | 0.00 1.98 0.00   | 0.10    0.11  | 0.00    0.00
Scenario 2  | 1.80 0.00  0.00  | 0.00 0.00 1.89   | 0.04    0.05  | 0.00    0.00
Scenario 3  | 1.80 0.00  0.00  | 1.93 0.00 0.00   | 0.08    0.09  | 0.00    0.00
```

A standard textual presentation of the output data is supported in recent optimization systems by graphical procedures, which provide different data views. An excellent example is AIMMS [21], which incorporates advanced report writing facilities. Its project entity includes a list of related pages — windows designed by a developer for the user.

Recent applications of stochastic optimization show that it is important to use more than one deterministic equivalent for a given underlying program. These equivalents may be compared, and the most suitable or easily solvable will be chosen. The approach of using more than one deterministic equivalent (model) is called *multimodeling*. The most important contribution of multimodeling is that the user learns more about reality using different viewpoints.

The experience with multimodeling applications leads us to the idea of automatic model building support. The reason is that model building is usually a time-consuming activity, which has no formal rules. Therefore, modeling languages have been developed to simplify the task of model changes, and it is now easier to change individual data items or to switch between data sets. In contrast, structural model changes must still be done by editing source codes, and this leads to difficulties when we need to 'play' with program's structural features. These difficulties motivate the building of special purpose programs designed to support easy changes of model structure. The approach, which tries to formalize it, is called *metamodeling*.

Therefore, model management systems can be understood as implementations of multimodeling and metamodeling ideas. There are only a few such systems under development. For stochastic programs, we refer to the management system SLP-IOR [113].

II.9 BIBLIOGRAPHICAL NOTES

Introduction and preliminaries. Part II exploits stochastic programming methodology whose motivation and the first simple applications come from the mid fifties and whose history, theoretical background and introduction to numerical approaches can be found in several recent textbooks and monographs, cf. [19], [91], [130]. An extensive list of books and collections on stochastic programming is contained in the preface of [169]. The basics in linear and nonlinear programming are assumed in Part II. It is briefly surveyed in the textbooks and monographs quoted above. For more details see e.g. [7], [8].

Multistage stochastic programs. Introduction to multistage problems of stochastic programming including survey of applications and an extensive list of references can be found in [45]. We focus here on multistage stochastic programs with recourse with reference to [130] for stochastic programs with probability constraints and to [99] for integer stochastic programs. For comparisons with stochastic dynamic programming see e.g. [73] or the recent paper [54] and references ib. The illustrative Example 2.4 was motivated by [29].

Multiple criteria. There are many tangent points between stochastic and multiobjective programming, see e.g. [72]. Chapter 3 deals with selected methods and results for multi-objective programming and their applications to static stochastic models for portfolio management, including the Markowitz model [112]. Criteria based on alternative definitions of risk were compared for example in [98]. We delineate in Section 3.3 how to use the concepts of multi-objective programming for formulation and interpretation of various types of stochastic programs. For introduction to financial optimization see [176] and for the state of the art in multiobjective optimization see [78].

Selected applications in finance and economics. The first large-scale advanced applications of stochastic programming in finance and economics can be found in [39], [128] and [179]. At that time, water resources planning and management was one of favorite applications areas, see [128]. Sophisticated approaches to portfolio management from the seventies, e.g., [24], [106], have become the cornerstones of the contemporary financial applications and have contributed also to modeling and software development for multistage stochastic programs. Section 4 reflects the fact that at present, the most popular seem to be financial applications of stochastic programming; see various survey papers and collections, e.g., [44], [119], [178], and descriptions of successful applications such as [28]. Our presentation of general features of multiperiod and multistage stochastic programs in portfolio optimization in 4.4 has been supported by ideas of [40], [119] and [178].

Applications in planning and management of energy generation and transmission are booming. Their presentation in 4.6 and 4.7 is based on [108], [130] and [157]. There is an increasing interest in stochastic programming applications for planning, allocation and management of resources, capacity expansion (see Chapter 12 in [19]), production planning and optimization of technological processes (see application 4.8 motivated by [126]), in design of networks, e.g. for telecommunications, and in logistics problems. In general, macroeconomic applications are rare. They concern growth models, socio-economic decisions supporting employment, etc.

Approximation via scenarios. In practise, one uses discrete probability distributions obtained mostly by approximation of a more complex and possibly not completely known underlying probability distribution. Chapter 5 devoted to generation of such sensible discrete probability distributions is based on several papers, cf. [2], [48], [53]. The critical item is how to create the required input for multistage stochastic programs and how to use the results of the approximated problem to draw inference about the sought results of the true one. The second mentioned problem area is usually called "what-if" or output analysis. Methods of output analysis have to be tailored to the structure of the problem and they should also reflect the source, character and precision of the input data. Accordingly, the suitable approaches are based on results of asymptotic and robust statistics (we use [144] as the reference book), moment problems, on simulation techniques and on general results of parametric programming, e.g. [5], [69], [62]; see [49] and references ib. The contamination method and its application to quantification of changes due to inclusion of out-of-sample scenarios was elaborated in several papers, see e.g. [46] and [52].

Case study: Bond portfolio management problem. The case study presented in Chapter 6 exploits results obtained within the contract "HPC-Finance" of the INCO '95 project founded by Directorate General III of the European Commission in a close collaboration with Professor Marida Bertocchi and her group at the University of Bergamo. The model is similar to the two-stage multiperiod model [70]. The need for output analysis is emphasized and various methods described in Chapter 5 are illustrated. The results are based on papers [15], [50], [51] and [52].

Incomplete input information. Selected techniques suitable for analysis of the results are applied also in Chapter 7. Except for stability of the European call price with respect to the estimated volatility, these results were presented in [47].

Numerical techniques and available software. In the eighties, the special care devoted to software development has resulted in the collection [58], in the recommended input format [20], in several monographs [77], [84], [113], software packages and test batteries.

The last Chapter written by Pavel Popela surveys and explains the available numerical approaches and software. It begins with a simple educational investment example solvable by simple calculations. Nevertheless, even this simple example motivates a need of numerical optimization algorithms. Traditional optimization techniques and software can be used for solution of medium size stochastic programs (see [117] for a historical overview, [66] for the GAMS description, [65] for the AMPL guide, and [21] for the AIMMS model development tool). The description of two main decomposition algorithms (LSHAPED and PHA) and their computer implementations follows the original papers [68] and [134]. Details and extensive references on other algorithms may be found in [19]. The brief exposition on problems of model management follows ideas of [21], [113] and [126].

Part III

STOCHASTIC ANALYSIS AND DIFFUSION FINANCE

III.1 MARTINGALES

stochastic processes, Brownian motion and martingales, Markov times and stopping theorem, local martingales and complete filtrations, L_2-martingales and density theorem, Doob-Meyer decomposition, quadratic variation of local martingales, helps to some exercises

1.1 Stochastic Processes

Recall first some basic definitions. Until further notice we shall fix a probability space $(\Omega, \mathcal{F}, P)$. A **stochastic process** is a collection

$$X = (X(t), t \geq 0), \qquad \text{also} \qquad X = (X_t, t \geq 0)$$

of real valued random variables. If only a collection $(X(t), t \in T)$, $T \subset \mathbb{R}^+$, is at our disposal, we shall speak about a stochastic process on T.

$X_1, X_2, \ldots X_d$ being stochastic processes, we shall call $X = (X_1, X_2, \ldots X_d)$ a **d-dimensional stochastic process**.

A trajectory $X(\omega)$ of a stochastic process X on a $T \subset \mathbb{R}^+$ is a function in $\mathbb{R}^T$ given by $t \to X(t, \omega)$ or by $t \to X_t(\omega)$ for a fixed random element $\omega \in \Omega$.

The basic equivalence relation on the set of stochastic processes is that given by the almost sure equality in the space of trajectories. More precisely, processes X and Y on T are said to be **equivalent** if

$$X(t, \omega) = Y(t, \omega) \quad \forall t \in T \quad \omega \notin N, \quad \text{for some } N \in \mathcal{F}, \ P(N) = 0 \text{ (a P-null set)}.$$

We write either $X \overset{as}{=} Y$ or $X(t) = Y(t)$, $t \in T$, a.s.. Observe that $X \overset{as}{=} Y$ implies that $X(t) \overset{as}{=} Y(t)$ for all $t \in T$ and if this holds we say that X is a **modification** of Y (and vice versa, of course).

The same definitions and observations apply to a collection of E-valued random variables where E is a metric space. Such a collection is called a **stochastic process with states in E** or simply **E-process**. Thus, a d-dimensional stochastic process is a process with states in $\mathbb{R}^d$.

1.1.1 Example. Let $U > 0$ be a random variable with an absolutely continuous distribution, put $X(t, \omega) := I_{\{t\}}(U(\omega))$ and $Y \equiv 0$. Obviously X is a modification of Y; on the other hand X and Y are not equivalent as in fact $P[\omega : X(\omega) \equiv 0] = 0$.

Almost sure properties and implications form an important and for a beginner not quite easy part of stochastic analysis. Read carefully the following probabilistic statements that refer to some properties of processes X, Y and a positive r.v. U: Obviously,

$$t \leq U \text{ implies almost surely that } X(t) = Y(t)$$
$$\text{Outside a P- null set} \quad [t \leq U(\omega) \Rightarrow X(t, \omega) = Y(t, \omega)]$$
$$X = Y \quad \text{on } [0, U] \text{ with probability one}$$

are equivalent statements and mean exactly that

$$\exists N \in \mathcal{F}, P(N) = 0 : \omega \notin N, t \leq U(\omega) \Rightarrow X(t, \omega) = Y(t, \omega).$$

We shall agree to call a stochastic process **continuous, decreasing,** ... if all its trajectories have the corresponding property and also to call it an **almost surely continuous process, almost surely decreasing process,** ... if the corresponding property is possessed by trajectories $X(\omega)$ outside a P-null set.

Our text will be centered mainly around continuous processes which assumption may in some cases simplify our arguments :

Recall that a function $B : \mathbb{R}^+ \to \mathbb{R}$ is of **finite variation** if for all $t \in \mathbb{R}^+$

$$B^v(t) \stackrel{\text{def}}{=} \sup_{\Delta(t)} V^{\Delta(t)}(B) < \infty, \quad \Delta(t) = \{0 = t_0 < \cdots < t_j < \cdots < t_k = t\},$$

where the $\Delta(t)$'s run through all finite partitions of the interval $[0, t]$ and where $V^{\Delta(t)}(B) \stackrel{\text{def}}{=} \sum_{j=1}^{k} |B(t_j) - B(t_{j-1})|$. The map $t \to B^v(t)$ will be called here the **variation of B.**

1.1.2 Lemma. *Let X and Y be either left or right continuous processes. Then*

(a) *X is a modification of Y iff $X \stackrel{\text{as}}{=} Y$.*

(b) *If X is a continuous process of finite variation then X^v is a continuous non decreasing stochastic process.*

For a continuous process X of finite variation, the process X^v will be referred to as the **variation of X.**

The first part follows easily by continuity of X and Y and by separability of $\mathbb{R}^+$, to verify **(b)**, namely the measurability of X^v's, you need to refresh your analysis by

1.1.3   Exercise. If $B : \mathbb{R}^+ \to \mathbb{R}$ is continuous and of finite variation then $B^v : \mathbb{R}^+ \to \mathbb{R}$ is continuous such that

$$|\Delta_n(t)| \stackrel{\text{def}}{=} \max_{1 \leq j \leq k_n} |t_j^n - t_{j-1}^n| \to 0 \Rightarrow B^v(t) = \lim_n V^{\Delta_n(t)}(B), \quad \forall t \geq 0,$$

where $\Delta_n(t) \stackrel{\text{def}}{=} \{0 = t_0^n < \cdots < t_j^n < \cdots < t_{k_n}^n = t\}$.

Besides the continuous processes of finite variation, Stochastic Analysis is inhabited mainly by continuous martingales. These processes, as we shall see later on, cannot be of finite variation unless they are constant on $[0, \infty)$ almost surely. Nevertheless, there is a concept of *variation* that makes a very good sense even for processes with such irregular trajectories as martingales:

We shall say that a stochastic process X is of **finite quadratic variation** if there is a process $(\langle X \rangle(t), t \in \mathbb{R}^+)$ such that

$$(1) \qquad \langle X \rangle(t) \overset{as}{=} p \lim_n Q^{\Delta_n(t)}(X) \quad \forall |\Delta_n(t)| \to 0 \quad \forall t \in \mathbb{R}^+,$$

where $Q^{\Delta_n(t)}(X) := \sum_{j=1}^{k_n} |X(t_j^n) - X(t_{j-1}^n)|^2$ and $\eta = p \lim_n \eta_n$ means that r.v.'s η_n converge to a r.v. η in probability as $n \to \infty$. We shall also write $\eta_n \overset{P}{\to} \eta$ in this case and $\eta_n - \eta_m \overset{P}{\to} 0$ if $\{\eta_n\}$ is a Cauchy sequence in probability.

The process $\langle X \rangle$ will be called the **quadratic variation of** X. Note the following facts:

(i) The correspondence $X \to \langle X \rangle$ respects the "modification classes", i.e. if X is a modification of Y then $\langle X \rangle$ is a modification of $\langle Y \rangle$.

(ii) Even though $s \leq t \Rightarrow \langle X \rangle(s) \leq \langle X \rangle(t)$ a.s., the quadratic variation need not be increasing almost surely due to a possible lack of continuity of $\langle X \rangle$.

(iii) The definition of $\langle X \rangle$ is a rather restrictive one: The process X in 1.1 where $U \equiv 1$ is not of finite quadratic variation.

1.1.4   Exercise. A stochastic process X is of finite quadratic variation iff

$$Q^{\Delta_n(t)}(X) - Q^{\Delta_m(t)}(X) \overset{P}{\to} 0, \quad \forall |\Delta_n(t)| \to 0 \quad \forall t \in \mathbb{R}^+, \quad n, m \to \infty.$$

As we have already observed, the concepts of variation and of the quadratic variation are not very friendly, indeed:

1.1.5   Lemma. *If X is a continuous process of finite quadratic variation then for any $t \in \mathbb{R}^+$ and outside a P-null set* $X^v(t, \omega) < \infty \quad \Rightarrow \quad \langle X \rangle(t, \omega) = 0.$

Proof. Fix $t > 0$ and choose $|\Delta_n(t)| \to 0$ such that $\lim_n Q^{\Delta_n(t)}(X) = \langle X \rangle(t)$ outside a P-null set N. Because

$$Q^{\Delta_n(t)}(X) \leq \sup\{|X(u) - X(v)|, \ |u - v| \leq |\Delta_n(t)|, \ u, v \leq t\} \times V^{\Delta_n(t)}(X),$$

it follows by continuity of X that $\omega \notin N$ and $X^v(t, \omega) < \infty \Rightarrow \langle X \rangle(t, \omega) = 0$. $\square$

If L is a linear space of processes of finite quadratic variation we may define a bilinear form $\langle X, Y \rangle$ on L by

$$\langle X, Y \rangle(t) \overset{def}{=} \frac{1}{4} (\langle X + Y \rangle(t) - \langle X - Y \rangle(t)) = p \lim_n Q^{\Delta_n(t)}(X, Y) \quad \forall t \in \mathbb{R}^+,$$

where $Q^{\Delta(t)}(X, Y)$ is defined by

$$\sum_{j=1}^k (X(t_j) - X(t_{j-1})) \, (Y(t_j) - Y(t_{j-1})) = \frac{1}{4} \left(Q^{\Delta(t)}(X + Y) - Q^{\Delta(t)}(X - Y) \right).$$

The process $\langle X, Y \rangle = (\langle X, Y \rangle(t), t \geq 0)$ will be called the **covariation of** X and Y.

1.1.6   Exercise. Let L be a linear space of processes of finite quadratic variation. Then
(a) $(X, Y) \rightarrow \langle X, Y \rangle$ is a symmetric bilinear form on L with values in the set of all processes on $(\Omega, \mathcal{F}, P)$ that respects the modification equivalence classes.

Moreover, for all $s \leq t$ and outside a P-null set the following inequalities hold :

(b) $|\langle X, Y \rangle(t) - \langle X, Y \rangle(s)|^2 \leq (\langle X \rangle(t) - \langle X \rangle(s)) (\langle Y \rangle(t) - \langle Y \rangle(s))$

(c) $|\langle X, Y \rangle(t) - \langle X, Y \rangle(s)| \leq \frac{1}{2} (\langle X \rangle(t) - \langle X \rangle(s)) + \frac{1}{2} (\langle Y \rangle(t) - \langle Y \rangle(s))$.

Frequently we meet problems that ignore the home space $(\Omega, \mathcal{F}, P)$ of a stochastic process X and ask for some properties of its probability distribution, only. Recall that for a measurable map $f : (F, \mathcal{F}) \rightarrow (G, \mathcal{G})^*$ and a measure μ on $\mathcal{F}$, $f \circ \mu$ denotes the measure on $\mathcal{G}$ defined by $(f \circ \mu)(G) := \mu(f^{-1}G)$ called the f-**image of** μ.

Recall also that f is called a random variable with values in a metric space E if it is defined on a probability space $(\Omega, \mathcal{F}, \mu)$, attains its values in E and is measurable in the sense $f : (\Omega, \mathcal{F}) \rightarrow (E, \mathcal{B}(E))$ where $\mathcal{B}(E)$ denotes the Borel σ-algebra of E. We shall abbreviate $\mathcal{B}^d := \mathcal{B}(\mathbb{R}^d)$, $\mathcal{B} := \mathcal{B}^1$ and $\mathcal{B}^+ := \mathcal{B}(\mathbb{R}^+)$. Agree to write $\mathcal{L}(f) := \mathcal{L}(f|\mu) := f \circ \mu$ and to call the image measure **the probability distribution of** f.

Having a continuous stochastic process $X = (X(t), t \in \mathbb{R}^+)$ we may observe the process as a map $\omega \rightarrow X(\omega)$ defined on Ω with values in the space $C(\mathbb{R}^+)$ of all continuous functions defined on $\mathbb{R}^+$. A suitable metrization of the space $C(\mathbb{R}^+)$ that makes X a random variable with values in $C(\mathbb{R}^+)$ is provided by

1.1.7   Exercise. Put $d(x, y) := \sum_{t=1}^{\infty} 2^{-t} (\max_{s \leq t} |x(s) - y(s)| \wedge 1)$ for $x, y \in C(\mathbb{R}^+)$ and prove:
   (a) d is a metric on $C(\mathbb{R}^+)$ such that $d(x_n, x) \rightarrow 0$ iff $x_n \rightarrow x$ uniformly on each $[0, t]$, $t \in \mathbb{R}^+$.
   (b) $(C(\mathbb{R}^+), d)$ is a separable complete metric space.
   (c) The Borel σ-algebra $\mathcal{B}(C) := \mathcal{B}(C(\mathbb{R}^+))$ is generated by closed neighbourhoods $\mathcal{O}(x_0, t, \epsilon)$ of the form

$$\{x \in C(\mathbb{R}^+) : \max_{s \leq t} |x(s) - x_0(s)| \leq \epsilon\}, \quad x_0 \in C(\mathbb{R}^+), t \in \mathbb{R}^+, \epsilon > 0$$

and therefore by the sets of the form

$$\{x \in C(\mathbb{R}^+) : |x(t) - x_0(t)| \leq \epsilon\}, \quad x_0 \in C(\mathbb{R}^+), t \in \mathbb{R}^+, \epsilon > 0.$$

Let us agree that speaking about a topology on $C(\mathbb{R}^+)$ we shall always mean the metric topology defined by (a).

Any stochastic process $X = (X(t), t \in \mathbb{R}^+)$ is naturally accompanied by the system of all its **finite dimensional distributions** $\mathcal{L}(X(t_1), .., X(t_n))$ where $(t_1 < t_2 < \cdots < t_n)$ goes through all finite increasing sequences in $\mathbb{R}^+$.

As expected, the probability distribution of a continuous process X is uniquely determined by its finite dimensional distributions.

---

*$\mathcal{F}$ and $\mathcal{G}$ are σ-algebras of subsets in F and G, respectively, such that $f^{-1}B \in \mathcal{F}$ for all $B \in \mathcal{G}$.

1.1.8   Lemma. *If X is a continuous process then $\omega \to X(\omega)$ is a $C(\mathbb{R}^+)$-valued random variable and such as, it has the probability distribution on $C(\mathbb{R}^+)$. Moreover, continuous processes X and Y are equally distributed on $C(\mathbb{R}^+)$ iff they have identical finite dimensional distributions.*

To prove such a statement we can not avoid a Dynkin argument: Consider a nonempty set Ω, a family of its subsets $\mathcal{S}$ and a family of maps $X_u : \Omega \to E_u$ where u goes through an index set U and each E_u is a metric space. We shall denote by $\sigma(\mathcal{S})$ the minimal σ-algebra that contains each $S \in \mathcal{S}$ and

$$\sigma(X_u) := \{[X_u \in B_u], B \in \mathcal{B}(E_u)\}, \quad \sigma(X_u, u \in U) := \sigma\left(\bigcup_{u \in U} \sigma(X_u)\right).$$

Obviously, having a measurable space $(\Omega, \mathcal{F})$ then X_u is an E_u-valued random variable defined on $(\Omega, \mathcal{F})$ if and only if $\sigma(X_u) \subset \mathcal{F}$.

The operator σ will appear in our proofs by means of promised

1.1.9   Dynkin Arguments.

(a) *If $X : \Omega \to E$ where E is a metric space and if $\mathcal{B}(E) = \sigma(\mathcal{H})$ then $\sigma(X) = \sigma([X \in H], H \in \mathcal{H})$.*

(b) *Let $\mathcal{X}$ be a vector space of functions $X : \Omega \to \mathbb{R}$ that contains constants, it is closed with respect to uniform convergence and posses the following property: If $X_n \in \mathcal{X}$ are nonnegative and uniformly bounded and $X_n \uparrow X$ on Ω then $X \in \mathcal{X}$.*
If $\mathcal{X}$ contains a subset $\mathcal{A}$ that is closed under multiplication, then any bounded $(\Omega, \sigma(\mathcal{A})) \to (\mathbb{R}, \mathcal{B}(\mathbb{R}))$-measurable function belongs to $\mathcal{X}$.
Weaker forms of (b) are:

(c) *Let $\mathcal{X}$ be a family of subsets of Ω such that $B_n \uparrow\downarrow B$ implies that $B \in \mathcal{X}$ for any sequence $(B_n) \subset \mathcal{X}$. If $\mathcal{A} \subset \mathcal{X}$ is an algebra then any $B \in \sigma(\mathcal{A})$ is a set in $\mathcal{X}$.*

(d) *Let $\mathcal{X}$ be a family of subsets of Ω such that*

$$\Omega \in \mathcal{X}, \quad B_1 \cap B_2 = \emptyset \Rightarrow B_1 \cup B_2 \in \mathcal{X}, \quad B_1 \subset B_2 \Rightarrow B_2 - B_1 \in \mathcal{X},$$

$$B_n \subset B_{n+1} \Rightarrow \bigcup_{n=1}^{\infty} B_n \in \mathcal{X}, \quad B_n \in \mathcal{X}$$

holds. If $\mathcal{A} \subset \mathcal{X}$ is a family closed under finite intersections then any $B \in \sigma(\mathcal{A})$ belongs to $\mathcal{X}$.

(e) *Let X and Y be random variables on $(\Omega, \mathcal{F})$ with values in a separable complete metric space E and an arbitrary metric space Z, respectively. Then X is a $\sigma(Y)$-measurable random variable if and only if there exists a Borel map $f : Z \to E$ such that $X = f(Y)$ holds everywhere on Ω.*

Proofs may be found in [135] or in [37].

Proof of 1.1.8. The first statement follows by 1.7 (c) and 1.9 (a). To prove the second one denote by $\mathcal{X}$ the family of all Borel sets $B \subset C(\mathbb{R}^+)$ such that $P[X \in B] = P[Y \in B]$ holds. Check that it owns all properties required by 1.9 (d). Denoting by $\mathcal{A}$ the family of all finite intersections of the sets $\{x \in C(\mathbb{R}^+) : |x(t) - x_0(t)| \le \epsilon\}$ it follows that $\mathcal{L}(X) = \mathcal{L}(Y)$ by 1.7 (c) and by 1.9 (d). $\square$

We will summarize the basic ingredients that allow to construct a (continuous) stochastic process $X = (X(t), t \in \mathbb{R}^+)$ with given finite dimensional marginals. In particular, consider a family of distribution functions

(2) $$\{F_{t_1,\ldots,t_n}(x_1,\ldots,x_n), \quad 0 \le t_1 < \cdots < t_n < \infty, \quad n \in \mathbb{N}\}$$

and ask if there is a (continuous) process X such that

(3) $$F_{t_1,\ldots,t_n}(x_1,\ldots,x_n) = P[X(t_1) \le x_1 \ldots, X(t_n) \le x_n],$$
$$0 \le t_1 < \cdots < t_n < \infty, \quad x_j \in \mathbb{R}, \quad n \in \mathbb{N}$$

holds. Note that if a process X exists such that (3) holds then the family of distribution functions (2) is **consistent**, i.e. such that $F_{s_1,\ldots,s_m}$ is the marginal distribution function of $F_{t_1,\ldots,t_n}$ whenever $\{s_1,\ldots,s_m\} \subset \{t_1,\ldots,t_n\}$.

The classical *Daniell-Kolmogorov Theorem* (see, Corollary 35.4 in [6], p.303) slightly simplified says

1.1.10. *For any consistent system of distribution functions (2) there exists a stochastic process X such that (3) holds.*

Once the existence of a process X with given finite dimensional distributions is established, a natural question arises: Under which conditions can X be modified to a continuous process Y? *Kolmogorov-Chenstov Theorem* (see, 39.3 in [6], p. 335) provides a partial answer.

1.1.11. *Let $X = (X(t), t \ge 0)$ be a stochastic process such that*

(4) $$E|X(t) - X(s)|^\alpha \le K |t - s|^{1+\beta}, \quad \forall t, s \in \mathbb{R}^+$$

holds for some α, β, $K > 0$. Then X has a continuous modification.

The natural state space for a continuous d-dimensional process $X = (X_1,\ldots,X_d)$ is the space of continuous maps from $\mathbb{R}^+$ to $\mathbb{R}^d$ denoted as $C(\mathbb{R}^+, \mathbb{R}^d)$ with the topology of uniform convergence on compact intervals in $\mathbb{R}^+$. Obviously, the space is metrizable to a separable complete metric space because it is homeomorphic to the separable complete product space $\bigotimes_1^d C(\mathbb{R}^+)$. We venture to identify both spaces and apply the notation $C(\mathbb{R}^+, \mathbb{R}^d) = \bigotimes_1^d C(\mathbb{R}^+)$.

Finite dimensional distributions of an d-dimensional stochastic process $X = (X_1,\ldots,X_d)$ are given as

$$\mathcal{L}(X(t_1),\ldots,X(t_n)), \quad 0 \le t_1 \le \ldots t_n < \infty, \quad n \in \mathbb{N},$$

where $\mathcal{L}(X(t_1), \ldots, X(t_n))$ is a Borel probability measure on $\mathbb{R}^{dn}$.

If $Y = (Y_1, \ldots, Y_d)$ is another d-dimensional process, defined perhaps on a different probability space, **agree to write** $X \stackrel{d}{=} Y$ if the processes X and Y have the identical finite dimensional distributions.

Since $C(\mathbb{R}^+, \mathbb{R}^d)$ is a separable metric space we may repeat arguments for 1.7 (c) to see that the Borel σ-algebra of $C(\mathbb{R}^+, \mathbb{R}^d)$ is generated by sets of the form $\{x \in C(\mathbb{R}^+, \mathbb{R}^d) : ||x(t) - x_0(t)|| \leq \epsilon\}$ where $x_0 \in C(\mathbb{R}^+, \mathbb{R}^d)$, $t \geq 0$, $\epsilon > 0$ and $||x||$ denotes the Eucleidian norm in $\mathbb{R}^d$. This and 1.9 (a),(d) yields

1.1.12 Lemma. *Let* $X = (X_1, \ldots, X_d)$ *be a d-dimensional continuous process. Then X is a $C(\mathbb{R}^+, \mathbb{R}^d)$-valued random variable such that*

$$\sigma(X) = \sigma(X(t), t \geq 0) = \sigma(X_j(t), t \geq 0, 1 \leq j \leq d)$$

holds. Moreover, $\mathcal{L}(X) = \mathcal{L}(Y)$ *if and only if* $X \stackrel{d}{=} Y$ *holds.*

We will say that $X = (X_1, \ldots, X_d)$ is a d-dimensional **Gaussian process** if all its finite dimensional distributions are normal. Thus, the $C(\mathbb{R}^+, \mathbb{R}^d)$-probability distribution of a **continuous** Gaussian process X is according to 1.12 uniquely determined by its **mean** and by its **covariance** matrix function given as

$$EX := (EX_1, \ldots, EX_n), \quad (s, t) \to \{\operatorname{cov}(X_i(s), X_j(t))\}_{i,j=1}^d.$$

A Gaussian process X with $EX(t) = 0$ for all $t \in \mathbb{R}^+$ will be called a **centered** Gaussian stochastic process.

The principal definitions of stochastic calculus are as follows:

Having a measurable space $(\Omega, \mathcal{F})$ we call $(\mathcal{F}_t, t \geq 0)$ a **filtration** of the measurable space if any $\mathcal{F}_t \subset \mathcal{F}$ is a σ-algebra and $\mathcal{F}_s \subset \mathcal{F}_t$ whenever $s \leq t$. We agree to denote $\mathcal{F}_\infty := \sigma\left(\bigcup_{t \geq 0} \mathcal{F}_t\right)$.

If X is an E-valued process on $(\Omega, \mathcal{F})$, where E is a metric space, we denote

$$\mathcal{F}_t^X := \sigma(X(s), s \leq t), \quad \mathcal{F}_\infty^X := \sigma(X(t), t \geq 0)$$

and call the filtration $(\mathcal{F}_t^X)$ the **canonical filtration** of X.

Having a filtration $(\mathcal{F}_t)$ of $(\Omega, \mathcal{F})$ we shall say that an E-valued process X on $(\Omega, \mathcal{F})$ is $\mathcal{F}_t$-**adapted process** if $\mathcal{F}_t^X \subset \mathcal{F}_t$ holds for all $t \geq 0$. This means that X is an $\mathcal{F}_t$-adapted process if and only if $X(t)$ is an $\mathcal{F}_t$-measurable E-valued random variable for all $t \geq 0$.

Observe that if X is a process on $\mathbb{R}^+$ whose trajectory records a random motion of a particle then the σ-algebra $\mathcal{F}_t^X$ may be interpreted as the history of the motion up to time t because it lists precisely those random events that may or may not happen to the particle on the time interval $[0, t]$ while $\mathcal{F}_\infty^X$ is the σ-algebra that records the complete motion of the particle.

A very natural continuous process lives on the measurable space $(C(\mathbb{R}^+), \mathcal{B}(C))$: It is defined by $\mathbf{x}(t, x) = x(t)$ for $x \in C(\mathbb{R}^+)$ and $t \in \mathbb{R}^+$ and called the **canonical or coordinate process**. Actually, the process is prepared to posses any probability distribution Q on $C(\mathbb{R}^+)$ we may choose just considering it as a stochastic process on $(C(\mathbb{R}^+), \mathcal{B}(C), Q)$. Having a continuous process $(\Omega, \mathcal{F}, P, X)$ we shall call the process $(C(\mathbb{R}^+), \mathcal{B}(C), \mathcal{L}(X|P), \mathbf{x})$ the **canonical representation of the process** X **on** $C(\mathbb{R}^+)$. Obviously,

$$(5) \qquad \mathcal{F}_t^\mathbf{x} = \left\{ p_t^{-1}(B_t), B_t \in \mathcal{B}(C[0,t]) \right\}, \quad \mathcal{F}_\infty^\mathbf{x} = \mathcal{B}(C(\mathbb{R}^+))$$

holds by 1.12. Here $C[0,t]$ denotes the space of continuous functions $[0,t] \to \mathbb{R}$ with usual topology of uniform convergence and $p_t : C(\mathbb{R}^+) \to C[0,t]$ the continuous projection.

A continuous time version to Daniell-Kolmogorov theorem we appreciate later on is provided by *Varadhan Theorem*, see [156], p.34.

1.1.13. For any $t \geq 0$ let Q^t be a probability measure on $\mathcal{F}_t^\mathbf{x}$ such that the family $(Q^t, t \in \mathbb{R}^+)$ is consistent in the sense $s \leq t \Rightarrow Q^t|\mathcal{F}_s^\mathbf{x} = Q^s$. Then there is a unique probability measure Q on $\mathcal{F}_\infty^\mathbf{x}$ such that $Q|\mathcal{F}_t^\mathbf{x} = Q^t$ holds for any $t \in \mathbb{R}^+$.

1.2   Brownian Motion and Martingales

Recall that a stochastic process X is said to have **independent increments** if the random variables

$$(6) \qquad X(t_1) - X(0), X(t_2) - X(t_1), \ldots, X(t_n) - X(t_{n-1}) \quad \text{are independent}$$

for all finite sequences $0 < t_1 < \cdots < t_n < \infty$. Both above properties are met by **Brownian motion** that is defined as a centered Gaussian process with $\operatorname{cov}(X(s), X(t)) = s \wedge t$ for $s, t \in \mathbb{R}^+$. The basic existence properties are provided by

1.2.1   Theorem.

(a) *A process* X *is a Brownian motion iff* $X(0) \overset{\text{as}}{=} 0$, *it has independent increments, and* $\mathcal{L}(X(t) - X(s)) = N(0, |t - s|)$ *holds for* $t, s \in \mathbb{R}^+$.

(b) *Brownian motion exists and each Brownian motion can be modified to a continuous process.*

A continuous Brownian motion will be called **Wiener process** and denoted mostly as $W = (W(t), t \geq 0)$. Observe that $W(0) \overset{\text{as}}{=} 0$ and $\mathcal{L}(W(t)) = N(0, t)$.

Proof. (a) If X is a Brownian motion then $\operatorname{cov}(X(t) - X(u), X(u) - X(s)) = 0$ for $s \leq u \leq t$, and the random vector (6) has a normal distribution. It follows easily that the increments are independent and $N(0, t_j - t_{j-1})$ random variables. Because the above reasoning is easily seen to be reversible the equivalence (a) is verified.

(b) Consider $0 \leq t_1 < \cdots < t_n < \infty$ and put $F_{t_1, \ldots, t_n} = N_n(\mathbf{0}, \|t_j \wedge t_k\|_{k,j=1}^n)$. Because $(s, t) \to s \wedge t$ is a positively semidefinite function on $\mathbb{R}^+ \times \mathbb{R}^+$ the $F_{t_1, \ldots, t_n}$'s are well defined normal distributions that obviously follow the consistency requirement of 1.10. It follows by the theorem that there is a stochastic process X with finite dimensional distributions given by (3), hence X is a Brownian motion. By (a) we get $E|X(t) - X(s)|^4 = 3|t - s|^2$ for all $t, s \geq 0$ and X possesses a continuous modification by 1.11. □

The motion as a physical phenomenon was first observed by English botanist R. Brown in 1897 as the motion of pollen particles in a liquid due to the incessant hitting of pollen by smaller liquid molecules. Amazingly, the first qualitative treatment of Brownian motion suggested by L. Bachelier in 1900 ([4]) was inspired by the stock price fluctuations. The physical and consequently mathematical theory of Brownian motion was set up by A. Einstein in 1906 ([56]) and by N. Wiener in 1923 ([171]). Wiener process, i.e. the continuous Brownian motion has since well established a claim to be one of the cornerstones of both theoretical and applied modern probability. Unquestionably, it is also the fundamental stochastic process for the present text due to its rôle of the stochastic driver of the stock prices dynamics in a model we shall present in Chapter 3.

1.2.2 Stability and the Quadratic Variation of W.

(a) For any $t > 0$ $(W_t(s) := W(t + s) - W(t), s \geq 0)$ is a Wiener process independent of the σ-algebra $\overleftarrow{\mathcal{F}_t^W} := \sigma\{W(u), u \leq t\}$.

(b) The process $\overleftarrow{W}$ defined by $\overleftarrow{W}(t) := tW(t^{-1})$, $\overleftarrow{W}(0) = 0$ is a Brownian motion with trajectories continuous on $(0, \infty)$.

(c) W is a process of finite quadratic variation with $\langle W \rangle(t) \overset{as}{=} t$ for all $t \in \mathbb{R}^+$.

Proof. There are two items of the above statements that need a finer treatment: Fix $t, s \in \mathbb{R}^+$ and denote

$$\mathcal{D}_n := \sigma \left\{ W((k+1)t2^{-n}) - W(kt2^{-n}), \quad 0 \leq k \leq 2^n - 1 \right\}, \quad \mathcal{D} := \bigcup_n \mathcal{D}_n.$$

It follows by definition of W that the random variable $W_t(s)$ is independent of the algebra $\mathcal{D}$, hence also of the σ-algebra $\sigma(\mathcal{D})$. By continuity of W we get $\mathcal{F}_t^W = \sigma(\mathcal{D})$ that establishes the independence stated by (a).

Fix $t \in \mathbb{R}^+$ and let $|\Delta_n(t)| \to 0$. Then

$$E(Q^{\Delta_n(t)}(W) - t)^2 = \mathrm{Var}(Q^{\Delta_n(t)}(W)) \leq \sum_{j=1}^{k_n} E\left(W(t_j^n) - W(t_{j-1}^n)\right)^4$$

$$= 3 \sum_j (t_j^n - t_{j-1}^n)^2 \leq 3|\Delta_n(t)| \, t \to 0 \quad \text{as} \quad n \to \infty$$

which proves (c) □

From the point of view of mathematical analysis the trajectories of W are fairly pathological:

1.2.3 Trajectories of Wiener Process W.

Denote by $Z_a(\omega)$ the set of $t \in \mathbb{R}^+$ such that $W(t, \omega) = a$ for $a \in \mathbb{R}$ and $\omega \in \Omega$. Then outside a P-null set

(a) $W^v(t, \omega) = \infty$ for $t > 0$.

(b) $\liminf_{t\to\infty} W(t,\omega) = -\infty$, and $\limsup_{t\to\infty} W(t,\omega) = +\infty$.

(c) $t^{-1}W(t,\omega) \to 0$ as $t \to \infty$.

(d) $\sup Z_a(\omega) = \infty$ $\quad \forall a \in \mathbb{R}$, $\quad \inf[Z_0(\omega) \cap (0,\infty)] = 0$.

We leave the proof to our reader who may prefer to consult any standard text book on modern probability, see [6], Chapter IX], for example. We complement 2.3 by

1.2.4 More on Trajectories of W.

(a) $\limsup_{t\to\infty} \dfrac{W(t)}{\sqrt{2t\ln\ln t}} \overset{as}{=} 1$, $\quad \liminf_{t\to\infty} \dfrac{W(t)}{\sqrt{2t\ln\ln t}} \overset{as}{=} -1$.

(b) *Outside a P-null set a trajectory of a Wiener process is nowhere differentiable on $(0,\infty)$.*

A very deep assertion **(a)** is called the **law of iterated logarithm**. We refer to [6], Chapter IX, again, for the proofs.

The natural state space for Brownian motion is of course $\mathbb{R}^3$. We shall say that a *continuous* d-dimensional stochastic process $W = (W_1, W_2, \ldots, W_d)$ with $W(0) \overset{as}{=} 0$ is a **d-dimensional Wiener process** if it has independent increments

$$W(t_1) - W(t_0),\ W(t_2) - W(t_1),\ldots, W(t_n) - W(t_{n-1}) \qquad \forall t_0 < t_1 < .. < t_n, n \in \mathbb{N}$$

and $\mathcal{L}(W(t) - W(s)) = N_d(0, |t - s| \cdot I_d)$ for $t, s \in \mathbb{R}^+$ where I_d denotes the unit $d \times d$ matrix. Hence, a d-dimensional Wiener process is a centered Gaussian process with the covariance function given as $(s,t) \to (s \wedge t) \cdot I_d$.

Only elementary arguments are needed to prove the following statements:

1.2.5 Theorem.

(a) $W = (W^1, \cdots, W^d)$ *is a d-dimensional Wiener process iff* $W^1, \cdots, W^d$ *are independent one dimensional Wiener processes.*

(b) *Let W be a d-dimensional Wiener process, fix $t > 0$. Then*
$$W_t := (W(s+t) - W(t),\ s \geq 0) \quad \text{is a d- dimensional Wiener process independent of the σ-algebra } \mathcal{F}_t^W \text{ where } (\mathcal{F}_u^W) \text{ denotes the canonical filtration of } W.$$

Wiener process is also a martingale.* The concept of the martingale was first introduced by J. Ville in 1939, see [166], and the pioneering fundamental results were discovered for the most part by P. Lévy and J.L. Doob in the period 1934-1950, see [41].

Recall that a stochastic process $X = (X(t), t \in T)$, where $T \subset \mathbb{R}^+$, is a **martingale** if for all $t \in T$

$$X(t) \in L_1 \quad \text{and} \quad s \leq t \Rightarrow E[X(t)|\mathcal{F}_s^X] = X(s).$$

Test your proficiency in the conditional expectations calculus, see 2.7 below, and verify the martingale property of W and some of its transformations.

---

*A strap connecting a horse's girth to the reins so as to hold down its head (the French language of the 15th century), later on a special gambling strategy.

1.2.6 Exercise. If $\lambda \in \mathbb{R}$ then $(W(t), t \in \mathbb{R}^+)$, $(W^2(t) - t, t \in \mathbb{R}^+)$ and
$\mathcal{E}(W, \lambda) \overset{\text{def}}{=} (\exp\{\lambda W(t) - \frac{\lambda^2}{2}t\}, t \in \mathbb{R}^+)$ are martingales.

1.2.7 Conditional Expectation Calculus. For any σ-algebra $\mathcal{G} \subset \mathcal{F}$ there exists an almost surely unique linear map $E^{\mathcal{G}} : L_1(\Omega, \mathcal{F}, P) \to L_1(\Omega, \mathcal{G}, P|\mathcal{G})$ such that

$$\int_G X \, dP = \int_G E^{\mathcal{G}} X \, dP, \quad G \in \mathcal{G}, X \in L_1(\mathcal{F}).$$

We shall also write $E[X|\mathcal{G}]$ instead of $E^{\mathcal{G}} X$. The following properties hold whenever the corresponding (conditional) expectations are defined:

(a) $X \geq 0$ a.s. $\Rightarrow$ $E^{\mathcal{G}} X \geq 0$ a.s.
(b) $g : \mathbb{R} \to \mathbb{R}$ convex $\Rightarrow$ $g \circ E^{\mathcal{G}} X \leq E^{\mathcal{G}} g(X)$ a.s., $|E^{\mathcal{G}} X| \leq E^{\mathcal{G}} |X|$ a.s.
(c) $E^{\mathcal{G}}(YX) = Y E^{\mathcal{G}} X$ a.s. if $Y \in \mathcal{G}$ and $YX \in L_1$.
(d) $E^{\mathcal{D}} E^{\mathcal{G}} X = E^{\mathcal{D}} X$ a.s. if $\mathcal{D} \subset \mathcal{G}$.
(e) $E[X|\mathcal{G} \vee \mathcal{D}] = E[X|\mathcal{G}]$ a.s. if $\sigma(X) \vee \mathcal{G}$ and $\mathcal{D}$ are independent.
(f) $X_n \overset{L_1}{\to} X$ $\Rightarrow$ $E^{\mathcal{G}} X_n \overset{L_1}{\to} E^{\mathcal{G}} X$.
(g) $(X_t, t \in T)$ uniformly integrable $\Rightarrow$ $(E[X_t|\mathcal{F}_t], t \in T)$ uniformly integrable.
(h) $0 \leq X_n \uparrow X$ a.s. $\Rightarrow$ $E^{\mathcal{G}} X_n \uparrow E^{\mathcal{G}} X$ a.s.
(i) $X_n \geq 0$ a.s. $\Rightarrow$ $E[\liminf_n X_n|\mathcal{G}] \leq \liminf_n E[X_n|\mathcal{G}]$ a.s.
(j) $\mathcal{F}_1 \supset \mathcal{F}_2 \supset \cdots \supset \mathcal{F}_n \supset \ldots,$ $X \in L_1$ $\Rightarrow$ $E[X|\mathcal{F}_n] \to E[X|\cap_n \mathcal{F}_n]$ both almost surely and in L_1.
(k) $\mathcal{F}_1 \subset \mathcal{F}_2 \subset \cdots \subset \mathcal{F}_n \subset \ldots,$ $X \in L_1$ $\Rightarrow$ $E[X|\mathcal{F}_n] \to E[X|\sigma \cup_n \mathcal{F}_n]$ both almost surely and in L_1.

By $Y \in \mathcal{G}$ we denoted and will continue to denote that a function $Y : \Omega \to \mathbb{R}$ is $\mathcal{G} \to \mathcal{B}$ -measurable random variable for a σ-algebra $\mathcal{G}$ of subsets of Ω. By $\mathcal{G} \vee \mathcal{D}$ we mean the σ-algebra $\sigma(\mathcal{G} \cup \mathcal{D})$.

See paragraph 15 in [6] for the proofs.

We are prepared to extend usefully the concept of Wiener process and that of martingale with respect to a given filtration $(\mathcal{F}_t)$.

We shall say that a d-dimensional Wiener process W is an $\mathcal{F}_t$-**Wiener process** if $W = (W_1, \ldots, W_d)$ is an $\mathcal{F}_t$ -adapted Wiener process such that

(7) $0 \leq s \leq t < \infty$ $\Rightarrow$ $W(t) - W(s)$ and $\mathcal{F}_s$ are independent.

We shall say that a one dimensional process X is an $\mathcal{F}_t$ - **martingale, (sub-martingale, supermartingale)** if X is an $\mathcal{F}_t$ -adapted process with $X(t) \in L_1$ for all $t \in \mathbb{R}^+$ such that

$$0 \leq s \leq t < \infty \quad \Rightarrow \quad E^{\mathcal{F}_s} X(t) \overset{\text{as}}{=} X(s), \quad (\geq \text{a.s.}, \ \leq \text{a.s.}),$$

respectively.

1.2.8 Stability of W w.r.t. a Filtration Change. *Let W be a d-dimensional $\mathcal{F}_t$-Wiener process. Put $\mathcal{F}_{t+} \overset{def}{=} \cap_{h>0}\mathcal{F}_{t+h}$. Then*

(a) *W is a $\mathcal{D}_t$ -Wiener process for any filtration $(\mathcal{D}_t)$ such that $\mathcal{F}_t^X \subset \mathcal{D}_t \subset \mathcal{F}_t$ for all $t \in \mathbb{R}$.*

(b) *W is an $\mathcal{F}_{t+}$ -Wiener process.*

(c) *W is an $\mathcal{F}_t \vee \mathcal{G}_t$ -Wiener process for any filtration $\mathcal{G}_t$ such that $\mathcal{F}_t$ and $\mathcal{G}_t$ are independent for all $t \in \mathbb{R}^+$ ($\iff$ $\mathcal{F}_\infty$ and $\mathcal{G}_\infty$ are independent σ-algebras).*

Thus, Wiener property (7) is safely preserved when moving the filtration $(\mathcal{F}_t)$ downwards as long as we keep the process W to be adapted. An enlargement of $(\mathcal{F}_t)$ is possible only in special cases as in **(b)** and **(c)**.

Proof. To verify **(b)** observe that if $t > s$ then $W(t) - W(s + h)$ and $\mathcal{F}_{s+}$ are independent for any $h > 0$, hence the assertion follows by continuity of W.

Having W, $(\mathcal{F}_t)$ and $(\mathcal{G}_t)$ as in **(c)**, the σ-algebras $\sigma(W(t) - W(s)) \vee \mathcal{F}_s \subset \mathcal{F}_\infty$ and $\mathcal{G}_s$ are independent for $t > s$. It follows immediately that $W(t) - W(s)$ and $\{F \cap G, F \in \mathcal{F}_s, G \in \mathcal{G}_s\}$ are independent also, which implies **(c)** because the latter family of sets is closed w.r.t. finite intersections and generates $\mathcal{F}_s \vee \mathcal{G}_s$. $\square$

1.2.9 Stability of Martingale Property.

(a) *If X is an $\mathcal{F}_t$ -martingale then it is also a $\mathcal{D}_t$ -martingale for any filtration $(\mathcal{D}_t)$ such that $\mathcal{F}_t^X \subset \mathcal{D}_t \subset \mathcal{F}_t$ holds for any $t \in \mathbb{R}^+$.*

(b) *Any continuous $\mathcal{F}_t$ -martingale X is $\mathcal{F}_{t+}$ -martingale.*

(c) *Let $(\mathcal{F}_t)$ and $(\mathcal{G}_t)$ be independent filtrations as in 2.8 (c), X an $\mathcal{F}_t$-martingale and Y a $\mathcal{G}_t$-martingale. Then X, Y and XY are $\mathcal{F}_t \vee \mathcal{G}_t$ -martingales.*

(d) *Let X be an $\mathcal{F}_t$ -martingale (a submartingale) such that $f(X_t)$ is in L_1 for $t \in \mathbb{R}^+$ for a convex (a convex non-decreasing) $f : \mathbb{R} \to \mathbb{R}$. Then $f(X) := (f(X_t), t \geq 0)$ is an $\mathcal{F}_t$ -submartingale.*

Proof. **(a)** follows easily when using **(d)** in 2.7. To prove **(b)** apply 2.7 (j) and the continuity of X to get for $s < t$

$$E[X(t)|\mathcal{F}_{s+}] = L_1 \lim_{h \to 0+} E^{\mathcal{F}_{s+h}}X(t) = L_1 \lim_{h \to 0+} X(s + h) \overset{as}{=} X(s).$$

X and Y in **(c)** are $\mathcal{F}_t \vee \mathcal{G}_t$ -martingales by 2.7 (e). To prove this for XY observe that the process is $\mathcal{F}_t \vee \mathcal{G}_t$ -adapted and $X(t)Y(t) \in L_1$ because $X(t)$ and $Y(t)$ are independent integrable variables. Fix $s < t$, consider $F_s \in \mathcal{F}_s$ and $G_s \in \mathcal{G}_s$. Because $X(t)I_{F_s}, Y(t)I_{G_s}$ and $X(s)I_{F_s}, Y(s)I_{G_s}$ are pairs of independent variables

$$\int_{F_sG_s} X(t)Y(t)\, dP = \int_{F_s} X(t)\, dP \cdot \int_{G_s} Y(t)\, dP = \int_{F_sG_s} X(s)Y(s)\, dP,$$

that obviously implies the martingale equality $E[X(t)Y(t)|\mathcal{F}_s \vee \mathcal{G}_s] \overset{as}{=} X(s)Y(s)$.

Finally, **(d)** is a direct consequence of Jensen Inequality 2.7 (b). $\square$

Next, we are going to establish a pair of important martingale inequalities for $X^*(t) \overset{def}{=} \sup_{0 \leq s \leq t} |X(s)|$ which quantity is easily seen to be a $[0, \infty]$-valued random variable if X is a right-continuous process. Observe also that $X^* \overset{def}{=} (X^*(t), t \in \mathbb{R}^+)$ is a continuous ($\mathcal{F}_t$ -adapted) process whenever the process X has the property.

1.2.10 Doob's Inequalities. *If X is a right-continuous martingale or a nonnegative submartingale, then*

(a) $p \geq 1 \quad \Rightarrow \quad P[\sup_{0 \leq s \leq t} |X(s)| \geq a] \leq a^{-p} E|X(t)|^p, \quad t \in \mathbb{R}^+, a > 0.$

(b) $p > 1 \quad \Rightarrow \quad E \sup_{0 \leq s \leq t} |X(s)|^p \leq (\frac{p}{p-1})^p E|X(t)|^p.$

A discrete version of the above inequalities provides

1.2.11 Lemma. *If $X_1, X_2, \ldots, X_n$ is a martingale or a nonnegative submartingale then*

(a) $p \geq 1 \quad \Rightarrow \quad P[\max_k |X_k| \geq a] \leq a^{-p} E|X_n|^p, \quad a > 0.$

(b) $p > 1 \quad \Rightarrow \quad E(\max_k |X_k|)^p \leq (\frac{p}{p-1})^p E|X_n|^p.$

Proof. Assume first that $X_1, \ldots, X_n$ is a nonnegative submartingale and denote $F_k := [X_1 < a, \ldots, X_{k-1} < a, X_k \geq a] \in \sigma(X_1, \ldots, X_k)$ for $1 \leq k \leq n$. Hence, by submartingale property $\int_{F_k} X_k \, dP \leq \int_{F_k} X_n \, dP$ and finally, if $Y := \max_k X_k$,

$$(8) \qquad aP[Y \geq a] = a \sum_k P(F_k) \leq \sum_k \int_{F_k} X_k \, dP \leq \int_{[Y \geq a]} X_n \, dP.$$

To prove **(b)** denote $Y := \max_k |X_k|$, fix $c > 0$ and apply (8) to the submartingale $|X_1|, \ldots, |X_n|$ to get

$$E(Y \wedge c)^p = E\left[\int_0^{Y \wedge c} pa^{p-1} \, da\right] = E \int_0^c pa^{p-1} I_{[Y \geq a]} \, da$$

$$= \int_0^c pa^{p-1} P[Y \geq a] \, da \leq \int_0^c pa^{p-2} E|X_n| I_{[Y \geq a]} \, da$$

$$= pE\left[|X_n| \int_0^{Y \wedge c} a^{p-2} \, da\right] = \frac{p}{p-1} E\left[|X_n|(Y \wedge c)^{p-1}\right].$$

Hölder inequality with the exponents p and $\frac{p}{p-1}$ yields

$$E(Y \wedge c)^p \leq \frac{p}{p-1} [E[(Y \wedge c)^p]^{(p-1)/p} [E|X_n|^p]^{1/p}$$

and finally

$$E[(Y \wedge c)^p] \leq \left(\frac{p}{p-1}\right)^p E|X_n|^p.$$

The proof of **(b)** is completed by letting $c \to \infty$. To prove **(a)** we may assume that $X_n \in L_p$ which according to **(b)** implies that $X_j \in L_p$ for any $1 \leq j \leq n$ and therefore $|X_1|^p, \ldots, |X_n|^p$ is a submartingale by 2.9 (d). The **(a)** inequality now follows in both cases by (8). □

Proof of 1.2.10. The right-continuity of X implies that

$$P[X^*(t) > a] = \lim_{n \to \infty} P\left[\max_{0 \le k \le 2^n} \left|X\left(\frac{kt}{2^n}\right)\right| > a\right] \le a^{-p} E|X(t)|^p, \quad \forall p \ge 1$$

and

$$E(X^*(t))^p = \lim_{n \to \infty} E\left[\max_{0 \le k \le 2^n} \left|X\left(\frac{kt}{2^n}\right)\right|\right]^p \le \left(\frac{p}{1-p}\right)^p E|X(t)|^p, \quad \forall p > 1$$

by 2.11 because each sequence $|X(0)|, |X(t2^{-n})|, \ldots, |X(kt2^{-n})|, \ldots, |X(t)|$ is a submartingale. $\square$

We shall say that a *martingale* X is an L_p-**martingale** if $X(t) \in L_p$ for $t \in \mathbb{R}^+$ and close this Section by a list of simple martingale properties

1.2.12 Exercise. Let X be an $(L_2, \mathcal{F}_t)$ -martingale and $a \le s \le b$. Then

(a) $E^{\mathcal{F}_s}(X(b) - X(a))^2 = E^{\mathcal{F}_s}(X(b) - X(s))^2 + (X(s) - X(a))^2$

(b) $E^{\mathcal{F}_a}(X(b) - X(s))^2 = E^{\mathcal{F}_a}[X^2(b) - X^2(s)]$

(c) $E(X(b) - X(a))^2 = E(X(b) - X(s))^2 + E(X(s) - X(a))^2$, i.e. L_2-martingale differences are L_2-orthogonal.

(d) An $\mathcal{F}_t$ -supermartingale X with constant $EX(t)$ is a martingale.
(e) If X is a right continuous nonnegative supermartingale on $[0, T]$ then

$$P[\max_{t \le T} X(t) \ge a] \le a^{-1} \cdot EX(0), \quad a > 0.$$

1.3   Markov Times and Stopping Theorem

We shall fix a measurable space $(\Omega, \mathcal{F})$ with a filtration $(\mathcal{F}_t, t \ge 0)$ and define a $\tau : \Omega \to [0, \infty]$ to be a $\mathcal{F}_t$ -**Markov time** if $[\tau \le t] \in \mathcal{F}_t$ holds for all $t \in \mathbb{R}^+$. We also enlarge the filtration $(\mathcal{F}_t)$ putting

$$\mathcal{F}_\tau \overset{\text{def}}{=} \{F \in \mathcal{F}_\infty : F \cap [\tau \le t] \in \mathcal{F}_t \quad \forall t \in \mathbb{R}^+\}$$

and call $\mathcal{F}_\tau$ the **pre-τ σ-algebra.** Observe that if $\tau \equiv t^0 \in \mathbb{R}^+$ then τ is an $\mathcal{F}_t$ -Markov time and $\mathcal{F}_\tau = \mathcal{F}_{t^0}$.

A Markov time τ is mostly looked upon as a random time in which some important event happens to an $\mathcal{F}_t$ -adapted stochastic process, see 3.6, for example. Note that the definition of Markov time supports this interpretation by saying that to determine whether the event appeared or not during $[0, t]$ time interval is always possible if we are familiar with the history $\mathcal{F}_t$ up to the time t. Thus the σ-algebra $\mathcal{F}_\tau$ may be interpreted as the history of the process up to the random time τ in which the event happens, hence the pre-τ σ-algebra.

First, let us summarize some almost obvious properties of Markov times.

1.3.1   Lemma. *Let τ and υ be $\mathcal{F}_t$-Markov times. Then*

(a) *τ is an $\mathcal{F}_\tau$-measurable random variable*
(b) *$\upsilon \wedge \tau$, $\upsilon \vee \tau$ and $\upsilon + \tau$ are $\mathcal{F}_t$-Markov times*
(c) *$\tau \wedge t$ is an $\mathcal{F}_t$-measurable random variable for any $t \in \mathbb{R}^+$*
(d) *$F \in \mathcal{F}_\upsilon \quad \Rightarrow \quad F \cap [\upsilon \le \tau] \in \mathcal{F}_\tau$*
(e) *$\upsilon \le \tau \quad \Rightarrow \quad \mathcal{F}_\upsilon \subset \mathcal{F}_\tau$*
(f) *$[\upsilon < \tau]$, $[\upsilon \le \tau]$ and $[\upsilon = \tau]$ are in $\mathcal{F}_\tau \cap \mathcal{F}_\upsilon$*
(g) *$\mathcal{F}_\upsilon \cap \mathcal{F}_\tau = \mathcal{F}_{\upsilon \wedge \tau}$*

Proof. The definitions yield **(a)**, **(b)**, **(c)** directly. As for **(d)** we observe that for an $F \in \mathcal{F}_\upsilon$ that (c) implies

$$F \cap [\upsilon \le \tau] \cap [\tau \le t] = F \cap [\upsilon \le t] \cap [\tau \le t] \cap [\upsilon \wedge t \le \tau \wedge t] \in \mathcal{F}_t.$$

By **(d)** we get **(e)**. As for **(f)** we have $[\upsilon \le \tau] \in \mathcal{F}_\tau$ again by **(d)** and $[\upsilon \ge \tau] = [\upsilon \wedge \tau = \tau] \in \mathcal{F}_\tau$ by **(a)** and **(c)** which is all we need to verify the statement. Finally,

$$F \in \mathcal{F}_\upsilon \cap \mathcal{F}_\tau \Rightarrow F \cap [\upsilon \wedge \tau \le t] = F \cap [\tau \le \upsilon] \cap [\tau \le t] \cup F \cap [\upsilon \le \tau] \cap [\upsilon \le t] \in \mathcal{F}_t$$

according to **(f)** which combined with **(e)** verifies **(g)**. $\square$

Also, we will find useful the techniques developed in the next four simple statements.

1.3.2   Lemma. *Let τ be an $\mathcal{F}_t$ -Markov time and*

$$\tau_n := +\infty \text{ if } \tau \ge n, \quad \tau_n := k 2^{-n} \text{ if } (k-1)2^{-n} \le \tau < k 2^{-n}, \quad 1 \le k \le 2^n n, \, n \in \mathbb{N}.$$

Then τ_n are $\mathcal{F}_t$-Markov times and $\tau_n \downarrow \tau$ on Ω as $n \to \infty$.

Proof. To see that τ_n's are Markov times fix $t \in \mathbb{R}^+$ and check that

$$(k-1)2^{-n} \le t < k 2^{-n} \Rightarrow [\tau_n \le t] = [\tau < (k-1)2^{-n}] \in \mathcal{F}_{(k-1)2^{-n}} \subset \mathcal{F}_t$$

and $t \ge n \Rightarrow [\tau_n \le t] = [\tau < n] \in \mathcal{F}_n \subset \mathcal{F}_t$. $\square$

1.3.2 may be obviously modified to

1.3.3   Exercise. *Let $\tau \le T < \infty$ be a bounded Markov time. Set*

$$\tau_n := k 2^{-n} T \quad \text{if} \quad (k-1)2^{-n} T \le \tau < k 2^{-n} T, \quad k \le 2^n, \quad \tau_n := T \quad \text{if} \quad \tau = T.$$

Then $\tau_n \downarrow \tau$ is a sequence of Markov times.

A **filtration** $(\mathcal{F}_t)$ is called **right continuous** if $\mathcal{F}_{t+} := \cap_{h>0} \mathcal{F}_{t+h} = \mathcal{F}_t$ for all $t \in \mathbb{R}^+$. The property may be used in some cases to simplify the verification that a random variable is a Markov time.

1.3.4   Exercise.

(a) If $(\mathcal{F}_t)$ is a right continuous filtration then τ is an $\mathcal{F}_t$-Markov time if and only if $[\tau < t] \in \mathcal{F}_t$ for any $t \geq 0$ and $\mathcal{F}_\tau = \{F \in \mathcal{F}_\infty : F \cap [\tau < t] \in \mathcal{F}_t, \forall t \geq 0\}$.

(b) For any filtration $\mathcal{F}_t$ the filtration $\mathcal{F}_{t+}$ is right continuous.

(c) Consider **the canonical filtration** $\mathcal{F}_t^x$ in $(C(\mathbb{R}^+), \mathcal{B}(C(\mathbb{R}^+)))$ defined by (5) and prove that $\mathcal{F}_0^x$ is a strictly smaller σ-algebra than $\mathcal{F}_{0+}^x$.

The following basic integration property requires a probability measure P to complete our model to $(\Omega, \mathcal{F}, P, \mathcal{F}_t)$.

1.3.5   Lemma.

Assume that $Z \in L_1$ and τ, υ are $\mathcal{F}_t$ -Markov times. Then

(a) $[\upsilon \leq \tau] \Rightarrow E[Z|\mathcal{F}_\upsilon] = E[Z|\mathcal{F}_{\upsilon \wedge \tau}]$ *outside a P-null set*

and

(b) $E^{\mathcal{F}_\upsilon} E^{\mathcal{F}_\tau} Z = E^{\mathcal{F}_{\upsilon \wedge \tau}} Z$ *almost surely.*

Proof. According to 3.1 (f) and (g), **(a)** is equivalent to

$$E^{\mathcal{F}_\upsilon} I_{[\upsilon \leq \tau]} Z \overset{as}{=} E^{\mathcal{F}_{\upsilon \wedge \tau}} I_{[\upsilon \leq \tau]} Z \quad \text{i.e.} \quad \int_F E^{\mathcal{F}_{\upsilon \wedge \tau}} Z I_{[\upsilon \leq \tau]} \, dP = \int_F Z I_{[\upsilon \leq \tau]} \, dP,$$

to be valid for all $F \in \mathcal{F}_\upsilon$. But this follows directly by the definition of the conditional expectation $E[.|\mathcal{F}_{\upsilon \wedge \tau}]$ because $F \cap [\upsilon \leq \tau] \in \mathcal{F}_{\upsilon \wedge \tau}$ according to (d),(f),(g) in 3.1.

For an $F \in \mathcal{F}_\upsilon$, **(a)** yields

$$\int_{F \cap [\upsilon \leq \tau]} E^{\mathcal{F}_{\upsilon \wedge \tau}} Z \, dP = \int_{F \cap [\upsilon \leq \tau]} E^{\mathcal{F}_\upsilon} Z \, dP = \int_{F \cap [\upsilon \leq \tau]} Z \, dP = \int_{F \cap [\upsilon \leq \tau]} E^{\mathcal{F}_\tau} Z \, dP,$$

because the integrand is in $\mathcal{F}_\upsilon \cap \mathcal{F}_\tau$, again according to 3.1 (d) and (f). Also by **(a)** we reason that

$$\int_{F \cap [\tau < \upsilon]} E^{\mathcal{F}_{\upsilon \wedge \tau}} Z \, dP = \int_{F \cap [\tau < \upsilon]} E^{\mathcal{F}_\tau} Z \, dP$$

and summing up both equalities we get

$$\int_F E^{\mathcal{F}_{\upsilon \wedge \tau}} Z \, dP = \int_F E^{\mathcal{F}_\tau} Z \, dP \quad \forall F \in \mathcal{F}_\upsilon,$$

which is equivalent to **(b)**. $\square$

We have already remarked that the concept of Markov time is designed mostly to denote the time in which a trajectory of a stochastic process X performs something very important for our purposes:

If $B \subset \mathbb{R}$ is a Borel set we denote $\varepsilon(X, B) := \inf\{t \geq 0 : X(t) \in B\}$, where $\inf \emptyset := +\infty$, and call the random time variable *the first entry of X into B*. Agree to write

$$\varepsilon_a(X) := \inf\{t \geq 0 : |X(t)| \geq a\}, \quad a \geq 0$$

for the first entry of $|X|$ into $[a, \infty)$. Speaking generally, each entry time of a measurable X into a Borel B can be forced to become a Markov time of a properly extended canonical filtration $(\mathcal{F}_t^X)$, see Theorem 6.7 in [93], for example. However, the following more elementary result covers all cases we shall treat.

1.3.6   Theorem. *Let X be a continuous $\mathcal{F}_t$-adapted process. Then $\varepsilon := \varepsilon(X, B)$ is an $\mathcal{F}_t$-Markov time under either of the following conditions:*

(a) *B is a closed set in $\mathbb{R}$;*
(b) *B is an open set in $\mathbb{R}$ and $(\mathcal{F}_t)$ is a right-continuous filtration.*

Proof. In (a) we have $\varepsilon = \min\{t \geq 0 : X(t) \in B\}$, hence

$$[\varepsilon \leq t] = [\exists s \leq t : X(s) \in B] = [\inf_{0 \leq s \leq t, s \in Q} \mathrm{dist}(X(s), B) = 0] \in \mathcal{F}_t.$$

In (b) we write

$$[\varepsilon < t] = [\exists s < t : X(s) \in B] = [\exists s < t, s \in Q : X(s) \in B] \in \mathcal{F}_t$$

that verifies (b) via 3.4 (a). $\square$

Note that (b) holds also if the continuity of X is weakened to its right-continuity.

Stochastic analysis embraces a lot of measurability concepts. The most important one is that of the progressive measurability: We shall say that a stochastic **process X is $\mathcal{F}_t$-progressively measurable** or simply **$\mathcal{F}_t$-progressive** if

$$(s, \omega) \to X(s, \omega) \quad \text{is} \quad \mathcal{B}[0, t] \otimes \mathcal{F}_t \to \mathcal{B}(\mathbb{R}) \text{ -measurable map} \quad \forall t \in \mathbb{R}^+.$$

Denote by $\mathrm{PM} := \mathrm{PM}(\mathcal{F}_t)$ the space of all $\mathcal{F}_t$-progressive processes and call a **set** $M \subset \mathbb{R}^+ \times \Omega$ **$\mathcal{F}_t$-progressive**   if $\mathrm{I}_M \in \mathrm{PM}(\mathcal{F}_t)$.

1.3.7   Exercise. Fix a filtration $\mathcal{F}_t$ and denote by $\mathcal{M}$ the set of all progressive subsets of $\mathbb{R}^+ \times \Omega$. Verify that

(a) $\mathcal{M} \subset \mathcal{B}^+ \otimes \mathcal{F}_\infty$ is a σ-algebra such that $X \in \mathrm{PM}$ iff X is an $\mathcal{M}$-measurable map.
(b) $X \in \mathrm{PM}(\mathcal{F}_t) \Rightarrow X$ is an $\mathcal{F}_t$-adapted process.
(c) According to (a) and (b), $X \in \mathrm{PM}$ is $\mathcal{F}_t$-adapted and $\mathcal{B}^+ \otimes \mathcal{F}$-measurable map $(\mathbb{R}^+ \times \Omega) \to \mathbb{R}$. Show that the implication can not be reversed.

As always, continuous processes are inclined to be handled more easily:

1.3.8   Theorem. *A right-continuous $\mathcal{F}_t$-adapted process X is $\mathcal{F}_t$-progressive.*

Proof. Fix a $t > 0$ and observe that

$$X_n(s) := X(0)\mathrm{I}_{\{0\}}(s) + \sum_{k=0}^{2^n - 1} X((k+1)2^{-n}t)\mathrm{I}_{(k2^{-n}t, (k+1)2^{-n}t]}(s), \quad s \leq t,$$

define $\mathcal{B}[0, t] \otimes \mathcal{F}_t$-measurable maps such that $\lim_{n \to \infty} X_n = X$ on $[0, t] \times \Omega$. $\square$

For a stochastic process X and a $[0, +\infty]$-valued random variable τ we shall denote

$$X^\tau(t) = X(t \wedge \tau), \quad X^\tau = (X^\tau(t), t \geq 0), \quad [0, \tau] = \{(t, \omega) \in \mathbb{R}^+ \times \Omega : t \leq \tau(\omega)\},$$
$$X(\tau)(\omega) = X(\tau(\omega)) \quad \text{for} \quad \tau(\omega) < \infty.$$

Obviously, we may interpret the variable $X(\tau)$ as the state of the process X at the time τ, hence, for example, $X(\varepsilon_a(X))$ is the state in which we find X at the moment in which its trajectory enters $\mathbb{R}\setminus(-a,a)$ for the first time. Thus, assuming that $X(0) = 0$ and that X is a continuous process we have $|X(\varepsilon_a)| = a$ on $[\varepsilon_a < \infty]$. For a progressive process X and Markov time τ, $X(\tau)$ is a random variable:

1.3.9 Lemma. Let X be $\mathcal{F}_t$-progressive and τ an $\mathcal{F}_t$-Markov time. Then

(a) $[0,\tau]$ is an $\mathcal{F}_t$-progressive set;

(b) X^τ is an $\mathcal{F}_t$-progressive stochastic process;

(c) $X(\tau)$ is an $\mathcal{F}_\tau$-measurable random variable defined on $[\tau < \infty]$ which means that $[X(\tau) \in B] \cap [\tau < \infty] \in \mathcal{F}_\tau$ for any $B \in \mathcal{B}$.

Proof. The continuous process $(t \wedge \tau, t \geq 0)$ is $\mathcal{F}_{t\wedge\tau}$-progressive, hence $\mathcal{F}_t$-progressive, according to 3.1.(a) and 3.8. This implies (a) because

$$[0,\tau] = \left\{ (t,\omega) \in \mathbb{R}^+ \times \Omega : t \wedge \tau(\omega) = t \right\}.$$

Denote $T(s,\omega) = (\tau(\omega) \wedge s, \omega)$ and observe that $\mathcal{F}_t$-progressivity of $(t \wedge \tau, t \geq 0)$ implies that for any fixed $t \in \mathbb{R}^+$ the map T is measurable as

$$T : ([0,t] \times \Omega, \mathcal{B}[0,t] \otimes \mathcal{F}_t) \to ([0,t] \times \Omega, \mathcal{B}[0,t] \otimes \mathcal{F}_t).$$

Hence, $X^\tau = X(T)$ is an $\mathcal{F}_t$-progressive process.

For a fixed t it follows by (b) that $X(t \wedge \tau) \in \mathcal{F}_t$ and therefore

$$[X(\tau) \in B, \tau < \infty, \tau \leq t] = [X(\tau \wedge t) \in B, \tau \leq t] \in \mathcal{F}_t$$

that proves (c). $\square$

The next pair of results is of a basic importance in what follows:

1.3.10 Optional Sampling. *If X is a right-continuous $\mathcal{F}_t$-martingale (submartingale) and $\upsilon \leq \tau \leq T < \infty$ a pair of bounded $\mathcal{F}_t$-Markov times, then*

$$X(\upsilon),\ X(\tau) \in L_1, \quad E[X(\tau)|\mathcal{F}_\upsilon] = X(\upsilon) \quad (E[X(\tau)|\mathcal{F}_\upsilon] \geq X(\upsilon))$$

holds almost surely.

Proof. Assume that X is a *submartingale* and that

(i) $\upsilon \leq \tau \leq T < \infty$ take their values in $\{k2^{-n}T, 0 \leq k \leq 2^n\}$ such that $\tau - \upsilon \leq 2^{-n}T$:

The first part of 3.10 follows easily by $|X(\tau)| \leq \max_{1 \leq k \leq 2^n} |X(t_k)|$ where $t_k := k2^{-n}T$. Choose $F \in \mathcal{F}_\upsilon$ and check that the submartingale property implies

$$\int_F X(\tau) - X(\upsilon)\, dP = \sum_k \int_{F \cap [\upsilon = t_k] \cap [\tau > t_k]} X(t_{k+1}) - X(t_k)\, dP \geq 0,$$

hence $E[X(\tau)|\mathcal{F}_\upsilon] \geq X(\upsilon)$ a.s. because $X(\upsilon) \in \mathcal{F}_\upsilon$ by 3.9 (c).

(ii) Assume that $v \leq \tau \leq T < \infty$ take their values in $\{k2^{-n}T, 0 \leq k \leq 2^n\}$:
Define Markov times v_j by $(v + j2^{-n}T) \wedge \tau$ to get a connecting sequence $v = v_0 \leq v_1 \leq .. \leq v_{2^n} = \tau$. It follows by (i) that

$$E[X(\tau)|\mathcal{F}_v] = E^{\mathcal{F}_{v_0}} E^{\mathcal{F}_{v_1}} \dots E^{\mathcal{F}_{v_{2^n}-1}} X(\tau) \geq X(v) \quad \text{a.s.}$$

(iii) Consider arbitrary Markov times $v \leq \tau \leq T < \infty$:
Accoding to 3.3 there are Markov times v_n and τ_n that satisfy (ii) for $n \in \mathbb{N}$ such that $v_n \leq \tau_n$, $v_n \downarrow v$ and $\tau_n \downarrow \tau$. According to (ii) we get

$$(9) \quad E^{\mathcal{F}_{v_n}} X(\tau_n) \geq X(v_n), \ E^{\mathcal{F}_{\tau_n+1}} X(\tau_n) \geq X(\tau_{n+1}), \ E^{\mathcal{F}_{v_n+1}} X(v_n) \geq X(v_{n+1}).$$

Hence, both $(X(v_n))$ and $(X(\tau_n))$ are inverse submartingales such that all expectations $EX(\tau_n)$ and $EX(v_n)$ are above $EX(0) > -\infty$ and therefore convergent in L_1 by Theorem 6.5.10 in [67], p.228, for example*. Thus, the first inequality in (9) and the right-continuity of X yield

$$E^{\mathcal{F}_v} X(\tau_n) \geq E^{\mathcal{F}_v} X(v_n), \quad X(v_n) \to X(v), \quad X(\tau_n) \to X(\tau) \quad \text{in } L_1.$$

Finally, according to 2.7 (f) and because $X(v) \in \mathcal{F}_v$ by 3.9 (c)

$$E^{\mathcal{F}_v} X(\tau) = \lim_{n \to \infty} E^{\mathcal{F}_v} X(\tau_n) \geq \lim_{n \to \infty} E^{\mathcal{F}_v} X(v_n) = E^{\mathcal{F}_v} X(v) = X(v),$$

where both limits are in L_1. $\square$

1.3.11 Stopping Theorem. *If X is a right-continuous $\mathcal{F}_t$-martingale (submartingale) and τ an $\mathcal{F}_t$-Markov time then X^τ inherits the corresponding property. If X is a right-continuous $\mathcal{F}_t$-martingale and $v \leq \tau$ $\mathcal{F}_t$-Markov times then*

$$E^{\mathcal{F}_v} X(\tau \wedge t) = X^v(t), \qquad \forall t \geq 0$$

and if moreover X is a bounded process and $v \leq \tau$ finite Markov times, then $E^{\mathcal{F}_v} X(\tau) = X(v)$.

Proof. The process X^τ is $\mathcal{F}_t$-adapted process by 3.9 (b), the random variables $X^\tau(t)$ are integrable by 3.10. Assume that X is submartingale. It follows by 3.10 and 3.5 (b) that

$$X^\tau(s) \leq E[X(\tau \wedge t)|\mathcal{F}_{\tau \wedge s}] = E^{\mathcal{F}_s} E^{\mathcal{F}_\tau} X(\tau \wedge t) = E[X^\tau(t)|\mathcal{F}_s]$$

holds for all $s < t$.

The latter part of **3.11** follows by 3.5 (b) and 3.10 as

$$E^{\mathcal{F}_v} X(\tau \wedge t) = E^{\mathcal{F}_v} E^{\mathcal{F}_t} X(\tau \wedge t) = E^{\mathcal{F}_{v \wedge t}} X(\tau \wedge t) = X(v \wedge t).$$

Letting $t \to \infty$ in the above equality we get $E^{\mathcal{F}_v} X(\tau) = X(v)$ by 2.7 (f). $\square$

---

*Note that the uniform integrability of $X(\tau_n)$'s and $X(v_n)$'s, that implies the L_1-convergence, follows simply by 1.2.7 (g) if X is a martingale or nonnegative submartingale

1.3.12 Exercise and Example. A right continuous $\mathcal{F}_t$-adapted process X is an $\mathcal{F}_t$-martingale iff $X(\tau) \in L_1$ and $EX(\tau) = EX(0)$ for all bounded $\mathcal{F}_t$-Markov times τ and iff $X(t) \in L_1$ for all $t \in \mathbb{R}^+$ and $EX(\tau) = EX(0)$ for all $\mathcal{F}_t$-Markov times τ with $\mathrm{card}(\tau(\Omega)) \leq 2$.

Setting $X := \mathcal{E}(W, 1)$ and $\tau := \varepsilon(X, \{a\})$ for an $a \in (0, 1)$ we get a martingale with $EX(\tau) \stackrel{as}{=} a < 1 = EX(0)$ because $\liminf_{t \to \infty} X(t) \stackrel{as}{=} 0$ (apply 2.6 and 2.3 (b)). Hence τ and $v := 0$ is a pair of Markov times such that 3.10 is not valid anymore.

Markov times provide indeed a mighty tool *to tame the continuum of time*:

1.3.13 Theorem. *Let* $X \geq 0$ *be a continuous supermartingale. Then outside a P-null set*

(10) $$t \geq \tau(\omega) \quad \Rightarrow \quad X(t, \omega) = 0$$

where $\tau := \varepsilon(X, \{0\})$ *denotes the first entry of X to 0.*

Hence, a continuous supermartingale X with $X(t) > 0$ almost surely for all $t \in \mathbb{R}^+$ has trajectories that are positive on $\mathbb{R}^+$ with probability one, i.e. the X is an almost surely positive supermartingale.

Proof. It follows by 3.6 (a) that τ is an $\mathcal{F}_t^X$-Markov time, hence we may use Stopping Theorem (or rather its supermartingale version): If $\tau \stackrel{as}{=} \infty$ we have nothing to prove. If the contrary is true let n_0 denote the minimal $n \in \mathbb{N}$ such that $P[\tau \leq n] > 0$. For a while fix $n \geq n_0$, denote $\tau_n := \tau \wedge n$ and choose a $0 < q \in Q$: Because $\tau_n \leq \tau_n + q \leq n + q$, 3.11 yields

$$E[X(\tau_n + q)|\mathcal{F}_{\tau_n}] \leq X(\tau_n) \quad \Rightarrow \quad \int_{[\tau \leq n]} X(\tau_n + q)\, dP \leq \int_{[\tau \leq n]} X(\tau_n)\, dP = 0$$

as $[\tau \leq n] \in \mathcal{F}_\tau \cap \mathcal{F}_n = \mathcal{F}_{\tau_n}$. It follows that $X(\tau + q) = 0$ for all $0 < q \in Q$ on $[\tau \leq n]$ outside a P-null set N_n, hence by continuity, $X(\tau + t) = 0$ for all $t \geq 0$ on $[\tau \leq n] \setminus N_n$. Putting $N := \cup_{n \geq n_0} N_n$ we exhibit a P-null set outside which the implication (10) operates.

As for the second part of our theorem suppose that $P[\tau < \infty] > 0$ which leads to a $T > 0$ such that $P[\tau \leq T] > 0$. According to the first part the trajectories of X vanish on $[T, \infty)$ with a positive probability, hence a contradiction and $\tau \stackrel{as}{=} \infty$ that is exactly our second assertion. $\square$

We close this section by extending the stability property of Wiener process given by 2.2 (a).

1.3.14 Strong Markov Property of W. *If W is an $\mathcal{F}_t$-Wiener process and τ an almost surely finite $\mathcal{F}_t$-Markov time, then*

$$W_\tau(t) := I_{[\tau < \infty]} \cdot (W(t + \tau) - W(\tau)) \quad \text{and} \quad B(t) := 2 \cdot W(t \wedge \tau) - W(t)$$

are Wiener processes, W_τ being a process independent of the σ-algebra $\mathcal{F}_\tau$.

Proof. Assume without loss of generality that τ is a finite Markov time. We need to verify

$$P\left([(W_\tau(t_1),\ldots,W_\tau(t_k)) \in B] \cap F\right) = P[(W(t_1),\ldots,W(t_k)) \in B] \cdot P(F),$$

(11)
$$0 \le t_1 < \ldots t_k < \infty, \quad B \in \mathcal{B}^k, \quad k \in \mathbb{N}, \quad F \in \mathcal{F}_\tau.$$

If $\tau(\Omega)$ is a countable set, the equality follows easily by its factorization to the atoms $[\tau = \alpha_j]$ where the α_j's are values of τ. Hence, (11) is proved for each Markov time τ_n defined in 3.2 and the validity of (11) in general is implied by $\tau_n \downarrow \tau$, $\mathcal{F}_\tau \subset \mathcal{F}_{\tau_n}$ and finally by the continuity of W and W_τ.

Since W_τ is a Wiener process independent of $\sigma(\tau, W^\tau) \subset \mathcal{F}_\tau$, it follows that (τ, W^τ, W_τ) and $(\tau, W^\tau, -W_\tau)$ are equally distributed $\mathbb{R}^+ \times C(\mathbb{R}^+) \times C(\mathbb{R}^+)$-valued random variables. Compute

$$W(t) = W^\tau(t) + W_\tau\left((t-\tau)^+\right), \quad B(t) = W^\tau(t) - W_\tau\left((t-\tau)^+\right)$$

to conclude that $W = f(\tau, W^\tau, W_\tau)$ and $B = f(\tau, W^\tau, -W_\tau)$ holds for a Borel map $f : \mathbb{R}^+ \times C(\mathbb{R}^+) \times C(\mathbb{R}^+) \to C(\mathbb{R}^+)$. Hence, W and B are equally distributed processes. $\square$

The strong Markov property may be applied to establish precisely

1.3.15 Reflection Principle. Let W be a Wiener process. For $t \ge 0$ and $a > 0$ put $S(t) := \max_{s \le t} W(s)$ and let τ_a denote the first entry of W to $[a, \infty)$. Then

$$P[S(t) \ge a] = P[\tau_a \le t] = 2P[W(t) \ge a] = P[|W(t)| \ge a].$$

Note that the Reflection Principle in fact says that

$$P[S(t) \ge a, W(t) < a] = P[S(t) \ge a, W(t) > a] = \frac{1}{2}P[S(t) \ge a],$$

which equality easily comes from by the obvious heuristic argument: Among the particles which enter a before time t "half" of them will be above a at time t.

Proof. Write $P[\tau_a \le t] = P[W(t) \ge a] + P[\tau_a \le t, W(t) < a]$. It is easy to check that $[\tau \le t, W(t) < a] = [B(t) > a]$ holds where $B := 2W^{\tau_a} - W$. Hence, $P[\tau_a \le t] = 2P[W(t) \ge a]$ by 3.14. $\square$

Thus, $\mathcal{L}(S(t)) = \mathcal{L}(|W(t)|)$ for each individual $t \in \mathbb{R}^+$ while obviously the distributions of S and W are singular probability measures in $C(\mathbb{R}^+)$.

1.3.16 Exercise. The Markov time $\tau_a := \varepsilon(W, \{a\})$ has for any $a \ne 0$ the probability distribution with the density

$$f_a(s) = |a|(2\pi s^3)^{-\frac{1}{2}} \exp\left(\frac{-a^2}{2s}\right), \quad s > 0.$$

1.4   Local Martingales and Complete Filtrations

Later on we shall appreciate the following extension of the martingale property:

Let X be an $\mathcal{F}_t$-adapted continuous stochastic process. We shall say that an $\mathcal{F}_t$-**Markov time** τ **stops** X **to martingale** if X^τ is an $\mathcal{F}_t$-martingale. Further, a sequence (τ_n) of $\mathcal{F}_t$-Markov times will be called an $\mathcal{F}_t$-**localization sequence for** X if $\tau_n \uparrow \infty$ almost surely and τ_n stops X to an $\mathcal{F}_t$-martingale for every $n \in \mathbb{N}$. Finally, X is a **local** $\mathcal{F}_t$-**martingale** if there is an $\mathcal{F}_t$-localization sequence (τ_n) for the shifted process $X - X(0)$, i.e. if

$$X^{\tau_n} - X(0) = (X - X(0))^{\tau_n} \quad \text{are} \quad \mathcal{F}_t\text{-martingales for all } n \in \mathbb{N}.$$

Fix a filtration $(\mathcal{F}_t)$ and verify as an exercise the following simple but frequently used properties:

1.4.1   Exercise. Let X be a continuous adapted process and τ, τ_n, υ Markov times. Then

(a) If $X(0) \in L_1$ then τ stops X to a martingale iff τ stops $X - X(0)$ to a martingale, i.e., X is a local martingale iff there is a localization sequence for X.

(b) X is a martingale $\Rightarrow$ X is a local martingale $\Rightarrow$ X^τ is a local martingale.

(c) If $\upsilon \leq \tau$ and τ stops X to a martingale then also υ stops X to a martingale.

(d) X is a local martingale, $\xi, \eta \in \mathcal{F}_0$ $\Rightarrow$ $\xi \cdot X + \eta$ is a local martingale.

Mostly we shall deal with continuous adapted processes X such that $X(0)$ is a constant. It follows by (a) that such a process is a local martingale if and only if there exists a localization sequence for the process X itself.

Warning. It is not true that local martingales need only be integrable in order to be martingales (see 2.3.7 for a counterexample).

Nevertheless, as we shall see later on, it is very important to recognize true martingales among local martingales. Agree to call a **process** X **bounded** if $|X(t,\omega)| \leq K < \infty$ for all t and ω, i.e. if it is a process with uniformly bounded trajectories.

1.4.2   Theorem. *For a fixed filtration $(\mathcal{F}_t)$ let X be a continuous adapted process. Then*

(a) *If X is a local martingale such that $|X(t)| \leq Z$ holds for all $t \geq 0$ and a random variable $Z \in L_1$ then X is a martingale, especially, bounded local martingales are martingales.*

(b) *Let ε_n denotes the first entry of $|X - X(0)|$ to $[n, +\infty)$. Then X is local martingale iff (ε_n) is a localization sequence for $X - X(0)$.*

(c) *If X is a nonnegative local $\mathcal{F}_t$-martingale with $X(0) \in L_1$ then it is a supermartingale.*

(d) *If X is a nonnegative local $\mathcal{F}_t$-martingale with $X(0) \in L_1$ such that $EX(t) = EX(0)$ holds for a $t > 0$ then X is a martingale on $[0, t]$.*

Proof. **(a)**: Choose a localization sequence for X, say (τ_n). Stopping Theorem implies

$$(11) \qquad E^{\mathcal{F}_s} X(\tau_n \wedge t) = X(\tau_n \wedge s), \qquad s \leq t,\, n \in \mathbb{N}.$$

Because $X(\tau_n \wedge t) \overset{\text{as}}{\to} X(t)$ and $X(\tau_n \wedge s) \overset{\text{as}}{\to} X(s)$ as $n \to \infty$, 1.2.7 (f) permits to pass $n \to \infty$ in (11) to receive $E[X(t)|\mathcal{F}_s] = X(s)$ for $s \leq t$.

(b): ε_n's are Markov times by 3.6, $\varepsilon_n \uparrow \infty$ on Ω because all trajectories of $X - X(0)$ are continuous, hence locally bounded functions on $\mathbb{R}^+$. If X is a local martingale then $(X - X(0))^{\varepsilon_n}$ is a local martingale by 4.1 (b), hence a martingale according to **(a)**.

(c): There is a localization sequence (τ_n) for X by 4.1 (a). Fix $s < t$ and check

$$E^{\mathcal{F}_s} X^{\tau_n}(t) = X^{\tau_n}(s), \quad \forall n \in \mathbb{N}, \quad \lim_{n \to \infty} X^{\tau_n}(t) \overset{\text{as}}{=} X(t).$$

By Fatou lemma we reason that $X(t) \in L_1$ as $EX(t) \leq \underline{\lim}_n EX^{\tau_n}(t) = EX(0) < \infty$ and finally, by Fatou lemma for conditional expectations 2.7 (i) we conclude the proof of **(c)**:

$$E[X(t)|\mathcal{F}_s] \leq \underline{\lim}_n E[X^{\tau_n}(t)|\mathcal{F}_s] = \lim_n X^{\tau_n}(s) = X(s), \quad \text{a.s..}$$

The assertion **(d)** is an immediate consequence of 2.12 (d) and our present **(c)**. $\square$

Note that 4.2 (b) says that given a local martingale X, the shifted local martingale $X - X(0)$ owns a localization sequence (τ_n) such that τ_n's are $\mathcal{F}_t^X$ -Markov times. This property enables to extend 2.9 to

1.4.3   Stability of Local Martingale Property.

(a) If X is a local $\mathcal{F}_t$-martingale then X is a local $\mathcal{D}_t$-martingale for any filtration $(\mathcal{D}_t)$ such that $\mathcal{F}_t^X \subset \mathcal{D}_t \subset \mathcal{F}_t$ holds for any $t \in \mathbb{R}^+$.
(b) If X is a local $\mathcal{F}_t$-martingale then it is a local $\mathcal{F}_{t+}$-martingale.
(c) If X is a local $\mathcal{F}_t$-martingale and Y a local $\mathcal{G}_t$-martingale then X, Y and XY are local $\mathcal{F}_t \vee \mathcal{G}_t$-martingales provided that filtrations $(\mathcal{F}_t), (\mathcal{G}_t)$ are independent.

Proof. The statements **(a)**, **(b)** follow easily by 2.9 and 4.2 (b). If X and Y are as in **(c)** assume $X(0) = Y(0) = 0$ without loss of generality (see, 4.1 (d)) and choose their corresponding localization sequences (τ_n) and (υ_n) such that τ_n's and υ_n's are $\mathcal{F}_t$ and $\mathcal{G}_t$ -Markov times, respectively. According to 2.9 (c) $X^{\tau_n}, Y^{\upsilon_n}$ and $X^{\tau_n} Y^{\upsilon_n}$ are $\mathcal{F}_t \vee \mathcal{G}_t$ -martingales. It follows by 4.1 (c) that $(XY)^{\tau_n \wedge \upsilon_n}$ are $\mathcal{F}_t \vee \mathcal{G}_t$ -martingales also and therefore $(\tau_n \wedge \upsilon_n)$ is an $\mathcal{F}_t \vee \mathcal{G}_t$ -localization sequence for XY. $\square$

Recall that by 1.5 the only continuous process with finite variation X^v and with the finite quadratic variation $\langle X \rangle$ is that for which $\langle X \rangle(t) \overset{\text{as}}{=} 0$ for all $t \in \mathbb{R}^+$. Continuous local martingales and those of finite variation behave in a similar fashion.

1.4.4   **Theorem.** *Each continuous local martingale X of finite variation is constant almost surely, i.e., $X(t, \omega) = X(0, \omega)$ for $t \geq 0$ outside a P-null set.*

Proof. Obviously we may restrict ourselves to the case that $X(0) \equiv 0$.

Assume first that the processes X and X^v are bounded, say $|X|, X^v \leq K < \infty$, fix a $t > 0$ and choose a sequence of partitions $|\Delta_n(t)| \to 0$. Then, because increments of L_2-martingale X are orthogonal by 2.12 (c),

$$EX^2(t) = E\left(\sum_{j=1}^{k_n}(X(t_j^n) - X(t_{j-1}^n))\right)^2$$

$$= E\sum_{j=1}^{k_n}(X(t_j^n) - X(t_{j-1}^n))^2 \leq E\left(X^v(t) \times \max_{|u-v| \leq |\Delta_n(t)|, u,v \leq t}|X(u) - X(v)|\right).$$

Hence, $EX^2(t) = 0$, by the dominated convergence theorem and the uniform continuity of each $X(\omega)$ on $[0, t]$. The assertion now easily follows by the continuity of X.

Let X be a continuous local martingale, say w.r.t. a filtration $(\mathcal{F}_t)$ and X^v its variation. Then X^v is continuous by 1.2 (b) and $\mathcal{F}_t$-adapted by 1.3. It follows by 4.1 (c) and 4.2 (b) that $\upsilon_n := \varepsilon_n(X) \wedge \varepsilon_n(X^v)$ defines an $\mathcal{F}_t$-localization sequence for X such that X^{υ_n} and $(X^{\upsilon_n})^v = (X^v)^{\upsilon_n}$ are bounded processes. Hence, by the first part of our proof, we get $X^{\upsilon_n} \overset{as}{=} 0$ for every $n \in \mathbb{N}$ and taking the limit $n \to \infty$ we conclude the proof. $\square$

The above theorem invites to combine the local martingales and the processes of finite variation by linear combinations:

Given a filtration $(\mathcal{F}_t)$ call a stochastic process X an $\mathcal{F}_t$-**semimartingale** if $X = B + M$ where B is a continuous $\mathcal{F}_t$-adapted process of finite variation and M a continuous local $\mathcal{F}_t$-martingale. Obviously, $\mathcal{F}_t$-semimartingales are continuous and $\mathcal{F}_t$-adapted processes.

For our future purposes our *notation* will be as follows:

$$\mathrm{CM}(\mathcal{F}_t) := \left\{\text{continuous } \mathcal{F}_t\text{-martingales } M \text{ with } M(0) \overset{as}{=} 0\right\}$$

$$\mathrm{CM}_p(\mathcal{F}_t) := \left\{M \in \mathrm{CM}(\mathcal{F}_t) : M(t) \in L_p, \, \forall \, t \in \mathbb{R}^+\right\}, \qquad p \geq 1$$

$$\mathrm{CM}_{loc}(\mathcal{F}_t) := \left\{[\text{continuous}] \text{ local } \mathcal{F}_t\text{-martingales } M \text{ with } M(0) \overset{as}{=} 0\right\}$$

$$\mathrm{CSM}(\mathcal{F}_t) := \{[\text{continuous}] \, \mathcal{F}_t\text{-semimartingales}\}$$

$$\mathrm{CFV}(\mathcal{F}_t) := \left\{B \text{ continuous } \mathcal{F}_t\text{-adapted with } B^v(t) < \infty \, \forall t \in \mathbb{R}^+, \, B(0) = 0\right\}$$

$$\mathrm{CI}(\mathcal{F}_t) := \{A \text{ continuous non decreasing } \mathcal{F}_t\text{-adapted}, \, A(0) = 0\}.$$

Sometimes we may prefer to be more specific about the home space of our processes, thus we may also write $M \in \mathrm{CM}(P, \mathcal{F}_t)$ or even $M \in \mathrm{CM}(\Omega, \mathcal{F}, P, \mathcal{F}_t)$. Also, we point out that unlike a martingale **we defined a local martingale and a semimartingale only as a continuous process.**

1.4.5. *Each semimartingale $X \in CSM(\mathcal{F}_t)$ can be written in the form*

$$(12) \qquad X = X(0) + B + M, \quad M \in CM_{loc}(\mathcal{F}_t), \quad B \in CFV(\mathcal{F}_t),$$

which decomposition is (according to 4.4) unique in the sense that B, B^0 and M, M^0, respectively, are pairs of equivalent processes if $X = X(0) + B^0 + M^0$ is another decomposition (12).

Instead of (12) we shall frequently write $dX = dB + dM$ and call the right-hand side the **stochastic differential of** X.

Practically everything we could do in favour of $\mathcal{F}_t$-martingales where $(\mathcal{F}_t)$ is a general filtration we have already done. For further purposes we lack the modification stability as

(13) If Y is a modification of an $\mathcal{F}_t$-adapted X then Y is $\mathcal{F}_t$-adapted.

It is obvious that we need filtrations $(\mathcal{F}_t)$ such that any $\mathcal{F}_t$ includes any P-null set in $\mathcal{F}$ where $(\Omega, \mathcal{F}, P)$ is the underlying probability space. In other words we force ourselves to believe that everything what *certainly will happen* (or certainly will not happen) to our particle is known at the time $t = 0$.

Let $(\Omega, \mathcal{F}, P)$ be a *complete probability space*. A filtration $(\mathcal{F}_t)$ of sub-σ-algebras of $\mathcal{F}$ will be called a **complete filtration** if $\mathcal{N}_P \subset \mathcal{F}_0$ where $\mathcal{N}_P := \{N \in \mathcal{F} : P(N) = 0\}$.

Note that a complete filtration is defined only as a filtration of a **complete probability space** which ensures that $\mathcal{N}_P$ is a hereditary class of sets, i.e., if N_1 is a subset of an $N \in \mathcal{N}_P$ then also $N_1 \in \mathcal{N}_P$. As an exercise solve the problems below to see how comfortable complete filtrations are:

1.4.6   **Exercise.** Fix a complete filtration $(\mathcal{F}_t)$. Then

(a) If X is an $\mathcal{F}_t$-adapted process and $Y(t) : \Omega \to \mathbb{R}$ are such that $Y(t) \overset{\text{as}}{=} X(t)$ holds for all $t \in \mathbb{R}^+$ then Y is an $\mathcal{F}_t$-adapted process. In particular, (13) holds.

(b) If X_n are $\mathcal{F}_t$-adapted processes and $X(t) : \Omega \to \mathbb{R}$ such that $\lim_n X_n(t) \overset{\text{as}}{=}$ $X(t)$ holds for every $t \geq 0$ then the process X is also $\mathcal{F}_t$-adapted.

A test for a process X to be adapted is provided by

1.4.7. Let $(\mathcal{F}_t)$ be a complete filtration. Consider $X(t) : \Omega \to \mathbb{R}$ for $t \geq 0$ and a sequence (τ_n) of $\mathcal{F}_t$-Markov times such that $\tau_n \uparrow +\infty$ almost surely.

(a) If X^{τ_n} are $\mathcal{F}_t$-adapted processes then X is a process that also is $\mathcal{F}_t$-adapted.

(b) If X^{τ_n} are $\mathcal{F}_t$-adapted continuous processes then X is a process that is $\mathcal{F}_t$-adapted and continuous almost surely.

(c) If $X^{\tau_n} \in CM_{loc}(\mathcal{F}_t)$ for all $n \in \mathbb{N}$ then X is a process equivalent to a process $Y \in CM_{loc}(\mathcal{F}_t)$.

For **(a)** *and* **(b)** *we do not need* τ_n's *to be Markov times.*

Proof.

(a) Observe that $X(t) \overset{\text{as}}{=} \lim_n X_n^{\tau_n}(t)$ and apply 4.6 (b).
(b) This follows by **(a)**, observing that $X(\omega)$ is a continuous function on $\mathbb{R}^+$ if $\lim_n \tau_n(\omega) = +\infty$.
(c) X is $\mathcal{F}_t$-adapted process that is continuous outside a set $N \in \mathcal{N}_P$ by **(a)** and **(b)**. Thus, $Y := I_{N^c} \cdot X$ is a continuous and $\mathcal{F}_t$-adapted process such that $Y \overset{\text{as}}{=} X$ and $Y^{\tau_n} \in \mathrm{CM}_{loc}(\mathcal{F}_t)$ for all n. It follows by 4.2 (b) that $\tau_n \wedge \varepsilon_n(Y)$ is a localization sequence for Y, hence $Y \in \mathrm{CM}_{loc}(\mathcal{F}_t)$.

$\square$

Adapted processes may be constructed from processes that are defined on $[0, \tau_n]$ only if $\tau_n \uparrow \infty$.

1.4.8. *Let* $(\mathcal{F}_t)$ *be complete filtration and* (τ_n) *a sequence of* $\mathcal{F}_t$-*Markov times such that* $\tau_n \uparrow \infty$ *almost surely. Further, let* X_n *be either a sequence of* $\mathcal{F}_t$-*adapted or continuous* $\mathcal{F}_t$-*adapted processes or processes in* $\mathrm{CM}_{loc}(\mathcal{F}_t)$ *such that*

$$(14) \qquad\qquad X_{n+1}^{\tau_n} \overset{\text{as}}{=} X_n^{\tau_n} \quad \text{for all } n \in \mathbb{N}.$$

Then there is a process X *that is* $\mathcal{F}_t$-*adapted, continuous* $\mathcal{F}_t$-*adapted and in* $\mathrm{CM}_{loc}(\mathcal{F}_t)$, *respectively, such that in all cases* $X^{\tau_n} \overset{\text{as}}{=} X_n^{\tau_n}$ *holds for all* n.
We need the τ_n's *to be Markov times only for the part that refers to the local martingale property.*

Proof. It follows by (14) that for $\omega \notin N$ where $N \in \mathcal{N}_P$ and for all $t \geq 0$ $X_n(t, \omega) \to X(t, \omega) \in \mathbb{R}$ as $n \to \infty$. Put $X(t, \omega) := 0$ for all $t \geq 0$ and $\omega \in N$. Obviously, again by (14), $X^{\tau_n} \overset{\text{as}}{=} X_n^{\tau_n}$ and all above assertions follow by 4.7 because processes $X_n^{\tau_n}$ are $\mathcal{F}_t$-adapted, continuous $\mathcal{F}_t$-adapted and in $\mathrm{CM}_{loc}(\mathcal{F}_t)$, respectively. $\square$

Whatever mathematical profit we may get working in the setting of complete filtrations we must build up a passage there from the uncompleted ones that would preserve Brownian motion and the (local) martingale property.

Let $(\Omega, \mathcal{F}, P)$ be an arbitrary probability space and $(\mathcal{F}_t)$ its arbitrary filtration. Denote by $\mathcal{N}_P^\circ$ the set of all $N_\circ \subset \Omega$ for which a P-null set $N \in \mathcal{F}$ exists such that $N_\circ \subset N$ holds.. Observe that $\mathcal{N}_P^\circ = \mathcal{N}_P$ if $(\Omega, \mathcal{F}, P)$ is a complete space. Denote $\mathcal{F}^P := \mathcal{F} \vee \sigma(\mathcal{N}_P^\circ)$ and stay to denote by P also the unique extension of P from $\mathcal{F}$ to $\mathcal{F}^P$. Note that $(\Omega, \mathcal{F}^P, P)$ is the standard completion of $(\Omega, \mathcal{F}, P)$. Finally denote $\mathcal{F}_t^P := \mathcal{F}_t \vee \sigma(\mathcal{N}_P^\circ)$ for $t \geq 0$ to construct a complete filtration of the complete probability space $(\Omega, \mathcal{F}^P, P)$. We shall say that $(\mathcal{F}_t^P)$ is the P-**completion** of $(\mathcal{F}_t)$.

1.4.9. *Let* $(\mathcal{F}_t)$ *be a filtration of a probability space* $(\Omega, \mathcal{F}, P)$. *Then*

(a) *If* W *is a* d-*dimensional* $\mathcal{F}_t$-*Wiener process then* W *is an* $\mathcal{F}_t^P$-*Wiener process.*
(b) *If* X *is either an* $\mathcal{F}_t$-*martingale or* $\mathcal{F}_t$-*local martingale or* $\mathcal{F}_t$-*semimartingale, then the same is true with* $\mathcal{F}_t$ *replaced by* $\mathcal{F}_t^P$.

Proof. Because any $\mathcal{F}_t$-martingale, $\mathcal{F}_t$-local martingale, $\mathcal{F}_t$-semimartingale and $\mathcal{F}_t$-Wiener process will keep the corresponding property if it is considered as a process on the completion $(\Omega, \mathcal{F}^P, P)$ we may, without loss of generality, assume that $(\Omega, \mathcal{F}, P)$ itself is complete. If it is so, then $\mathcal{F}_t^P = \mathcal{F}_t \vee \mathcal{G}_t$ where $\mathcal{G}_t := \sigma(\mathcal{N}_P)$ and $\mathcal{F}_t$ are independent σ-algebras. The statements (a), (b) now follow by the (c) statements in 2.8, 2.9 and 4.3. $\square$

1.5 L_2-Martingales and Density Theorem

Denote

$$\mathrm{CP}_2(\mathcal{F}_t) := \left\{ X \text{ an } \mathcal{F}_t\text{-adapted continuous process, } E \max_{s \le t} X^2(s) < \infty \; \forall \, t \ge 0 \right\},$$

$\|X\|_t := \sqrt{EX^{*2}(t)}$ for $t \in \mathbb{R}^+$ and finally $\|X\| := \sum_{t=1}^{\infty} 2^{-t}(\|X\|_t \wedge 1)$ for $X \in \mathrm{CP}_2(\mathcal{F}_t)$ where, as you may remember, $X^*(t) := \max_{s \le t} |X(s)|$.

The space $\mathrm{CP}_2 := \mathrm{CP}_2(\mathcal{F}_t)$ with convergence induced by the pseudometric $\|X - Y\|$ provides a suitable *operation field* for handling continuous processes as the convergence combines the uniform convergence on bounded intervals with that in L_2. Next Exercise summarizes simple properties of $(\mathrm{CP}_2, \|\cdot\|)$:

1.5.1   Exercise. Verify that

(a) $\|X - Y\|$ defines a pseudometric on CP_2 such that $\|X - Y\| = 0$ iff $X \overset{\mathrm{as}}{=} Y$.

(b) $\|X_n - X\| \to 0 \quad$ iff $\quad \max_{s \le t} |X_n(s) - X(s)| \to 0$ in L_2 for all $t \ge 0$
$\Rightarrow X_n(t) \to X(t)$ in L_2 for all $t \ge 0$.

(c) $\mathrm{CM}_2 \subset \mathrm{CP}_2$ and if M, M_n are in CM_2 then $\|M_n - M\| \to 0$ iff $M_n(t) \to M(t)$ in L_2 for each $t \in \mathbb{R}^+$.

A deeper quality of CP_2 is given by

1.5.2   Theorem. *For a fixed complete filtration $(\mathcal{F}_t)$ the pseudometric space CP_2 is complete and $\mathrm{CM}_2 \subset \mathrm{CP}_2$ is a closed subspace.*

Proof. Let (X_n) be a Cauchy sequence in CP_2, observe that $\|X_n\|_t \le C_t < \infty$ for all $n \in \mathbb{N}$ and $t \ge 0$. Choose $n_1 < n_2 < \dots$ such that $\|X_{n_{k+1}} - X_{n_k}\| \le 2^{-\frac{k}{2}}$ holds for all $k \in \mathbb{N}$. Then for any $t \in \mathbb{N}$

$$\left(E \max_{s \le t} |X_{n_{k+1}} - X_{n_k}| \right)^2 \le \|X_{n_{k+1}} - X_{n_k}\|_t^2 \le 2^{2t} \cdot 2^{-k}$$

if $k \ge k(t) \in \mathbb{N}$. Hence, for any $t \ge 0$

$$E \sum_{k=1}^{\infty} \max_{s \le t} |X_{n_{k+1}}(s) - X_{n_k}(s)| < \infty \Rightarrow \sum_{k=1}^{\infty} \max_{s \le t} |X_{n_{k+1}}(s) - X_{n_k}(s)| < \infty \text{ a.s..}$$

It follows that there is an $F \in \mathcal{F}$ with $P(F) = 1$ such that $X_{n_k}(\omega)$ is a Cauchy, hence a convergent sequence in $C(\mathbb{R}^+)$ for $\omega \in F$ by 1.7. Denote its limit by $X(\omega)$ and put $X(\omega) \equiv 0$ for $\omega \notin F$. Thus, we have constructed a continuous process X such that

$$(15) \qquad \max_{s \leq t} |X_{n_k}(s) - X(s)| \to 0 \quad \forall\, t \geq 0 \text{ a.s.}$$

and therefore the X is an adapted process by 4.6 (b). Also, Fatou lemma and (15) yield for a fixed $t \in \mathbb{R}^+$

$$E \max_{s \leq t} |X(s)|^2 \leq \liminf_{k \to \infty} E \max_{s \leq t} |X_{n_k}(s)|^2 \leq C_t^2 < \infty,$$

hence $X \in \mathrm{CP}_2$. Finally, fix $t \geq 0, \epsilon > 0$ and choose $n_t \in \mathbb{N}$ such that $n, m \geq n_t, \Rightarrow$ $\|X_n - X_m\|_t^2 \leq \epsilon$. By Fatou lemma and (15) again

$$E \max_{s \leq t} |X_n(s) - X(s)|^2 \leq \liminf_{k \to \infty} E \max_{s \leq t} |X_n(s) - X_{n_k}(s)|^2 \leq \epsilon$$

if $n \geq n_t$, therefore $\|X_n - X\|_t \to 0$ and finally $\|X_n - X\| \to 0$.

To prove that CM_2 is a closed subspace consider $M_n \in \mathrm{CM}_2$ and $M \in \mathrm{CP}_2$ such that $\|M_n - M\| \to 0$ holds. The convergence implies that $M_n(t) \to M(t)$ in L_2 for any fixed $t \geq 0$. Hence,

$$E[M(t)|\mathcal{F}_s] = \lim_n E[M_n(t)|\mathcal{F}_s] = \lim_n M_n(s) = M(s), \qquad s \leq t,$$

where all limits are in L_1. □

Before we shall proceed further test your proficiency in the following elements of measure theory:

Let $B : \mathbb{R}^+ \to \mathbb{R}$ be a continuous function of finite variation such that $B(0) = 0$. Recall that the variation of B denoted by B^v is an increasing continuous function $\mathbb{R}^+ \to \mathbb{R}$ with $B^v(0) = 0$ by 1.3. Also denote $B^+ := \frac{1}{2}(B^v + B)$ and $B^- := \frac{1}{2}(B^v - B)$. Agree to write dB or μ_B for the locally finite Lebesgue-Stieltjes signed measure defined on $\mathcal{B}^+$ by $\mu_B([0, t]) = B(t)$ for $t \geq 0$. Let $\mu_B = (\mu_B)^+ - (\mu_B)^-$ be **Hahn-Jordan decomposition** of μ_B and $|\mu_B| = (\mu_B)^+ + (\mu_B)^-$ its **total variation**. Prove that

1.5.3   Exercise. B^+ and B^- are increasing (continuous, starting from 0) functions such that $(\mu_B)^+ = \mu_{B^+}$, $(\mu_B)^- = \mu_{B^-}$ and $|\mu_B| = \mu_{B^v}$ holds.

Recall that $\mathrm{CFV} := \mathrm{CFV}(\mathcal{F}_t)$ and $\mathrm{CI} := \mathrm{CI}(\mathcal{F}_t)$ denote the spaces of all $\mathcal{F}_t$-adapted continuous processes B with $B(0) = 0$ of finite variation and increasing on $\mathbb{R}^+$, respectively. It follows by 5.3 that $B \in \mathrm{CFV} \Rightarrow B^v, B^+, B^- \in \mathrm{CI}$ where the latter processes are $\mathcal{F}_t$-adapted by 1.3. Thus, $B = B^+ - B^-$ provides the decomposition $\mathrm{CFV} = \mathrm{CI} \ominus \mathrm{CI}$ that is equivalent to Hahn-Jordan decomposition of corresponding induced measures.

Our aim now will be to give a sense to $\int G\, dB$ where G is an $\mathcal{F}_t$-progressive process and $B \in \mathrm{CFV}(\mathcal{F}_t)$. An obvious idea is to integrate separately, if possible, each trajectory $G(t, \omega)$ by Lebesgue-Stieltjes signed measure $\mu_{B(\omega)}$ induced by the trajectory $B(\omega)$. Denote

$$(16) \qquad T(G, B) := \left\{ \omega \in \Omega : \int_0^t |G(s, \omega)| \, dB^v(s, \omega) < \infty \qquad \forall t \geq 0 \right\}$$

and

$$(17) \quad \left(\int_0^t G(s) \, dB(s) \right)(\omega) := I_{T(G,B)}(\omega) \cdot \int_0^t G(s, \omega) \, dB(s, \omega), \quad (t, \omega) \in \mathbb{R}^+ \times \Omega.$$

The process $\int G \, dB = \left(\int_0^t G \, dB(s), t \geq 0 \right)$ will be called **the integral of a process** $G \in \mathrm{PM}(\mathcal{F}_t)$ **with respect to a process** $B \in \mathrm{CFV}(\mathcal{F}_t)$ where by $\mathrm{PM}(\mathcal{F}_t)$ we have denoted the space of all $\mathcal{F}_t$-progressive processes. Note that $\int G \, dB$ is defined as a continuous process whose trajectories are of finite variation on $\mathbb{R}^+$ and we may ask under what conditions it will also define an $\mathcal{F}_t$-adapted process. For a pair $G \in \mathrm{PM}(\mathcal{F}_t)$, $B \in \mathrm{CFV}(\mathcal{F}_t)$ and $p \geq 1$ also denote

$$(18) \qquad \mathrm{PM}_p(B, \mathcal{F}_t) = \{ G \in \mathrm{PM}(\mathcal{F}_t) : \int_0^t |G(s)|^p \, dB^v(s) < \infty \text{ a.s. } \forall t \geq 0 \}.$$

1.5.4. Let G be an $\mathcal{F}_t$-progressive process and B a process in $\mathrm{CFV}(\mathcal{F}_t)$. Then $T(G, B) \in \mathcal{F}_\infty$ and

(i)

$$G \geq 0, \ B \in \mathrm{CI}(\mathcal{F}_t) \Rightarrow \omega \to \int G(\omega) \, dB(\omega) \text{ is an } \mathcal{F}_t\text{-adapted } [0, \infty]\text{-process,}$$

(ii)

$$T(G, B) = \Omega \ \Rightarrow \ \int G \, dB \in \mathrm{CFV}(\mathcal{F}_t),$$

(iii)

$$(\mathcal{F}_t) \text{ a complete filtration, } G \in \mathrm{PM}_1(B, \mathcal{F}_t) \ \Rightarrow \ \int G \, dB \in \mathrm{CFV}(\mathcal{F}_t).$$

Having a complete filtration $(\mathcal{F}_t)$, $B \in \mathrm{CFV}(\mathcal{F}_t)$ and $G \in \mathrm{PM}_1(B, \mathcal{F}_t)$ **agree to write** $dX = G \, dB$ or $dX \stackrel{\mathrm{as}}{=} G \, dB$ if X is a process in $\mathrm{CFV}(\mathcal{F}_t)$ such that $X \stackrel{\mathrm{as}}{=} \int G \, dB$. **Note** that the latter differential shorthand means exactly that outside a P-null set $\mu_{X(\omega)} \ll \mu_{B(\omega)}$ and the $G(\omega)$ is the **Radon-Nikodym derivative** of $\mu_{X(\omega)}$ with respect to $\mu_{B(\omega)}$.

We leave our reader to apply 1.11 (b) and 5.3 to get a neat proof of 5.4.

In some instances we may avoid the progressive measurability of the integrand G:

1.5.5. For a fixed complete filtration $(\mathcal{F}_t)$ let G be a process that is $\mathcal{F}_t$-adapted and $(\mathcal{B}^+ \otimes \mathcal{F})$-measurable. Moreover assume that all its trajectories are locally *integrable with respect to the Lebesgue measure on* $\mathbb{R}^+$. *Then* $(\int_0^t G\,ds, t \geq 0)$ *is a continuous and* $\mathcal{F}_t$ *-adapted process.*

Proof. Fix $t > 0$ and consider a sequence (ξ_n) of i.i.d. random variables uniformly distributed on $[0, t]$ and defined on a probability space $(\bar{\Omega}, \bar{\mathcal{F}}, \bar{P})$. The joint measurability of G implies that

(19) $\qquad (\bar{\omega}, \omega) \to \dfrac{1}{n} \sum_1^n G\left(\xi_j(\bar{\omega}), \omega\right)$ is an $(\bar{\mathcal{F}} \otimes \mathcal{F})$-measurable map.

Hence, it follows by *strong law of large numbers* that if $n \to \infty$ then

$$\frac{1}{n} \sum_1^n G\left(\xi_j, \omega\right) \to EG(\xi_1, \omega) = t^{-1} \int_0^t G(s, \omega)\,ds \quad \bar{P}\text{-a.s.}$$

for all $\omega \in \Omega$. Combined with (19) it yields $(\bar{P} \otimes P)(A) = 1$ where

$$A := \left\{ (\bar{\omega}, \omega) : \lim_n \frac{1}{n} \sum_1^n G\left(\xi_j(\bar{\omega}), \omega\right) = t^{-1} \int_0^t G(s, \omega)\,ds \right\} \in \bar{\mathcal{F}} \otimes \mathcal{F}.$$

It follows that there is at least one $\bar{\omega} \in \bar{\Omega}$ with $P(A_{\bar{\omega}}) = 1$. Denoting $t_j := \xi_j(\bar{\omega})$ for $j \in \mathbb{N}$,

$$\lim_n \frac{1}{n} \sum_1^n G\left(t_j, \omega\right) = t^{-1} \int_0^t G(s, \omega)\,ds \quad \forall \omega \in A_{\bar{\omega}},$$

hence P-almost surely. The random variable $t^{-1} \int_0^t G(s)\,ds$ is $\mathcal{F}_t$ -measurable because $\frac{1}{n} \sum_1^n G(t_j)$ are $\mathcal{F}_t$ -measurable and the filtration $(\mathcal{F}_t)$ is complete. $\square$

For the rest of the present section **we will fix a complete probability space** $(\Omega, \mathcal{F}, P)$ **endowed by a complete filtration** $(\mathcal{F}_t)$ and close it by establishing a suitable approximation method to be used in our construction of stochastic integral.

Consider $A \in \mathrm{CI}(\mathcal{F}_t)$ and define

$$\mathrm{EPM}_2(A, \mathcal{F}_t) = \mathrm{EPM}_2 := \left\{ G \in \mathrm{PM}(\mathcal{F}_t) : E \int_0^t G^2(s)\,dA(s) < \infty, \quad \forall t \geq 0 \right\},$$

$$[G]_t^A = [G]_t := \sqrt{E \int_0^t G^2\,dA}, \quad [G]^A = [G] := \sum_{t=1}^{\infty} 2^{-t}([G]_t \wedge 1)$$

for $G \in \mathrm{EPM}_2(A, \mathcal{F}_t)$. Note that $\mathrm{EPM}_2 \subset \mathrm{PM}_2 \subset \mathrm{PM}_1$.

As in 5.1 it is easy to verify the following simple properties of EPM_2:

1.5.6   Exercise. Let $A \in \mathrm{CI}$ be a process such that $EA(t) < \infty$ for all $t \geq 0$, denote again $\mathrm{EPM}_2 = \mathrm{EPM}_2(A)$ and prove:

(a) $[G - H]$ defines a pseudometric on EPM_2.

(b) $[G - H] = 0$ iff $[G - H]_t = 0$ for all $t \geq 0$ and iff outside a P-null set $G(\omega) = H(\omega)$ on $\mathbb{R}^+$ almost everywhere with respect to the measure associated with $A(\omega)$.

(c) $[G_n - G] \to 0$ iff $[G_n - G]_t \to 0$ for all $t \in \mathbb{R}^+$ and if and only if $E \int_0^t (G_n - G)^2 \, dA \to 0$ for all $t \in \mathbb{R}^+$.

(d) If (G_n) is a sequence in EPM$_2$ such that $[G_n - G_m] \to 0$ as $n, m \to \infty$ then there is a $G \in$ EPM$_2$ such that $[G_n - G] \to 0$.

We are interested in finding a suitable simple dense subset of EPM$_2$: Let $\Delta = \{0 = t_1 < t_2 < \ldots\}$ be a locally finite partition of $\mathbb{R}^+$ and K_j a bounded $\mathcal{F}_{t_j}$-measurable random variable for $j \in \mathbb{N}$. Then

$$K(t) := K_0 I_{\{0\}}(t) + \sum_{j=0}^{\infty} K_j I_{(t_j, t_{j+1}]}, \qquad t \geq 0$$

will be called an $\mathcal{F}_t$-**simple stochastic process**. The set of $\mathcal{F}_t$-simple processes will be denoted by $J(\mathcal{F}_t) =: J$ and we note that simple processes are $\mathcal{F}_t$-adapted (the destiny of any trajectory $K(\omega)$ on $(t_j, t_{j+1}]$ is determined by $K_j(\omega) \in \mathcal{F}_{t_j}$) and left-continuous, hence $\mathcal{F}_t$-progressive and obviously $J(\mathcal{F}_t) \subset$ EPM$_2(A, \mathcal{F}_t)$ for any $A \in$ CI$(\mathcal{F}_t)$ with $EA(t) < \infty$ for all $t \geq 0$.

1.5.7   Density Theorem. *Let $A \in$ CI be a process such that $EA(t) < \infty$ for $t \geq 0$. Then for any $G \in$ EPM$_2(A)$ there are simple processes G_n such that*

$$(20) \qquad \left([G_n - G]_t^A\right)^2 = E \int_0^t (G_n - G)^2 \, dA \to 0, \quad \forall \, t \geq 0.$$

Observe that the completeness of $(\mathcal{F}_t)$ is crucial in 5.7, again.

When constructing a sequence $(K_n) \subset J$ such that $[K_n - G] \to 0$ where $G \in$ EPM$_2$ we may restrict ourselves to bounded G's.

If $G \in$ EPM$_2$ is a **bounded continuous process**, say $|G| \leq C < \infty$, then

$$K_n(t) := G(0) \cdot I_{\{0\}}(t) + \sum_{j=1}^{\infty} G(t_j^n) \cdot I_{(t_j^n, t_{j+1}^n]}(t),$$

where $\Delta_n = \{0 = t_0^n < t_1^n < \ldots\}$ is a sequence of locally finite partitions of $\mathbb{R}^+$ with $|\Delta_n| \to 0$, defines $K_n \in J$ such that $|K_n| \leq C$ and $\lim_n K_n = G$ everywhere on $\mathbb{R}^+ \times \Omega$.

Thus it remains to prove that for every bounded process $G \in$ EPM$_2$ there are bounded continuous adapted processes G_n such that (20) holds. The first step is provided by

1.5.8   Lemma. *If G is a bounded adapted $\mathcal{B}(\mathbb{R}^+) \otimes \mathcal{F}$-measurable process, then*

$$G_n(t) := n \cdot \int_{(t-n^{-1})+}^{t} G(u) \, du, \qquad t \geq 0 \quad n \in \mathbb{N}$$

*defines a sequence of uniformly bounded adapted and continuous processes such
that*

$$(21) \qquad \int_0^t (G_n(s, \omega) - G(s, \omega))^2 \, ds \to 0, \quad \forall (t, \omega) \in \mathbb{R}^+ \otimes \Omega.$$

Note that 5.8 does not require the progressivity of G.

Proof. Obviously, if $|G| \leq C < \infty$ then $|G_n| \leq C$ also and G_n's are continuous
$\mathcal{F}_t$-adapted by 5.5. Moreover, everywhere on Ω

$$G_n(t) = \frac{\int_0^t G(u) \, du - \int_0^{(t-n^{-1})^+} G(u) \, du}{n^{-1}} \to G(t) \quad \text{almost everywhere on } \mathbb{R}^+$$

with respect to Lebesgue measure λ as $n \to \infty$ by the Fundamental Theorem of
Calculus. The Dominated Convergence Theorem now proves (21). $\square$

To conclude the proof of 5.7 we may assume that

$$(22) \qquad A \text{ is strictly increasing with } A(t) \geq t \text{ for all } t \in \mathbb{R}^+,$$

(hence with $A(\infty) = \infty$), because if A does not posses the additional properties,
the $\bar{A}(t) := A(t) + t$ does, and

$$E \int_0^t G^2 \, dA \leq E \int_0^t G^2 \, d\bar{A}, \quad \forall t \geq 0$$

holds for any bounded progressive process G.

1.5.9   Exercise. Let $A \in \mathrm{CI}$ be a process such that (22) holds. Put

$$\tau(t) := \inf\{s \geq 0 : A(s) \geq t\} \quad \text{and} \quad \tau := (\tau(t), t \geq 0) = A^{-1}$$

and consider an $\mathcal{F}_t$-progressive process G. Then

  (i) $\mathcal{F}_{\tau(t)} \subset \mathcal{F}_t$ is a complete filtration such that $G(\tau) = (G(\tau(t)), t \geq 0)$ is an
      $\mathcal{F}_{\tau(t)}$-adapted process that is measurable with respect to $\mathcal{B}(\mathbb{R}^+) \otimes \mathcal{F}$.

  (ii) If H is a continuous and $\mathcal{F}_{\tau(t)}$-adapted process then $H(A)$ is a continuous
       process that is $\mathcal{F}_t$-adapted.

Consider A as in (22), a bounded progressive process G and define τ as in 5.9.
As a consequence of 5.9 (i) we are entitled to employ 5.8 substituting there simply
the complete filtration $(\mathcal{F}_{\tau(t)}) \to (\mathcal{F}_t)$ and $G(\tau) \to G$. Hence, there are continuous
uniformly bounded $\mathcal{F}_{\tau(t)}$-adapted processes H_n such that $\int_0^t (H_n - G(\tau))^2 \, ds \to 0$
for all $t \geq 0$. By 5.9 (ii) we argue that $G_n := H_n(A)$ are continuous uniformly
bounded $\mathcal{F}_t$-adapted processes such that for all $t \geq 0$

$$\int_0^t |G_n - G|^2 \, dA = \int_0^t |H_n(A) - G(A \circ \tau)|^2 \, dA = \int_0^{A(t)} |H_n(u) - G(\tau(u))|^2 \, du \to 0$$

holds. Because obviously

$$\int_0^t (G_n - G)^2 \, dA \leq 4C^2 \cdot A(t) \in L_1(P)$$

we arrive at (20) and 5.7 is proved.

1.5.10   Corollary. *Let A be a process in CI. Then for arbitrary $G \in PM_2(A)$ a sequence (G_n) of simple processes exists such that*

$$(23) \qquad \int_0^t |G_n - G|^2 \, dA \xrightarrow{p} 0, \quad \forall t \in \mathbb{R}^+.$$

Proof. Assume first that A is as in 5.7: If G is a bounded process then there are $G_n \in J$ such that $E \int_0^t |G_n - G|^2 \, dA \to 0$, for all $t \geq 0$ according to 5.7, hence such as in (23). Further, for any $G \in PM_2(A)$ we put $G_n := G \cdot I_{[|G| \leq n]}$ to get a sequence of bounded progressive processes such that (23) holds. Observing that the convergence (23) is generated by the pseudometric $[[G - H]] := \sum_t 2^{-t} E \min(\int_0^t |G - H|^2 \, dA, 1)$ we combine the above pair of convergences to verify 5.10 for our special choice of A.

Consider A and G as in 5.10. Denote $A_n := A^{\tau_n}$ where τ_n is the first entry of A to n. According to the first part of our proof there are $G_n \in J$ such that $P[\int_0^n |G_n - G|^2 \, dA_n \geq n^{-1}] \leq n^{-1}$ holds for all $n \in \mathbb{N}$. Because $P[\tau_n < t] \to 0$ and $A_n(t) = A(t)$ if $\tau_n \geq t$, it follows that $G_n \to G$ as in (23). $\square$

1.6   Doob-Meyer Decomposition

Throughout this section we shall **fix a complete probability space** $(\Omega, \mathcal{F}, P)$ endowed by a **complete filtration** $(\mathcal{F}_t)$.

The following theorem provides the principal step towards the construction of stochastic integral and shows that martingale and its quadratic variation are concepts with a deep and rich relation.

Let first $\Delta = \{0 = t_0 < t_1 < t_2 < \dots\}$ be a locally finite partition of $\mathbb{R}^+$, i.e., with $t_n \uparrow \infty$ and put

$$S_t^\Delta(M) := \sum_{j=1}^k (M(t_j) - M(t_{j-1}))^2 + (M(t) - M(t_k))^2, \quad t_k \leq t < t_{k+1}, \quad k \in \mathbb{N}$$

and $S_0^\Delta(M) := 0$ for arbitrary stochastic process M. Observe that $S^\Delta(M)$ is continuous adapted provided that M possesses the properties and M is a process of finite quadratic variation iff $(S_t^{\Delta_n})$ is a Cauchy sequence in probability for any $t \geq 0$ and any sequence of partitions (Δ_n) with $|\Delta_n| \to 0$. Moreover,

$$|\Delta_n| \to 0 \quad \Rightarrow \quad \langle M \rangle(t) = p\text{-}\lim_{n \to \infty} S_t^{\Delta_n}, \quad t \in \mathbb{R}^+.$$

1.6.1   Doob-Meyer Decomposition. *Let $M \in CM_{loc}$. Then M is of finite quadratic variation and $\langle M \rangle$ possesses a modification that belongs to CI.*

Agree to choose $\langle M \rangle \in CI$ any time we have $M \in CM_{loc}$.

Further,

$$(24) \qquad M^2 \overset{\text{as}}{=} \langle M \rangle + R, \quad \text{where} \quad R \in CM_{loc}$$

and if $M^2 - A \in CM_{loc}$ holds for some other $A \in CI$ then $A \overset{as}{=} \langle M \rangle$. Moreover,

(25) $\qquad |\Delta_n| \to 0 \quad \Rightarrow \quad \max_{s \le t} |S_s^{\Delta_n}(M) - \langle M \rangle(s)| \overset{P}{\to} 0 \qquad \forall t \in \mathbb{R}^+.$

If M is a bounded martingale then $R = M^2 - \langle M \rangle$ is an $\mathcal{F}_t$-martingale.

The relation (24) will be referred to as **Doob-Mayer decomposition of M^2.**

Before starting a rather lengthy and complex proof we shall make some preparatory computations for $M \in CM_2$, a partition Δ and $s < t$ such that

$$t_i \le s < t_{i+1} \le t_k < t \le t_{k+1} \quad \text{for some} \quad i < k.$$

Apply 2.12 (a) and (b) to get

(26)

$$E^{\mathcal{F}_s}\left(S_t^{\Delta}(M) - S_s^{\Delta}(M)\right) =$$

$$= E^{\mathcal{F}_s}\left(M(t) - M(t_k)\right)^2 + \cdots + E^{\mathcal{F}_s}\left(M(t_{i+1}) - M(t_i)\right)^2 - (M(s) - M(t_i))^2 =$$

$$= E^{\mathcal{F}_s}\left(M^2(t) - M^2(t_k)\right) + \cdots + E^{\mathcal{F}_s}\left(M^2(t_{i+1}) - M^2(s)\right) =$$

$$= E^{\mathcal{F}_s}\left(M^2(t) - M^2(s)\right) = E^{\mathcal{F}_s}\left(M(t) - M(s)\right)^2.$$

Hence, if Δ, Δ^* are locally finite partitions, then

(27) $\qquad M^2 - S^{\Delta}(M), \qquad S^{\Delta}(M) - S^{\Delta^*}(M) \quad \text{are in} \quad CM.$

We shall also need the following upper boundary independent of the choice of $\Delta = \{0 = t_0 < t_1 \cdots < t_n < \ldots\}$:

(28) $\qquad M \in CM, \quad |M| \le C < \infty \Longrightarrow E\left(S_t^{\Delta}(M)\right)^2 \le 12C^4, \quad t \ge 0.$

Because $S_t^{\Delta} = S_t^{\Delta \cup \{t\}}$, we may assume that $t \in \Delta$, say $t = t_n$. Observe that

$$\left(S_t^{\Delta}(M)\right)^2 =$$

$$= \sum_{k=1}^{n} (M(t_k) - M(t_{k-1}))^4 + 2 \sum_{k=1}^{n} (M(t_k) - M(t_{k-1}))^2 \sum_{j=k}^{n-1} (M(t_{j+1}) - M(t_j))^2 =$$

$$= \sum_{k=1}^{n} (M(t_k) - M(t_{k-1}))^4 + 2 \sum_{k=1}^{n} \left(S_{t_k}^{\Delta}(M) - S_{t_{k-1}}^{\Delta}(M)\right) \left(S_t^{\Delta}(M) - S_{t_k}^{\Delta}(M)\right)$$

and use (26) to get

$$E\left(S_t^{\Delta}(M)\right)^2 - E \sum_{k=1}^{n} (M(t_k) - M(t_{k-1}))^4 =$$

$$= 2 \sum_{k=1}^{n} E\left(S_{t_k}^{\Delta}(M) - S_{t_{k-1}}^{\Delta}(M)\right) E^{\mathcal{F}_{t_k}}\left((S_t^{\Delta}(M) - S_{t_k}^{\Delta}(M))\right) =$$

$$= 2 \sum_{k=1}^{n} E\left(S_{t_k}^{\Delta}(M) - S_{t_{k-1}}^{\Delta}(M)\right) E^{\mathcal{F}_{t_k}}\left(M(t) - M(t_k)\right)^2 =$$

$$= 2E \sum_{k=1}^{n} \left(S_{t_k}^{\Delta}(M) - S_{t_{k-1}}^{\Delta}(M)\right) (M(t) - M(t_k))^2.$$

Hence,

$$E\left(S_t^{\Delta}(M)\right)^2 \le 4C^2 ES_t^{\Delta}(M) + 8C^2 ES_t^{\Delta}(M) = 12C^2 EM^2(t) \le 12C^4.$$

The uniqueness of the decomposition (24) is a direct consequence of 4.4.

Proof for bounded M. Let $|M| \le C < \infty$. Our plan is to verify

(29) $|\Delta| \to 0, \quad |\Delta^*| \to 0 \Rightarrow E|S_t^{\Delta}(M) - S_t^{\Delta^*}(M)|^2 \to 0, \quad \forall t \ge 0$

as it implies directly that M is a process of finite quadratic variation and supported by (27) it says that

(30) $|\Delta_n| \to 0 \Rightarrow M^2 - S^{\Delta_n}(M)$ is a Cauchy sequence in CM_2.

By 5.2 there is a martingale $R \in CM_2$ such that denoting $R_n := M^2 - S^{\Delta_n}(M)$

(31) $|\Delta_n| \to 0 \Rightarrow E\max_{s \le t}|R_n(s) - R(s)|^2 \to 0, \quad \forall t \ge 0.$

Check that (30) implies that the limit R is independent of the choice of $|\Delta_n| \to 0$: If also $|\Delta_n^*| \to 0$ observe the *mixed sequence* $\{\Delta_1, \Delta_1^*, \Delta_2, \Delta_2^*, \dots\}$.
    The convergence (31) yields

(32) $|\Delta_n| \to 0 \Rightarrow E\max_{s \le t}|S_s^{\Delta_n}(M) - [M^2(s) - R(s)]|^2 \to 0, \quad \forall t \ge 0.$

Thus, the quadratic variation $\langle M \rangle$ is a modification of $M^2 - R$ which implies (24) and (25) directly. To conclude the proof for bounded M it remains to show that $M^2 - R$ is an increasing process almost surely:
    In (32) choose (Δ_n) such that Δ_{n+1} is a finer partition than Δ_n for all $n \in \mathbb{N}$ and such that $\Delta := \cup\Delta_n$ is a (countable) dense set in $\mathbb{R}^+$. Fix $s \le t$ a pair of points in Δ, observe that $S_s^{\Delta_n}(M) \le S_t^{\Delta_n}(M)$ for sufficiently large $n \in \mathbb{N}$ which yields

$$M^2(s) - R(s) = L_1 - \lim_n S_s^{\Delta_n}(M) \le L_1 - \lim_n S_t^{\Delta_n}(M) = M^2(t) - R(t), \quad \text{a.s..}$$

It follows that outside a P-null set the process $M^2 - R$ is increasing on the set Δ, hence on $\mathbb{R}^+$ as $\bar{\Delta} = \mathbb{R}^+$. We proved that $\langle M \rangle$ can be modified to a process in CI.
    To prove (29) consider partitions $\Delta = \{0 = t_0 < t_1 < \dots\}, \Delta^* = \{0 = t_0^* < t_1^* < \dots\}$ and $\Delta\Delta^* = \{0 = s_0 < s_1 < \dots\}$ the partition generated by $\Delta \cup \Delta^*$. For a fixed $s_k \in \Delta\Delta^*$ define $t_{l(k)} := t_l$ as the maximal point of Δ such that $t_{l(k)} \le s_k < s_{k+1} \le t_{l(k)+1}$ and compute

(33)
$$S_{s_{k+1}}^{\Delta}(M) - S_{s_k}^{\Delta}(M) = (M(s_{k+1}) - M(t_l))^2 - (M(s_k) - M(t_l))^2$$
$$= (M(s_{k+1}) - M(s_k))(M(s_{k+1}) + M(s_k) - 2M(t_l)).$$

Note that $S := S^\Delta(M) - S^{\Delta^*}(M)$ is a martingale by (27), hence $S \in CM_2$ as $|M| \leq C$. Apply (26) and a simple inequality $(a + b)^2 \leq 2a^2 + 2b^2$ to see that for any $t \geq 0$

$$(34) \quad E|S_t^\Delta(M) - S_t^{\Delta^*}(M)|^2 = ES_t^2 = ES_t^{\Delta\Delta^*}(S)$$
$$\leq 2ES_t^{\Delta\Delta^*}(S^\Delta(M)) + 2ES_t^{\Delta\Delta^*}(S^{\Delta^*}(M)).$$

Denote

$$Z_t^\Delta := 2\max\{|M(u) - M(v)|, |u - v| \leq |\Delta|, u, v \leq t\}, \quad K_t^\Delta := \sqrt{E(Z_t^\Delta)^4}.$$

Fix a $t \in \Delta\Delta^*$, say $t = s_{j+1}$, and apply (33) to verify

$$S_t^{\Delta\Delta^*}(S^\Delta(M)) = \sum_{k=0}^{j}(S_{s_{k+1}}^\Delta(M) - S_{s_k}^\Delta(M))^2$$
$$\leq \max_{0 \leq k \leq j}|M(s_{k+1}) + M(s_k) - 2M(t_{l(k)})|^2 \times S_t^{\Delta\Delta^*}(M) \leq (Z_t^\Delta)^2 \times S_t^{\Delta\Delta^*}(M).$$

It follows by (28) that

$$ES_t^{\Delta\Delta^*}(S^\Delta(M)) \leq K_t^\Delta \times \sqrt{E(S_t^{\Delta\Delta^*}(M))^2} \leq K_t^\Delta \times \sqrt{12}C^2.$$

By (34) we conclude that

$$(35) \quad E|S_t^\Delta(M) - S_t^{\Delta^*}(M)|^2 \leq 2\sqrt{12}C^2(K_t^\Delta + K_t^{\Delta^*}), \quad \forall t \in \Delta\Delta^*.$$

Since $S^2 = |S_t^\Delta(M) - S_t^{\Delta^*}(M)|^2$ is a submartingale and the partition $\Delta\Delta^*$ cofinal in $\mathbb{R}^+$, the inequality (35) implies that

$$E|S_t^\Delta(M) - S_t^{\Delta^*}(M)|^2 \leq 2\sqrt{12}C^2(K_{t+1}^\Delta + K_{t+1}^{\Delta^*})$$

holds for all $t \geq 0$ if $|\Delta\Delta^*| \leq 1$. It follows easily by the continuity and boundedness of M that $K_{t+1}^\Delta + K_{t+1}^{\Delta^*} \to 0$ as $|\Delta|, |\Delta^*| \to 0$ and the convergence (29) is proved. $\square$

Proof for $M \in CM_{loc}$. Let $\tau_n \uparrow \infty$ be a localization sequence for M such that $M_n := M^{\tau_n}$ are all bounded martingales. According the first part of our proof

$$(36) \quad M_n^2 \overset{as}{=} \langle M_n \rangle + R_n, \quad \text{where} \quad R_n \in CM, \quad n \in \mathbb{N}.$$

We can also write

$$M_n^2 = (M_{n+1}^2)^{\tau_n} \overset{as}{=} (\langle M_{n+1} \rangle + R_{n+1})^{\tau_n} = \langle M_{n+1} \rangle^{\tau_n} + R_{n+1}^{\tau_n}, \quad n \in \mathbb{N},$$

that implies

$$\langle M_n \rangle \overset{as}{=} \langle M_{n+1} \rangle^{\tau_n}, \quad R_n \overset{as}{=} R_{n+1}^{\tau_n}, \quad n \in \mathbb{N}$$

because of the already proved uniqueness of the decompositions (36). By 4.8 there are continuous adapted processes A and $R \in CM_{loc}$ such that

$$(37) \qquad A^{\tau_n} \overset{as}{=} \langle M_n \rangle, \qquad R^{\tau_n} \overset{as}{=} R_n, \qquad n \in \mathbb{N}.$$

The process A can be obviously chosen in CI. Combining (36) and (37) we arrive to the equalities

$$(M^2)^{\tau_n} \overset{as}{=} A^{\tau_n} + R^{\tau_n}, \qquad n \in \mathbb{N}$$

and taking limit $n \to \infty$ there we exhibit the decomposition $M^2 \overset{as}{=} A + R$. Everything we need to conclude the proof is to verify

$$(38) \qquad |\Delta_n| \to 0 \Rightarrow \max_{s \leq t} |S_s^{\Delta_n}(M) - A(s)| \overset{P}{\to} 0, \qquad \forall t \geq 0,$$

because this would imply that M is a process of finite quadratic variation, $A \in CI$ is a modification of $\langle M \rangle$ and also the convergence (25), of course.

For these purposes fix $t \geq 0$, $\epsilon > 0$ and $\delta > 0$ and let $k \in \mathbb{N}$ be such that $P[\tau_k \leq t] < \frac{\epsilon}{2}$. Since $A^{\tau_k} \overset{as}{=} \langle M^{\tau_k} \rangle$ by (37), we have

$$P[\max_{s \leq t} |S_s^{\Delta_n}(M) - A(s)| \geq \delta] \leq \frac{\epsilon}{2} + P[\max_{s \leq t} |S_s^{\Delta_n}(M) - A(s)| \geq \delta, \tau_k > t]$$

$$\leq \frac{\epsilon}{2} + P[\max_{s \leq t} |S_{s \wedge \tau_k}^{\Delta_n}(M) - A(s \wedge \tau_k)| \geq \delta]$$

$$= \frac{\epsilon}{2} + P[\max_{s \leq t} |S_s^{\Delta_n}(M^{\tau_k}) - \langle M^{\tau_k} \rangle(s)| \geq \delta] \to \frac{\epsilon}{2},$$

as $n \to \infty$, because we have already proved (25) for bounded martingales. $\square$

Recall that we have agreed that writing $\langle M \rangle$ for a local $\mathcal{F}_t$-martingale we mean the modification which belongs to $CI(\mathcal{F}_t)$.

Consider processes M, N and $\Delta = \{0 = t_0 < t_1 < \ldots\}$ a locally finite partition of $\mathbb{R}^+$. For $t_k \leq t < t_{k+1}$ denote

$$S_t^\Delta(M, N) :=$$

$$= \sum_{j=1}^{k} (M(t_j) - M(t_{j-1})) (N(t_j) - N(t_{j-1})) + (M(t) - M(t_k)) (N(t) - N(t_k)).$$

1.6.2   Theorem. Let M and N be in CM_{loc}. Then the covariation of these processes

$$(39) \qquad \langle M, N \rangle := \frac{1}{4} ((\langle M + N \rangle) - \langle M - N \rangle)$$

defines a process in CFV such that $M \cdot N - \langle M, N \rangle \in CM_{loc}$.

If $M \cdot N - B \in CM_{loc}$ for some other $B \in CFV$ then $B \overset{as}{=} \langle M, N \rangle$. Moreover,

$$(40) \qquad |\Delta_n| \to 0 \Rightarrow \max_{s \leq t} |S_s^{\Delta_n}(M, N) - \langle M, N \rangle(s)| \overset{P}{\to} 0.$$

Proof. The uniqueness follows again easily by 4.4, for the rest apply Doob-Mayer theorem and the equalities (39) jointly with

$$S_t^\Delta(M,N) = \frac{1}{4}\left(S_t^\Delta(M+N) - S_t^\Delta(M-N)\right), MN = \frac{1}{4}\left((M+N)^2 - (M-N)^2\right)$$

□

In many cases our reasoning will depend entirely on our ability to compute $\langle M \rangle$ and $\langle M, N \rangle$. Below we will present techniques that may smooth the computations.

We will say that **processes** of finite quadratic variation M, N **are orthogonal** if $\langle M, N \rangle \overset{as}{=} 0$. Observe that $M, N \in CM_{loc}$ are orthogonal iff their product MN is in CM_{loc} and that their orthogonality implies $\langle M + N \rangle \overset{as}{=} \langle M \rangle + \langle N \rangle$.

1.6.3 Corollary. *Independent local martingales X and Y are orthogonal. In particular, if $(W_1, W_2, \ldots, W_d)$ is an d-dimensional Wiener process then*

$$\langle W_i, W_j \rangle(t) \overset{as}{=} t\delta_{ij} \qquad \forall t \in \mathbb{R}^+.$$

According to 4.3 (c) and 4.9 (b) XY is a local $(\mathcal{F}_t^X \vee \mathcal{F}_t^Y)^P$-martingale, hence $\langle X, Y \rangle \overset{as}{=} 0$ by the uniqueness part of 6.2 as the $\langle X, Y \rangle$ does not depend on the choice of a filtration by (25).

1.6.4 Corollary. *Any continuous semimartingale $dX = dB + dM$, i.e.,*

$$X = X(0) + B + M, \qquad B \in CFV, \quad M \in CM_{loc}$$

is a process of finite quadratic variation with $\langle X \rangle = \langle M \rangle$. If $dY = dC + dN$ is also a process in CSM, then $\langle X, Y \rangle = \langle M, N \rangle$.

Proof. We need to prove that for any $t \in \mathbb{R}^+$ and any sequence of (Δ_n) where Δ_n is a locally finite partition of $\mathbb{R}^+$

(41)
$$|\Delta_n| \to 0 \Rightarrow S_t^{\Delta_n}(X, Y) \overset{P}{\to} \langle M, N \rangle.$$

Because

$$S_t^{\Delta_n}(X, Y) = S_t^{\Delta_n}(B, C) + S_t^{\Delta_n}(B, N) + S_t^{\Delta_n}(C, M) + S_t^{\Delta_n}(M, N),$$

(41) follows by (40) observing that the first three sequences tend to 0 in probability. For example

$$S_t^{\Delta_n}(C, M) \leq \max_{|u-v| \leq \Delta_n, u, v \leq t} |M(u) - M(v)| \times B^v(t) \to 0$$

everywhere on Ω because M is continuous and C of finite variation. □

Important Remark A. Let $X \in CSM(\Omega, \mathcal{F}, P, \mathcal{F}_t)$. Since $\langle X \rangle$ is the limit in probability of $S^{\Delta_n}(X)$, it does not change if we replace $(\mathcal{F}_t, P)$ by $(\mathcal{G}_t, Q)$ such that $(\Omega, \mathcal{F}, Q)$ is a complete space, $(\mathcal{G}_t)$ its complete filtration, $X \in CSM(\Omega, \mathcal{F}, Q, \mathcal{G}_t)$ and $Q \ll P$ $(P(N) = 0 \Rightarrow Q(N) = 0)$.

Very frequently we shall use

1.6.5   Theorem. *Let* M, N *be processes in* CM_{loc} *and consider a Markov time* τ. *Then outside a P-null set*

$$(42) \qquad \langle M^\tau \rangle = \langle M \rangle^\tau, \qquad \langle M^\tau, N^\tau \rangle = \langle M, N \rangle^\tau = \langle M, N^\tau \rangle.$$

Proof. Since

$$M^\tau N^\tau - \langle M, N \rangle^\tau = (MN - \langle M, N \rangle)^\tau$$

is in CM_{loc}, by 6.2 and 4.1 (b), it follows again by the uniqueness part of 6.2 that $\langle M^\tau, N^\tau \rangle = \langle M, N \rangle^\tau$. The last equality in (42) is less obvious:
Verify first that for all $(s, \omega) \in \mathbb{R}^+ \times \Omega$ there is a constant $C_s = C_s(\omega)$ such that

$$|S_s^\Delta(M, N^\tau) - S_{s \wedge \tau}^\Delta(M, N)| \le C_s \cdot \max \{|M(u) - M(v)|, \, |u - v| \le |\Delta|, \, u, v \le s\},$$

whatever we may have a partition Δ. Hence, for any fixed $t \ge 0$

$$|\Delta_n| \to 0 \Rightarrow \max_{s \le t} |S_s^{\Delta_n}(M, N^\tau) - \langle M, N \rangle^\tau(s)| \le \max_{s \le t} |S_s^{\Delta_n}(M, N) - \langle M, N \rangle(s)| \to 0$$

in probability by (40). Thus, $\langle M, N \rangle^\tau = \langle M, N^\tau \rangle$ again by (40). $\square$

1.6.6   Exercise. *If* $M \in CM_{loc}(\mathcal{F}_t)$ *and* $r > 0$ *then* $M_r(t) := M(t + r) - M(r)$ *is a process in* $CM_{loc}(\mathcal{F}_{t+r})$ *and* $\langle M_r \rangle(t) = \langle M \rangle(t + r) - \langle M \rangle(r)$.
Moreover, if $N \in CM_{loc}(\mathcal{F}_{t+r})$ *define* $N_o = (N_o(t), t \ge 0)$ *by* $N_o(t) = 0$ *if* $t \le r$ *and* $N_o(t + r) = N(t)$ *if* $t \ge 0$. *Prove that*

$$N_o \in CM_{loc}(\mathcal{F}_t), \qquad \langle M_r, N \rangle(t) \overset{as}{=} \langle M, N_o \rangle(t + r) \quad \forall t \ge 0.$$

1.7   Quadratic Variation of Local Martingales

A complete probability space $(\Omega, \mathcal{F}, P)$ and a **complete filtration** $\mathcal{F}_t \subset \mathcal{F}$ will be **fixed** again. We shall study interactions between a local martingale M and its quadratic variation $\langle M \rangle$. The first observation is that the quadratic variation and covariation maps help us to recognize when $M, N \in CM_{loc}$ are equivalent processes. Indeed,

$$(43) \quad M \overset{as}{=} N \quad \Longleftrightarrow \quad \langle M - N \rangle \overset{as}{=} 0 \quad \Longleftrightarrow \quad \langle L, M - N \rangle \overset{as}{=} 0 \quad \forall L \in CM_{loc}.$$

We are left to prove only that $\langle M \rangle \overset{as}{=} 0 \Rightarrow M \overset{as}{=} 0$:
If $\langle M \rangle \overset{as}{=} 0$ then $M^2 \in CM_{loc}$ and by a localization, $(M^{\tau_n})^2$ and M^{τ_n} are true martingales for a sequence of Markov times $\tau_n \uparrow \infty$ almost surely. It follows by 2.12. (b) that $M^{\tau_n} \overset{as}{=} 0$ for all $n \in \mathbb{N}$ that completes the verification of $M \overset{as}{=} 0$.
The simple device (43) is extended by

1.7.1 Mutual Continuity of M and $\langle M \rangle$. If $M \in CM_{loc}$ then outside a P-null set and for all $0 \le a < b < \infty$

(44) $\qquad M(\cdot, \omega)$ is a constant on $[a, b] \quad \Longleftrightarrow \quad \langle M \rangle (b, \omega) = \langle M \rangle (a, \omega)$.

holds. In other words, M and $\langle M \rangle$ have common intervals of constancy with probability one.

Proof. **(a):** It follows by the definition of $\langle M \rangle$ that for each pair of rationals $0 \le r < q < \infty$ there exists a P-null set $N_{r,q}$ such that for $\omega \notin N_{r,q}$

$$M(\cdot, \omega) \quad \text{is a constant on} \quad [r, q] \quad \Rightarrow \quad \langle M \rangle (q, \omega) = \langle M \rangle (r, \omega).$$

The continuity of M and $\langle M \rangle$ implies that for $\omega \notin \cup N_{r,q}$ and any interval $[a, b] \subset \mathbb{R}^+$

$$M(\cdot, \omega) \quad \text{is a constant on} \quad [a, b] \quad \Rightarrow \quad \langle M \rangle (a, \omega) = \langle M \rangle (b, \omega),$$

that proves "$\Rightarrow$" in (44).

(b): We shall prove that there is a P-null set N such that outside N and for any $t \ge 0$

$$\langle M \rangle (\cdot, \omega) = 0 \quad \text{on} \quad [0, t] \quad \Rightarrow \quad M(\cdot, \omega) = 0 \quad \text{on} \quad [0, t].$$

According to 4.3 (b) we may assume without loss of generality that $(\mathcal{F}_t)$ is a right-continuous filtration. Denote $v := \inf\{t \ge 0 : \langle M \rangle (t) > 0\}$ and recall 3.6 (b) to confirm that v is an $\mathcal{F}_t$-Markov time. It follows that $\langle M^v \rangle \overset{as}{=} \langle M \rangle^v \overset{as}{=} 0$, hence $M^v \overset{as}{=} 0$ by (43) that is exactly what we promised to prove.

(c): Choose $r \in Q^+$, denote $M^r(t) := M(r + t) - M(r)$, observe that $M^r \in CM_{loc}(\mathcal{F}_{r+t})$ with

$$\langle M^r \rangle (t) = \langle M \rangle (r + t) - \langle M \rangle (r), \quad \forall t \ge 0 \quad \text{almost surely}$$

by 6.6. The part **(b)** of the present proof applied to M^r provides a P-null set N_r such that outside N_r and for any $t \ge 0$

$$\langle M \rangle (r + t, \omega) = \langle M \rangle (r, \omega) \quad \Rightarrow \quad M(\cdot, \omega) \quad \text{is a constant on} \quad [r, r + t].$$

Using again the continuity of M and $\langle M \rangle$ we verify "$\Leftarrow$" in (44) for $\omega \notin \cup N_r$ and any interval $[a, b] \subset \mathbb{R}^+$. $\square$

We may deepen 7.1 a little combining it with 1.5 to get a local version of 4.4:

1.7.2 Corollary. Let $dX = dB + dM$ be semimartingale in CSM. Then outside a P-null set and for any interval $[a, b]$

$X(\cdot, \omega)$ is of finite variation on $[a, b] \quad \Longleftrightarrow \quad M(\cdot, \omega)$ is a constant on $[a, b]$.

Proof. The X is of finite quadratic variation with the **continuous** $\langle X \rangle$ by 6.4, hence we may apply 1.5 to exhibit a P-null set N such that for $\omega \notin N$ and all intervals $[a, b] \subset \mathbb{R}$

$$X(\cdot, \omega) \quad \text{is of finite variation on} \quad [a, b] \Longrightarrow \langle M \rangle (b, \omega) = \langle M \rangle (a, \omega).$$

Combining the above implication with "$\Leftarrow$" in (44) we conclude the proof. $\square$

The integrability of the quadratic variation of a local martingale forces it to be a true martingale:

1.7.3   Theorem. Let $M \in CM_{loc}$. Then $M \in CM_2$ if and only if $E\langle M \rangle(t) < \infty$ for any $t \geq 0$. If $M \in CM_2$ then $R := M^2 - \langle M \rangle \in CM$.

Proof. Let $\tau_n \uparrow \infty$ almost surely be a localization sequence for M such that M^{τ_n}'s are bounded martingales. Then $(M^{\tau_n})^2 - \langle M^{\tau_n} \rangle$ is a true martingale by the last statement of 6.1 and therefore

$$(45) \qquad EM^2(\tau_n \wedge t) = E\langle M \rangle(\tau_n \wedge t), \qquad \forall t \geq 0.$$

First assume that $M \in CM_2$: Apply the monotone convergence theorem, (45) and finally the submartingale part of Stopping Theorem 3.11 to see that

$$E\langle M \rangle(t) = \lim_{n \to \infty} E\langle M \rangle(\tau_n \wedge t) = \lim_{n \to \infty} EM^2(\tau_n \wedge t) \leq EM^2(t) < \infty.$$

If $E\langle M \rangle(t) < \infty$ then by Fatou lemma, (45) and the Monotone Convergence Theorem

$$EM^2(t) \leq \lim_{n \to \infty} EM^2(\tau_n \wedge t) = \lim_{n \to \infty} E\langle M \rangle(\tau_n \wedge t) = E\langle M \rangle(t) < \infty,$$

hence $M(t) \in L_2$ for all $t \in \mathbb{R}^+$. Having fixed $t \geq 0$ we note that $\{M(\tau_n \wedge t)\}$ is a sequence of uniformly integrable r.v.'s because

$$EM^2(\tau_n \wedge t) = E\langle M \rangle(\tau_n \wedge t) \leq E\langle M \rangle(t) < \infty$$

again by (45). Hence, $M(\tau_n \wedge t) \to M(t)$ in L_1 and if $s \leq t$, it follows by 2.7 (f) that

$$E^{\mathcal{F}_s} M(t) = \lim_n E^{\mathcal{F}_s} M(\tau_n \wedge t) = \lim_n M(\tau_n \wedge s) = M(s),$$

where both limits are in L_1. Hence $M \in CM_2$.

Finally, if $M \in CM_2$ fix $t \geq 0$ and note that

$$\max_{s \leq t} |R(t)| \leq \max_{s \leq t} M^2(s) + \langle M \rangle(t) \in L_1$$

by Doob inequality 2.10 (b). Hence, it follows by 4.2 (a) that R^t is a martingale for any $t \in \mathbb{R}^+$ which is exactly as to say that $R \in CM(\mathcal{F}_t)$. $\square$

1.7.4   Exercise. Use the polarization $\langle M, N \rangle = \frac{1}{4}(\langle M + N \rangle - \langle M - N \rangle)$ and 7.3 to prove

(i) $M, N \in CM_2 \implies E|\langle M, N \rangle(t)| < \infty, \quad t \geq 0, \quad MN - \langle M, N \rangle \in CM.$

Hence,

(ii) $M, N \in CM_2, \quad \langle M, N \rangle = 0 \quad \Rightarrow \quad MN \in CM.$

Prove also

(iii) $M, N \in CM, \quad N \text{ bounded}, \langle M, N \rangle = 0 \quad \Rightarrow \quad MN \in CM.$

1.7.5 Exercise. If $M, N \in CM_2$ and $v \le \tau$ Markov times then for any $t \ge 0$

$$E^{\mathcal{F}_v}(M(\tau \wedge t) - M(v \wedge t))^2 = E^{\mathcal{F}_v}(M^2(\tau \wedge t) - M^2(v \wedge t))^2$$

(46)
$$= E^{\mathcal{F}_v}(\langle M \rangle(\tau \wedge t) - \langle M \rangle(v \wedge t))$$

and by polarization of (46)

$$E^{\mathcal{F}_v}(M(\tau \wedge t) - M(v \wedge t))(N(\tau \wedge t) - N(v \wedge t))$$

$$= E^{\mathcal{F}_v}(M(\tau \wedge t)N(\tau \wedge t) - M(v \wedge t)N(v \wedge t))$$

(47)
$$= E^{\mathcal{F}_v}(\langle M, N \rangle(\tau \wedge t) - \langle M, N \rangle(v \wedge t)).$$

The equalities (46) and (47) may be applied to determine the quadratic variation and covariation:

1.7.6 Exercise. Let M, N be martingales in CM_2. Then $\langle M \rangle$ is the only integrable process in CI for which

(48) $$E^{\mathcal{F}_s}(M(t) - M(s))^2 = E^{\mathcal{F}_s}(\langle M \rangle(t) - \langle M \rangle(s)), \qquad \forall s \le t$$

holds. Similarly, $\langle M, N \rangle$ is the only integrable process in CFV with the property

(49) $$E^{\mathcal{F}_s}(M(t) - M(s))(N(t) - N(s)) = E^{\mathcal{F}_s}(\langle M, N \rangle(t) - \langle M, N \rangle(s)), \qquad \forall s \le t.$$

Recall that CM_2 as a closed subspace in CP_2 is a complete pseudometric space by 5.2 with convergence $M_n \to M$ that is equivalently defined by

$$E \max_{s \le t} |M_n(s) - M(s)|^2 \to 0 \quad \forall t \ge 0 \iff E|M_n(t) - M(t)|^2 \to 0 \quad \forall t \ge 0$$

$$\iff E \langle M_n - M \rangle(t) \to 0 \quad \forall t \ge 0$$

according to 5.1 and 7.3.

For local martingales the above equivalence has the following simple form:

1.7.7 Continuity Theorem. If $M_n, M \in CM_{loc}$ and a $t > 0$ fixed then

$$\langle M_n - M \rangle(t) \xrightarrow{P} 0 \iff \max_{s \le t} |M_n(s) - M(s)| \xrightarrow{P} 0.$$

For the proof we need

1.7.8 Lenglart Inequality. If $M \in CM_{loc}$ and $t, \delta, \epsilon > 0$, then

$$P := P[\max_{s \le t} |M(s)| \ge \epsilon, \langle M \rangle(t) < \delta] \le 4\epsilon^{-2} E \min(\delta, \langle M \rangle(t)).$$

To prove the inequality denote by τ the first entry of $\langle M \rangle$ into $[\delta, \infty)$. Then $\langle M^\tau \rangle \leq \delta$ implies that $M^\tau \in CM_2$ and by 7.3 we arrive at

$$E(M^\tau)^2(t) = E\langle M^\tau \rangle(t) \leq E \min(\delta, \langle M \rangle(t)).$$

Consequently,

$$P = P[\max_{s \leq t} |M(s)| \geq \epsilon, \tau > t] \leq \epsilon^{-2} \int_{[\tau > t]} \max_{s \leq t} |M(s)|^2 \, dP \leq 4\epsilon^{-2} E(M^\tau(t))^2$$

by Doob inequality 2.10 (b) which completes the proof of the inequality.

Proof of 7.7. By Lenglart inequality, $\langle M_n - M \rangle(t) \xrightarrow{P} 0$ implies

$$P[\max_{s \leq t} |M_n(s) - M(s)| \geq \epsilon]$$
$$\leq P[\langle M_n - M \rangle(t) \geq \epsilon] + P[\max_{s \leq t} |M_n(s) - M(s)| \geq \epsilon, \langle M_n - M \rangle(t) < \epsilon]$$
$$\leq P[\langle M_n - M \rangle(t) \geq \epsilon] + 4\epsilon^{-2} E \min\{\epsilon, \langle M_n - M \rangle(t)\} \to 0.$$

To prove the" $\Leftarrow$ " part consider $\epsilon > 0$ and denote by τ_n the first entry of $|M_n - M|$ into $[\epsilon, \infty)$. Because $(M_n - M)^{\tau_n}$ is a bounded martingale, it follows by 7.3 that $E\langle M_n - M \rangle^{\tau_n}(t) = E(M_n - M)^2(\tau_n \wedge t) \leq \epsilon^2$ for any $t \geq 0$ and

$$P[\langle M_n - M \rangle(t) \geq \epsilon] \leq P[\tau_n < t] + \epsilon^{-1} \int_{[\tau_n \geq t]} \langle M_n - M \rangle(t) \, dP$$
$$\leq P[\tau_n < t] + \epsilon^{-1} E\langle M_n - M \rangle(\tau_n \wedge t) \leq P[\max_{s \leq t} |M_n(s) - M(s)| \geq \epsilon] + \epsilon,$$

that verifies " $\Leftarrow$ ". $\square$

We shall also need

1.7.9 Kunita-Watanabe Inequality. *For* $M, N \in CM_{loc}$, $K \in PM_2(M)$ *and* $H \in PM_2(N)$

$$\int_0^t |K \cdot H| \, d\langle M, N \rangle^v \leq \sqrt{\int_0^t K^2 \, d\langle M \rangle} \cdot \sqrt{\int_0^t H^2 \, d\langle N \rangle}$$

holds almost surely for any $t \geq 0$.

Remark. Recall that for a $B \in CFV(\mathcal{F}_t)$ we defined the space $PM_2(B, \mathcal{F}_t)$ as the set of all $\mathcal{F}_t$-progressive processes K such that $\int_0^t K^2 \, dB^v < \infty$ almost surely for any $t \geq 0$. Having $M, N \in CM_{loc}(\mathcal{F}_t)$ we agree to relax our notation letting $PM_2(M, \mathcal{F}_t) = PM_2(M) := PM_2(\langle M \rangle, \mathcal{F}_t)$ and $PM_1(M, N) := PM_1(\langle M, N \rangle)$.

Note that Kunita-Watanabe inequality also says that

$$K \in PM_2(M), \ H \in PM_2(N) \implies K \cdot H \in PM_1(M, N).$$

Proof. Note that the proof may be relaxed to K and H bounded by the Monotone Convergence Theorem.

Assume first that K and H are $\mathcal{F}_t$-simple processes, without loss of generality such that

$$K(t) = K_0 \cdot I_{\{0\}}(t) + \sum_{i=0}^{\infty} K_i \cdot I_{(t_i, t_{i+1}]}(t), \quad H(t) = H_0 \cdot I_{\{0\}}(t) + \sum_{i=0}^{\infty} H_i \cdot I_{(t_i, t_{i+1}]}(t),$$

where $\Delta = \{0 = t_0 < t_1 < \ldots\}$ is a locally finite partition of $\mathbb{R}^+$ and K_i, H_i bounded $\mathcal{F}_{t_i}$-measurable random variables. Fix $t > 0$ and assume, again without loss of generality, that $t = t_{k+1}$. Let $\Delta_n(t) = \{0 = t_0^n < t_1^n < \cdots < t_{k_n}^n = t\}$ be a sequence of partitions of $[0, t]$ with $|\Delta_n(t)| \to 0$ such that $\{t_0, t_1, \ldots, t_{k+1}\} \subset \Delta_n(t)$ for all $n \in \mathbb{N}$. Applying successively 1.3 and 1.6 (b) we prove the inequality:

$$\int_0^t |K \cdot H| \, d\langle M, N \rangle^v = \sum_{j=0}^{k} |K_j H_j| \{\langle M, N \rangle^v(t_{j+1}) - \langle M, N \rangle^v(t_j)\}$$

$$= \lim_{n \to \infty} \sum_{j=0}^{k} |K_j H_j| \sum_{t_j < t_i^n < t_{i+1}^n \leq t_{j+1}} |\langle M, N \rangle(t_{i+1}^n) - \langle M, N \rangle(t_i^n)|$$

$$\leq \lim_{n \to \infty} \sum_{j=0}^{k} |K_j H_j| \sum_{t_j < t_i^n < t_{i+1}^n \leq t_{j+1}} |\langle M \rangle(t_{i+1}^n) - \langle M \rangle(t_i^n)|^{\frac{1}{2}} |\langle N \rangle(t_{i+1}^n) - \langle N \rangle(t_i^n)|^{\frac{1}{2}}$$

$$\leq \lim_{n \to \infty} \sqrt{\sum_{j=0}^{k} \sum_{t_j < t_i^n < t_{i+1}^n \leq t_{j+1}} K_j^2 \{\langle M \rangle(t_{i+1}^n) - \langle M \rangle(t_i^n)\}} \times$$

$$\times \lim_{n \to \infty} \sqrt{\sum_{j=0}^{k} \sum_{t_j < t_i^n < t_{i+1}^n \leq t_{j+1}} H_j^2 \{\langle N \rangle(t_{i+1}^n) - \langle N \rangle(t_i^n)\}}$$

$$= \sqrt{\sum_{j=0}^{k} K_j^2 \{\langle M \rangle(t_{j+1}) - \langle M \rangle(t_j)\}} \cdot \sqrt{\sum_{j=0}^{k} H_j^2 \{\langle N \rangle(t_{j+1}) - \langle N \rangle(t_j)\}},$$

the last term being exactly the right-hand side of Kunita-Watanabe inequality.

If K and H are bounded processes then $K, H \in \text{PM}_2(A)$ where $A := \langle M \rangle + \langle N \rangle + \langle M, N \rangle^v \in \text{CI}$ and 5.10 applies to exhibit for a fixed $t > 0$ approximations

$$\int_0^t (K_n - K)^2 \, dA \to 0, \qquad \int_0^t (H_n - H)^2 \, dA \to 0, \qquad \text{almost surely},$$

where K_n, H_n are simple processes. Hence, outside a P-null set the integrals

$$\int_0^t (K_n - K)^2 \, d\langle M \rangle, \qquad \int_0^t (H_n - H)^2 \, d\langle N \rangle, \qquad \int_0^t |K_n \cdot H_n| - |K \cdot H| \, d\langle M, N \rangle^v$$

tend to zero as $n \to \infty$. The inequality for K and H follows now easily applying the inequality already proved for K_n, H_n's. $\square$

1.8   Helps to Some Exercises

1.4   To verify the ($\Leftarrow$) part of the equivalence you need to prove that the probability limit in (1) does not depend on the choice of $\{\Delta_n(t)\}$. You will achieve this considering a mixed sequence $\Delta_1(t), \Delta_1^*(t), \Delta_2(t), \Delta_2^*(t), \Delta_3(t) \ldots$, where $|\Delta_n(t)| \to 0$ and $|\Delta_n^*(t)| \to 0$.

1.6   Use the "limit definition" of $\langle X, Y \rangle$ and the trivial inequalities $\left(\sum_i a_i b_i \right)^2 \le \sum_i a_i^2 \sum_i b_i^2$ and $ab \le \frac{1}{2}(a^2 + b^2)$.

1.7   To prove (c) apply *Lindelöf Theorem* to prove that any open set in $C(\mathbb{R}^+)$ is a countable union of open neighbourhoods $\{x \in C(\mathbb{R}^+) : \max_{s \le t} |x(s) - x_0(s)|\} < \epsilon$.

2.12   In **(d)** observe that $E|X(s) - E^{\mathcal{F}_s} X(b)| = 0$. As for **(e)** compute as in (8)

$$a \cdot P[\max_{0 \le k \le 2^n} X\left(\frac{kt}{2^n}\right) \ge a] = a \cdot \sum_k P(F_k) \le \sum_k \int_{F_k} X\left(\frac{kt}{2^n}\right) dP \le EX(0),$$

where $F_k := [X(\frac{t}{2^n}) < a, \ldots, X(\frac{(k-1)t}{2^n}) < a, X(\frac{kt}{2^n}) \ge a]$ and perform $n \to \infty$.

3.12   Assume that all $X(t)$'s are integrable and that $EX(\tau) = EX(0)$ for all at most two valued Markov times τ. Fix $s < t$, $F \in \mathcal{F}_s$, define $\tau := s \cdot I_{F^c} + t \cdot I_F$ and check that τ is a Markov time. Hence,

$$\int_F X(t) \, dP = EX(\tau) - \int_{F^c} X(s) \, dP = EX(\tau) - EX(s) + \int_F X(s) \, dP = \int_F X(s) \, dP,$$

which proves that X is a martingale. The rest follows by 3.10.

3.16   Use the reflection principle and differentiate

$$P[\tau_a \le t] = 2 \int_a^\infty \frac{1}{\sqrt{2\pi t}} \exp(-y^2/2t) \, dy, \quad a > 0$$

to obtain $f_a(s)$ in the form

$$\frac{1}{\sqrt{2\pi}} \left(-\frac{1}{s^{3/2}} \int_a^\infty \exp(-y^2/2s) \, dy + \frac{1}{s^{5/2}} \int_a^\infty y \cdot y \exp(-y^2/2s) \, dy \right)$$

and integrate the right hand integral by parts.

5.6   Put $Q(M) := \int_\Omega \mu_{A(\omega)}(M_\omega) \, dP$ for $M \in \mathcal{M}$. Because $\omega \to \mu_{A(\omega)}(M_\omega)$ is a random variable on $(\Omega, \mathcal{F}, P)$ by 5.4 (i), the Q is defined correctly as a measure on $\mathcal{M}$ such that $Q([0, t] \times \Omega) < \infty$ for any $t \ge 0$. A standard procedure 1.11 (b) shows that $\int_{\mathbb{R}^+ \times \Omega} G \, dQ = E \int_{\mathbb{R}^+} G \, dA$ for any progressive $G \ge 0$. In particular, $[G]_t = \sqrt{\int G^2 \cdot I_{[0,t]} \, dQ}$ for all $t \ge 0$ and all $G \in \mathrm{EPM}_2$. Hence,

$$\mathrm{EPM}_2 = \{G \in \mathrm{PM} : G \cdot I_{[0,t]} \in L_2(\mathbb{R}^+ \times \Omega, \mathcal{M}, Q), \quad \forall t \ge 0\}.$$

This interpretation proves all four statements. As for (d) recall that $L_2(\mathcal{M}) := L_2(\mathbb{R}^+ \times \Omega, \mathcal{M}, Q)$ is known as a complete pseudometric space provided that the distance of G_1 and G_2 is given by $\sqrt{\int (G_1 - G_2)^2 \, dQ}$. This and a simple *projectivity* argument imply that EPM$_2$ is also a complete pseudometric space if the distance of G_1 and G_2 is given as $[G_1 - G_2]$.

5.9 $G(\tau)$ is an $\mathcal{F}_{\tau(t)}$-adapted process by 1.3.9 because $\tau(t)$'s are finite $\mathcal{F}_t$-Markov times. Obviously

$$A(t) \geq t \Rightarrow \tau(t) \leq t \Rightarrow \mathcal{F}_{\tau(t)} \subset \mathcal{F}_t \quad \forall t \geq 0$$

holds. It follows that τ is an $\mathcal{F}_t$-progressive process (continuous and $\mathcal{F}_t$-adapted) and therefore measurable as

$$\tau : (\mathbb{R}^+ \times \Omega, \mathcal{B}(\mathbb{R}^+) \otimes \mathcal{F}) \to (\mathbb{R}, \mathcal{B}(\mathbb{R})).$$

It follows that $T(s, \omega) := (\tau(s, \omega), \omega)$ defines a process that is measurable as

$$(\mathbb{R}^+ \times \Omega, \mathcal{B}(\mathbb{R}^+) \otimes \mathcal{F}) \to (\mathbb{R}^+ \times \Omega, \mathcal{B}(\mathbb{R}^+) \otimes \mathcal{F})$$

which verifies that $G(\tau) = G \circ T \in \mathcal{B}(\mathbb{R}^+) \otimes \mathcal{F}$ because $G \in \mathcal{B}(\mathbb{R}^+) \otimes \mathcal{F}$.

Let H be continuous and without loss of generality bounded $\mathcal{F}_{\tau(t)}$-adapted process. Because there are $\mathcal{F}_{\tau(t)}$-simple processes J_n such that $H = \lim_n J_n$ on $\mathbb{R}^+ \times \Omega$ the problem of proving (ii) reduces to verify that $G := I_F \cdot I_{(a,b]}$, where $F \in \mathcal{F}_{\tau(a)}$ and $0 \leq a < b < \infty$, is an $\mathcal{F}_t$-adapted process. But this follows easily as

$$[G(A)(t) = 1] = F \cap [\tau(a) < t] \cap [\tau(b) \geq t] \in \mathcal{F}_t, \quad \forall t \geq 0.$$

6.6 If an $\mathcal{F}_t$-Markov τ time stops M to $\mathcal{F}_t$-martingale then $\upsilon := (\tau - r)^+$ is an $\mathcal{F}_{t+r}$-Markov time such that

$$M_r^\upsilon(t) = I_{[\tau \geq r]} \cdot [M^\tau(t + r) - M(r)] =: L(t)$$

holds because $M_r^\upsilon(t) = M((t + r) \wedge (\upsilon + r)) - M(r)$. Check that $L \in \mathrm{CM}(\mathcal{F}_{t+r})$ and compute $\langle M_r \rangle(t) = p - \lim Q^{\Delta_n(t)}(M_r)$ for a sequence $|\Delta_n(t)| \to 0$.

As for the second part observe that if an $\mathcal{F}_{t+r}$-Markov time τ stops N to an $\mathcal{F}_{t+r}$-martingale then $\tau + r$ is an $\mathcal{F}_t$-Markov time that stops $N_\circ$ to an $\mathcal{F}_t$-martingale. Compute $\langle M, N_\circ \rangle(t + r)$ as the probability limit of $Q^{\Delta_n(t)}(M, N_\circ)(t + r)$ where $|\Delta_n(t)| \to 0$ and $r \in \Delta_n(t)$ for all n.

7.4 In (iii) let (τ_n) be a localization sequence such that $M^{\tau_n} \in \mathrm{CM}_2$ for all n and $s < t$. Then $\langle M^{\tau_n}, N \rangle = 0$ and according to (ii)

$$E^{\mathcal{F}_s} M(t) N(t) = \lim_n E^{\mathcal{F}_s} M^{\tau_n}(t) N(t) = \lim_n M^{\tau_n}(s) N(s) = M(s) N(s),$$

where the $[\lim_n E^{\mathcal{F}_s} = E^{\mathcal{F}_s} \lim_n]$ interchange follows by 2.7 (g) as $|M(t \wedge \tau_n) N(t)| \leq C \cdot |E^{\mathcal{F}_{t \wedge \tau_n}} M(t)|$ by 3.10.

III.2 Stochastic Integration

stochastic integral, stochastic per partes and Itô formula, exponential martingales and Lévy theorem, Girsanov theorem, integral and Brownian representations, helps to some exercises

Through the present chapter, if not stated otherwise explicitly, we fix a complete probability space $(\Omega, \mathcal{F}, P)$ with a complete filtration $\mathcal{F}_t \subset \mathcal{F}$.

2.1   Stochastic Integral

We begin by a very standard procedure and define the **stochastic integral** of an $\mathcal{F}_t$-**simple process** $K = (K(t), t \geq 0) \in J := J(\mathcal{F}_t)$, i.e.

$$(1) \qquad K(t) = K_0 \cdot I_{\{0\}}(t) + \sum_{i=0}^{\infty} K_i \cdot I_{(t_i, t_{i+1}]}(t), \quad 0 = t_0 < t_1 < t_2 \ldots \uparrow \infty,$$

$$K_i \in \mathcal{F}_{t_i} \quad \text{bounded},$$

with respect to a **local martingale** $M \in \mathrm{CM}_{loc}(\mathcal{F}_t)$ by

$$(2) \qquad I_t^M(K) := \int_0^t K \, dM := \sum_{i=0}^{k-1} K_i(M(t_{i+1}) - M(t_i)) + K_k(M(t) - M(t_k))$$

where $k \in \mathbb{N}$ and $t_k \leq t < t_{k+1}$. Observe that our definition is independent of the choice of the partition $\Delta = \{0 = t_0 < t_1 < t_2 \ldots\}$ and that $I^M(K)$ is a continuous $\mathcal{F}_t$-adapted process. Even more is true:

2.1.1   Stochastic Integration - a Junior Grade. Let $M, N \in CM_{loc}$ and $K \in J$. Then $I^M(K) \in CM_{loc}$ and

$$(3) \qquad \langle I^M(K), N \rangle(t) \stackrel{as}{=} \int_0^t K \, d\langle M, N \rangle \quad \text{holds for all} \quad t \geq 0.$$

Moreover, if $M \in CM_2(CM)$ then also $I^M(K) \in CM_2(CM)$.

Proof. We shall perform the proof in several steps:

(a) If $M \in \mathrm{CM}_2(\mathrm{CM})$ then $I^M(K) \in \mathrm{CM}_2(\mathrm{CM})$: Obviously, $I_t^M(K)$ is $\mathcal{F}_t$-measurable and in $L_2(L_1)$. If K is as in (1) let $s < t$ and assume without loss of generality that $s = t_i$ and $t = t_{k+1}$. Then

$$E^{\mathcal{F}_s}(I_t^M(K) - I_s^M(K)) = E^{\mathcal{F}_s} \sum_{j=i}^{k} K_j(M(t_{j+1}) - M(t_j))$$

$$= E^{\mathcal{F}_s} \sum_{j=i}^{k} E^{\mathcal{F}_{t_j}} K_j(M(t_{j+1}) - M(t_j)) = 0$$

because $K_j \in \mathcal{F}_{t_j}$. It follows that $I^M(K) \in \mathrm{CM}_2$ (CM).

(b) $I^M(K) \in \mathrm{CM}_{loc}$ for arbitrary $M \in \mathrm{CM}_{loc}$: If τ is a Markov time such that M^τ is in CM_2, then according to **(a)** $(I^M(K))^\tau = I^{M^\tau}(K) \in \mathrm{CM}_2$ and $I^M(K) \in \mathrm{CM}_{loc}$ by an obvious localization.

(c) The relation (3) holds for any M and N in CM_2: Let $s = t_i < t = t_{k+1}$ as before. Then straightforward computations via 1.7.6 provide

$$E^{\mathcal{F}_s}(I_t^M(K) - I_s^M(K))(N(t) - N(s))$$

$$= E^{\mathcal{F}_s} \sum_{j=i}^{k} K_j(M(t_{j+1}) - M(t_j)) \cdot \sum_{j=i}^{k}(N(t_{j+1}) - N(t_j))$$

$$= E^{\mathcal{F}_s} \sum_{j=i}^{k} K_j E^{\mathcal{F}_{t_j}}(\langle M, N\rangle(t_{j+1}) - \langle M, N\rangle(t_j))$$

$$= E^{\mathcal{F}_s}\left\{\int_0^t K\, d\langle M, N\rangle - \int_0^s K\, d\langle M, N\rangle\right\}$$

which proves (3) again by 1.7.6 because $t \to \int_0^t K\, d\langle M, N\rangle$ is in CFV.

(d) The equality (3) holds for arbitrary M and N: This follows by a localization of **(c)**, since for any Markov time τ such that $M^\tau, N^\tau \in \mathrm{CM}_2$ we may compute, using repeatedly 1.6.5,

$$\langle I^M(K), N\rangle(\tau \wedge t) = \langle (I^M(K))^\tau, N^\tau\rangle(t) = \langle I^{M^\tau}(K), N^\tau\rangle(t) = \int_0^t K\, d\langle M^\tau, N^\tau\rangle$$

$$= \int_0^{\tau \wedge t} K\, d\langle M, N\rangle.$$

$\square$

A good idea for a general definition of the integral seems to be offered by (3): For $M \in \mathrm{CM}_{loc}$ and $G \in \mathrm{PM}_2(M)$ a stochastic process $I^M(G) \in \mathrm{CM}_{loc}$ shall be called the **stochastic integral of G with respect to M** if

$$(4) \qquad \langle I^M(G), N\rangle(t) \overset{\mathrm{as}}{=} \int_0^t G\, d\langle M, N\rangle, \quad t \geq 0, \qquad \forall N \in \mathrm{CM}_{loc}(\mathcal{F}_t).$$

We shall also use notations as: $\int G\, dM := I^M(G)$ and $\int_0^t G(s)\, dM(s) = \int_0^t G\, dM := I_t^M(G)$ and relax our notation to $\mathrm{PM}_2(M) := \mathrm{PM}_2(\langle M\rangle, \mathcal{F}_t)$, $\mathrm{PM}_1(M, N) := \mathrm{PM}_1(\langle M, N\rangle)$, or $\mathrm{EPM}_2(M) := \mathrm{EPM}_2(\langle M\rangle)$, etc. as before.

The integral $\int G\, d\langle M, N\rangle$ in (4) is a continuous and adapted process by 1.5.4, since $\mathrm{PM}_2(M) \subset \mathrm{PM}_1(M, N)$ according to Kunita-Watanabe inequality and therefore we may modify the definition of $I^M(G)$ as

2.1.2. *The stochastic integral $I^M(G)$ is almost surely uniquely determined by the requirements*

$$(5) \qquad I^M(G) \in \mathrm{CM}_{loc}, \qquad \langle I^M(G), N\rangle \overset{\mathrm{as}}{=} \int G\, d\langle M, N\rangle \quad \forall N \in \mathrm{CM}_{loc}$$

or equivalently by

(6) $I^M(G) \in CM_{loc}$, $\langle I^M(G), N \rangle \overset{as}{=} \int G \, d\langle M, N \rangle$ $\forall N \in CM_2$.

Indeed, if a process $I \in CM_{loc}$ is another one that satisfies the requirements (5), then $\langle I^M(G) - I \rangle \overset{as}{=} 0$ and therefore $I \overset{as}{=} I^M(G)$ by 1.7.1.

If there is an $I \in CM_{loc}$ such that $\langle I, N \rangle \overset{as}{=} \int G \, d\langle M, N \rangle$ holds for all $N \in CM_2$ take $N \in CM_{loc}$ arbitrary and consider a sequence of Markov times $\tau_n \uparrow \infty$ almost surely such that $N^{\tau_n} \in CM_2$ for all $n \in \mathbb{N}$. Then outside a P-null set and for all $n \in \mathbb{N}$

$$\langle I, N \rangle^{\tau_n} = \langle I, N^{\tau_n} \rangle = \int G \, d\langle M, N^{\tau_n} \rangle = \left(\int G \, d\langle M, N \rangle \right)^{\tau_n}$$

holds. Letting $n \to \infty$ we get by continuity that I satisfies (5), hence $I \overset{as}{=} I^M(G)$ according to the first part of 1.2.

Denoting

$$\mathbf{A} := \{(M, G), M \in CM_{loc}, G \in PM_2(M) \text{ such that } I^M(G) \text{ exists}\}$$

we observe that **A is a nonempty set**. Indeed, any pair (M, K), where $M \in CM_{loc}$ and K is a simple process, belongs to **A** and the corresponding stochastic integral $I^M(K)$ is given by (2).

2.1.3   Theorem. *The stochastic integral $I^M(G)$ exists for any $M \in CM_{loc}$ and $G \in PM_2(M)$.*

Here are **the principal properties of the stochastic integral.**

2.1.4   Quadratic Variation. *Consider $M, N \in CM_{loc}, G \in PM_2(M), H \in PM_2(N)$. Then*

(7) $\langle I^M(G), I^N(H) \rangle \overset{as}{=} \int GH \, d\langle M, N \rangle$, $\langle I^M(G) \rangle \overset{as}{=} \int G^2 \, d\langle M \rangle$.

2.1.5   Localization Lemma. *Let $M \in CM_{loc}, G \in PM_2(M)$ and consider a Markov time τ. Then $G \cdot I_{[0,\tau]} \in PM_2(M), G \in PM_2(M^\tau)$ and*

(8) $$I^M(G \cdot I_{[0,\tau]}) \overset{as}{=} (I^M(G))^\tau \overset{as}{=} I^{M^\tau}(G)$$

holds. In particular, $M^\tau \overset{as}{=} I^M(I_{[0,\tau]})$.

2.1.6   Linearity of the Integral. *If $G_i \in PM_2(M), a_i \in \mathbb{R}$ and $M \in CM_{loc}$ then*

(9) $$I^M(a_1 G_1 + a_2 G_2) \overset{as}{=} a_1 I^M(G_1) + a_2 I^M(G_2).$$

2.1.7 L_2-Integral. If $M \in CM_2$ and $G \in EPM_2(M)$ then $I^M(G) \in CM_2$ and it is a process with $EI_t^M(G) = 0$ such that

$$(10) \qquad E[I_t^M(G) \cdot I_s^M(G)] = E \int_0^{t \wedge s} G^2 \, d\langle M \rangle, \quad E[I_t^M(G)]^2 = E \int_0^t G^2 \, d\langle M \rangle$$

hold for all $t, s \geq 0$.

Remark. Recall that the spaces CM_2 and $EPM_2(M)$ (see 1.5.1 and 1.5.6) are endowed by the pseudometrics $\|M_1 - M_2\|$ and $[G_1 - G_2]^M := [G_1 - G_2]^{\langle M \rangle}$, respectively. Denoting

$$(11) \quad \||M\|| := \sum_{t=1}^{\infty} 2^{-t} \min(\sqrt{EM^2(t)}, 1), \quad \text{we get} \quad \||M\|| \leq \|M\| \leq 2 \cdot \||M\||$$

and thus $\||M_1 - M_2\||$ defines a pseudometric on CM_2 equivalent to that given by $\|M_1 - M_2\|$ (the inequalities in (11) follow by 1.2.10 (b)). The statements of 1.6 and 1.7 now can be read as follows:

For any fixed $M \in CM_2$ the map $G \to I^M(G)$ from $EPM_2(M)$ to CM_2 is a linear map such that $\||I^M(G)\|| = [G]^M$ holds for all $G \in EPM_2(M)$.

The above **isometry** is the cornerstone of all definitions of the stochastic integral.

2.1.8   Localization of Integrands and Integrators. Let τ be a Markov time. Then for any $M_i \in CM_{loc}$ and $G_i \in PM_2(M_i)$

$$(12) \qquad G_1^\tau \stackrel{as}{=} G_2^\tau, \quad M_1^\tau \stackrel{as}{=} M_2^\tau \quad \Rightarrow \quad I^{M_1}(G_1)^\tau \stackrel{as}{=} I^{M_2}(G_2)^\tau.$$

Proof of 1.3, 1.4, 1.5, 1.6 and 1.8. The next steps will constitute the proof of the above statements:

(a)   1.4 holds for all pairs (M, G) and (N, H) in **A**: It follows by 1.2 that

$$d\langle I^M(G), I^N(H) \rangle \stackrel{as}{=} G \cdot d\langle M, I^N(H) \rangle, \qquad d\langle M, I^N(H) \rangle \stackrel{as}{=} H \cdot d\langle M, N \rangle,$$

hence by the chain rule for the Radon-Nikodym derivations we simply compute that $d\langle I_M(G), I^N(H) \rangle \stackrel{as}{=} GH \, d\langle M, N \rangle$, which verifies 1.4 on **A**.

(b)   1.5 holds for $(M, G) \in \mathbf{A}$ and arbitrary Markov time τ: Consider any $N \in CM_{loc}$ and $t \geq 0$. According to 1.6.5 and 1.2, we get that

$$\langle I^M(G)^\tau, N \rangle(t) = \langle I^M(G), N \rangle(\tau \wedge t) = \int_0^{\tau \wedge t} G \, d\langle M, N \rangle =$$

$$= \int_0^t \mathrm{I}_{[0,\tau]} \cdot G \, d\langle M, N \rangle = \int_0^t G \, d\langle M^\tau, N \rangle$$

holds almost surely. This proves that $(M, I_{[0,\tau]}G)$ and consequently (M^τ, G) are in **A** and both equalities in (8).

(c)   1.6 holds for $(M, G_i) \in \mathbf{A}$ because $\mathrm{PM}_2(M)$ is a linear space and the covariation is a bilinear form on CM_{loc}.

(d)   1.7 holds for any $(M, G) \in \mathbf{A}$ such that $M \in \mathrm{CM}_2$ and $G \in \mathrm{EPM}_2(M)$: It follows by **(a)** that $\langle I^M(G)\rangle(t) \overset{as}{=} \int_0^t G^2 \, d\langle M\rangle$ holds for all $t \geq 0$, the latter integral being an integrable random variable by the definition of the space $\mathrm{EPM}_2(M)$. Hence $I^M(G) \in \mathrm{CM}_2$ by 1.7.3. If $s < t$ then $I_s^M(G) = (I_t^M(G))^s$ and it follows by 1.7.4, 1.6.5 and by **(a)** that

$$E[I_t^M(G) \cdot I_s^M(G)] = E\langle I^M(G), I^M(G)^s\rangle(t) = E \int_0^{t \wedge s} G^2 \, d\langle M\rangle.$$

(e) $(M, G) \in \mathbf{A}$ for any $M \in \mathrm{CM}_2$ and $G \in \mathrm{EPM}_2(M)$: Because $E\langle M\rangle(t) < \infty$ by 1.7.3, it follows by 1.5.7 that there are simple processes G_n such that $[G_n - G]^M \to 0$. Hence, $\|I^M(G_n) - I^M(G_m)\| \leq 2 \cdot [G_n - G_m]^M \to 0$ as $n, m \to \infty$ by 1.7 and the subsequent Remark. It follows by 1.5.2 that $\|I^M(G_n) - I\| \to 0$ where I is a process in CM_2. Denote $I_n := I^M(G_n)$, choose $N \in \mathrm{CM}_2$ and $t \geq 0$. It follows by 1.1.6 (b) and 1.7.3 that

$$\{E|\langle I, N\rangle(t) - \langle I_n, N\rangle(t)|\}^2 \leq E(I_n - I)^2(t) \cdot EN^2(t) \to 0.$$

Further, by Kunita-Watanabe inequality,

$$E|\langle I_n, N\rangle(t) - \int_0^t G \, d\langle M, N\rangle| = E\left|\int_0^t (G_n - G) \, d\langle M, N\rangle\right| \leq$$

$$\leq E \int_0^t |G_n - G| \, d\langle M, N\rangle^v \leq E\left\{\int_0^t (G_n - G)^2 \, d\langle M\rangle\right\}^{\frac{1}{2}} \cdot \{\langle N\rangle(t)\}^{\frac{1}{2}} \leq$$

$$\leq \left\{E \int_0^t (G_n - G)^2 \, d\langle M\rangle\right\}^{\frac{1}{2}} \cdot \{E\langle N\rangle(t)\}^{\frac{1}{2}} \to 0.$$

Combining both of the above convergences we arrive at $\langle I, N\rangle(t) \overset{as}{=} \int_0^t G \, d\langle M, N\rangle$ for any $N \in \mathrm{CM}_2$ and $t \geq 0$ which shows that $I^M(G)$ exists by 1.2 (6).

(f)   1.8 is true for all (M, G) and (N, H) in **A** and arbitrary Markov time τ: Choose $Z \in \mathrm{CM}_{loc}$ and $t \geq 0$. Then by **(a)** and **(b)** we get

$$\langle I^M(G)^\tau - I^N(H)^\tau, Z\rangle(t) \overset{as}{=} \int_0^t G^\tau \, d\langle M^\tau, Z\rangle - \int_0^t H^\tau \, d\langle N^\tau, Z\rangle \overset{as}{=} 0.$$

This, of course, implies that $I^M(G)^\tau \overset{as}{=} I^N(H)^\tau$.

(g)   Any (M, G) with $M \in \mathrm{CM}_{loc}$ and $G \in \mathrm{PM}_2(M)$ belongs to **A**: Choose such a pair and a sequence of Markov times $\tau_n \uparrow \infty$ such that $M_n := M^{\tau_n}$ and $(\int_0^{\tau_n \wedge t} G^2 \, d\langle M\rangle, t \geq 0)$ are bounded processes for any $n \in \mathbb{N}$. Consequently, $M_n \in$

CM$_2$ and $G \in$ EPM$_2(M_n)$ because $\int_0^t G^2 \, d\langle M_n \rangle \overset{as}{=} \int_0^{\tau_n \wedge t} G^2 \, d\langle M \rangle \leq C_n < \infty$. Hence, according to **(e)** $(M_n, G) \in \mathbf{A}$ with $I^{M_n}(G) \in$ CM$_2$ and $I^{M_n}(G)^{\tau_n} \overset{as}{=} I^{M_{n+1}}(G)^{\tau_n}$ according to **(f)**. It follows by 1.4.8 that there is a process $I \in$ CM$_{loc}$ such that $I^{\tau_n} \overset{as}{=} I^{M_n}(G)^{\tau_n}$ holds for any $n \in \mathbb{N}$. Further, for any $N \in$ CM$_{loc}$ and $t \geq 0$

$$\langle I, N \rangle(\tau_n \wedge t) = \langle I^{\tau_n}, N \rangle(t) = \langle I^{M_n}(G)^{\tau_n}, N \rangle(t) =$$

$$= \langle I^{M_n}(G), N \rangle(\tau_n \wedge t) = \int_0^{\tau_n \wedge t} G \, d\langle M, N \rangle$$

holds almost surely. Letting $n \to \infty$ in the above equality we get $\langle I, N \rangle(t) \overset{as}{=} \int_0^t G \, d\langle M, N \rangle$ for all $N \in$ CM$_{loc}$ and $t \geq 0$ which, by definition, means that $(M, G) \in \mathbf{A}$. $\square$

For $0 \leq a < b < \infty$ denote $\int_a^b G \, dM := \int_0^b G \, dM - \int_0^a G \, dM$.

2.1.9. *If* $M \in$ CM$_{loc}$ *and* $G \in$ PM$_2(M)$ *then outside P-null set and for all intervals* $[a, b] \subset \mathbb{R}^+$ *either* $G = 0$ *almost everywhere with respect to* $\mu_{\langle M \rangle}$ *on* $[a, b]$ *or* $\langle M \rangle(b) = \langle M \rangle(a)$ *implies that* $\int_a^t G \, dM = 0$, $a \leq t \leq b$.

The implication is an obvious consequence of 1.7.1 applied to the local martingale $I^M(G)$ whose quadratic variation is equal to $\int_0^t G^2 \, d\langle M \rangle$.

If $N \in$ CM$_{loc}$ is such that $N \overset{as}{=} \int G \, dM$, then we shall agree to write $dN = G \, dM$ or $dN \overset{as}{=} G \, dM$.

2.1.10 Stochastic Chain Rule. *If* $M \in$ CM$_{loc}$, $G \in$ PM$_2(M)$, $dN = G \, dM$ *and* $H \in$ PM$_2(N)$ *then* $GH \in$ PM$_2(M)$ *and* $I^N(H) \overset{as}{=} I^M(GH)$ *which equality reads in the differentials as*

$$dN = G \, dM \qquad \Rightarrow \qquad H \, dN = H \cdot G \, dM.$$

Proof. According to 1.4 $d\langle N \rangle \overset{as}{=} G^2 \, d\langle M \rangle$, hence $(HG)^2 \, d\langle M \rangle \overset{as}{=} H^2 \, d\langle N \rangle$ and therefore $GH \in$ PM$_2(M)$. Consider finally a $Z \in$ CM$_{loc}$ and apply (7) in 1.4 twice to get

$$\langle I^N(H), Z \rangle \overset{as}{=} \int H \, d\langle N, Z \rangle \overset{as}{=} \int HG \, d\langle M, Z \rangle,$$

what, by definition, yields $I^N(H) \overset{as}{=} I^M(GH)$. $\square$

2.1.11 Exercise. *If* $G \in$ PM$_2(M) \cap$ PM$_2(N)$ *then* $G \in$ PM$_2(aM + bN)$ *and* $I^{aM+bN}(G) \overset{as}{=} aI^M(G) + bI^N(G)$ *holds for all* $a, b \in \mathbb{R}$.

The continuity of $G \to I^M(G)$ that maps PM$_2(M)$ into CM$_{loc}$ is most precisely expressed by

2.1.12 Continuity Theorem. *For any* $M \in CM_{loc}$, G *and* G_n *in* $PM_2(M)$

$$\max_{s \leq t} \left| \int_0^s G_n \, dM - \int_0^s G \, dM \right| \xrightarrow{p} 0 \qquad \Longleftrightarrow \qquad \int_0^t (G_n - G)^2 \, d\langle M \rangle \xrightarrow{p} 0$$

and in particular,

$$I_s^M(G) \overset{as}{=} 0 \quad \forall s \leq t \qquad \Longleftrightarrow \qquad \int_0^t G^2 \, d\langle M \rangle \overset{as}{=} 0$$

holds for any fixed $t \geq 0$.

The equivalence follows directly by 1.4, 1.6 and 1.7.7 and permits to transfer a variety of Lebesgue-Stieltjes integral convergence theorems to the theory of the stochastic integral. We shall manage with

2.1.13 Dominated Convergence. *Let* $M \in CM_{loc}$. *Assume that* G_n, G *and* H *are processes in* $PM_2(M)$ *such that* $\lim_{n \to \infty} G_n = G$ *on* $[0, t] \times \Omega$ *for a fixed* $t > 0$. *Then either* $|G_n| \leq H$ *on* $[0, t] \times \Omega$ *or* $G_n(\cdot, \omega) \to G(\cdot, \omega)$ *uniformly on* $[0, t]$ *for all* $\omega \in \Omega$ *implies that* $\max_{s \leq t} \left| \int_0^s G_n \, dM - \int_0^s G \, dM \right| \xrightarrow{p} 0$.

Thus, we get

2.1.14 Riemann Integration. *Let* M *be a process in* CM_{loc}.

(a) *If* K *is a process given by (1) with random variables* $K_i \in \mathcal{F}_{t_i}$ *that need not be bounded then* $K \in PM_2(M)$ *and the stochastic integral* $I^M(K)$ *is a process given by (2).*

(b) *If* F *is an adapted continuous process then for any fixed* $t \geq 0$ *and* $|\Delta_n(t)| \to 0$

$$\int_0^t F \, dM = p - \lim_{n \to \infty} \sum_{j=0}^{k_n - 1} F(t_j^n) \left(M(t_{j+1}^n) - M(t_j^n) \right).$$

Proof. Denote $K^n := K \cdot I_{[|K| \leq n]}$, observe that K^n's are simple processes and therefore

$$\int_0^t K^n \, dM = \sum_{i=0}^{k-1} K_i^n \left(M(t_{i+1}) - M(t_i) \right) + K_k^n \left(M(t) - M(t_k) \right)$$

if $t_k \leq t < t_{k+1}$ and $K_i^n := K_i \cdot I_{[|K_i| \leq n]}$. Letting $n \to \infty$ we verify **(a)** by means of 1.13.

According to **(a)** we have $\sum_{j=0}^{k_n - 1} F(t_j^n) \left(M(t_{j+1}^n) - M(t_j^n) \right) \overset{as}{=} \int_0^t F_n \, dM$ where $F_n := \sum_{j=0}^{k_n - 1} F(t_j^n) I_{(t_j^n, t_{j+1}^n]}$. Obviously $F_n(\cdot, \omega) \to F(\cdot, \omega)$ uniformly on $[0, t]$ for all $\omega \in \Omega$ and 1.13 applies to prove **(b)**. $\square$

Corollary 1.5.10 and 1.12 apply to verify

2.1.15   Limit Definition of Stochastic Integral. Let $M \in CM_{loc}$ and $G \in PM_2(M)$. Then $I^M(G)$ is almost surely a unique process in CM_{loc} such that for arbitrary sequence of simple processes G_n

$$\int_0^t |G_n - G|^2 \, d\langle M \rangle \xrightarrow{\mathrm{P}} 0 \quad \forall t \quad \Rightarrow \quad \int_0^t G_n \, dM \xrightarrow{\mathrm{P}} I_t^M(G) \quad \forall t,$$

where $I^M(G_n)$ is a process defined by (2).

Important Remark B. The limit definition provides an argument for the following statements: Let $M \in CM_{loc}(\mathcal{F}_t)$ and $G \in PM_2(M, \mathcal{F}_t)$. Recall the Important Remark A we put forward at the end of 1.6 Section of our text.

(1) The stochastic integral $I^M(G)$ does not change if we legally change the underlying filtration.

   More precisely, if $\mathcal{F}_t^M \subset \mathcal{G}_t \subset \mathcal{F}_t$ is another complete filtration such that G is a $\mathcal{G}_t$-progressive process then $M \in CM_{loc}(\mathcal{F}_t) \cap CM_{loc}(\mathcal{G}_t)$ and $G \in PM_2(M, \mathcal{F}_t) \cap PM_2(M, \mathcal{G}_t)$. It follows by 1.5.10 and 1.12 that the stochastic integrals of G with respect to M on the filtrations $(\mathcal{F}_t)$ and $(\mathcal{G}_t)$, respectively, are equivalent processes.

(2) The stochastic integral does not change if we replace the underlying probability measure P by a $Q \sim P$ ($P(N) = 0 \iff Q(N) = 0$) and $M \in CM_{loc}(Q, \mathcal{F}_t)$.

   Indeed, $(\Omega, \mathcal{F}, Q, (\mathcal{F}_t))$ continues to be a complete filtration and because obviously $G \in PM_2(P, \mathcal{F}_t) = PM_2(Q, \mathcal{F}_t)$, the integral $I^M(G)$ is defined also in the setting of the filtered space $(\Omega, \mathcal{F}, Q, (\mathcal{F}_t))$. It follows again by 1.5.10 and 1.12 that both the P-integral and the Q-integral $I^M(G)$ are equivalent processes with respect to both P and Q.

Solve the next pair of exercises to accommodate the properties of the stochastic integral.

2.1.16   Exercise. Consider an $\mathcal{F}_t$-Wiener process W and a Borel measurable function $g : \mathbb{R}^+ \to \mathbb{R}$ such that $\int_0^t g^2(s) \, ds < \infty$ for all $t \geq 0$. Then $\int_0^t g \, dW$ is a centered Gaussian process with independent increments such that for all $s, t \in \mathbb{R}^+$

$$\mathrm{cov}\left(\int_0^t g \, dW, \int_0^s g \, dW\right) = \int_0^{t \wedge s} g^2(u) \, du, \quad \mathrm{var}\left(\int_0^t g \, dW\right) = \int_0^t g^2(u) \, du$$

holds.

2.1.17   Exercise.

(i) Consider $M \in CM_{loc}$, $r > 0$, $\xi \in \mathcal{F}_r$ and finally a $G \in PM_2(M)$ such that $G = 0$ on $[0, r]$. Obviously, $\xi \cdot G \in PM_2(M)$, prove that $\xi \cdot \int G \, dM \stackrel{\mathrm{as}}{=} \int \xi \cdot G \, dM$.

(ii) If $M \in \mathrm{CM}_{loc}$ and $r > 0$, denote $M_r(t) := M(t+r) - M(r)$ and recall that $M_r \in \mathrm{CM}_{loc}(\mathcal{F}_{r+t})$ by 1.6.6. If, moreover $G \in \mathrm{PM}_2(M)$ then obviously $G_r := (G(r+t), t \geq 0) \in \mathrm{PM}_2(M_r)$. Prove that outside a P-null set $\int_0^t G_r \, dM_r \overset{as}{=} \int_r^{r+t} G \, dM$ holds for all $t \geq 0$.

Let $X \in \mathrm{CSM}$ be **a continuous semimartingale** with the stochastic differential $dX = dB + dM$ for a $B \in \mathrm{CFV}$ and an $M \in \mathrm{CM}_{loc}$. Recall that the differential formula stays for $X = X(0) + B + M$ and the decomposition is unique almost surely. Denote $\mathrm{PM}_{12}(X) := \mathrm{PM}_1(B) \cap \mathrm{PM}_2(M)$ and define **the stochastic integral** of a process $G \in \mathrm{PM}_{12}(X)$ with respect to X by

$$(13) \qquad \int G \, dX = I^X(G) := \int G \, dB + \int G \, dM.$$

If $Y \in \mathrm{CSM}$ is such that $Y \overset{as}{=} \int G \, dX$ holds, we write $dY = G \, dX$ or $dY \overset{as}{=} G \, dX$.

The process $\int G \, dX$ is a semimartingale in CSM according to 1.5.4 (iii) and according to the definition of the stochastic integral. It also means that $\int G \, dX \overset{as}{=} \int G \, dB + \int G \, dM$ is the unique decomposition of the left-hand side semimartingale into the continuous finite variation part and the continuous local martingale one. Hence, in differentials our definition reads as $GdX = GdB + GdM$.

It should be stressed that (13) defines the stochastic integral with property $\int G \, dX = \int G \, d(X + \xi_0)$ for any $\xi_0 \in \mathcal{F}_0$ and the integral of a G with respect to a continuous local martingale M as $\int G \, dM = \int G \, d(M - M(0))$.

The domain of this integral enables us to apply the common fundamental properties both of the Lebesgue-Stieltjes and the stochastic integral, the domain $\mathrm{PM}_{12}(X)$ being importantly large enough for any X to include continuous adapted processes or more generally locally bounded progressive processes. Recall that a **process G is locally bounded** if all its trajectories are bounded on all bounded intervals.

2.1.18 Quadratic Variation. *Let X and Y be processes in CSM with stochastic differentials $dX = dB + dM$ and $dY = dC + dN$. Consider also processes $G \in \mathrm{PM}_{12}(X)$ and $H \in \mathrm{PM}_{12}(Y)$. Then outside a P-null set*

$$d\langle I^X(G), I^Y(H) \rangle = GH \, d\langle M, N \rangle$$

holds.

2.1.19 Dominated Convergence. *Let $X \in \mathrm{CSM}$ and assume that G_n, G and F are processes in $\mathrm{PM}_{12}(X)$ such that $G_n \to G$ on $[0,t] \times \Omega$ for a fixed t. If either $|G_n| \leq F$ on $[0,t] \times \Omega$ or $G_n(\omega) \to G(\omega)$ uniformly on $[0,t]$ for all $\omega \in \Omega$ then $\max_{s \leq t} |I_s^X(G_n) - I_s^X(G)| \overset{P}{\to} 0$.*

2.1.20 Chain Rule. *Consider $X \in \mathrm{CSM}$ and $G \in \mathrm{PM}_{12}(X)$, denote $Y := \int G \, dX$. Then for any $H \in \mathrm{PM}_{12}(Y)$ we get $HG \in \mathrm{PM}_{12}(X)$ and $I^Y(H) \overset{as}{=} I^X(HG)$ or equivalently*

$$\boxed{dY = G \, dX \quad \Rightarrow \quad H \, dY = HG \, dX.}$$

Moreover, if both G and H are processes in $PM_{12}(X)$ and either G or H is a locally bounded process then $H \in PM_{12}(Y)$ and the above Chain rule formula holds.

The first assertion follows by a combination of Radon-Nikodym Theorem and the Chain rule formula 1.10.

As for the *moreover part* chose X as $dX = dB + dM$ and note in both cases $\int_0^t |HG|\,dB^v < \infty$ and $\int_0^t H^2G^2\,d\langle M \rangle < \infty$ hold almost surely for arbitrary t. Hence, Radon-Nikodym Theorem implies that $H \in PM_{12}(Y)$ and Chain rule formula follows by the first part of 1.20.

2.1.21   Riemann Integration. If $X \in CSM$ and F is a continuous adapted process then for any $t \geq 0$ and $|\Delta_n(t)| \to 0$

$$\int_0^t F\,dX = p - \lim_{n \to \infty} \sum_{j=1}^{k_n} F(t_{j-1}^n) \left(X(t_j^n) - X(t_{j-1}^n) \right).$$

2.1.22   Localization of Integrands and Integrators. Let τ be a Markov time. Then for any $X_i \in CSM$ and any $G_i \in PM_{12}(X_i)$

$$G_1^\tau \overset{as}{=} G_2^\tau, \quad X_1^\tau \overset{as}{=} X_2^\tau \quad \Rightarrow \quad I^{X_1}(G_1)^\tau \overset{as}{=} I^{X_2}(G_2)^\tau.$$

2.2   Stochastic per partes and Itô formula

The following summation equality yields both the **Lebesgue-Stieltjes** and **stochastic per partes** integration formulas:

Let X and Y be continuous processes, $\Delta(t) = \{0 = t_0 < t_1 < \ldots t_k = t\}$ a partition of $[0, t]$. Then

(14)

$$X(t)Y(t) = X(0)Y(0) + \sum_{j=1}^{k} \left(X(t_j) - X(t_{j-1}) \right) \left(Y(t_j) - Y(t_{j-1}) \right) +$$

$$+ \sum_{j=1}^{k} X(t_{j-1}) \left(Y(t_j) - Y(t_{j-1}) \right) + \sum_{j=1}^{k} Y(t_{j-1}) \left(X(t_j) - X(t_{j-1}) \right).$$

2.2.1   Lebesgue-Stieltjes Per Partes. If X and Y are processes in *CFV* then XY is also a process in *CFV* such that

(15) $$X(t)Y(t) = X(0)Y(0) + \int_0^t X\,dY + \int_0^t Y\,dX.$$

holds for all $t \geq 0$. In particular, $X^2(t) = X^2(0) + 2\int_0^t X\,dX$.

In differentials we read 2.1 as

$$\boxed{dXY = X\,dY + Y\,dX} \qquad \boxed{dX^2 = 2X\,dX}.$$

We get (15), hence 2.1 as a whole, by passing $|\Delta(t)| \to 0$ in (14) where the second term tends to 0 because the trajectories of X are uniformly continuous on $[0, t]$ and because $Y^v(t) < \infty$.

2.2.2   Stochastic Per Partes. *Let X and Y be processes in CSM. Then XY is also a process in CSM such that outside a P-null set and for any $t \geq 0$*

$$(16) \qquad X(t)Y(t) = X(0)Y(0) + \int_0^t X\,dY + \int_0^t Y\,dX + \langle X, Y\rangle(t)$$

holds. In particular, $X^2(t) = X^2(0) + 2\int_0^t X\,dX + \langle X\rangle(t)$.

In differentials we read 2.2 as

$$\boxed{dXY = X\,dY + Y\,dX + d\langle X, Y\rangle} \qquad \boxed{dX^2 = 2X\,dX + d\langle X\rangle}.$$

To verify (16), which equality already implies that $XY \in \text{CSM}$, consider (14) for a sequence of partitions $|\Delta_n(t)| \to 0$, then apply the definition of $\langle X, Y\rangle$ and 1.21.

Writing the formula (16) for local martingales $N, N \in \text{CM}_{loc}$ only, we specify the local martingale term in Doob-Meyer decomposition of M^2 and of MN, respectively: $M^2 - \langle M\rangle = 2I^M(M)$ and $MN - \langle M, N\rangle = I^M(N) + I^N(M)$.

Next pair of examples suggests the advantages of the differential symbolic:

2.2.3   Example. Let us establish the Doob-Meyer decomposition of X^2 where $X(t) = t + W(t)$ is the Wiener process with a linear trend:

$$dX^2(t) = 2(t + W(t))\,d(t + W(t)) + dt = (2t + 1 + 2W(t))\,dt + 2(t + W(t))\,dW(t),$$

i.e., the decomposition reads as $X^2(t) = \int_0^t 2s + 1 + 2W(s)\,ds + 2\int_0^t s + W(s)\,dW(s)$.

2.2.4   Example. Let X and Y be processes in CSM. Then XY^2 is a process in CSM, too, by 2.2 and according to (16), 1.6.4, 1.10 and 1.4 the stochastic differential dXY^2 equals to

$$(17)$$
$$Y^2\,dX + X\,dY^2 + d\langle X, Y^2\rangle = Y^2\,dX + X\,d\left(\langle Y\rangle + 2I^Y(Y)\right) + d\langle X, 2I^Y(Y)\rangle$$
$$= Y^2\,dX + 2XY\,dY + X\,d\langle Y\rangle + 2Y\,d\langle X, Y\rangle.$$

It is only natural to consider a d-**dimensional semimartigale** X where

$$(18) \qquad X = (X_1, X_2, \ldots, X_d), \quad X_j \in \text{CSM}, \quad dX_j = dB_j + dM_j, \quad 1 \leq j \leq d$$

and a continuous $f : \mathbb{R}^d \to \mathbb{R}$ asking a question under which circumstances $f(X)$ is also a semimartingale in CSM and what form its stochastic differential $df(X)$ has. A complete answer is available for $f \in C^2 := C^2(\mathbb{R}^d)$ where, as usual, we denote by $C^2(G)$ the set of those $f : \mathbb{R}^d \to \mathbb{R}$ that have continuous derivatives

$$f_i(x_1, \ldots, x_d) := \frac{\partial f}{\partial x_i}(x_1, \ldots, x_d), \qquad f_{ij}(x_1, \ldots, x_d) := \frac{\partial f}{\partial x_i \partial x_j}(x_1, \ldots, x_d)$$

on an open set $G \subset \mathbb{R}^d$ for all $1 \leq i, j \leq d$.

Agree to read (18) as $X \in \text{CSM}^d$ and observe that putting $f(x_1, x_2) = x_1 \cdot x_2^2$ we may rewrite (17) as

$$(19) \qquad \boxed{df(X) = \sum_i f_i(X)\,dX_i + \frac{1}{2}\sum_{i,j} f_{ij}(X)\,d\langle X_i, X_j\rangle}$$

to be true for any $X = (X_1, X_2) \in \text{CSM}^2$. It is easy to extend 2.4 by induction.

2.2.5 **Exercise.** Let X be a process in CSM^d and $f(x_1, x_2, \ldots, x_d)$ a polynomial $\mathbb{R}^d \to \mathbb{R}$. Then the formula (19) holds, and in particular, $f(X)$ is a process in CSM^d.

To extend 2.5 further, we obviously need

2.2.6 **Weierstrass Theorem.** *Let $G \subset \mathbb{R}^d$ be an open set. Then for any $f \in C^2(G)$ there are polynomials $p^n : \mathbb{R}^d \to \mathbb{R}$ such that for all compacts $K \subset G$*

$$(20) \quad p^n \to f, \quad p_i^n \to f_i \quad \text{and} \quad p_{ij}^n \to f_{ij} \quad \text{uniformly on } K \quad \forall 1 \le i, j \le d.$$

Proof. First fix an $f \in C^2(G)$, a compact $K \subset G$ and $\epsilon > 0$. According to a deep extension of the classical Weierstrass theorem, see [142], pp. 596/2-596/3, there is a polynomial $p = p_{\epsilon, K}$ such that

$$\|p - f\|_K < \epsilon, \quad \|p_i - f_i\|_K < \epsilon, \quad \|p_{ij} - f_{ij}\|_K < \epsilon \quad 1 \le i, j \le d,$$

where $\|g\|_K := \max_{x \in K} |g(x)|$. Because G is a separable and locally compact metric space there is an increasing sequence of compacts $K_n \subset G$ such that $\cup K_n = G$ and such that any compact $K \subset G$ is already subset of some K_n. Putting $p^n := p_{\epsilon_n, K_n}$ for a sequence $\epsilon_n \to 0$ we complete the construction. $\square$

The case $d = 1$ in 2.6 is the only one which does not require a sophisticated treatment.

2.2.7 **Exercise.** Prove 2.6 for $d = 1$.

The next theorem is the true cornerstone of Stochastic Analysis.

2.2.8 **Itô Formula.** *Let G be an open set in $\mathbb{R}^d$, $f \in C^2(G)$ and $X \in \mathrm{CSM}^d$ such that $X \in G$ everywhere on $\mathbb{R}^+ \times \Omega$. Then the process $f(X)$ is in CSM and its stochastic differential is given by (19) which means equivalently that*

(21)

$$f(X(t)) = f(X(0)) + \sum_{i=1}^d \int_0^t f_i(X)\, dX_i + \frac{1}{2} \sum_{i,j=1}^d \int_0^t f_{i,j}(X)\, d\langle X_i, X_j \rangle$$

holds almost surely for any $t \ge 0$. In particular,

$$\boxed{df(X) = f'(X)\, dX + \frac{1}{2} f''(X)\, d\langle X \rangle}$$

holds if we choose $d = 1$.

Proof. Note that the integrands in (21) are continuous adapted processes, hence $f(X)$ is a process in CSM if the formulas are true.

Choose p^n's as in (20). It follows by 2.5 that for any fixed $t > 0$ and $n \in \mathbb{N}$

(22)

$$p^n(X(t)) \overset{as}{=} p^n(X(0)) + \sum_{i=1}^{d} \int_0^t p_i^n(X)\, dX_i + \frac{1}{2} \sum_{i,j=1}^{d} \int_0^t p_{i,j}^n(X)\, d\langle X_i, X_j \rangle$$

holds almost surely. Fix $t > 0$ and $\omega \in \Omega$ and consider compacts

$$K_\omega := \{X(s,\omega), s \leq t\} \subset G.$$

Obviously, $p^n(X(0,\omega)) \to f(X(0,\omega))$ and because all the convergences stated by 2.6 are uniform on arbitrary K_ω we conclude that for all $1 \leq i,j \leq d$ and uniformly for $s \in [0,t]$

$$p_i^n(X(s,\omega)) \to f_i(X(s,\omega)), \quad p_{ij}^n(X(s,\omega)) \to f_{ij}(X(s,\omega))$$

holds as $n \to \infty$. Itô formula (21) follows if we let $n \to \infty$ in (22) and apply 1.19. $\square$

Later on we shall appreciate the following local form of Itô formula: Having a function $f(t,x)$ defined on a subset of $\mathbb{R}^+ \times \mathbb{R}$ we denote

$$f' := \frac{\partial f}{\partial t}, \quad f_x := \frac{\partial f}{\partial x} \quad \text{and} \quad f_{xx} := \frac{\partial^2 f}{\partial x^2}$$

at those $(t,x) \in \mathbb{R}^+ \times \mathbb{R}$ where the derivatives exist.

2.2.9   Corollary. *Consider $T > 0$ and $-\infty \leq a < b \leq +\infty$. Let f be a continuous function defined on $\mathbb{R}^+ \times (a,b)$ such that f', f_x and f_{xx} exist continuous on $[0,T) \times (a,b)$. Let $X \in CSM$ be a process such that $X(t,\omega) \in (a,b)$ for all $(t,\omega) \in \mathbb{R}^+ \times \Omega$. Denote $Y(t) := f(t, X(t))$ for $t \geq 0$ and choose $0 \leq U < T$ arbitrary. Then Y^U is a process in CSM with the stochastic differential*

(23) $$dY^U = f'(X)\, dt + f_x(X)\, dX + \frac{1}{2} f_{xx}(X)\, d\langle X \rangle,$$

where

$$f'(X)(t) := f'(t, X(t)) \cdot I_{[0,U]}(t), \quad f_x(X)(t) := f_x(t, X(t)) \cdot I_{[0,U]}(t),$$

(24)

$$f_{xx}(X)(t) := f_{xx}(t, X(t)) \cdot I_{[0,U]}(t), \quad t \geq 0.$$

Proof. Note that Y is a continuous adapted process, that processes $f'(X)$, $f_x(X)$ and $f_{xx}(X)$ are in $\text{PM}_{12}(X)$ because they are adapted and locally bounded. Hence, (23) defines a continuous semimartingale if the formula is true.

Fix $U < T$ and $t \le U$. Let $\Delta_n(t) \to 0$ and apply 2.8 to get

(25)

$$
\begin{aligned}
Y^U(t) - Y^U(0) &= f(t, X(t)) - f(0, X(0)) \\
&= \sum_k f(t_k^n, X(t_k^n)) - f(t_{k-1}^n, X(t_k^n)) + \sum_k f(t_{k-1}^n, X(t_k^n)) - f(t_{k-1}^n, X(t_{k-1}^n)) \\
&\overset{as}{=} \sum_k \int_{t_{k-1}^n}^{t_k^n} f'(s, X(t_k^n))\, ds + \sum_k \int_{t_{k-1}^n}^{t_k^n} f_x(t_{k-1}^n, X(s))\, dX(s) + \\
&\quad + \frac{1}{2} \sum_k \int_{t_{k-1}^n}^{t_k^n} f_{xx}(t_{k-1}^n, X(s))\, d\langle X\rangle(s) \\
&\overset{as}{=} \int_0^t A_n(s)\, ds + \int_0^t B_n(s)\, dX(s) + \frac{1}{2}\int_0^t C_n(s)\, d\langle X\rangle(s).
\end{aligned}
$$

In (25) we have denoted

$$
A_n := f'(0, X(0)) \cdot I_{\{0\}} + \sum_k f'(t_k^n, X)) \cdot I_{(t_{k-1}^n, t_k^n]},
$$

$$
B_n := f_x(0, X(0)) \cdot I_{\{0\}} + \sum_k f_x(t_{k-1}^n, X)) \cdot I_{(t_{k-1}^n, t_k^n]},
$$

$$
C_n := f_{xx}(0, X(0)) \cdot I_{\{0\}} + \sum_k f_{xx}(t_{k-1}^n, X)) \cdot I_{(t_{k-1}^n, t_k^n]}
$$

and observe that these are processes in $\mathrm{PM}_{12}(X)$ such that $A_n \to f'(X)$, $B_n \to f_x(X)$ and $C_n \to f_{xx}(X)$ uniformly on $[0, T]$ for each $\omega \in \Omega$ where $f'(X)$, $f_x(X)$ and $f_{xx}(X)$ are processes defined by (24). Theorem 1.19 and the equality (25) finally prove (23). $\square$

2.2.10   Example and Exercise. Consider two models for a two-dimensional motion (X, Y) with $X(0) = Y(0) = 0$:

$$
\begin{aligned}
dX(t) = t\, dW(t) \qquad & dY(t) = X^2(t)\, dW(t) \qquad && W - \text{model} \\
dX(t) = t\, dW_1(t) \qquad & dY(t) = X^2(t)\, dW_2(t) \qquad && (W_1, W_2) - \text{model},
\end{aligned}
$$

where W and (W_1, W_2) is Wiener and two-dimensional Wiener process, respectively.

In both models $X, Y \in \mathrm{CM}_2$, X is a centered Gaussian process with $EX^2(t) = E\langle X\rangle(t) = \int_0^t s^2\, ds = \frac{t^3}{3}$ by 1.16 and $EY^2(t) = E\langle Y\rangle(t) = \int_0^t EX^4(s)\, ds = \mathrm{c} \cdot t^7$.

Define $C(t) := X(t) \cdot Y(t)$, compute the stochastic differential of C and the expectation $EC(t)$ in both models.

Stochastic differential equations, shortly SDE, provide a method how to generate a rich class of **diffusion processes** $X \in \mathrm{CSM}$ that satisfy the equation

(26) $dX(t) = b(X)\, dt + \sigma(X)\, dW(t), \quad X(0) \overset{as}{=} x_0, \quad x_0 \in \mathbb{R},$

where W is an $\mathcal{F}_t$-Wiener process, $\sigma, b : \mathbb{R} \to \mathbb{R}$ Borel measurable functions such that $b(X) \in \mathrm{PM}_1(W)$, $\sigma(X) \in \mathrm{PM}_2(W)$, i.e. such that

(27) $$\int_0^t \left(|b(X(s))| + \sigma^2(X(s)) \right) ds < \infty \quad \text{holds almost surely} \quad \forall t \geq 0.$$

The equation (26) is called (b, σ)-SDE, any $X \in \mathrm{CSM}$ such that (26) and (27) hold is called a **solution to** (26) with the initial condition x_0. To meet purposes of our treatment of the diffusion financial mathematics in Chapter 3 we do not need more than to be able to solve a general linear equation 2.13. Next two examples show its possible applications.

2.2.11   Example and Exercise.   Langevin Equation. The velocity of the motion of a Brownian particle in a viscous environment is a solution to *Langevin equation*:

(28) $$dX(t) = bX(t)\, dt + \sigma\, dW(t), \quad X(0) = x_0 > 0, \quad \sigma > 0,$$

where x_0 represents the starting position of the particle, $-b$ the viscosity coefficient, in physics making a sense only if $b < 0$.

Modify the deterministic solution of $dX(t) = bX(t)\, dt$ and prove that

$$X(t) := \sigma e^{bt} \left(x_0 + \int_0^t e^{-bs}\, dW(s) \right), \quad t \geq 0.$$

is up to a P-null set the unique solution of (28). The process X is a Gaussian process by 1.16, called **Ornstein-Uhlenbeck process.** Compute $EX(t)X(s)$.

2.2.12   Example and Exercise.   Geometrical Brownian Motion. Solve the SDE

(29) $$dX(t) = bX(t)\, dt + \sigma X(t)\, dW(t), \quad X(0) = x_0,$$

where W is a Wiener process and x_0, b real numbers. Again, we easily guess the solution

(30) $$X(t) = x_0 \exp\left\{ (b - \frac{\sigma^2}{2})t + \sigma W(t) \right\}$$

and verify the validity of our guess by Itô formula.

According to 2.13, the solution (30) is determined almost surely and the process $X(t) = x_0 e^{bt} \mathcal{E}_t(W, \sigma)$ is called the **geometrical Brownian motion.** Also observe that the latter multiplicator is a continuous martingale by 1.2.6 and therefore $EX(t) = x_0 e^{bt}$.

2.2.13 Doléans Equation. *Let V and U be processes in CSM and $\xi \in \mathcal{F}_0$. Put*

$$Z := \exp\left\{V - V(0) - \frac{1}{2}\langle V\rangle\right\}, \quad Y := \xi + I^{U - \langle U, V\rangle}(Z^{-1}).$$

Then $X := ZY$ is the almost surely unique solution to the equation

$$X \stackrel{as}{=} \xi + U + \int X \, dV.$$

Proof. Obviously $Z \in$ CSM and $dZ = Z\,dV$ by 2.8 and therefore by 1.18 $d\langle Y, Z\rangle = Z\,d\langle Y, V\rangle$.

First we shall prove that the X solves the equation: X is a continuous adapted process and therefore the right hand side of the equation is a process in CSM. It follows by 1.20 and by 1.18 that $Z\,dY = d(U - \langle U, V\rangle)$ and that $d\langle Y, V\rangle = Z^{-1}\,d\langle U, V\rangle$ and therefore $\langle Y, Z\rangle = \langle U, V\rangle$. Integrating per partes and substituting the above information we get

$$dX = Z\,dY + Y\,dZ + d\langle Y, Z\rangle = d(U - \langle U, V\rangle) + X\,dV + d\langle U, V\rangle = dU + X\,dV.$$

Since $X(0) = \xi$, we proved that X is a solution to the equation.

If a process X solves the equation then $X \in$ CSM and putting $Y := XZ^{-1}$ we get again $Y \in$ CSM. Integrating per partes and substituting $dZ = Z\,dV$ we get

$$dU = dX - X\,dV = Y\,dZ + Z\,dY + d\langle Z, Y\rangle - X\,dV = Z\,dY + d\langle Z, Y\rangle.$$

It follows by 1. 18 that $\langle U, V\rangle = \langle Y, Z\rangle$, hence $dY = Z^{-1}\,d(U - \langle U, V\rangle)$ and since $Y(0) = \xi$ we get $Y = \xi + \int Z^{-1}\,d(U - \langle U, V\rangle)$. $\square$

Choosing in 2.13 $dV(t) := b\,dt$, $dU(t) := \sigma\,dW(t)$ and $dV(t) := b\,dt + \sigma\,dW(t)$, $U(t) := 0$, respectively we get the equations in 2.11 and in 2.12 as special cases.

Some strange things may happen when diffusion processes are involved (see also [153]):

2.2.14 Exercise. Consider a planar motion given by

$$dX(t) = -Y(t)\,dW(t), \quad dY(t) = X(t)\,dW(t), \quad X(0) = x_0, Y(0) = y_0$$

where W is a Wiener process and $r_0^2 := x_0^2 + y_0^2 > 0$. Let $R(t) := \sqrt{X^2(t) + Y^2(t)}$ be the distance of the particle from the origin at time t. Compute dR^2 and prove that R is a deterministic process.

Our reader may need some more exercises to master the calculus offered by Itô formula. A variety of examples and exercises is provided by [111].

To close the present section we introduce a fragment of **higher dimension technologies** we shall need later on.

Having a matrix $A = (A_{ij})$ we denote by $A^\top = (A_{ji})$ the transposition of the A, by $A \cdot B$ its product with another matrix B, by A^{-1} its inverse, if defined. **Agree** any r- dimensional vector C to be looked upon as an $(r \times 1)$-matrix if involved in a matrix algebra computation. Also agree that if A is an $(r \times d)$-matrix of integrable functions then $\int A\, dt$ denotes the $(r \times d)$-matrix the elements of which are given as $\int A_{ij}\, dt$.

Consider a d-dimensional semimartingale $X = (X_1, \ldots, X_d)^\top \in \mathrm{CSM}^d$ and an $(r \times d)$-matrix $G = (G_{ij})$ of progressive processes such that $G_{ij} \in \mathrm{PM}_{12}(X_j)$ for all i, j involved. Further we write $G \in \mathrm{PM}_{12}(X)$ and define **the stochastic integral** of G with respect to X as an r-dimensional semimartingale $Y = (Y_1, \ldots, Y_r)^\top \in \mathrm{CSM}^r$ where

$$dY_i(t) := \sum_{j=1}^{d} \int_0^t G_{ij}\, dX_j, \quad t \geq 0, \quad 1 \leq i \leq r.$$

The integral will be denoted by $\int G\, dX$ or as $I^X(G)$, as usual we shall also write $dY = G\, dX$. We will also simplify our notation writing $\mathrm{PM}_2(X) := \mathrm{PM}_{12}(X)$ if all X_j's are local martingales.

Recall the definition of the integrals $\int G_{ij}\, dX_j$ and compute that

$$Y_i = \sum_{j=1}^{d} G_{ij}\, dB_j + \sum_{j=1}^{d} G_{ij}\, dM_j, \quad 1 \leq i \leq r,$$

if $dX_i = dB_i + dM_i$ where $B_i \in \mathrm{CFV}$ and $M_i \in \mathrm{CM}_{loc}$. **Note** that the integration of an $(r \times d)$-matrix G by a semimartingale $X \in \mathrm{CSM}^d$ results in an r-dimensional process in CSM^r while $\int G\, ds$ results in an $(r \times d)$-dimensional matrix of progressive processes if the integrals $\int G_{ij}\, ds$ are defined. It may help you to observe the differential dX as the column vector $(dX_1, \cdots, dX_d)^\top$ and the differential ds as a scalar.

Multidimensional Chain Rule Formula needs a more refined treatment, see also 2.19. The assertion 1.10 applies to prove

2.2.15 Chain Rule. Let $X \in \mathrm{CSM}^d$ and $G \in \mathrm{PM}_{12}(X)$ be an $(r \times d)$-matrix, denote $Y := I^X(G)$. Let H be an $(m \times r)$-matrix of progressive processes such that

$$H_{ki} \in \mathrm{PM}_{12}\left(I^{X_j}(G_{ij})\right) \quad \text{holds for all} \quad 1 \leq k \leq m, 1 \leq i \leq r, 1 \leq j \leq d.$$

Then $H \in \mathrm{PM}_{12}(Y)$ and $H \cdot G \in \mathrm{PM}_{12}(X)$ are such that

$$I^Y(H) \overset{\mathrm{as}}{=} I^X(H \cdot G) \quad \text{or equivalently} \quad H\, dY = (H \cdot G)\, dX$$

is true. Especially, the Chain Rule Formula holds if either of the following conditions holds:

(a) H is a matrix of locally bounded progressive processes.

(b) G is a matrix of locally bounded processes and processes $H_{k,i}$ are in $\mathrm{PM}_{12}(X_j)$ for all k, i, j.

Indeed, if the integrability requirements of the first statement are satisfied then we argue by 1.11 and 1.10 that

$$\left(I^Y(H)\right)_k = \sum_{i=1}^r I^{Y_i}(H_{ki}) = \sum_{i=1}^r \sum_{j=1}^d \int H_{ki}\, dI^{X_j}(G_{ij})$$

$$= \sum_{j=1}^d \sum_{i=1}^r \int H_{ki} G_{ij}\, dX_j = \left(I^X(H \cdot G)\right)_k, \quad 1 \le k \le m.$$

Consider finally a d-dimensional $\mathcal{F}_t$-Wiener process $W = (W_1, \cdots, W_d)$, an r-dimensional vector $C = (C_1, \ldots, C_r)^\top$ and an $(r \times d)$-matrix $S = (S_{ij})$ of progressive processes such that

$$\int_0^t |C_i| + |S_{ij}|^2\, ds < \infty, \quad t \ge 0, \quad 1 \le i \le r, 1 \le j \le d$$

holds almost surely. We define a semimartingale $Y = (Y_1, \ldots, Y_r) \in \mathrm{CSM}^r$ by

$$Y_i(t) := x_i + \int_0^t C_i\, ds + \sum_{j=1}^d \int_0^t S_{ij}\, dW_j, \quad t \ge 0, \quad 1 \le i \le r$$

where $x = (x_1, \ldots, x_r)^\top$, writing also,

$$(31) \quad Y(t) = x + \int_0^t C\, ds + \int_0^t S\, dW \quad \text{or} \quad dY(t) = C\, dt + S\, dW, \quad Y(0) = x$$

to reflect the symbolism we have just introduced. Agree to call such a process Y an **r-dimensional Itô semimartingale** with coefficients C and S.

2.2.16 Gaussian Itô Semimartingales. *If C_i and S_{ij} in (31) are deterministic processes then Y is an r-dimensional Gaussian process such that*

$$EY(t) = x + \int_0^t C\, du, \quad K(t,s) = \int_0^{t \wedge s} S \cdot S^\top\, du, \quad t, s \ge 0,$$

where $K(t,s)$ denotes the covariance matrix function of Y.

Y is a Gaussian process due to 1.16 because $\left(I^{W_j}(S_{ij}), i \le r, j \le d\right)$ are independent processes. Apply 1.7.4, note that $\langle W_i, W_j \rangle \overset{\text{as}}{=} 0$ if $i \ne j$ and finally 1.4 to verify that

$$\mathrm{cov}(Y_k(t), Y_l(s)) = E \sum_{j=1}^d \langle I^{W_j}(S_{kj}), I^{W_j}(S_{lj}) \rangle (t \wedge s) = \sum_{j=1}^d \int_0^{s \wedge t} S_{kj} S_{lj}\, du.$$

Call a $(d \times d)$-matrix S of progressive processes S_{ij} **uniformly positive definite** if there is an $\epsilon > 0$ such that

$$x^\top \cdot S(t) \cdot x \ge \epsilon \cdot x^\top \cdot x = \epsilon \cdot ||x||^2, \quad \forall t \ge 0, \quad \forall x \in \mathbb{R}^d$$

holds.

2.2.17 Exercise. Let A be a $d \times d$ matrix such that $x^\top \cdot A \cdot A^\top \cdot x \geq \epsilon \cdot ||x||^2$ holds for an $\epsilon > 0$ and all $x \in \mathbb{R}^d$. Then A is a regular matrix such that $||A^{-1} \cdot x|| \leq \epsilon^{-\frac{1}{2}} \cdot ||x||$ holds for all $x \in \mathbb{R}^d$.

2.2.18 Orthogonalization of Itô Semimartingales. *Assume that Y is a d-dimensional Itô semimartingale (31) such that the matrix $S \cdot S^\top$ is uniformly positive definite and such that all S_{ij} are bounded processes. Then*

$$Y_{\text{ort}} := \int S^{-1} \, dY = \int S^{-1} \cdot C \, dt + W$$

correctly defines a d-dimensional Itô semimartingale such that $Y = \int S \, dY_{\text{ort}}$ holds almost surely.

It follows by 2.17 that all processes $(S^{-1})_{ij}$ are bounded and of course progressive. Hence, $S^{-1} \in \mathrm{PM}_{12}(Y)$ and therefore the stochastic integral $I^Y(S^{-1})$ is well defined. Further, the processes S_{ij} are supposed to be bounded, hence $S \in \mathrm{PM}_{12}(Y_{\text{ort}})$ and the relation between Y and Y_{ort} follows by 2.15.

2.2.19 Chain Rule. *Let Y be an Itô process (31) with locally bounded coefficients C and S. Then for any $m \times r$-matrix $H \in \mathrm{PM}_2(W)$ the chain rule formula reads as follows:*

$$H \in \mathrm{PM}_{12}(Y), \quad H \cdot S \in \mathrm{PM}_2(W) \quad \text{and} \quad H \, dY = (H \cdot C) \, ds + (H \cdot S) \, dW.$$

In other words $I^Y(H)$ is an m-dimensional Itô process with coefficients $H \cdot C$ and $H \cdot S$.

Read the definition of the set $\mathrm{PM}_2(W)$ to see that a matrix H belongs to the set iff $\int_0^t H_{ij}^2 \, ds < \infty$ is true almost surely for all i, j, t.

Denote $dY_1 = C \, ds$ and $dY_2 = S \, dW$ and note that both Y_1 and Y_2, satisfy requirements of 2.15 **(b)** and then apply 2.11.

2.3 Exponential Martingales and Lévy Theorem

We start with a simple problem: What functions $f(x, y) \in C^2(\mathbb{R}^2)$ are such that $f(M, \langle M \rangle)$ is a local martingale whatever may be given $M \in \mathrm{CM}_{loc}$? Because $\langle M, \langle M \rangle \rangle \overset{\text{as}}{=} 0$ and also $\langle \langle M \rangle, \langle M \rangle \rangle \overset{\text{as}}{=} 0$, we get

$$(32) \qquad df(M(t), \langle M \rangle(t)) = \frac{\partial f}{\partial x}(M, \langle M \rangle) \, dM$$

if $\frac{\partial f}{\partial y} + \frac{1}{2} \frac{\partial^2 f}{\partial x^2} = 0$ holds on $\mathbb{R}^2$ and therefore $f(M, \langle M \rangle)$ is a local martingale.

Choosing $f(x, y) = \exp\{\lambda x - \frac{\lambda^2}{2} y\}$ where $\lambda \in \mathbb{C}$ (the set of all complex numbers) we get

2.3.1. *If $M \in CM_{loc}$ and $\lambda \in \mathbb{C}$ then*

$$\mathcal{E}_t(M, \lambda) := \exp\{\lambda M(t) - \frac{\lambda^2}{2} \langle M \rangle(t)\}, \quad t \geq 0$$

is a **complex local** *$\mathcal{F}_t$-martingale, i.e. a process with states in $\mathbb{C}$ such that both $\mathfrak{R}\mathcal{E}$ and $\mathfrak{I}\mathcal{E}$ are local $\mathcal{F}_t$-martingales.*

If $M \in CM_{loc}$, we shall denote $\mathcal{E}(M) := \mathcal{E}(M, 1) \in CM_{loc}$ and call it the **exponential of** M. It follows by 2.13 that $\mathcal{E}(M)$ is the only solution to the equation $dX(t) = X(t)\, dM(t)$ with the initial condition $X(0) = 1$ which provides an alternative proof that $\mathcal{E}(M)$ is a local martingale.

Note that if W is an $\mathcal{F}_t$-Wiener process then, according to 1.2.6, $\mathcal{E}(W, \lambda)$ is a true martingale for any $\lambda \in \mathbb{R}$.

Important Remark C.

(a) If $\lambda \in \mathbb{R}$ and $M \in CM_{loc}$ then $\mathcal{E}(M, \lambda)$ is a local martingale and therefore a supermartingale by 1.4.2 (c). More importantly, if $T > 0$ is arbitrary then $\mathcal{E}(M, \lambda)$ is an $\mathcal{F}_t$-martingale on $[0, T]$ if and only if $E\mathcal{E}_T(M, \lambda) = 1$ holds by 1.4.2 (d).

In 3.7 we shall present an example of $\mathcal{E}(M)$ that does not posses the martingale property on $[0, 1]$.

(b) If $\lambda \in C$ and $M \in CM_{loc}$ are such that $|\mathcal{E}(M, \lambda)| \leq K < \infty$ then, $\mathcal{E}(M, \lambda)$ is a true complex martingale, i.e. $\mathfrak{R}\mathcal{E}$ and $\mathfrak{I}\mathcal{E}$ are in CM by 1.4.2 (a).

Apply 1.16 to prove:

2.3.2   Exercise. *If $g : \mathbb{R} \to \mathbb{R}$ is a Borel measurable function with $\int_0^t g^2\, ds < \infty$ for all $t > 0$ and W an $\mathcal{F}_t$-Wiener process, then*

$$\mathcal{E}_t(I^W(g)) = \exp\left\{ \int_0^t g\, dW - \frac{1}{2} \int_0^t g^2\, ds \right\}$$

is an $\mathcal{F}_t$-martingale or equivalently, $E\mathcal{E}_t(I^W(g)) = 1$ holds for all $t \geq 0$.

Now we shall state a theorem that is, by its importance, the next to Itô formula and is, indeed, one of the cornerstones of Stochastic analysis.

2.3.3   Lévy Theorem. *Let $X = (X_1, X_2, \ldots, X_d)$ be a d-dimensional continuous $\mathcal{F}_t$-adapted process. Then X is an $\mathcal{F}_t$-Wiener process if and only if*

$$X_j \in CM_{loc}(\mathcal{F}_t) \quad \text{and} \quad \langle X_j, X_k \rangle(t) = t\delta_{jk}, \quad 1 \leq j, k \leq d, \quad t \geq 0.$$

A local version of the statement is as follows:

If, for a fixed T, the stopped process X_j^T is in $CM_{loc}(\mathcal{F}_t)$ with $\langle X_j, X_k \rangle(t) = t\delta_{jk}$ for $t \in [0, T]$ and $1 \leq j \leq d$ then X is an $\mathcal{F}_t$-Wiener process on the interval $[0, T]$.

Applying Doob-Meyer Theorem we get as a corollary

2.3.4   One Dimensional Lévy Theorem.

(a) $\mathcal{F}_t$-Wiener process is the unique local martingale $M \in CM_{loc}(\mathcal{F}_t)$ with $\langle M \rangle(t) \equiv t$.

(b) A one dimensional process M is $\mathcal{F}_t$-Wiener iff M and $(M^2(t) - t, t \geq 0)$ are processes in $CM_{loc}(\mathcal{F}_t)$.

What an exciting property of the normal probability distribution. To verify that a continuous motion M is Brownian you need only to test the martingale property of the motion itself and of its transformation $M^2(t) - t$.

Remark. We cannot avoid the assumption on the continuity of M in (b):

If $N(t)$ is the right-continuous Poisson process with the intensity $\lambda = 1$ then $N(t) - t$ and $(N(t) - t)^2 - t$ are martingales.

Agree to call a continuous and adapted process X a **local $\mathcal{F}_t$-martingale on** $[0, T]$ if, as in 3.3, $(X - X(0))^T \in CM_{loc}(\mathcal{F}_t)$. Prove

2.3.5   Exercise.   A continuous adapted process X is a local martingale on $[0, T]$ iff there is an increasing sequence $\tau_n \leq T$ of Markov times with $P[\tau_n = T] \to 1$ such that $(X - X(0))^{\tau_n}$ is a martingale for all $n \in \mathbb{N}$.

Proof of 3.3. Taking in consideration 1.2.2 (c) and 1.6.3 it is enough to verify the implication stated by the local version of the theorem.

For any $u = (u_1, u_2, \ldots, u_d) \in \mathbb{R}^d$ the linear combination defined by $M(t) := \sum_{k=1}^d u_k X_k(t \wedge T)$ is a process in CM_{loc} with

$$\langle M \rangle(t) = \sum_{k,l=1}^d u_k u_l \langle X_k, X_l \rangle(t \wedge T) = \sum_{k=1}^d u_k^2(t \wedge T) \qquad \text{and with}$$

$$\mathcal{E}_t(M, i) = \exp\left\{ i \sum_{k=1}^d u_k X_k(t \wedge T) + \frac{1}{2} \sum_{k=1}^d u_k^2(t \wedge T) \right\}.$$

Because $|\mathcal{E}_t(M, i)| \leq \exp\{2^{-1} T \sum_k u_k^2\} =: K < \infty$ for any $t \geq 0$, the local martingale $\mathcal{E}(M, i)$ is a true complex $\mathcal{F}_t$-martingale as we have already stressed in (b) of the preceding Important Remark. If $s < t \leq T$ and $Y \in \mathcal{F}_s$ is a bounded random variable, then

$$E\left[Y \exp\left\{ i \sum_{k=1}^d u_k(X_k(t) - X_k(s)) \right\} \right]$$

$$= e^{-\frac{1}{2} \sum_k u_k^2(t-s)} \cdot E\left[Y E^{\mathcal{F}_s} \exp\left\{ i \sum_{k=1}^d u_k(X_k(t) - X_k(s)) + \frac{1}{2} u_k^2 t - \frac{1}{2} u_k^2 s \right\} \right]$$

$$= e^{-\frac{1}{2} \sum_k u_k^2(t-s)} \cdot E\left[Y E^{\mathcal{F}_s} \mathcal{E}_t(M, i) \cdot \mathcal{E}_s^{-1}(M, i) \right] = e^{-\frac{1}{2} \sum_k u_k^2(t-s)} \cdot EY.$$

Read first the equality with $Y = 1$ to see that

(33) $$\Phi_{X(t)-X(s)}(u_1, \ldots, u_d) = e^{-\frac{1}{2} \sum_k u_k^2(t-s)},$$

where $\Phi_{X(t)-X(s)}$ denotes the characteristic function of the increment $X(t) - X(s)$. Putting there $Y = e^{iu_{d+1}I_F}$ for a $u_{d+1} \in \mathbb{R}$ and $F_s \in \mathcal{F}_s$ we get

$$\Phi_{X(t)-X(s),I_F}(u_1,\ldots,u_d,u_{d+1}) = \exp\{-\frac{1}{2}\sum_k^d u_k^2(t-s)\} \cdot E\exp\{iu_{d+1}I_F\}$$

(34)
$$= \Phi_{X(t)-X(s)}(u_1,\ldots,u_d) \cdot \Phi_{I_F}(u_{d+1}),$$

where $\Phi_{X(t)-X(s),I_F}$ is the characteristic function of $d+1$-dimensional random vector $(X(t) - X(s), I_F)$ and Φ_{I_F} the characteristic function of I_F. Because (33) and (34) are proved for arbitrary $(u_1,\ldots,u_d,u_{d+1}) \in \mathbb{R}^{d+1}$, (33) implies that $X(t) - X(s)$ has the normal distribution $N_d(0, (t-s) \cdot I_d)$ and (34) that it is independent of F. It follows that X is an $\mathcal{F}_t$-Wiener process on $[0,T]$. $\square$

2.3.6   Exercise. Let (W_1, W_2) be a two-dimensional $\mathcal{F}_t$-Wiener process and $0 \le G \le 1$ an $\mathcal{F}_t$-progressive process. Then

$$dW(t) = G(t)\,dW_1(t) + \sqrt{1 - G^2(t)}\,dW_2(t), \qquad W(0) = 0$$

defines another $\mathcal{F}_t$-Wiener process on the underlying probability space.

We promised to exhibit an example of a nonnegative integrable local martingale that is not a true martingale.

2.3.7   Example and Exercise. Consider an $\mathcal{F}_t$-Wiener process W and a sequence $0 < u_n < 1$ such that $u_n \uparrow 1$. Put

$$F_n(u) := \frac{1}{1-u} \cdot I_{[0,u_n]}(u), \quad 0 \ge u \ge 1, \quad X_n(t) := \int_0^t F_n\,dW, \quad t \ge 0.$$

Obviously, $X_n \in CM_2$ and since $X_{n+1}^{u_n} \overset{as}{=} X_n^{u_n}$, by 1.8 there is a continuous $\mathcal{F}_t$-adapted process $X = (X(t), 0 \le t < 1)$ such that $X^{u_n} \overset{as}{=} X_n^{u_n}$ hods for any $n \in \mathbb{N}$. Observe the process $Y = (Y(s), s \ge 0)$ defined by $Y(s) = X(\frac{s}{s+1})$ and apply 3.4.(a) to prove:

(i) $\tau := \inf\{0 \le t < 1 : X(t) = -1\}$ is an $\mathcal{F}_t$-Markov time such that $\tau < 1$ almost surely ($\inf \emptyset \overset{def}{=} +\infty$).

(ii)   If $F(u) := \frac{1}{1-u} \cdot I_{[0,\tau]}(u) \in PM_2(W)$, then $\tau \overset{as}{=} \inf\{t \ge 0 : I_t^W(F) = -1\}$. Especially, $I^W(F) \ge -1$ almost surely on $\mathbb{R}^+$ and $I^W(F) \overset{as}{=} -1$ on $[1, \infty]$.

(iii) $1 + I^W(F)$ is an nonnegative local martingale, hence a supermartingale, hence an integrable local martingale. Obviously it is not a true martingale.

(iv) $I^W(F)$ is local martingale whose exponential $\mathcal{E}(I^W(F))$ is not a true martingale.

The next theorem is a sophisticated inverse to 3.1

2.3.8   Theorem. *Let M be a continuous adapted process with $M(0) \stackrel{as}{=} 0$ and $B \in CFV$. Denote*

$$Z^\lambda(t) := \exp\{\lambda M(t) - \frac{\lambda^2}{2}B(t)\}, \quad t \in \mathbb{R}^+, \quad \lambda \in \mathbb{R}.$$

Then $M \in CM_{loc}$ with $\langle M \rangle \stackrel{as}{=} B$ if and only if $Z^\lambda - 1 \in CM_{loc}$ for all $\lambda \in \mathbb{R}$.

Proof. The "$\Rightarrow$" part is provided by 3.1. As for "$\Leftarrow$" assume first that M and B are bounded processes, say $|M|, |B| \leq K < \infty$. Then $|Z^\lambda(t)| \leq \exp\{K(|\lambda| + \frac{\lambda^2}{2})\} =: D(\lambda)$. Hence, Z^λ is a true martingale for each $\lambda \in \mathbb{R}$ by 1.4.2 (a). If $s < t$ and $F \in \mathcal{F}_s$ then

(35)
$$\int_F Z^\lambda(t)\, dP = \int_F Z^\lambda(s)\, dP.$$

It follows from

$$\frac{\partial Z^\lambda(t)}{\partial \lambda} = Z^\lambda(t)\,(M(t) - \lambda B(t)), \quad \frac{\partial^2 Z^\lambda(t)}{\partial \lambda^2} = Z^\lambda(t)\,(M(t) - \lambda B(t))^2 - Z^\lambda(t)B(t),$$

that

$$\left| \frac{\partial Z^\lambda(t)}{\partial \lambda} \right| \leq K D_\lambda(1 + |\lambda|), \quad \left| \frac{\partial^2 Z^\lambda(t)}{\partial \lambda^2} \right| \leq D_\lambda[2K^2 + 2\lambda^2 K^2 + K].$$

Therefore we may differentiate twice (35) and put $\lambda = 0$ to get

$$\int_F M(t)\, dP = \int_F M(s)\, dP, \quad \int_F M^2(t) - B(t)\, dP = \int_F M^2(s) - B(s)\, dP,$$

for all $s < t$ and $F \in \mathcal{F}_s$. This is exactly as to say that M and $M^2 - B$ are both true $\mathcal{F}_t$-martingales, it follows by 1.6.2 that $\langle M \rangle \stackrel{as}{=} B$.

Finally, let $\tau_n \uparrow \infty$ almost surely be a sequence of $\mathcal{F}_t$-Markov times such that $|M^{\tau_n}|, |B^{\tau_n}| \leq K_n < \infty$. It follows that

$$(Z^\lambda)^{\tau_n}(t) = \exp\{\lambda M^{\tau_n}(t) - \frac{\lambda^2}{2}B^{\tau_n}(t)\}, \quad t \geq 0$$

is a local martingale for any $\lambda \in \mathbb{R}$ according to 1.4.1 (b). Hence, for any $n \in \mathbb{N}$ the process M^{τ_n} is a true $\mathcal{F}_t$-martingale with $\langle M^{\tau_n} \rangle \stackrel{as}{=} B^{\tau_n}$ by the first part of the proof. It follows by definition that $M \in CM_{loc}$ and letting $n \to \infty$ we get $\langle M \rangle \stackrel{as}{=} B$. $\square$

2.4   Girsanov Theory

We shall study what happens with a local martingale $M \in \mathrm{CM}_{loc}(P, \mathcal{F}_t)$ and with a $(P, \mathcal{F}_t)$-Wiener process W if we replace the probability measure P by a probability measure Q such that $P \sim Q$.

Assume first that $Q \ll P$ and denote by $P^t := P|\mathcal{F}_t$ and $Q^t := Q|\mathcal{F}_t$ the restriction of P and of Q to the σ-algebra $\mathcal{F}_t$, respectively. Obviously, $Q^t \ll P^t$ for $t \geq 0$. Denote $D_t := \dfrac{dQ^t}{dP^t}$ and $D := (D_t, t \geq 0)$. Observe that if N is a $(Q, \mathcal{F}_t)$-martingale and then $D_t N(t) \in L_1(P)$ for all $t \geq 0$. Moreover, if $t > s$ and $F_s \in \mathcal{F}_s$, then

$$\int_{F_s} D_t N(t)\, dP = \int_{F_s} N(t)\, dQ = \int_{F_s} N(s)\, dQ = \int_{F_s} D_s N(s)\, dP$$

and therefore,

(36) D is a $(P, \mathcal{F}_t)$-martingale.

Even more generally:

(37)   If N is a $(Q, \mathcal{F}_t)$-martingale, then $N \cdot D$ is a $(P, \mathcal{F}_t)$-martingale.

Note that there is no reason to expect that D can be modified to a continuous process.

Denoting

$$\mathcal{G}(P) := \{Q : Q \sim P \quad \wedge \quad D \quad \text{can be modified to a continuous process}\}$$

we are in a better position because if $Q \in \mathcal{G}(P)$ then the derivative process $D_t = (\frac{dQ^t}{dP^t})$ is a continuous $\mathcal{F}_t$-martingale with $D_t > 0$ almost surely for all $t \geq 0$. It follows by 1.3.13 that D is a continuous martingale with trajectories that are outside a P-null set positive functions on $\mathbb{R}^+$. Without loss of generality **we always choose**

$$D = \left(\frac{dQ^t}{dP^t}, t \geq 0\right) \text{ as a positive continuous } \mathcal{F}_t\text{-martingale denoted by } D(Q|P)$$

and call it **the derivative process of a $Q \in \mathcal{G}(P)$ with respect to P.**

Important Remark D. Let $Q \in \mathcal{G}(P)$.

(a) Since Q-null and P-null sets coincide, $(\Omega, \mathcal{F}, Q)$ is a complete probability space and $(\Omega, \mathcal{F}, Q, \mathcal{F}_t)$ a **complete filtration**.

(b) The Important Remark A in Section 1.6 says that if $Q \sim P$ then a stochastic process X is of a finite quadratic variation on $(\Omega, \mathcal{F}, P)$ if and only if it has the property on $(\Omega, \mathcal{F}, Q)$ and the P-quadratic variation of X equals almost surely to its Q-quadratic variation.

(c) Note that $Q \in \mathcal{G}(P)$ implies $P \in \mathcal{G}(Q)$. Because for any $t \geq 0$ we have $\frac{dP^t}{dQ^t} \overset{as}{=} (\frac{dQ^t}{dP^t})^{-1}$, we are always allowed to choose $D(P|Q)$ as $D(Q|P)^{-1}$.

Also remark that the logarithm of the derivation process $D := D(Q|P)$ is by 2.8 a semimartingale in $\mathrm{CSM}(P)$ with the stochastic differential given by

$$(38) \qquad d(\ln D) = D^{-1}\, dD - \frac{1}{2} D^{-2}\, d\langle D \rangle = dL - \frac{1}{2} d\langle L \rangle,$$

$$(39) \qquad \text{where } dL := D^{-1}\, dD \text{ defines an } L \in \mathrm{CM}_{loc}(P) \text{ with } d\langle L \rangle = D^{-2}\, d\langle D \rangle.$$

We are now prepared to prove

2.4.1   Theorem. *Let $Q \in \mathcal{G}(P)$ be a probability measure with the derivative process $D := D(Q|P)$. Consider an $\mathcal{F}_t$-adapted process N and $M \in \mathrm{CM}_{loc}(P, \mathcal{F}_t)$. Then*

(a)   *N is a $(Q, \mathcal{F}_t)$-(local) martingale if and only if $D \cdot N$ is a $(P, \mathcal{F}_t)$-(local) martingale.*

(b)   *$\widehat{M} := M - \langle M, \ln D \rangle$ is a process in $\mathrm{CM}_{loc}(Q, \mathcal{F}_t)$.*

(c)   *If M is a $(P, \mathcal{F}_t)$-Wiener process then $\widehat{M}$ is $(Q, \mathcal{F}_t)$-Wiener process.*

Proof. Due to the symmetry stated by (c) in the Important Remark D it is enough in **(a)** to verify only ($\Rightarrow$) implications. For a Q-martingale N we have already proved and stated in (37) that DN is a P-martingale. So assume that N is a local Q-martingale. According to (37) there is a sequence of Markov times $\tau_n \uparrow \infty$ such that $D(N - N(0))^{\tau_n}$ is a P-martingale for all $n \in \mathbb{N}$. Thus, $D(N - N(0))$ is a P-local martingale and DN has also the property because $DN(0)$ is a P-local martingale by 1.4.1 (d).

We apply 3.8 to prove **(b)**. We shall consider a $\lambda \in \mathbb{R}$ and prove that $Z^\lambda := \exp\{\lambda \widehat{M} - \frac{\lambda^2}{2} \langle M \rangle\}$ is a continuous local Q-martingale:

Because $\langle M, \ln D \rangle \overset{as}{=} \langle M, L \rangle$ it follows by (38) and (39) that outside a P-null set

$$(40) \qquad D_0^{-1} \cdot Z^\lambda \cdot D = \exp\{\lambda M - \lambda \langle M, L \rangle - \frac{\lambda^2}{2} \langle M \rangle + L - \frac{1}{2} \langle L \rangle\}$$

$$= \exp\{\lambda M + L - \frac{1}{2} \langle \lambda M + L \rangle\}$$

holds. Hence a combination of 3.1 and (40) proves that $Z^\lambda D$ is a continuous local P-martingale and finally that Z^λ is local Q-martingale by **(a)**.

If M is a $(P, \mathcal{F}_t)$-Wiener process, then $\widehat{M} \in \mathrm{CM}_{loc}(Q)$ by **(b)**. Since $\langle \widehat{M} \rangle(t) = \langle M \rangle(t) = t$ for every $t \geq 0$ Q-almost surely by 1.6.4 and by **(b)** in the Important Remark D, 3.4 (a) implies that $\widehat{M}$ is a $(Q, \mathcal{F}_t)$-Wiener process. $\square$

Let P, Q and D be as in 4.1. It follows by **(b)** that for any $X \in \mathrm{CSM}(P)$ with the stochastic differential $dX = dB + dM$

$$(41) \qquad d\overline{X} = d\left(B + \langle M, \ln D \rangle\right) + d\left(M - \langle M, \ln D \rangle\right), \quad \overline{X}(0) = X(0).$$

defines a process $\overline{X} \in \mathrm{CSM}(Q)$. It is important to note that even though the processes X and $\overline{X}$ are identical as continuous adapted processes, they have own separate identities as semimartingales in $\mathrm{CSM}(P)$ and $\mathrm{CSM}(Q)$. To understand the following important theorem properly, keep in mind that the identity on the space CSM is given by

$$X_1 = X_2 \quad \Longleftrightarrow \quad X_1(0) \stackrel{as}{=} X_2(0), \quad B_1 \stackrel{as}{=} B_2, \quad M_1 \stackrel{as}{=} M_2,$$

where $dX_i = dB_i + dM_i$.

2.4.2   Theorem. *Let P, Q and D be as in 4.1, X_i and X processes in $\mathrm{CSM}(P)$ with the stochastic differentials $dX_i = dB_i + dM_i$ and $dX = dB + dM$, respectively. In what follows the identities are almost surely with respect both to P and Q.*

(a) *The map $X \to \overline{X}$ defined by (41) is a bijective map $\mathrm{CSM}(P) \to \mathrm{CSM}(Q)$, especially $\mathrm{CSM}(Q) = \mathrm{CSM}(P)$ as sets of continuous adapted processes. The map $M \to \widehat{M}$ defined in (b) is bijective map $\mathrm{CM}_{loc}(P) \to \mathrm{CM}_{loc}(Q)$.*

(b) *$\langle X_1, X_2 \rangle = \langle \overline{X_1}, \overline{X_2} \rangle$ and $\langle X_1, X_2 \rangle = \langle \widehat{X_1}, \widehat{X_2} \rangle$ if $X_1, X_2 \in \mathrm{CM}_{loc}(P)$.*

(c) *If $G \in \mathrm{PM}_{12}(X)$ then $G \in \mathrm{PM}_{12}(\overline{X})$ and $\overline{I^X(G)} = I^{\overline{X}}(G)$, especially $I^X(G) = I^{\overline{X}}(G)$ holds. If moreover $X \in \mathrm{CM}_{loc}(P)$ then $\widehat{I^X(G)} = I^{\widehat{X}}(G)$ holds almost surely.*

Proof. The assertion (b) is obvious, see Important Remark (b).

(a)   If $\overline{X_1} = \overline{X_2}$, an easy computation shows that $M_1 - M_2 \in \mathrm{CFV} \cap \mathrm{CM}_{loc}(P)$ and therefore $M_1 = M_2$ by 1.4.4. Since $\overline{X_1} = \overline{X_2}$, we have also $B_1 = B_2$, hence $X_1 = X_2$.

If $Y \in \mathrm{CSM}(Q)$ is a process with $dY = dA + dN$ apply 4.1 for $P \in \mathcal{G}(Q)$ and derivation process D^{-1} to see that the process X defined by

$$(A - \langle N, \ln D \rangle) + (N + \langle N, \ln D \rangle) = (A + \langle N, \ln D^{-1} \rangle) + (N - \langle N, \ln D^{-1} \rangle)$$

is a process in $\mathrm{CSM}(P)$ such that $\overline{X} = Y$. This, of course, also proves the bijectivity of $M \to \widehat{M}$.

(c)   If G is a process in $\mathrm{PM}_{12}(X) := \mathrm{PM}_1(B) \cap \mathrm{PM}_2(M)$, then the integrals

$$\int_0^t G^2 \, d\langle M - \langle M, \ln D \rangle \rangle = \int_0^t G^2 \, d\langle M \rangle \quad \text{and} \quad \int_0^t |G| \, d \left(B + \langle M, \ln D \rangle \right)^v$$

are finite almost surely for all $t \geq 0$ by 1.7.9 and therefore $G \in \mathrm{PM}_{12}(\overline{X})$. To verify the first equality in (c) it is sufficient to prove that

$$\int G \, dB + \langle I^M(G), \ln D \rangle = \int G \, d \left(B + \langle M, \ln D \rangle \right),$$

(this follows directly by 1.4), and that

$$\widehat{I^M(G)} := \int G \, dM - \langle I^M(G), \ln D \rangle = \int G \, d \left(M - \langle M, \ln D \rangle \right) =: I^{\widehat{M}}(G).$$

If $Y \in CM_{loc}(Q)$ is an arbitrary local martingale then $Y = \widehat{Z}$ for a $Z \in CM_{loc}(P)$. It follows by **(a)** that

$$\langle \widehat{I^M(G)}, Y \rangle = \langle I^M(G), Z \rangle = \int G \, d\langle M, Z \rangle = \int G \, d\langle \widehat{M}, Y \rangle,$$

which proves that $\widehat{I^M(G)} = \int G \, d\widehat{M}$. $\square$

Theorem 4.1 is not easy to apply. Mostly we are in a possession of

a positive continuous $\mathcal{F}_t$-martingale $D = (D_t, t \geq 0)$

such that each D_t is a probability density w.r.t. measure P

rather than of a probability measure $Q \in \mathcal{G}(P)$ which would yield a derivation process $D(Q|P)$. Having a D as in (41) we are able to construct a $Q \in \mathcal{G}(P)$ such that $D = D(Q|P)$ if, for example,

(42)   there is a positive probability density w.r.t. P, say D_∞,

such that the martingale D is closed by D_∞,

which means that $D_t = E_P[D_\infty | \mathcal{F}_t]$ holds for all t,

choosing simply $dQ = D_\infty \, dP$. Indeed, $Q^t(F_t) = \int_{F_t} D_\infty \, dP = \int_{F_t} D_t \, dP^t$ for all $F_t \in \mathcal{F}_t$ and $t \geq 0$ and therefore $Q \in \mathcal{G}(P)$ and $D = D(Q|P)$.

The most obvious exhibition of a process D such as in (42) is as follows: Choose an $L \in CM_{loc}(P)$ and $T > 0$ to fulfil

(43) $$E\mathcal{E}_T(L) = E_P \exp\{L(T) - \frac{1}{2}\langle L \rangle(T)\} = 1,$$

recall **(a)** in Important Remark C, we made before stating 3.3, to argue that $\mathcal{E}(L)$ is an $\mathcal{F}_t$-martingale on $[0, T]$ which obviously means that

$$D_t := \mathcal{E}_t(L^T) := \exp\{L(t \wedge T) - \frac{1}{2}\langle L \rangle(t \wedge T)\}$$

defines a process D as in (42) with $D_\infty := D_T = \mathcal{E}_T(L)$.

Given an $L \in CM_{loc}(P)$ such that (43) holds, the above considerations deliver a setting for 4.1 and 4.2 specified as

$$dQ = \mathcal{E}_T(L) \, dP, \quad \ln D = L^T - \frac{1}{2}\langle L^T \rangle, \quad \widehat{M} = M - \langle M, L^T \rangle, \quad M \in CM_{loc}(P).$$

Thus, as a direct corollary to 4.1, we have proved

2.4.3   Girsanov Theorem. *If $L \in CM_{loc}(P)$ and $T > 0$ satisfy (43) then $dQ = \mathcal{E}_T(L) \, dP$ defines a probability measure such that*

$$\widehat{M} = M - \langle M, L^T \rangle \in CM_{loc}(Q) \quad \text{for all} \quad M \in CM_{loc}(P).$$

In particular, if M is a $(P, \mathcal{F}_t)$-Wiener process then $\widehat{M}$ defines a $(Q, \mathcal{F}_t)$-Wiener process.

A classical version of Girsanov theorem reads as

2.4.4   Theorem. Let $W = (W_1,\ldots,W_d)$ be a d-dimensional $(P,\mathcal{F}_t)$-Wiener process. Further consider $T > 0$ and $G = (G_1\ldots,G_d)^\top$ a process in $PM_2(W)$ such that $E\mathcal{E}_T(I^W(G^\top)) = 1$. Then $dQ = \mathcal{E}_T(I^W(G^\top))\,dP$ defines a probability measure such that

$$\widehat{W} := W - \int G^\top \cdot I_{[0,T]}\,ds = \left(W_1 - \int G_1 \cdot I_{[0,T]}\,ds,\ldots,W_d - \int G_d \cdot I_{[0,T]}\,ds \right)$$

is a d-dimensional $(Q,\mathcal{F}_t)$-Wiener process.

Recall that

$$I^W(G^\top) := \sum_{i=1}^d I^{W_i}(G_i), \qquad \langle I^W(G^\top) \rangle = \int \sum_{i=1}^d G_i^2\,ds,$$

$$\mathcal{E}_T\left(I^W(G^\top)\right) := \exp\left\{ I_T^W(G^\top) - \frac{1}{2}\langle I^W(G^\top)\rangle(T) \right\}.$$

Proof. We compute

$$\widehat{W}_i := W_i - \int G_i \cdot I_{[0,T]}\,ds = W_i - \langle W_i, I^W(G^\top \cdot I_{[0,T]})\rangle = W_i - \langle W_i, L^T\rangle$$

where $L := I^W(G^\top)$. It follows by 4.3 that each $\widehat{W}_i$ is a process in $\mathrm{CM}_{loc}(Q)$ and because

$$\langle \widehat{W}_i, \widehat{W}_j \rangle \overset{\mathrm{as}}{=} \langle W_i, W_j \rangle, \quad 1 \leq i,j \leq d$$

by 4.1, we may apply 3.3 to prove that $\widehat{W}$ is a d-dimensional $(Q,\mathcal{F}_t)$-process. $\square$

Observing the condition (43) we should be worried how to verify it. Recall Important Remark C we put forward at the beginning of 2.3 to remind that given $L \in \mathrm{CM}_{loc}$ we get $\mathcal{E}(L)$ as a local martingale and supermartingale that is a martingale on $[0,T]$ iff (43) holds. Example 3.7 presents an M in CM_{loc} with $\mathcal{E}(M)(1) < 1$.

A handy and a very sophisticated sufficient condition for (53) is delivered by

Novikov Theorem. Let L be a process in CM_{loc} and $T > 0$. Then

$$(44) \qquad\qquad E\exp\{\frac{1}{2}\langle L\rangle(T)\} < \infty \quad \Rightarrow \quad E\mathcal{E}_T(L) = 1.$$

The proofs may be found in [131], (1.15) p.318, (1.16) p.319 or in [95], 5.12 p. 198). The condition (44) presents a very strong integrability requirement, indeed.

2.4.5   Exercise. If $E\exp\{\frac{1}{2}\langle L\rangle(T)\} < \infty$ then L is an L_p-martingale on $[0,T]$ for all $p \in \mathbb{N}$ and $\exp\{\frac{1}{2}L(t)\}$ is a submartingale on $[0,T]$.

Next pair of results will serve all our purposes.

2.4.6   Lemma. For an $L \in CM_{loc}$ with $E \exp\{\max_{t \leq T} L(t)\} < \infty$ (43) holds.

Proof. $\mathcal{E}(L)$ is a local martingale with $\mathcal{E}_o(L) = 1$ and therefore there are Markov times $\tau_n \uparrow \infty$ almost surely such that $\mathcal{E}(L)^{\tau_n}$ is a martingale for each $n \in \mathbb{N}$ and therefore $E\mathcal{E}_{T \wedge \tau_n}(L) = 1$ for arbitrary $T > 0$. Because

$$\mathcal{E}_{T \wedge \tau_n}(L) = \exp\{L(T \wedge \tau_n) - \frac{1}{2}\langle L\rangle(T \wedge \tau_n)\} \leq \exp\{\max_{t \leq T} L(t)\},$$

we may send $n \to \infty$ to get $E\mathcal{E}_T(L) = 1$. $\square$

2.4.7   Theorem. If $L \in CM_{loc}$ and $T > 0$ are such that $\langle L\rangle(T) \leq c$ almost surely for a $c \in (0, \infty)$, then (43) holds.

Proof. Denote $Y := \max_{t \leq T} L(t)$, consider $y > 0$ and put $\lambda := \frac{y}{c}$. Then

$$(45) \qquad P[Y \geq y] \leq P[\max_{t \leq T} \mathcal{E}_t(\lambda L) \geq \exp\{\lambda y - \frac{1}{2}\langle \lambda L\rangle(T)\}]$$

$$\leq P[\max_{t \leq T} \mathcal{E}_t(\lambda L) \geq \exp\{\lambda y - \frac{1}{2}\lambda^2 c\}]$$

$$\leq \exp\{\frac{1}{2}\lambda^2 c - \lambda y\} \cdot E\mathcal{E}_0(\lambda L) = \exp\left\{-\frac{1}{2}\frac{y^2}{c}\right\},$$

where the last inequality follows by 1.2.12 (e) because $\mathcal{E}(\lambda L)$ is a positive super-martingale.

The rest follows easily as

$$Ee^Y \leq 1 + \int_{[Y>0]} e^Y \, dP = 1 + P[Y > 0] + \int_{[Y>0]} (e^Y - 1) \, dP$$

$$= K + \int_{[Y>0]} \int_0^\infty e^y \cdot I_{[Y \geq y]} \, dy \, dP = K + \int_0^\infty e^y \cdot P[Y \geq y] \, dy$$

holds for a constant $K < \infty$. It follows from (45) that $Ee^Y < \infty$ and according to 4.6 that $\mathcal{E}_T(L) = 1$. $\square$

2.4.8   Corollary. Let W be an $\mathcal{F}_t$-Wiener process, $G \in PM_2(W)$ such that $|G| \leq g$ on $\mathbb{R}^+ \times \Omega$ where g is a locally square integrable function $\mathbb{R}^+ \to \mathbb{R}$. Define

$$R := W - \int G \, ds \quad \text{and} \quad dQ_t := \mathcal{E}_t(I^W(G)) \, dP.$$

Then Q_t is a probability measure such that R is $(Q_t, \mathcal{F}_s)$-Wiener process on $[0, t]$

Our program to exhibit a probability measure Q for which a P-semimartingale $W - \int G \, ds$ would become a Q-Wiener process has been successful on bounded intervals $[0, T]$. Next theorem extends the procedure to $[0, \infty)$ in some special cases.

2.4.9   Theorem. *Let* $b : \mathbb{R} \to \mathbb{R}$ *be a Borel and locally bounded function. Then there exists a complete filtration* $(\Omega, \mathcal{F}, P, \mathcal{F}_t)$, *an* $\mathcal{F}_t$-*Wiener process* W *and a probability measure* Q *on* $\mathcal{F}_\infty^W$ *such that*

$$N := W - \int b(W)\, ds \text{ is an } (\Omega, \mathcal{F}_\infty^W, Q, \mathcal{F}_t^W)\text{-Wiener process on } [0, \infty),$$

where Q *is defined uniquely by*

(46) $Q|\mathcal{F}_t^W = Q_t|\mathcal{F}_t^W,$ *where* $dQ_t = \mathcal{E}_t[I^W(b(W))]\, dP,$ $t \geq 0.$

Note that $b(W) \in \mathrm{PM}_2(W)$ if b is as in 4.9.

2.4.10   Exercise. Let $b \neq 0$ in 4.9 be a constant function and define Q by (46). Then $N(t) := W(t) - bt$ is a drifted Wiener process under P, a Wiener process under Q and P, Q are singular measures ($P(N) = 0$, $Q(N) = 1$ for some $N \in \mathcal{F}$, denoting $P \perp Q$).

Proof of 4.9. Denote by ν the probability distribution of the Wiener process and by $(\Omega, \mathcal{F}, P)$ the completion of the space $(C(\mathbb{R}^+), \mathcal{B}(C(\mathbb{R}^+)), \nu)$. According to 1.4.9 (a), denoting $\mathcal{F}_t := \mathcal{F}_t^{\mathbf{x},P}$ and $W := \mathbf{x}$, where $\mathbf{x}$ is the canonical process on $C(\mathbb{R}^+)$, we define by $(\Omega, \mathcal{F}, P, \mathcal{F}_t)$ a complete filtration, by W an $\mathcal{F}_t$-Wiener process. It follows by 1.5.4 (ii) that

(47) N is a continuous and $\mathcal{F}_t^W$-adapted process on the space $(\Omega, \mathcal{F}_\infty^W)$.

Our aim is to find a probability measure Q on $\mathcal{F}_\infty^W$ such that N will become an $(\Omega, \mathcal{F}_\infty^W, Q, \mathcal{F}_t^W)$-Wiener process:

Apply 4.8 with $G := b(W)$ to construct for any $t > 0$ a probability measure Q_t on $\mathcal{F}$ by $dQ_t = \mathcal{E}_t[I^W(b(W))]\, dP$ and put $Q^t := Q_t|\mathcal{F}_t^W$. Because $\mathcal{E}[I^W(b(W))]$ is a $(P, \mathcal{F}_t)$-martingale, it follows by a direct computation that $Q^s = Q^t|\mathcal{F}_s^W$ is true if $s \leq t$. According to 1.1.13 there is a unique probability measure Q on $\mathcal{F}_\infty^W = \mathcal{B}(C(\mathbb{R}^+))$ such that $Q|\mathcal{F}_t^W = Q^t$ holds for all t. Thus, the Q is constructed as required by (46) and to prove that N is an $(\Omega, \mathcal{F}_\infty^W, Q, \mathcal{F}_t^W)$-Wiener process fix $s < t$, $B \subset \mathbb{R}$ a Borel set and $F \in \mathcal{F}_s^W$. Since, according to 4.8, $N(u)$ is a $(Q_t, \mathcal{F}_u)$-Wiener process on $[0, t]$, (47) implies that

$$Q\left(F \cap [N(t) - N(s) \in B]\right) = Q_t\left(F \cap [N(t) - N(s) \in B]\right)$$
$$= Q_t(F) \cdot Q_t\left[N(t) - N(s) \in B\right] = Q(F) \cdot Q\left[N(t) - N(s) \in B\right]$$

and therefore the increment $N(t) - N(s)$ is independent of the σ-algebra $\mathcal{F}_s^W$. Choosing $F = \Omega$ we get $Q[N(t) - N(s) \in B] = Q_t[N(t) - N(s) \in B]$ which yields $\mathcal{L}(N(t) - N(s)|Q) = N(0, t - s)$. This, assisted by (47) proves that N is a $(Q, \mathcal{F}_t^W)$-Wiener process as also $N(0) \overset{\mathrm{as}}{=} 0.$ $\square$

2.4.11 Upcrossing a Level by Drifted Wiener Process. Consider a shifted Wiener process $Y(t) := W(t) + \mu t$ and establish the probability distribution of the first entry $\upsilon_{a,\mu} := \inf\{t \geq 0 : Y(t) = a\}$ where $a \neq 0$.

Put $\tau_a := \upsilon_{a,0}$ and recall 1.3.16 that provides the density $f_a(s)$ of the distribution of the Markov time τ_a. Our reasoning will go along the lines suggested by 4.9 with $b(x) \equiv \mu$. Let $(\Omega, \mathcal{F}, P, \mathcal{F}_t)$, W and Q be such as in 4.9. Then $N(t) := W(t) - \mu t$ is a $(Q, \mathcal{F}_t^W)$-Wiener process. Obviously,

$$\mathcal{L}(Y|P) = \mathcal{L}(W|Q) \quad \text{and therefore} \quad \mathcal{L}(\upsilon_{a,\mu}|P) = \mathcal{L}(\tau_a|Q)$$

holds. Hence,

$$P[\upsilon_{a,\mu} \leq t] \overset{(1)}{=} Q[\tau_a \leq t] = E_P I_{[\tau_a \leq t]} \cdot \mathcal{E}_t(\mu W)$$

$$\overset{(2)}{=} E_P I_{[\tau_a \leq t]} E_P \left[\mathcal{E}_t(\mu W) | \mathcal{F}_{\tau_a \wedge t} \right]$$

$$\overset{(3)}{=} E_P I_{[\tau_a \leq t]} \cdot \mathcal{E}_{\tau_a \wedge t}(\mu W) = E_P I_{[\tau_a \leq t]} \exp\left\{ \mu W(\tau_a) - \frac{1}{2}\mu^2 \tau_a \right\}$$

$$= E_P I_{[\tau_a \leq t]} \cdot \exp\left\{ \mu a - \frac{1}{2}\mu^2 \tau_a \right\},$$

where the equality (1) follows from the fact that $[\tau_a \leq t] \in \mathcal{F}_t^W$, the equality (2) holds because $[\tau_a \leq t] \in \mathcal{F}_{t \wedge \tau_a}$ and finally (3) is implied by 1.3.11 because $\mathcal{E}(\mu W)$ is a true $\mathcal{F}_t$-martingale. Hence, we get

$$P[\upsilon_{a,\mu} \leq t] = \int_0^t \exp\left\{ \mu a - \frac{1}{2}\mu^2 s \right\} f_a(s)\, ds = \frac{|a|}{\sqrt{2\pi}} \int_0^t s^{-\frac{3}{2}} \exp\left\{ -\frac{(a - \mu s)^2}{2s} \right\} ds$$

and letting $t \to \infty$ in

$$P[\upsilon_{a,\mu} \leq t] = E_P I_{[\tau_a \leq t]} \exp\left\{ \mu a - \frac{1}{2}\mu^2 \tau_a \right\},$$

we arrive at

$$P[\upsilon_{a,\mu} < \infty] = E_P \exp\left\{ \mu a - \frac{1}{2}\mu^2 \tau_a \right\} = \exp\{\mu a - |\mu a|\}.$$

We conclude that for $\mu \neq 0$

$$P[\upsilon_{a,\mu} < \infty] = 1 \quad \Leftrightarrow \quad a > 0, \mu > 0 \quad \vee \quad a < 0, \mu < 0.$$

2.5   Integral and Brownian Representations

Having fixed again a complete filtration $(\mathcal{F}_t)$ of a complete probability space $(\Omega, \mathcal{F}, P)$ and choosing an $M \in CM_{loc}$ we denote

$$SI(M) := SI(\Omega, \mathcal{F}, P, \mathcal{F}_t, M) := \{I^M(G), \, G \in PM_2(M)\}.$$

Natural questions are obviously offered: How voluminous subset of CM_{loc} is the set of all M-stochastic integrals $SI(M)$? What might be a property of M or of $(\mathcal{F}_t)$ to imply that $SI(M) = CM_{loc}$?

Later on in the developments of our proofs we shall be able to appreciate the fact that the set $SI(M)$ is *projectively closed* as it is stated by 1.4.8 for the sets of adapted continuous processes and local martingales, respectively. Recall that we have denoted by $\mathcal{M} := \mathcal{M}(\mathcal{F}_t)$ the σ-algebra of all $\mathcal{F}_t$-progressive sets in $\mathbb{R}^+ \times \Omega$.

2.5.1   Lemma. *Let M be a process in CM_{loc}, (τ_n) a sequence of Markov times such that $\tau_n \uparrow \infty$ almost surely and finally $G_n \in PM_2(M)$ processes such that*

$$(48) \qquad \left(I^M(G_{n+1})\right)^{\tau_n} \overset{as}{=} \left(I^M(G_n)\right)^{\tau_n}, \quad n \in \mathbb{N}.$$

Then there exists a $G \in PM_2(M)$ such that

$$(49) \qquad \left(I^M(G)\right)^{\tau_n} \overset{as}{=} \left(I^M(G_n)\right)^{\tau_n}, \quad n \in \mathbb{N}.$$

Proof. The projectivity (48) assisted by 1.12 and by 1.5 implies that outside a P-null set N

$$\int_0^{t \wedge \tau_n} (G_r - G_n)^2 \, d\langle M \rangle = 0 \quad \text{holds for all} \quad t \geq 0, \, r \geq n.$$

Hence, for any $\omega \notin N$ there is a Borel set $D_\omega \subset \mathbb{R}^+$ such that $\mu_{\langle M \rangle(\omega)}(D_\omega) = 0$ and such that

$$(50) \qquad G_r(t, \omega) \cdot I_{[0, \tau_n(\omega)]}(t) = G_n(t, \omega) \cdot I_{[0, \tau_n(\omega)]}(t), \quad \forall t \notin D_\omega, \, r \geq n$$

holds. Define

$$G := \liminf_{n \to \infty} G_n \cdot I_{[\liminf_n G_n \in \mathbb{R}]},$$

observe that it is a progressive process for all $n \in \mathbb{N}$ and check that according to (50) we get

$$(51) \qquad G(t, \omega) = \lim_{n \to \infty} G_n(t, \omega) \quad \text{for all } \omega \notin N \text{ and } t \notin D_\omega.$$

To prove that $G \in PM_2(M)$ recall 1.4.8 and (49) to exhibit a local martingale $I \in CM_{loc}$ such that $I^{\tau_n} \overset{as}{=} \left(I^M(G_n)\right)^{\tau_n} \overset{as}{=} \left(I^M(G_r)\right)^{\tau_n}$ holds for all $r \geq n$. It follows by Fatou Lemma and by (51) that outside a P-null set for all $t \geq 0$ and $n \in \mathbb{N}$

$$\int_0^{t \wedge \tau_n} G^2 \, d\langle M \rangle \leq \liminf_{r \to \infty} \int_0^{t \wedge \tau_n} G_r^2 \, d\langle M \rangle = \langle I \rangle (t \wedge \tau_n).$$

Sending $n \to \infty$ in the above inequality we get that $G \in PM_2(M)$.

Finally, choose an $N \in CM_{loc}$, apply 1.6.5, (50) and compute that

$$\langle I, N \rangle (t \wedge \tau_n) = \langle I^M(G_n), N \rangle (t \wedge \tau_n) = \int_0^{t \wedge \tau_n} G \, d\langle M, N \rangle$$

holds for all $t \geq 0$ and $n \in \mathbb{N}$ almost surely. Letting $n \to \infty$ we arrive at $\langle I, N \rangle \overset{as}{=}$ $\int G \, d\langle M, N \rangle$ to be true for all $N \in CM_{loc}$ and therefore $I \overset{as}{=} I^M(G)$. Hence, the G is the process the existence of which is stated by 5.1. $\square$

Recall that having processes M and L in CM_{loc} we call the local martingales orthogonal if $\langle L, M \rangle = 0$, agree to write $L \perp M$ if it is so. From the definition of the stochastic integral it follows that

(52) $L \perp M \quad \Longleftrightarrow \quad L \perp SI(M).$

Any $N \in CM_{loc}$ owns a uniquely determined projection to the set of stochastic integrals $SI(M)$ as specified by

2.5.2   Projection Theorem. *Let M and N be arbitrary processes in CM_{loc}. Then there exist an $L \in CM_{loc}$ and a $G \in PM_2(M)$ such that*

(53) $N \overset{as}{=} L + I^M(G) \quad \text{and } L \perp M \text{ hold.}$

If $N = L_1 + I^M(G_1)$ is another decomposition (53) then $L_1 \overset{as}{=} L$ (and therefore $\int_0^t (G_1 - G)^2 \, d\langle M \rangle \overset{as}{=} 0$ for all $t \geq 0$ by 1.12).

Proof. Verify first the uniqueness statements: It follows by (53) that $\langle L - L_1 \rangle = \langle L - L_1, I^M(G_1 - G) \rangle = 0$, hence $L_1 \overset{as}{=} L$.

Further, consider $\tau \leq T < \infty$ a **bounded** Markov time such that $R := M^\tau \in CM_2$ and denote

$$\mathcal{H} := \{ I_\tau^R(F), \quad F \in EPM_2(R) \} \subset L_2(\mathcal{F}_\tau),$$

where the ($\subset$) statement is proved as follows: If $F \in EPM_2(R)$ then $I^R(F) \in CM_2$ by 1.7 and therefore $I_\tau^R(F) \in L_2(\mathcal{F}_\tau)$ by 1.2.10 (b) combined with 1.3.9 (c) because $|I_\tau^R(F)|^2 \leq \max_{t \leq T} |I_t^R(F)|^2$.

Next we shall prove that $\mathcal{H}$ is a closed subset of $L_2(\mathcal{F}_\tau)$: If $Y = I_\tau^R(F) \in \mathcal{H}$ then $E[Y|\mathcal{F}_{t \wedge \tau}] = I_{t \wedge \tau}^R(F)$ for all $t \geq 0$ as may be seen recalling 1.3.10. Hence, we get for all $t \geq 0$ that

(54) $E \int_0^t F^2 \, d\langle R \rangle = E \left(I_t^R(F) \right)^2 = E \left(I_{t \wedge \tau}^R(F) \right)^2 = E \left(E[Y|\mathcal{F}_{t \wedge \tau}] \right)^2 \leq EY^2$

as $R^\tau = R$. Now, let $Y_n = I_\tau^R(G_n) \in \mathcal{H}$ and $Y \in L_2(\mathcal{F}_\tau)$ be such that $E(Y_n - Y)^2 \to 0$ holds. If it is so, (54) yields that $E \int_0^t (G_n - G_m)^2 \, d\langle R \rangle \to 0$ as $n, m \to \infty$ for all $t \geq 0$, hence (G_n) is a Cauchy sequence in $EPM_2(R)$. From 1.5.6 (d) it follows that there is a $G \in \mathrm{EPM}_2(R)$ such that $E \int_0^t (G_n - G)^2 \, d\langle R \rangle \to 0$ for all $t \geq 0$; consequently $I^R(G_n - G) \to 0$ in the space CM_2 by 1.7. The latter convergence implies that

$$E\left(I_\tau^R(G) - Y\right)^2 = \lim_{n \to \infty} E\left(I_\tau^R(G) - I_\tau^R(G_n)\right)^2 \leq \lim_{n \to \infty} E \max_{t \leq T} |I_t^R(G - G_n)|^2 = 0,$$

thus $Y \in \mathcal{H}$.

Choose finally a Markov time τ with the properties

(55) τ is bounded and both M^τ and N^τ are processes in CM_2.

Observe that $N(\tau) \in L_2(\mathcal{F}_\tau)$. Since $\mathcal{H}$ is proved to be a **nonempty** subspace of the pseudo-Hilbert space $L_2(\mathcal{F}_\tau)$, there is a random variable $Z \in L_2(\mathcal{F}_\tau)$ that is L_2-orthogonal to $\mathcal{H}$ such that $N(\tau) = Z + I_\tau^R(G)$ holds for a process G in $\mathrm{EPM}_2(R)$. Denoting $L := N^\tau - I^R(G)$ we get $L = L^\tau \in \mathrm{CM}_2$ such that $L(\tau) = Z$.

We shall prove that L and R are orthogonal processes: If λ is an arbitrary bounded Markov time, then

$$E[L(\tau) \cdot R(\tau \wedge \lambda)] = E[Z \cdot I_\tau^R(\mathrm{I}_{[0,\lambda]})] = 0,$$

because obviously $\mathrm{I}_{[0,\lambda]} \in \mathrm{EPM}_2(R)$. It follows by 1.3.10, as also $L^\tau = L$ and $R^\tau = R$, that

$$EL(\lambda) \cdot R(\lambda) = EL(\tau \wedge \lambda) \cdot R(\tau \wedge \lambda) = ER(\tau \wedge \lambda) \cdot E^{\mathcal{F}_{\tau \wedge \lambda}} L(\tau) = ER(\tau \wedge \lambda) \cdot L(\tau) = 0$$

holds. Thus, $L \cdot R$ is an $\mathcal{F}_t$-martingale according to 1.3.12, hence $L \perp R$.

Summarize what we have proved. If τ is as in (55), then there is a local martingale L and a $G \in \mathrm{PM}_2(M^\tau)$ such that $L \perp M^\tau$ and

$$N^\tau = L + I^{M^\tau}(G) = L^\tau + I^M(G \cdot \mathrm{I}_{[0,\tau]})$$

holds almost surely. As $\langle L^\tau, M \rangle = \langle L, M^\tau \rangle = 0$, our summary reads as follows: If τ_n is as in (55) then there is an $L_n \in \mathrm{CM}_{loc}$ and $G_n \in \mathrm{PM}_2(M)$ such that

(56) $$N^{\tau_n} \overset{\mathrm{as}}{=} L_n + I^M(G_n), \quad L_n \perp M$$

holds. Finally, choose a sequence (τ_n) of Markov times such that $\tau_n \uparrow \infty$ holds almost surely, such that (55), and therefore (56) also, hold for arbitrary $n \in \mathbb{N}$. Because (56) can also be written as $N^{\tau_n} = L_{n+1}^{\tau_n} + (I^M(G_{n+1}))^{\tau_n}$ and $\langle L_{n+1}^{\tau_n}, M \rangle = \langle L_{n+1}, M^{\tau_n} \rangle = 0$, it follows from the uniqueness of the decomposition (53) proved above, that $L_{n+1}^{\tau_n} \overset{\mathrm{as}}{=} L_n$ holds for all $n \in \mathbb{N}$. This yields the projectivity (48) and therefore by 5.1 the existence of a $G \in \mathrm{PM}_2(M)$ such that (48) is true. Thus, if $L := N - I^M(G)$ then according to (49)

$$\langle L, M \rangle(t \wedge \tau_n) = \langle N^{\tau_n} - I^M(G_n)^{\tau_n}, M \rangle(t) = \langle L_n^{\tau_n}, M \rangle(t) = 0$$

holds almost surely for all $t \geq 0$ and $n \in \mathbb{N}$. Passing $n \to \infty$ we get that $\langle L, M \rangle \overset{\mathrm{as}}{=} 0$ which verifies that $N = L + I^M(G)$ is the decomposition (53). $\square$

2.5.3 Exercise. For any $M \in \mathrm{CM}_2$ the set $\{I^M(G),\, G \in \mathrm{EPM}_2(M)\}$ is closed in the space CM_2.

2.5.4 Exercise. For any $M \in \mathrm{CM}_{loc}$ the set $\mathrm{SI}(M)$ is closed in the space CM_{loc} with respect to the convergence $N_n \to N$ defined by

$$\langle N_n - N \rangle(t) \xrightarrow{p} 0 \quad \forall t, \quad N_n,\, N \in \mathrm{CM}_{loc}.$$

To execute the program postulated at the beginning of the present section fix a d-dimensional *continuous process* $X : (X_1, \ldots, X_d) : \Omega \to C(\mathbb{R}^+) \times \cdots \times C(\mathbb{R}^+)$ and put

$$\mathcal{F}_t^X := \sigma\left(X_j(s),\, s \le t,\, 1 \le j \le d\right), \quad \mathcal{F}_\infty^X := \sigma \bigcup_{t \ge 0} \mathcal{F}_t^X.$$

Having a probability measure P defined on the σ-algebra $\mathcal{F}_\infty^X$, we will adapt our notation as

$$\mathrm{CM} - \mathrm{CM}_2 - \mathrm{CM}_{loc}(P, X) := \mathrm{CM} - \mathrm{CM}_2 - \mathrm{CM}_{loc}\left(\Omega, (\mathcal{F}_\infty^X)^P, P, (\mathcal{F}_t^X)^P\right).$$

and

$$\mathrm{PM}_2(P, X_j) := \mathrm{PM}_2\left(\Omega, (\mathcal{F}_\infty^X)^P, P, (\mathcal{F}_t^X)^P, X_j\right), \quad 1 \le j \le d.$$

Recall that $\left(\Omega, (\mathcal{F}_\infty^X)^P, P\right)$ denotes the P-completion of $(\Omega, \mathcal{F}_\infty^X, P)$ and $(\mathcal{F}_t^X)^P$ the P-completion of $(\mathcal{F}_t^X)$.

We also agree to denote by $\mathrm{MM}(X)$ the set of all probability measures P defined on the σ-algebra $\mathcal{F}_\infty^X$ such that any X_j is a $(P, \mathcal{F}_t^X)$-martingale and such that $P[X_j = 0] = 1$. Equivalently, $P \in \mathrm{MM}(X)$ is defined by

$$X_j \in \mathrm{CM}(\Omega, \mathcal{F}_\infty^X, P, \mathcal{F}_t^X), \quad 1 \le j \le d \quad \Longleftrightarrow \quad X_j \in \mathrm{CM}(P, X), \quad 1 \le j \le d$$

by 1.4.9 (b). Later on we shall see why we should be interested in the extremal points, called the **extremal measures**, of the convex set $\mathrm{MM}(X)$. Recall that a measure P in a convex set of probability measures $\mathcal{K}$ is said to be **extremal in $\mathcal{K}$** if

$$(57) \qquad P = \alpha \cdot P_1 + (1 - \alpha) \cdot P_2, \quad P_i \in \mathcal{K},\, 0 < \alpha < 1 \quad \Rightarrow \quad P = P_1 = P_2.$$

Denote by $\mathrm{ex}(\mathcal{K})$ the set of all extremal measures in $\mathcal{K}$. Generally, $\mathrm{MM}(X)$ need not own any extremal points but a possibility to discover a Brownian particle with trajectories X makes all the difference.

2.5.5 Theorem. *If P is a probability measure defined on $\mathcal{F}_\infty^X$ such that $X = (X_1, \ldots, X_d)$ is a d-dimensional P-Wiener process, then P is an extremal measure in $MM(X)$.*

Proof. Obviously $P \in \mathrm{MM}(X)$. Assume that P is a convex combination of P_i's in $\mathrm{MM}(X)$ as in (57). Applying 1.4.9 we can see that X is a d-dimensional

$(\mathcal{F}_t^X)^P$-Wiener process on $(\Omega,(\mathcal{F}_\infty^X)^P,P)$ and of course $X_j \in CM(P_i,X)$ for all i,j. Since $P_i \ll P$ it follows by the definition that $\langle X_j, X_k \rangle(t) = \delta_{jk}t$ holds almost surely with respect to any of the measures P, P_1 and P_2 for all j,k,t. Theorem 3.3 therefore says that X is simultaneously $(\mathcal{F}_t^X)^P$, $(\mathcal{F}_t^X)^{P_1}$ and $(\mathcal{F}_t^X)^{P_2}$-Wiener process. Hence $P_1 = P_2 = P$ on the σ-algebra $\mathcal{F}_\infty^X$ as the σ-algebra equals to $\sigma(X) := \{[X \in B], B$ a Borel subset of $C(\mathbb{R}^+,\mathbb{R}^d)\}$ according to 1.1.12. $\square$

2.5.6 Exercise. If P is a probability measure on $\mathcal{F}_\infty^X$ such that a **one** dimensional continuous process X is under the P a centered Gaussian process with independent increments with $\sigma^2(\infty) = \infty$ and $\sigma^2(0) = 0$, where $\sigma^2(t) := EX^2(t)$, then P is an extremal measure in $MM(X)$.

The extremal martingale measures in $MM(X)$ are endowed with extremal properties, indeed. For example, they know only continuous martingales.

2.5.7   Theorem. *If P is an extremal measure in $MM(X)$, then any $(\mathcal{F}_t^X)^P$-martingale on $(\Omega,(\mathcal{F}_\infty)^P,P)$ can be modified to continuous $(\mathcal{F}_t^X)^P$-martingale M, hence to a process M such that $M - M(0) \overset{as}{=} M - EM(0) \in CM(P,X)$.*

Especially, if $W = (W_1,\ldots,W_d)$ is a d-dimensional Wiener process on a probability space $(\Omega,\mathcal{F},P)$, then any $(\mathcal{F}_t^W)^P$-martingale has a continuous modification according to 5.5.

To prove 5.7 we shall need the following corollary to Hahn-Banach theorem known as *Douglas Density Theorem*:

2.5.8   Exercise. Let $(\Omega,\mathcal{F})$ be a measurable space and $\mathbf{L}$ a set of $\mathcal{F}$-measurable functions. Denote by $\mathcal{K}_\mathbf{L}$ the set of all probability measures P defined on $\mathcal{F}$ such that

$$(58) \qquad \mathbf{L} \subset L_1(P) \quad \text{and} \quad \int_\Omega f \, dP = 0 \quad \forall f \in \mathbf{L}.$$

Then P is an extremal measure of the convex set $\mathcal{K}_\mathbf{L}$ if and only if the linear space generated by $f \equiv 1$ and by $\mathbf{L}$ is a dense set in $L_1(P)$.

Observe that if we choose $\mathcal{F} := \mathcal{F}_\infty^X$ in 5.8, where $X = (X_1,\ldots,X_d)$ is an d-dimensional continuous process, and define $\mathbf{L}$ by

$$(59) \quad \{I_{F_s} \cdot (X_j(t) - X_j(s)), 0 \le s < t, F_s \in \mathcal{F}_s^X, I_{F_0} \cdot X(0), F_0 \in \mathcal{F}_0^X, j \le d\}$$

we get $\mathcal{K}_\mathbf{L} = MM(X)$ because $P \in \mathcal{K}_\mathbf{L}$ implies that $\int_{F_0} X_j(0) \, dP = 0$ for all $F_0 \in \mathcal{F}_0^X$ and therefore $P[X_j(0) = 0] = 1$.

2.5.9   Lemma. *Let P be an extremal measure in $MM(X)$. Consider $T > 0$ and $H(T) \in (\mathcal{F}_T^X)^P$ an integrable random variable with $EH(T) = 0$. Then there* is a sequence $G_n^\top = (G_{1n}, \ldots, G_{dn})$ where $G_{jn} \in PM_2(P, X_j)$ and $I^{X_j}(G_{jn}) \in CM(P, X)$ such that

(60)
$$\int_0^T G_n^\top \, dX = \sum_{j=1}^d \int_0^T G_{jn} \, dX_j \to H(T) \quad \text{in} \quad L_1, \quad n \to \infty.$$

Moreover,

$$H \in CM_{loc}(P, X), \quad \langle H, X_j \rangle \overset{as}{=} 0, \quad 1 \le j \le d \quad \Rightarrow \quad H \overset{as}{=} 0.$$

Proof. Note that for any fixed $s < t$ and $F_s \in \mathcal{F}_s^X$ the random variable $I_{F_s} \cdot (X_j(t) - X_j(s))$ in (59) equals almost surely to $I_t^{X_j}(G)$ where $G := I_{(s,\infty] \times F_s}$ is a simple process and therefore $I^{X_j}(G) \in CM(P, X)$ by 1.1. This and 5.8 yields that the linear span of

$$\mathbf{L}^* := \left\{ I_t^{X_j}(G) \; : \; t \ge 0, G \in PM_2(P, X_j), I^{X_j}(G) \in CM(P, X), j \le d \right\}$$

is a dense set in the subspace $\{H \in L_1(P), EH = 0\}$. Thus, there are

$$t_{jn} \ge 0, \, G_{jn} \in PM_2(P, X_j), \, I^{X_j}(G_{jn}) \in CM(P, X)$$

such that $\quad H_n := \sum_{j=1}^d \int_0^{t_{jn}} G_{jn} \, dX_j \overset{L_1}{\to} H(T), \quad n \to \infty.$

Further, if $t_{jn} < T$, replace G_{jn} by $G_{jn} \cdot I_{[0, t_{jn}]}$ and put $t_{jn} = T$ for such a j and n, apply 1.9 to check that that you will get a random variable that equals H_n almost surely. Thus $E[H_n | (\mathcal{F}_T^X)^P] \overset{as}{=} \int_0^T G_n^\top \, dX$ and therefore (60) holds.

To prove the *moreover* statement assume without loss of generality that H is a bounded martingale. Choose $T \ge 0$ and let $\int_0^T G_n^\top \, dX \to H(T)$ as in (60). Note that $H \cdot I^X(G_n^\top) \in CM(P, X)$ by 1.7.4 (iii). Hence, therefore

$$E\left(H(T) \cdot I_T^X(G_n^\top)\right) = E \langle H, I^X(G_n^\top) \rangle (T) = \sum_{j=1}^d E \langle H, I^{X_j}(G_{jn}) \rangle (T) = 0$$

holds because $H \perp X_j$ for all $1 \le j \le d$. It follows by (60) that $EH(T)^2 = 0$. $\square$

Proof of 5.7. Let $P \in exMM(X)$ and H an $(\mathcal{F}_t^X)^P$-martingale, assume without loss of generality that $H(0) \overset{as}{=} 0$ and fix $T > 0$. It follows by 5.9 that there are continuous martingales $M_n \in CM(P, X)$ such that $E|M_n(T) - H(T)| \to 0$ as $n \to \infty$. It follows by 1.2.10 (a) that for any $\epsilon > 0$

$$P[\max_{t \le T} |M_n(t) - M_m(t)| \ge \epsilon] \le \epsilon^{-1} E |M_n(T) - M_m(T)| \to 0, \quad n, m \to \infty.$$

Hence, we may pick up a sequence (n_k) such that

$$P[\max_{t \le T} |M_{n_{k+1}}(t) - M_{n_k}(t)| \ge 2^{-k}] \le 2^{-k}, \quad k \in \mathbb{N}$$

and therefore outside a P-null set $(M_{n_k}(t), t \leq T)$ is a sequence convergent in $C[0, T]$. According 1.4.6 (b) there is a continuous $(\mathcal{F}_t^X)^P$-adapted process M on $[0, T]$ such that $\max_{t \leq T} |M_{n_k}(t) - M(t)| \to 0$ almost surely as $k \to \infty$. Because M_n's are constructed to satisfy $E|H(T) - M_n(T)| \to 0$ we get for any $t \geq 0$ the equalities

$$H(t) \overset{\text{as}}{=} E\left[H(T)|(\mathcal{F}_t^X)^P\right] \overset{\text{as}}{=} \lim_n E\left[M_n(T)|(\mathcal{F}_t^X)^P\right] = \lim_n M_n(t) \overset{\text{as}}{=} M(t),$$

where both limits are in $L_1(P)$. Thus, $(M(t), t \leq T)$ is a continuous $(\mathcal{F}_t^X)^P$-martingale that defines a modification of $(H(t), t \leq T)$. Because $T > 0$ was chosen arbitrary the proof is complete. $\square$

What follows is the most important achievement of the present section.

2.5.10 Brownian Representation. Let $W = (W_1, \ldots, W_d)$ be a Wiener process on $(\Omega, \mathcal{F}, P)$ and H an arbitrary $(\mathcal{F}_t^W)^P$-local martingale. Then there is a process $G = (G_1, \ldots, G_d)^\top \in PM_2(P, W)$ and such that

$$(61) \qquad H \overset{\text{as}}{=} H(0) + \int G^\top \, dW := H(0) + \sum_{j=1}^d \int G_j \, dW_j$$

holds. The process G in (61) is determined on $\mathbb{R}^+ \times \Omega$ up to a $\lambda \otimes P$-null set.

Proof. Assume without loss of generality that $H(0) \overset{\text{as}}{=} 0$. According to 5.2 we get

$$(62) \quad N \overset{\text{as}}{=} L_1 + I^{W_1}(G_1), \quad L_1 \in CM_{loc}(P, W), \quad G_1 \in PM_2(P, W_1), \quad L_1 \perp W_1.$$

Assume that for a $k < d$

$$(63) \qquad N \overset{\text{as}}{=} L_k + \sum_{j=1}^k I^{W_j}(G_j), \quad L_k \in CM_{loc}(P, W),$$

$$G_j \in PM_2(P, W_j), \quad L_k \perp W_j, \quad j \leq k.$$

Then, according 5.2 again,

$$L_k \overset{\text{as}}{=} L_{k+1} + I^{W_{k+1}}(G_{k+1}), \quad L_{k+1} \in CM_{loc}(P, W),$$

$$G_{k+1} \in PM_2(P, W_{k+1}), \quad L_{k+1} \perp W_{k+1}$$

holds. If $j \leq k$ then

$$\langle L_{k+1}, W_j \rangle \overset{\text{as}}{=} \langle L_{k+1} + I^{W_{k+1}}(G_{k+1}), W_j \rangle \overset{\text{as}}{=} \langle L_k, W_j \rangle \overset{\text{as}}{=} 0$$

and therefore, (62) yields by induction that (63) holds for $k = d$. It follows by 5.9 that $L_d \overset{\text{as}}{=} 0$, hence (61) is true.

If $G^* \in PM_2(P, W)$ is a process such that $\int G^\top \, dW \overset{\text{as}}{=} \int G^{*\top} \, dW$ holds, then

$$\int \sum_{j=1}^d (G_j - G_j^*)^2 \, ds \overset{\text{as}}{=} \langle I^W(G^\top) - I^W(G^{*\top}) \rangle \overset{\text{as}}{=} 0$$

on $\mathbb{R}^+$. This and Fubini Theorem imply that $G = G^*$ almost everywhere on $\mathbb{R}^+ \times \Omega$ with respect to $\lambda \otimes P$. $\square$

2.5.11 Corollary. *Let W be as in 5.10, $T > 0$ and $H_T \in (\mathcal{F}_T^W)^P$ an integrable random variable. Then there is a process $G = (G_1, \ldots, G_d)^\top \in PM_2(P, W)$ such that*

$$(64) \qquad H_T \overset{as}{=} EH_T + \int_0^T G^\top \, dW := EH_T + \sum_{j=1}^d \int_0^T G_j \, dW_j$$

holds.

   If, moreover, $EH_T^2 < \infty$, then G_j in (64) satisfy $E\left(\int_0^T \sum_{j=1}^d G_j^2 \, ds\right) < \infty$. This property of a G in (64) determines the G on $[0, T] \times \Omega$ up to a $\lambda \otimes P$-null set.

   Proof. Consider an $(\mathcal{F}_t^W)^P$-martingale M defined as $M(t) := E[H_T | (\mathcal{F}_t^W)^P]$. According to 5.7 M can be modified to a continuous $(\mathcal{F}_t^W)^P$-martingale H. Thus, a representation (64) always exists by 5.10.

   If H_T in (64) is a square integrable random variable then $I^W(G^\top)^T$ is a continuous L_2-martingale and therefore $E \int_0^T \sum_1^d G_j^2 \, ds = E\langle I^W(G^\top)\rangle(T) < \infty$ by 1.7.3. If $G^* = (G_1^*, \ldots, G_d^*)^\top$ is another subject for (64) such that $E\left(\int_0^T \sum_{j=1}^d G_j^{*2} \, ds\right) < \infty$ holds, then also $I^W(G^{*\top})^T$ is a continuous L_2-martingale. Hence,

$$E\left(\sum_{j=1}^d \int_0^T (G_j - G_j^*)^2 \, ds\right) = E\langle I^W(G^\top) - I^W(G^{*\top})\rangle(T) = 0$$

and $G = G^*$ holds almost everywhere on $[0, T] \times \Omega$ with respect to $\lambda \otimes P$. $\square$

   Having a (P, X) such that $P \in MM(X)$ where $X = (X_1, \ldots, X_d)$ is a d-dimensional continuous process we shall say that (P, X) has the **integral representation property (IRP)** if arbitrary $(\mathcal{F}_t^X)^P$-martingale H can be represented as

$$(65) \qquad H(t) \overset{as}{=} EH(0) + \int_0^t G^\top \, dX = EH(0) + \sum_{j=1}^d \int_0^t G_j \, dX_j, \quad t \geq 0,$$

where $G = (G_1, \ldots, G_d)^\top \in PM_2(X)$. In particular, for a (P, X) with IRP any $(\mathcal{F}_t^X)^P$-martingale has a continuous modification.

   If X is a **one** dimensional continuous process then the IRP characterizes the extremal measures in $MM(X)$. For such a process X denote

$$SI(P, X) := SI\left(\Omega, (\mathcal{F}_\infty^X)^P, P, (\mathcal{F}_t^X)^P, X\right).$$

2.5.12 Theorem. *Let X be a one dimensional continuous process and P a measure in $MM(X)$. Then P is an extremal measure in $MM(X)$ if and only if (P, X) has the integral representation property. Moreover, if P is an extremal measure in $MM(X)$ then $CM_{loc}(P, X) = SI(P, X)$.*

   Proof. To verify $\Rightarrow$ and the *moreover* part apply the 5.7 to see that any $(\mathcal{F}_t^X)^P$-martingale has a continuous modification and therefore it is sufficient to prove that any $H \in \mathrm{CM}(P, X)$ can be represented as $H \stackrel{\mathrm{as}}{=} \int G \, dX$ where G is a process in $\mathrm{PM}_2(X)$. But it follows directly by 5.2 and by 5.9.

Assume that the IRP is present for (P, X). Then (65) says that any $N \in \mathrm{CM}(P, X)$ can be represented as $N \stackrel{\mathrm{as}}{=} I^X(G)$ where G is a process in $\mathrm{PM}_2(P, X)$. It follows by 5.1 that it remains to be true also for any $N \in \mathrm{CM}_{loc}(P, X)$ and therefore $\mathrm{CM}_{loc}(P, X) = \mathrm{SI}(P, X)$. Further assume that $P = \alpha P_1 + (1 - \alpha) P_2$ holds for an $\alpha \in (0, 1)$ and some $P_1, P_2 \in \mathrm{MM}(X)$. Put

$$H(t) := E_P^{\mathcal{F}_t^X} \left[\frac{dP_1}{dP} \right] = \frac{d\left(P_1 | \mathcal{F}_t^X\right)}{d\left(P | \mathcal{F}_t^X\right)}, \quad t \geq 0.$$

It follows easily that H is a bounded $(P, \mathcal{F}_t^X)$-martingale such that $P[H(0) = 1] = 1$ as $\mathcal{F}_0^X$ is a trivial σ-algebra under P. Because of the IRP property the H owns a continuous modification L such that $L - 1 \in \mathrm{CM}(P, X)$. The process X is also a $(P_1, \mathcal{F}_t^X)$-martingale and therefore we get for any $s < t$ and $F \in \mathcal{F}_s^X$ that

$$\int_F X(t) L(t) \, dP = \int_F X(t) \, dP_1 = \int_F X(s) \, dP_1 = \int_F X(s) L(s) \, dP$$

holds. It follows that $X(L - 1) \in \mathrm{CM}(P, X)$ and therefore $\langle X, L - 1 \rangle = 0$. The equivalence (52) further yields that $L - 1 \perp \mathrm{SI}(P, X) = \mathrm{CM}_{loc}(P, X)$. Hence, $L \stackrel{\mathrm{as}}{=} 1$, therefore $P = P_1$ and P is proved to be an extremal measure in $\mathrm{MM}(X)$. $\square$

An obvious choice of a *process* $X : \Omega \to C(\mathbb{R}^+)$ is the canonical process $\mathbf{x} : C(\mathbb{R}^+) \to C(\mathbb{R}^+)$. Having chosen such a process we translate our definitions and notations as follows:

$\mathrm{MM} := \mathrm{MM}(\mathbf{x})$ is the set of Borel probability measures P in $C(\mathbb{R}^+)$, called the **martingale measures**, such that the canonical process $\mathbf{x}$ is a $(P, \mathcal{F}_t^{\mathbf{x}})$-martingale and such that $P[\mathbf{x}(0) = 0]$ holds. Equivalently, the set of martingale measures is defined as the set of $\mathcal{L}(X)$ where X goes through all continuous martingales with $X(0) \stackrel{\mathrm{as}}{=} 0$ whenever they may be defined.

The set of martingale measures is a nice set. Its **extremal boundary** $\mathrm{ex}(\mathrm{MM})$ is according to 5.12 the set of Borel probability measures P on $C(\mathbb{R}^+)$ such that $(P, \mathbf{x})$ has the integral representation property. It is a set rich enough not only to include $\mathcal{L}(W)$ but also to generate the convex set MM via *Krein-Milmann* and *Choquet Theorem*. See III.4 (Bibliographical notes) for references.

2.6   Helps to Some Exercises

1.11   Let $a = b = 1$ without loss of generality and consider a $G \in \mathrm{PM}_2(M) \cap \mathrm{PM}_2(N)$. By Kunita-Watanabe inequality, we get

$$\int_0^t G^2 \, d\langle M + N \rangle \leq \int_0^t G^2 \, d\langle M \rangle + \int_0^t G^2 \, d\langle N \rangle + 2 \left[\int_0^t G^2 \, d\langle M \rangle \cdot \int_0^t G^2 \, d\langle N \rangle \right]^{\frac{1}{2}},$$

which shows that $G \in \mathrm{PM}_2(M + N)$. The linearity formula 1.11 is then proved directly by the definition of the stochastic integral.

1.16   The process $I^M(g_n)$ is obviously a Gaussian process with independent increments for any simple **deterministic** process g_n. A general $I^M(g)$ inherits the above properties because according to 1.5.10 and 1.12 there are deterministic simple processes such that $I^M(g) \overset{as}{=} \lim I^M(g_n)$. Apply 1.7 to complete your proof.

1.17 (i)   Apply subsequently 1.9 and 1.10 to verify that for any $t > r$

$$\xi \int_0^t G\,dM = \xi \int_r^t G\,dM = \xi \cdot [I_t^M(G) - I_r^M(G)]$$

$$= \int_0^t \xi \cdot I_{(r,\infty)}\,dI^M(G) = \int_0^t \xi \cdot G\,dM$$

holds almost surely.

1.17 (ii)   Denote $J_t := \int_r^{r+t} G\,dM$, it follows by 1.6.6 that $J \in \mathrm{CM}_{loc}(\mathcal{F}_{r+t})$. If $N \in \mathrm{CM}_{loc}(\mathcal{F}_{r+t})$ then, again by 1.6.6,

$$\langle J, N \rangle(t) = \langle I^M(G), N_\circ \rangle(r+t) = \int_0^{r+t} G\,d\langle M, N_\circ \rangle$$

$$= \int_0^t G(r+s)\,d\langle M, N_\circ \rangle(r+s) = \int_0^t G_r\,d\langle M_r, N \rangle(s) = \langle I^{M_r}(G_r), N \rangle(t),$$

where $N_\circ \in \mathrm{CM}_{loc}(\mathcal{F}_t)$ is a process defined in 1.6.6.

2.17   If A is a singular matrix, then there is an $x \neq 0$ such that $A^\top \cdot x = 0$, hence a contradiction.

Take x arbitrary and put $y := A^{-1} \cdot x$. Compute

$$||x||^2 = x^\top \cdot x = (A \cdot y)^\top \cdot (A \cdot y) = y^\top \cdot A \cdot A^\top \cdot y \geq \epsilon ||y||^2 = \epsilon ||A^{-1} \cdot x||^2.$$

3.7 (i)   Obviously, $Y \in \mathrm{CM}_2(\mathcal{G}_t)$ where $\mathcal{G}_s := \mathcal{F}_{\frac{s}{s+1}}$ defines a complete filtration. If $u_n > \frac{s}{s+1}$, then by the limit definition of the quadratic variation

$$\langle Y \rangle(s) = \langle X_n \rangle \left(\frac{s}{s+1} \right) = \int_0^{\frac{s}{s+1}} \left(\frac{1}{1-u} \right)^2 du = s,$$

i.e., Y is a $\mathcal{G}_t$-Wiener process. If v is its first entrance to -1 then $v < \infty$ almost surely and $\tau = \frac{v}{1+v} < 1$ almost surely.

3.7 (ii) F is in $\mathrm{PM}_2(W)$ because $\tau < 1$ implies that $\int_0^\infty F^2\,du = \int_0^\tau F^2\,du = \frac{\tau}{1-\tau} < \infty$ almost surely.

By definition and by 1.8 we prove

$$(I^W(F))^{\tau \wedge u_n} \overset{as}{=} (I^W(F_n))^{\tau \wedge u_n} \overset{as}{=} X_n^{\tau \wedge u_n} \overset{as}{=} X^{\tau \wedge u_n}.$$

Letting $n \to \infty$ we get $I^W(F) \overset{as}{=} (I^W(F))^\tau \overset{as}{=} X^\tau$ by 1.5, and therefore **(ii)**.

3.7 (iv)   Note that $\mathcal{E}(I^W(F)) \leq e^{-1}$ almost surely.

5.8   Assume that the linear space L^c generated by L and by all constants in $\mathbb{R}$ is a dense set in $L_1(P)$ and that P_i in $\mathcal{K}_L$ are such that $P = \alpha P_1 + (1 - \alpha)P_2$ holds for an $\alpha \in (0, 1)$. Because $dP_i = g_i\, dP$ where g_i's are bounded densities, we get that L^c is a dense set in both spaces $L_1(P_i)$, too. Hence $P_1 = P_2$ because $\int g\, dP_i = \int g\, dP$ for all $g \in L^c$.

If L^c is not a dense set in $L_1(P)$ then it follows by Hahn-Banach Theorem that there is a bounded $\mathcal{F}$-measurable function h such that $P[h \neq 0] > 0$ and $\int hg\, dP = 0$ for any $g \in L^c$. Without loss of generality we assume that $|h| \leq \frac{1}{2}$ and putting $dP_\pm := (1 \pm h)\, dP$ we define probability measures $P_+ \neq P_-$ in $\mathcal{K}_L$ such that $P = \frac{1}{2}(P_+ + P_-)$.

III.3 Diffusion Financial Mathematics

Black-Scholes calculus, Girsanov calculus, market regulations and option pricing, helps to some exercises

Throughout the present chapter, if not stated otherwise explicitly, $(\Omega, \mathcal{F}, P)$ will be a complete probability space, $W = (W_1, \ldots, W_n)$ an n-dimensional Wiener process and $(\mathcal{F}_t)$ the completed canonical filtration of W, that is the filtration $(\mathcal{F}_t)$ defined as $\mathcal{F}_t := (\mathcal{F}_t^W)^P \subset \mathcal{F}$ for all $t \geq 0$.

3.1   Black-Scholes Calculus

This calculus considers a market where $n + 1$ assets, say $0, 1, \ldots, n$, are traded continuously at any time $t \geq 0$ or, more frequently, at any time $0 \leq t \leq T$ where $T < \infty$ is the trading expiration time. Denote $X_0(t), X_1(t), \ldots, X_n(t)$ the price of the corresponding security at time t and assume that 0 is a **bond** whose price $X_0(t)$ is not exposed to the diffusion perturbations generated by the Wiener process W. Its price is evolved by the equation

$$(1) \qquad X_0(t) = x_0 + \int_0^t r(s) X_0(s)\, ds, \quad t \geq 0, \quad x_0 > 0,$$

i.e., by $dX_0(t) = X_0(t) r(t)\, dt$ and $X(0) = x_0$, where

$$(2) \qquad\qquad r \text{ is a bounded progressive process}$$

called **the interest rate process**. It follows that the price process X_0 of the bond is a continuous adapted process of finite variation (hence $X_0 \in \mathrm{CFV}$) determined almost surely by

$$(3) \qquad X_0(t) = x_0 \cdot \exp\left\{ \int_0^t r(s)\, ds \right\}, \quad t \geq 0.$$

Thus, $d(t) X_0(t) = x_0$ for all $t \geq 0$ where $d(t) := \exp\{-\int_0^t r\, ds\}$ and therefore the process d enters our calculations as the **discount process** associated with the interest rate process r. Assuming that r is a positive deterministic process we get the asset 0 as a security that offers a riskless possibility of investments. If it is so, then the discount process d, denoted as v, is called in the **discount function** in Part I. If moreover, $r(s) \equiv r > 0$ is a constant process, r being called the force of interest in this case, we simply arrive to a continuous time model for a typical savings bank account.

On the other hand, the assets $1, 2, \ldots, n$ are assumed to be **stocks**, or other risky securities, that of course accompanied by a risk, may offer a more substantial profit than the safe bond. Their prices exposed to the diffusion W are modeled by the stochastic differential equations

$$(4) \quad X_i(t) = x_i + \int_0^t X_i(s) b_i(s)\, ds + \sum_{j=1}^n \int_0^t X_i(s) \sigma_{ij}(s)\, dW_j, \quad t \geq 0, \quad 1 \leq i \leq n,$$

where $x_i \geq 0$ for all $1 \leq i \leq n$ and

$$b = (b_1, \ldots, b_n)^\top \text{ and } \sigma = (\sigma_{ij}) \text{ is an } n\text{-dimensional vector and an } n \times n\text{-matrix}$$
(5)

of bounded progressive processes, respectively.

The vector process b is called the **rate of return** and the matrix process σ the **volatility matrix**. Hence, the vector $X = (X_1, \ldots X_n)^\top$ of prices of risky securities is correctly defined as a n-dimensional Itô process with locally bounded coefficients.

Denoting by $D(X)$ the diagonal $n \times n$-matrix defined by $D(X)_{ii} := X_i$ we get (4) equivalently as

$$(6) \qquad X(t) = x + \int_0^t D(X) \cdot b \, ds + \int_0^t D(X) \cdot \sigma \, dW, \quad t \geq 0,$$

where $x = (x_1, \ldots, x_n)^\top$. Also define an Itô process $R = (R_1, \ldots, R_n)^\top$ by

$$(7) \qquad R(t) = \int_0^t b(s) \, ds + \int_0^t \sigma(s) \, dW(s), \quad t \geq 0,$$

or equivalently by

$$(8) \qquad R_i(t) := \int_0^t b_i \, ds + \sum_{j=1}^n \int_0^t \sigma_{ij} \, dW_j, \quad t \geq 0, \quad 1 \leq i \leq n$$

and call R the **return process**. Since R is an n-dimensional Itô process with bounded coefficients and X is a continuous process, 2.2.19 implies that X is a process further equivalently defined by

$$(9) \qquad X \overset{as}{=} x + \int D(X) \, dR, \quad \text{or by} \quad X_i \overset{as}{=} x_i + \int X_i \, dR_i, \quad 1 \leq i \leq n$$

where $x = (x_1, \ldots, x_n)^\top$.

Before trying to defend the equations (4) as a suitable model for stock prices stochastic dynamics we should convince ourselves that they have a unique solution. To this end observe that $\langle R_i \rangle(t) = \int_0^t \sum_{j=1}^n \sigma_{ij}^2(s) \, ds$ holds and apply 2.2.13 to each equation $dX_i = X_i \, dR_i$ separately to prove

3.1.1   Stock Prices. *Under (5) the equations (4) have almost surely unique solutions given by*

$$(10) \qquad X_i(t) = x_i \exp\left\{ \sum_{j=1}^n \int_0^t \sigma_{ij} \, dW_j + \int_0^t \left(b_i - \frac{1}{2} \sum_{j=1}^n \sigma_{ij}^2 \right) ds \right\}, \quad t \geq 0$$

for all $1 \leq i \leq n$. Assuming that all

$$(11) \qquad \textit{initial prices } x_1, \ldots, x_n \textit{ are positive numbers,}$$

and denoting $\ln X := (\ln X_1, \ldots, \ln X_n)^\top$ *and* $\ln x := (\ln x_1, \ldots, \ln x_n)^\top$ *we get (10) as*

$$(12) \qquad \ln X(t) = \ln x + \int_0^t \left(b - \frac{1}{2} d(\sigma \cdot \sigma^\top) \right) ds + \int_0^t \sigma \, dW, \quad t \geq 0$$

where $d(A) := (a_{11}, \ldots, a_{nn})^\top$ *denotes the diagonal of an* $n \times n$-*matrix* A.

Since $X_i \overset{as}{=} 0$ for any stock i with initial price $x_i = 0$, we may assume (11) and therefore also that **each price process** X_i **is a positive semimartingale** without loss of generality. Thus, we may write (9), and therefore also (4), equivalently as

$$(13) \qquad\qquad D(X)^{-1} \, dX = dR, \quad X(0) = x$$

by 2.2.19, since both processes $D(X)$ and $D(X)^{-1}$ are continuous adapted and both X and R are Itô processes with locally bounded coefficients.

In case that b and σ are deterministic processes we may be very specific about the probability distribution $\mathcal{L}(X)$ in the space $C(\mathbb{R}^+, \mathbb{R}^n)$. By 2.2.16 we verify directly

3.1.2   Log-Normal Distribution. *Assume that all processes* b_i *and* σ_{ij} *in (5) are deterministic, further assume (11). Then* $\ln X$ *is an* n-*dimensional continuous Gaussian process whose probability distribution in the space* $C(\mathbb{R}^+, \mathbb{R}^n)$ *is uniquely determined by*

$$E \ln X_i(t) = \ln x_i + \int_0^t \left(b_i - \frac{1}{2} \sum_{j=1}^n \sigma_{ij}^2 \right) ds, \quad t \geq 0, \quad 1 \leq i \leq n,$$

(14)

$$cov\,(\ln X_i(t), \ln X_k(s)) = \int_0^{t \wedge s} \sum_{j=1}^n \sigma_{ij}\sigma_{kj} \, ds, \quad t, s \geq 0, \quad 1 \leq i, k \leq n.$$

The arguments in favour of the model and terminology are as follows: Obviously, the $dX_i = X_i \, dR_i$ equations in (9) justify the return process term chosen to label the process R. Note that (9) and 2.1.21 imply for any $t \geq 0$, $1 \leq i \leq n$ and $|\Delta_m(t)| \to 0$ that

$$x_i + \sum_{k=1}^{k_m} X_i(t_{k-1}^m) \left(R_i(t_k^m) - R_i(t_{k-1}^m) \right) \overset{P}{\to} X_i(t), \quad m \to \infty$$

holds, hence a more precise justification.

3.1.3   Example and Exercise. The role of the rate of return process b and that of the volatility matrix σ is easily understood if the coefficients b and σ are bounded deterministic. Apply 2.2.16 and the Fundamental Theorem of Calculus to prove:

The return process R is a Gaussian process such that almost everywhere on $\mathbb{R}^+$ and for all $1 \leq i, k \leq n$

$$\frac{d}{dt}(ER_i(t)) = b_i(t), \quad \frac{d}{dt}(\text{cov}(R_i(t), R_k(t))) = \sum_{j=1}^{n} \sigma_{ij}(t)\sigma_{kj}(t)$$

holds in this case.

Thus, the rate process b controls infinitesimal changes of the mean $ER(t)$ of the return process while the volatility matrix σ is designed to control local changes of its covariance matrix function. Processes b and σ jointly influence the infinitesimal behavior of the probability distribution of the random variable $R(t)$.

Even a more deep insight provides

3.1.4 Exercise. Under the assumption (5)

$$\lim_{h \to 0+} E \left| \frac{R_i(t+h) - R_i(t)}{h} - b_i(t) \right|^2 = 0$$

holds λ-almost everywhere on $\mathbb{R}^+$ for all i.

Hence, the rate of return b may also be interpreted as an L_2-derivative of the return process R.

Let us come back to Part I where the **random walk hypothesis**, supported by empirical experience about the log-normal distribution of a single stock's price $X(t)$, was applied to model the price by means of the equation

(15) $$X^{-1}(t)\, dX(t) = b\, dt + \sigma dW(t), \quad X(0) = x > 0,$$

where W is a one dimensional process. Observing that (15) coincides with (13) if $n = 1$ and the return process R is given by $dR(t) = b\, dt + \sigma\, dW(t)$ we are encouraged to believe that our more complex model (9) approximates the genesis of stock's prices with a acceptable precision.

To summarize what has been said up to now agree to call

$$(X_0, X) \text{ where } X_0 \in \text{CSM and } X = (X_1, \cdots, X_n)^\top \in \text{CSM}^n$$

the **Black-Scholes model (BS-model)** if the following hypotheses are satisfied:

(BS1) X_0 is a process defined by (1) or equivalently by (3) where r is a bounded progressive process and $x_0 > 0$.
(BS2) Processes X_i are defined by (4), also by (8), (9), (10), (12), (13), processes b_i and σ_{ij} are bounded and progressive, $x_i > 0$ for all $1 \leq i, j \leq n$.
(BS3) The matrix $\sigma \cdot \sigma^\top$ is uniformly positive definite.

In what follows **we will fix a BS model** (X_0, X).

Remark that the hypothesis (BS3) has not been applied yet, however it will appear to be crucial in what follows.

The bond and stocks are traded: The investor may continuously change the structure of his holdings by selling some assets $0, 1, \ldots, n$ and buying others. Any fraction (share) of any asset can be traded at any time $t \geq 0$. No commissions are paid for the transactions.

Denoting by $\phi_i(t)$ *the number* of shares of a security $i = 0, 1, \ldots, n$ held by the investor at time t, we do not exclude a possibility of a negative position $\phi_i(t) < 0$.

A negative position $\phi_0(t) < 0$ for the bond may obviously be interpreted as a credit received by the investor to support his operations with the stocks $i = 1, \ldots, n$. A negative position $\phi_i(t) < 0$ for a stock $1 \leq i \leq n$ models a short sale operation. This means that the investor is allowed to sell shares of the security i he does not possess and, of course, has to buy back or pay for in a near future. A typical situation for a short selling operation is that the investor suspects that the price X_i of the stock i is about to fall rapidly.

Denote by

$$(16) \qquad Y(t) = \sum_{i=0}^{n} \phi_i(t) X_i(t), \quad t \geq 0$$

the total **wealth**-capital invested in the assets $i = 0, 1 \ldots, n$ at time t and **postulate that**:

- The investor has no prior knowledge of future prices $X_i(t)$ to help him design a winning **trading strategy** $(\phi_0, \phi_1, \ldots, \phi_n)$.

Since the price processes are positive, trading strategy is equivalently but more comfortably expressed in terms of the **portfolio** process given as $\pi_i(t) = \phi_i(t) X_i(t)$ for each asset i which quantity declares the money value of the shares of i possessed by the investor at time t.

- The investor owns an **initial endowment** – capital $Y(0) = y \geq 0$. However, later on, we shall prove that a void initial endowment $y = 0$ excludes any possibility to acquire a future positive capital $Y(t)$ in decent markets.

- The investor's gains and losses, i.e., positive or negative profits, are entirely due to his trading strategies ϕ_i given the underlying prices X_i. Hence, denoting by $F(t)$ the global **profit** achieved by the investor up to time t, we have

$$(17) \qquad dF(t) = \sum_{i=0}^{n} \phi_i(t) \, dX_i(t), \quad F(0) = 0.$$

Be careful, please, to distinguish the stochastic differential dF from the product $d \cdot F$ of the discount process d with a process F.

- The investor is of course allowed to consume continuously a part of the profit. A **consumption strategy** is defined by the rate $c(s) \geq 0$ of his cumulative spendings $\int_0^t c(s) \, ds$ up to time t.

- The investor's combined **trading-consumption strategy** $(\phi_0, \phi_1 \ldots, \phi_n, c)$ has to be **self-financing** which means that any increase and any decline $dY(t)$ of his wealth $Y(t)$ is entirely caused by gains and losses coming from his investments and consumption. From the point of view of a single investor the market is closed, there is no way for money in and no way for money out. The obvious mathematization of this requirement combined with (17) is given as

$$(18) \qquad\qquad dY(t) = dF(t) - c(t)\, dt, \quad Y(0) = y.$$

Thus, having already fixed a BS-model (X_0, X) we add a **continuous time market model** whose principal components are given by the following definitions:

An $n + 1$-dimensional stochastic process

$$\phi = (\phi_0, \phi_1, \dots, \phi_n)^\top \text{ with } \phi_0 \in \mathrm{PM}_1(W_1) \text{ and } (\phi_1, \dots, \phi_n)^\top \in \mathrm{PM}_2(W)$$

will be called a **trading strategy** or T-strategy if $\sum_{i=0}^n \phi_i(0)X_i(0) \geq 0$ almost surely.

A n-dimensional stochastic process $\pi = (\pi_1, \dots, \pi_n)^\top \in \mathrm{PM}_2(W)$ will be called a **portfolio** process. The portfolio process defined by

$$\pi(\phi) := (\pi_1(\phi), \dots, \pi_n(\phi))^\top := (\phi_1 X_1, \dots, \phi_n X_n)^\top$$

will be called the **portfolio** process **associated with a T-strategy** ϕ. We will also denote $\pi_0(\phi) := \phi_0 X_0$.

The stochastic process $Y(\phi)$ defined by

$$Y(\phi) := \sum_{i=0}^n \phi_i X_i = \sum_{i=0}^n \pi_i(\phi)$$

will be called the **wealth process associated with a T-strategy** ϕ. Since $Y(0) \in \mathcal{F}_0$ and $\mathcal{F}_0 := (\mathcal{F}_0^W)^P$ is a trivial σ-algebra we argue that there is $y \geq 0$ such that

$$Y(0, \phi) = \sum_{i=0}^n \phi_i(0)X_i(0) = \sum_{i=0}^n \pi_i(0, \phi) \overset{as}{=} y.$$

The number y will be called the **initial endowment** (associated with ϕ).

The stochastic process $F(\phi)$ defined as

$$F(t, \phi) := \sum_{i=0}^n \int_0^t \phi_i \, dX_i, \quad t \geq 0$$

will be called the **profit process associated with a T-strategy** ϕ.

A stochastic process $c \in \mathrm{PM}_1(W_1)$, where $c(t) \geq 0$ for all $t \geq 0$ will be referred to as a **consumption process**. We shall write $C := \int c\, ds$ and will refer to C as to an accumulated consumption process.

A pair (ϕ, c) where ϕ is a T-strategy and c a consumption process will be called a **trading-consumption strategy** or TC-strategy. Any pair (π, c) where π is a portfolio and c a consumption process will be called a **portfolio-consumption strategy** or PC-strategy.

Note that the integrability requirements on a T-strategy ϕ say that

$$(19) \qquad \int_0^t |\phi_0|\, ds < \infty, \quad \int_0^t |\phi_i|^2\, ds < \infty \quad \text{almost surely}, \quad t \geq 0, \quad 1 \leq i \leq n$$

or equivalently that

$$(20) \qquad \pi_0(\phi) \in \mathrm{PM}_1(W_1) \quad \text{and} \quad \pi(\phi) \in \mathrm{PM}_2(W),$$

as the processes X_0 and X are continuous adapted and positive. The stochastic process $F(\phi)$ is defined correctly for arbitrary T-strategy ϕ because (19) and (5) are easily seen to imply that $\phi_i \in \mathrm{PM}_{12}(X_i)$ holds for each $0 \leq i \leq n$.

Let (ϕ, c) be a TC-strategy. We shall say that it is a **self-financing strategy** if

$$(21) \qquad Y(t, \phi) \overset{\text{as}}{=} y + \sum_{i=0}^n \int_0^t \phi_i\, dX_i - \int_0^t c\, ds, \quad t \geq 0$$

holds.

Note that $Y(\phi)$ is defined by (16) for a general T-strategy ϕ as a progressive process. However, if the ϕ can be accompanied by a consumption process c such that (ϕ, c) is a self-financing TC-strategy, we get $Y(\phi)$ as a process that can be modified to a process Y in CSM, in particular, to a continuous adapted process.

We define a self-financing strategy as one that satisfies the requirements (18) and (17). This definition not quite precisely written as $Y(\phi) = y + F(\phi) - C$, simply states the accounting balance equation

$$(22) \qquad \text{current wealth=initial wealth + profit} - \text{accumulated consumption.}$$

Remark. If we apply a TC-strategy (ϕ, c) such that all processes ϕ_i and c are $\mathcal{F}_t$-simple, we in fact assume that individual trading operations take place at discrete time points $0 = t_0 < t_1 < \cdots < \infty$ and that the rate of consumption is constant on each interval $(t_k, t_{k+1}]$. Note that (ϕ, c) is defined as a self-financing strategy precisely by (22) in this case.

If we apply a TC-strategy (ϕ, c) such that all processes ϕ_i and c are continuous and adapted, then according to 2.1.14,

$$|\Delta_m| \to 0 \quad \Rightarrow \quad Y_m(t) \overset{\mathrm{P}}{\to} Y(t, \phi), \quad t \geq 0,$$

where $Y_m(t)$ is the wealth accumulated, according to (22), by trading in discrete time points $\Delta_m = \{0 = t_0^m < t_1^m < \cdots < \infty\}$ while the rate of consumption remains constant on each interval $(t_k^m, t_{k+1}^m]$.

We are able to recover the profit process $F(\phi)$, where ϕ is a T-strategy, in terms of the return process $R = (R_1, \ldots, R_n)^\top$ that is defined equivalently by (7),(8) or (9) and of the associated portfolio $\pi := \pi(\phi)$: It follows by (1), (9) and 2.1.10 that for arbitrary $t \geq 0$

$$F(t, \phi) = \int_0^t \phi_0 \cdot X_0 \cdot r\, ds + \sum_{j=1}^n \int_0^t \phi_j \cdot X_j\, dR_j = \int_0^t \pi_0(\phi) \cdot r\, ds + \sum_{j=1}^n \int_0^t \pi_j(\phi)\, dR_j$$

holds almost surely because $\phi_j \in PM_{12}(R_j)$ for $1 \leq j \leq n$ by (19) and R_j is an Itô process with bounded coefficients.

Hence, if Y is a continuous adapted modification of $Y(\phi)$, then

$$(23) \qquad F(\phi) \overset{\text{as}}{=} \int (Y - \pi(\phi)^\top \cdot 1_n) \cdot r \, ds + \int \pi(\phi)^\top \, dR,$$

where $1_n := (1, \ldots, 1)^\top$ is an n-dimensional vector. It follows further by (7) and 2.2.19 that for such a modification Y also

$$(24) \qquad F(\phi) \overset{\text{as}}{=} \int (Y - \pi(\phi)^\top \cdot 1_n) \cdot r \, ds + \int \pi(\phi)^\top \cdot b \, ds + \int \pi(\phi)^\top \cdot \sigma \, dW$$

is true as $\pi \in PM_2(W)$, $\pi_0 \in PM_1(W_1)$ and R is a d-dimensional Itô process with bounded coefficients. Hence, a TC-strategy (ϕ, c) with an initial endowment $Y(0, \phi) \overset{\text{as}}{=} y \geq 0$ is self-financing if and only if the wealth process $Y(\phi)$ has a continuous adapted modification Y that satisfies

$$(25) \quad Y \overset{\text{as}}{=} y + \int r \cdot Y \, ds - \int c \, ds + \int \pi(\phi)^\top \cdot (b - r \cdot 1_n) \, ds + \int \pi(\phi)^\top \cdot \sigma \, dW.$$

Thus we get

3.1.5   Theorem. *Consider the stochastic differential equation*

$$(26) \qquad \boxed{dY(t) = (Y \cdot r - c) \, dt + \pi^\top \cdot (b - r \cdot 1_n) \, dt + \pi^\top \cdot \sigma \, dW, \quad Y(0) = y,}$$

where (π, c) is a PC-strategy and $y \geq 0$. Then (26) has an almost surely unique solution Y given for $t \geq 0$ by

$$(27) \qquad \boxed{Y(t) = d^{-1}(t) \cdot \left\{ y + \int_0^t d \cdot [\pi^\top \cdot (b - r \cdot 1_n) - c] \, ds + \int_0^t d \cdot \pi^\top \cdot \sigma \, dW \right\}.}$$

Let (ϕ, c) be a TC-strategy with an initial endowment $y \geq 0$. Then (ϕ, c) is self-financing if and only if $Y(\phi)$ can be modified to a continuous adapted process Y that solves (26) with $\pi := \pi(\phi)$.

Recall that $d(t) := \exp\{-\int_0^t r \, ds\}$ denotes the discount process associated with the interest rate process r.

Proof. The second statement is already proved by arguments (23), (24) and (25). Apply 2.2.13 with U and V that are such that $U(0) = V(0) = 0$ and

$$dV(t) = r(t) \, dt, \quad \xi = y, \quad U(t) = \int_0^t \pi^\top \cdot (b - r \cdot 1_n) - c \, ds + \int_0^t \pi^\top \cdot \sigma \, dW,$$

which gives $\langle V \rangle = \langle V, U \rangle = 0$, to prove that (28) has an almost surely unique solution given as

$$Y = \exp\{V\} \cdot \{y + I^U(\exp\{-V\})\}.$$

The process Y equals almost surely to the right hand side of (27) by 2.1.20. $\square$

The assertion invites to formulate the following definition: The process Y defined for $t \geq 0$ equivalently by (26) or (27) or by

$$\textbf{(WE)} \quad Y(t) = y + \int_0^t (Y \cdot r - c)\, ds + \int_0^t \pi^\top \cdot (b - r \cdot 1_n)\, ds + \int_0^t \pi^\top \cdot \sigma\, dW,$$

where (π, c) is a PC-strategy and $y \geq 0$, will be called the **wealth process** generated by (π, c, y), Agree to call the equation (WE) the **wealth equation** for (π, c, y).

The wealth equation expresses the requirement on self-financing behavior of the investor in the following manner:

3.1.6   Theorem. *Let Y be the wealth process generated by (π, c, y). Then*

$$(28) \qquad \phi := (\phi_0, \phi_1, \ldots, \phi_n)^\top, \quad \phi_0 := \frac{Y - \sum_{j=1}^n \pi_j}{X_0}, \quad \phi_j := \frac{\pi_j}{X_j}, \quad 1 \leq j \leq n$$

defines a T-strategy ϕ such that $Y(\phi) = Y$ and the following holds true:

$$(29) \qquad (\phi, c) \text{ is a self-financing strategy,} \quad \pi(\phi) = \pi, \quad Y(0, \phi) \overset{\text{as}}{=} y.$$

If $\phi^* = (\phi_0^*, \phi_1^*, \ldots, \phi_n^*)^\top$ is another T-strategy that satisfies (29) then

$$\phi_0^*(t) \overset{\text{as}}{=} \phi_0(t), \quad t \geq 0 \qquad \text{and} \qquad \phi_j^* = \phi_j, \quad 1 \leq j \leq n$$

is true.

Especially, $\phi_0 \overset{\text{ae}}{=} \phi_0^*$ on $\mathbb{R}^+$ by Fubini Theorem because ϕ_0 and ϕ_0^* are progressive processes. By $f \overset{\text{ae}}{=} g$ on a Borel set $B \subset \mathbb{R}^+$ we mean that f equals to g almost everywhere on B with respect to the Lebesgue measure λ.

Proof. The T-strategy ϕ defined by (28) is designed to satisfy $Y(\phi) = Y$ and $\pi(\phi) = \pi$. Hence, $Y(\phi)$ solves (WE) and (ϕ, c) is a self-financing strategy by 1.5. If ϕ^* has properties (29) it follows by 1.5 again that $Y(\phi^*)$ is a modification of Y. Hence,

$$\phi_0^*(t) = \frac{Y(t, \phi^*) - \sum_{j=1}^n \pi_j(t)}{X_0(t)} \overset{\text{as}}{=} \frac{Y(t) - \sum_{j=1}^n \pi_j(t)}{X_0(t)} = \phi_0(t), \quad t \geq 0$$

holds. $\square$

There is a very precise one to one correspondence between the trajectories of the wealth process Y and that of the PC-process (π, c) in (WE).

3.1.7 Uniqueness Theorem. *Let Y be the wealth generated by (π, c, y). Then there is a P-null set N such that for all $\omega \notin N$ and $0 \le a < b < \infty$*

(30)

$\qquad Y(\omega)$ *is a function of finite variation on* $[a, b] \iff \pi(\omega) \overset{\text{ae}}{=} 0$ *on* $[a, b]$,

(31)

$\qquad d(\omega)Y(\omega)$ *is constant on* $[a, b] \iff \pi(\omega) \overset{\text{ae}}{=} 0, \quad c(\omega) \overset{\text{ae}}{=} 0$ *on* $[a, b]$,

(32)

$\qquad Y(\omega) = 0$ *on* $[a, b] \iff \pi(\omega) \overset{\text{ae}}{=} 0, \quad c(\omega) \overset{\text{ae}}{=} 0$ *on* $[a, b], \quad Y(a, \omega) = 0$.

Proof. Put $M := \int \pi^{\mathsf{T}} \cdot \sigma \, dW$. Obviously, $M \in \mathrm{CM}_{loc}$ and a straightforward computation gives $\langle M \rangle = \int \pi^{\mathsf{T}} \cdot \sigma \cdot \sigma^{\mathsf{T}} \cdot \pi \, ds$. According to 1.7.1 and 1.7.2 there is a P-null set N such that for $\omega \notin N$ and arbitrary $0 \le a < b < \infty$

$Y(\omega)$ is a function of finite variation on $[a, b] \iff M(\omega)$ is a constant on $[a, b]$

$$\iff \int_a^b \pi(\omega)^{\mathsf{T}} \cdot \sigma(\omega) \cdot \sigma(\omega)^{\mathsf{T}} \cdot \pi(\omega) \, ds = 0.$$

It follows by (BS3) that the last integral equals to 0 iff $\pi(\omega) \overset{\text{ae}}{=} 0$ on $[a, b]$; (30) is proved.

Take an arbitrary $\omega \notin N$ where N is the P-null set constructed for (30) and arbitrary $0 \le a < b < \infty$. If $d(\omega)Y(\omega)$ is a constant on $[a, b]$ then $\pi(\omega) \overset{\text{ae}}{=} 0$ on $[a, b]$ according to (30) and therefore by (27)

$$d(t, \omega)Y(t, \omega) = d(a, \omega)Y(a, \omega) - \int_a^t d(\omega)c(\omega) \, ds, \quad a \le t \le b.$$

Hence, $\int_a^t d(\omega)c(\omega) \, ds = 0$ for all $t \in [a, b]$ which implies that $c(\omega) \overset{\text{ae}}{=} 0$ on $[a, b]$ because $d > 0$.

To verify ($\Leftarrow$) in (31) assume without loss of generality that the P-null set N we constructed for (30) is large enough to ensure that for $\omega \notin N$ and for all $a < b$

$$\pi(\omega) \overset{\text{ae}}{=} 0 \text{ on } [a, b] \Rightarrow \left(\int_a^t d \cdot \pi^{\mathsf{T}} \cdot \sigma \, dW \right)(\omega) = 0, \quad a \le t \le b$$

holds. It follows by (27) that outside N and for all $a < b$

$$\pi(\omega) \overset{\text{ae}}{=} 0, \, c(\omega) \overset{\text{ae}}{=} 0 \text{ on } [a, b] \Rightarrow d(t, \omega)Y(t, \omega) = d(a, \omega)Y(a, \omega), \quad a \le t \le b.$$

The equivalence (32) is valid as a consequence of (31). $\square$

Remark. It easy to check that 1.7 will stay to be true even in the case of (WE) with bounded not necessarily nonnegative consumption c and for arbitrary $y \in \mathbb{R}$.

3.1.8   Corollary. *Let* Y *and* Y^* *be wealth processes generated by* (π, c, y) *and* (π^*, c^*, y^*), *respectively. Then outside a* P*-null set* N *and for all* $0 \le a < b < \infty$

(33)
$$Y(\omega) = Y^*(\omega) \text{ on } [a,b] \iff \pi(\omega) \stackrel{ae}{=} \pi^*(\omega), \quad c(\omega) \stackrel{ae}{=} c^*(\omega) \text{ on } [a,b]$$
$$\text{and } Y(a,\omega) = Y^*(a,\omega).$$

Especially,

$$Y \stackrel{as}{=} Y^* \iff y = y^* \quad \text{and} \quad \pi \stackrel{ae}{=} \pi^*, \quad c \stackrel{ae}{=} c^* \quad \text{on } \mathbb{R}^+ \text{ holds almost surely.}$$

The generalized wealth process $Y - Y^*$ is generated by $(\pi - \pi^*, c - c^*, y - y^*)$. The equivalence (33) is therefore verified by (32) in 1.7 and by the subsequent Remark.

Note that if π and π^* are progressive processes, then

$$\pi \stackrel{ae}{=} \pi^* \text{ on } \mathbb{R}^+ \text{ almost surely} \iff \pi = \pi^* \quad \lambda \otimes P - \text{almost everywhere on } \mathbb{R}^+ \times \Omega.$$

Theorem 1.7 states properties of self-financing strategies that are almost obvious if termed in the language of finance:

The equivalence (30) states that a trajectory of the wealth is of finite variation during a time period $[a, b]$ if and only if we are sure that the investor ceased entirely to possess risky assets during the period. His wealth is generated only by trading the bonds, hence by means of the ordinary differential equation $dY(t) = (Yr - c)\, dt$ during the time interval.

Further, the equivalence (31) says that the discounted investor's wealth becomes constant on an interval if and only if he owns only bonds and does not consume anything during the period.

Finally, (32) informs us that the investor goes bankrupt at time a and will never recover from this state until b if and only if he owns no bonds, no stocks and consumes nothing between times a and b.

Most of the problems of (diffusion) financial mathematics we shall encounter in the present text may be formulated as below.

Given a requirement on the wealth Y and given an initial endowment y we seek answers to the following questions:

• Is there a (unique) PC-strategy (π, c) such that (π, c, y) generates a wealth process Y fulfilling that requirement?

• If this is the case, is there an effective construction of a suitable PC-strategy (π, c) to get a wealth Y with the desired property ?

One of the principal requirements on the wealth Y is to keep it as a nonnegative process or even as a positive one. Being ready to adapt our consumption c to our current wealth Y the problem is simple.

3.1.9   Positive Wealth Process. *Let* $p = (p_1, \dots, p_n)^\top \in PM_2(W)$ *and* $q \in PM_1(W_1)$ *be an* n*-dimensional and a one-dimensional process, respectively;* $y > 0$.

Then there is a PC-strategy (π, c) such that (π, c, y) generates a wealth process Y having the following property:

$$(34) \qquad Y > 0 \text{ on } \mathbb{R}^+ \times \Omega, \quad c = q \cdot Y, \quad \pi_j = p_j \cdot Y, \quad 1 \le j \le n.$$

Proof. Consider the stochastic differential equation

$$(35) \quad dY(t) = [Y \cdot (r - q)] \, dt + [Y \cdot p^\top \cdot (b - r \cdot 1_n)] \, dt + [Y \cdot p^\top \cdot \sigma] \, dW, \quad Y(0) = y.$$

According to 1.10 below it has an almost surely unique solution Y given by (36) and (37) that is obviously such that $Y > 0$ holds. If we define (π, c) by (34), i.e. as $\pi^\top := Y \cdot p^\top \in \mathrm{PM}_2(W)$ and $c := q \cdot Y \in \mathrm{PM}_1(W_1)$, the equation (35) is translated to (WE) and therefore Y is the wealth process generated by (π, c, y). $\square$

3.1.10 Exercise. For any p, q that are as in 1.9 and for arbitrary $y \ge 0$ equation (35) has an almost surely unique solution given by

$$(36) \qquad Y = y \cdot \exp\left\{ V - \frac{1}{2} \int p^\top \cdot \sigma \cdot \sigma^\top \cdot p \, ds \right\},$$

where

$$(37) \qquad dV = [p^\top \cdot (b - r \cdot 1_n) + r - q] \, ds + p^\top \cdot \sigma \, dW, \quad V(0) = 0.$$

3.1.11   Exercise. The price X_i of an arbitrary stock $1 \le i \le n$ is a wealth process generated by $\pi^{(i)} = X_i \cdot e_i$, $c = 0$ and $y = x_i$, where $e_i \in \mathbb{R}^n$ denotes the i-th unit vector. In other words, the trading-consumption strategy $(\phi^{(i)}, c)$ defined by $\phi_0^{(i)} := 0$, $\phi_j^{(i)} := \delta_{ij}$ for $1 \le j \le n$ and $c := 0$ is self-financing.

If we are not willing to accommodate our consumption to our wealth, i.e. if we fix c, we may confront a problem of bankruptcy.

Let Y be the wealth process generated by a (π, c, y) or, more generally, an arbitrary process such that $Y(0) \ge 0$ holds almost surely. Denote by

$$\tau_0 := \inf\{t \ge 0 : Y(t) = 0\}, \quad \text{i.e., time of the first entry of } Y \text{ to } \{0\}$$

and call it the **bankruptcy time** of the (wealth) process Y.

We feel that a decent model for continuous trading should be such that if the investor goes bankrupt at $\tau_0 < \infty$ then he will never recover from this position. In other words, we require that only such strategies (π, c, y) should be permitted whose wealth process Y satisfies the equality $Y \stackrel{as}{=} Y^{\tau_0}$. Note that according to (32) in 1.7 the equality says that, outside a P-null set, $\tau_0(\omega) < \infty$ implies

$$Y(\omega) = 0, \quad \pi(\omega) \stackrel{ae}{=} 0, \quad c(\omega) \stackrel{ae}{=} 0 \quad \text{on } [\tau_0(\omega), \infty),$$

i.e. that both trading and consumption are *dead* after the bankruptcy. In particular,

(38) $Y(t) \geq 0$ almost surely for all $t \geq 0$ $\iff$ $Y \geq 0$ almost surely on $\mathbb{R}^+ \times \Omega$

if $Y^{\tau_0} \overset{as}{=} Y$.

We shall say that a PC-strategy (π, c) is **admissible** for $y \geq 0$ if (38) holds for the wealth process Y generated by (π, c, y). If Y is such a process and (38) is true only for $0 \leq t \leq T < \infty$ we shall say that the strategy (π, c) is **admissible up to time T** for the initial endowment y.

In the forthcoming section we shall prove that a PC-strategy is admissible for an $y \geq 0$ if and only if $Y^{\tau_0} \overset{as}{=} Y$ where Y is the wealth process generated by (π, c, y), i.e., that the trivial implication ($\Leftarrow$) can be reversed. Given an arbitrary PC-strategy (π, c) and an arbitrary initial endowment y there is an obvious way how to stop (π, c) to an admissible strategy for y.

3.1.12   Exercise. Let Y be the wealth process generated by (π, c, y). Then Y^{τ_0} is the wealth process generated by (π^*, c^*, y), where $\pi^* := \pi I_{[0,\tau_0]}$ and $c^* := c I_{[0,\tau_0]}$. Obviously, (π^*, c^*) is an admissible strategy for y.

3.1.13   Example and Exercise. Consider a portfolio $\pi^{(i)}$ introduced in 1.11 that suggests to own forever only one risky security i. This time, however, generate the investor wealth by $(\pi^{(i)}, c, y)$ where c is a general, perhaps positive, consumption and $y = x_i$. The wealth equation (27) says that

$$Y(t) \overset{as}{=} d^{-1}(t) \cdot \left\{ x_i + \int_0^t d \cdot X_i \cdot (b_i - r) \, ds + \sum_{j=1}^n \int_0^t d \cdot X_i \cdot \sigma_{ij} \, dW_j \right\}$$

$$- d^{-1}(t) \cdot \int_0^t d \cdot c \, ds$$

$$\overset{as}{=} X_i(t) - d^{-1}(t) \cdot \int_0^t d \cdot c \, ds = X_i(t) - \frac{\int_0^t d \cdot c \, ds}{X_0(t) \cdot d(t)} \cdot X_0(t) = X_i(t) - k(t) \cdot X_0(t)$$

according to 1.11.

Thus, $(\pi^{(i)}, c, x_i)$ generates a wealth that equals almost surely to the price X_i iff $c \overset{ae}{=} 0$. In order to satisfy his consumption the investor needs to keep selling $k(t)$ shares of the bond short at any time t. This, however, can not be performed forever if only admissible PC-strategies Y^{τ_0}, generated by $(\pi^{(i)} \cdot I_{[0,\tau_0]}, c \cdot I_{[0,\tau_0]}, x_1)$, where τ_0 is the bankruptcy time of Y, are allowed. Note that τ_0 is the first time when the discounted price $d \cdot X_i$ gets on top of the accumulated discounted consumption $\int d \cdot c \, ds$. Thus, having been forced to trade along the wealth process Y^{τ_0} the investor has no means to cover his consumption after time τ_0 that will happen almost surely if $\tau_0 < \infty$ with probability one.

Prove that if b and σ are deterministic processes such that $b_i < \frac{1}{2} \sum_{j=1}^n \sigma_{ij}^2$, if r and c are positive processes, then $\tau_0 < \infty$ holds almost surely.

Indeed, to get a reliable model for a secure market we can not ignore the admissibility requirement or equivalently that given by the equality $Y^{\tau_0} \stackrel{as}{=} Y$ because otherwise the investor would be allowed to perform the following risky financial operations:

3.1.14 Example and Exercise. We consider the Black-Scholes model for one bond and one stock with the following simplifications:

$$r = 0, \quad x_0 = 1, \quad x_1 = 1, \quad \sigma = 1, \quad b \in \mathbb{R} \quad \text{a constant.}$$

This bears the discount factor $d = 1$ and the bond and stock prices X_0 and X_1, respectively, given as

$$X_0 \equiv 1, \quad X_1(t) = \exp\left\{\left(b - \frac{1}{2}\right)t + W(t)\right\}, \quad t \geq 0,$$

where W is a one dimensional Wiener process. Further, consider the trading consumption model with no consumption, i.e. with $c = 0$, where the wealth process Y generated by a portfolio process π and by an initial endowment $y \geq 0$ is

$$(39) \qquad\qquad Y = y + \int b\pi \, ds + \int \pi \, dW.$$

Choose $y = 1$ and a constant trading strategy $\phi = (\phi_0, \phi_1)^\top$ defined by

$$\phi_0 := 1 + K, \quad \phi_1 := -K, \qquad \text{where } K > 0 \text{ is a fixed number.}$$

Check that $Y(\phi) = 1 + F(\phi)$, where $Y(\phi)$ and $F(\phi)$ is the wealth process and the profit process associated with ϕ, respectively, to see that ϕ defines a self-financing strategy with the initial endowment $Y(0, \phi) = 1$. The following strategy, called also a *suicide strategy* is an example of a short sale financial operation: The investor starts at time $t = 0$ with one dollar initial endowment, sells K shares of the stock short and buys $1 + K$ bonds.

Thus, putting $\pi := \pi(\phi)$ in (39) we get the wealth process Y generated by π and by $y = 1$ as

(40)

$$Y(t) = Y(t, \phi) = 1 + K - KX_1(t) = 1 + K - K \exp\left\{\left(b - \frac{1}{2}\right)t + W(t)\right\}, \quad t \geq 0$$

and therefore $(\pi, 0)$ is not an admissible strategy for the initial endowment $y = 1$. Prove that the following assertions hold:

(a) For any $t > 0$ the probability distribution of the wealth $Y(t)$ is supported by the interval $(-\infty, K + 1]$, i.e., that the distribution function $P[Y(t) \leq y]$ is increasing on the interval and equals to 1 on $[K + 1, \infty)$.

(b) Letting $t \to \infty$ we get $\lim Y(t) \stackrel{as}{=} -\infty$ if $b > \frac{1}{2}$ and $\lim Y(t) \stackrel{as}{=} 1 + K$ if $b < \frac{1}{2}$.

(c) If $b = \frac{1}{2}$, then $\overline{\lim}_{t \to \infty} Y(t) \stackrel{as}{=} 1 + K$ and $\underline{\lim}_{t \to \infty} Y(t) \stackrel{as}{=} -\infty$.

(d) $EY(t) = 1 + K - Ke^{bt}$ for arbitrary $t \geq 0$, $b \in \mathbb{R}$ and $EY(t) < 1$ for $b > 0$ and $t > 0$.

Thus, if the rate of return b is grater than one half of the volatility, the trading strategy ϕ is really a *suicide one* if the infinite trading expiration time is considered. With $0 < b < \frac{1}{2}$ we might be inclined to acquire the strategy as a lucrative one as $\lim_{t\to\infty} Y(t) \overset{\text{as}}{=} K + 1$ in this case, if we had not been warned by **(d)** that the expected wealth at any time $t > 0$ is strictly below the initial wealth $y = 1$ and that $EY(t) \downarrow -\infty$ as $t \to \infty$. Perhaps even more discouraging is the fact that the bankruptcy has a positive probability:

The bankruptcy time τ_0 of Y is easily computed as

$$\inf\left\{t \geq 0 : X_1(t) = 1 + K^{-1}\right\} = \inf\left\{t \geq 0 : \left(b - \frac{1}{2}\right)t + W(t) = \ln(1 + K^{-1})\right\}.$$

Girsanov theorem, together with 2.4.11, offers probabilities for the bankruptcy before time $t > 0$ as

$$P[\tau_0 \leq t] = \frac{\ln(1 + K^{-1})}{\sqrt{2\pi}} \int_0^t s^{-\frac{3}{2}} \exp\left\{-\frac{(\ln(1 + K^{-1}) - (b - \frac{1}{2})s)^2}{2s}\right\} ds =: q_K(t)$$

with $q_K(t) \downarrow 0$ as $K \downarrow 0$ and $q_K(t) \uparrow 1$ as $K \uparrow \infty$.

More importantly, by 2.4.11, $b \geq \frac{1}{2} \Rightarrow P[\tau_0 < \infty] = 1$ and $b < \frac{1}{2}$ implies that

$$P[\tau_0 = +\infty] = 1 - \exp\left\{2\left(b - \frac{1}{2}\right)\ln(1 + K^{-1})\right\} = 1 - \left(1 + K^{-1}\right)^{2b-1} =: p_K.$$

Hence, there is no escape from the bankruptcy if the rate of return is greater than one half of the volatility. Of course, this is something that also follows directly by **(b)**. If $b < \frac{1}{2}$ then the investor may avoid the bankruptcy with a positive probability p_K which decreases from 1 to 0 with K increasing in $(0, \infty)$. Indeed, a high rate of return b if compared with the volatility σ supports the risk coming from the negative position $\phi_1(t) = -K$ for the stock.

Finally, according to 1.12, $\pi^* := \pi(\phi)I_{[0,\tau_0]}$ defines an admissible strategy π^* for the initial endowment $y = 1$ which jointly generate the wealth process given by Y^{τ_0} where the process Y is defined by (40). Let $t \to \infty$ and prove:

(e) $b \geq \frac{1}{2} \Rightarrow Y^{\tau_0}(t) \to 0$, $\quad b < \frac{1}{2} \Rightarrow Y^{\tau_0}(t) \to (K + 1) \cdot I_{[\tau_0 = +\infty]}$ a.s.

(f) $b \geq \frac{1}{2} \Rightarrow EY^{\tau_0}(t) \to 0$.

(g) $b < \frac{1}{2} \Rightarrow EY^{\tau_0}(t) \to (K + 1) \cdot [1 - (1 + K^{-1})^{2b-1}]$.

We shall revisit the example in 2.14.

3.2   Girsanov Calculus

We shall keep a fixed Black-Scholes model for bond-stocks prices (X_0, X) that is specified by hypotheses (BS1), (BS2) and (BS3). Agree to denote by

(41) $$G = (G_1, \ldots, G_n)^\top := \sigma^{-1} \cdot (r \cdot 1_n - b)$$

a bounded progressive process that is defined correctly due to (BS3) and 2.2.17. The process G, as we shall see later on, combines the interest rate process r, the rate of return b and the volatility σ to a very useful characteristic of our BS-model. The process also yields another important *discount* process D associated with the model, the process that is defined by

$$(42) \qquad D_t := \mathcal{E}_t \left(I^W(G^\top) \right) = \exp \left\{ I_t^W(G^\top) - \frac{1}{2} \int_0^t \|G^\top\|^2 \, ds \right\}$$

$$= \exp \left\{ \sum_{j=1}^n \int_0^t G_j \, dW_j - \frac{1}{2} \int_0^t \sum_{j=1}^n G_j^2 \, ds \right\}, \quad t \geq 0.$$

Hence, D is defined as the exponential of the stochastic integral $I^W(G^\top)$ whose quadratic variation is obviously given as $\int \|G^\top\|^2 \, ds$. Further define n-dimensional Itô processes $\widehat{W}$ and $\widehat{W}^{(T)}$ in $\mathrm{CSM}(P)$ by

$$(43) \qquad \widehat{W} := W - \int G \, ds, \qquad \widehat{W}^{(T)} := W - \int G \cdot \mathrm{I}_{[0,T]} \, ds, \quad T \geq 0.$$

Section 2.4 on **Girsanov calculus** provides mathematical tools that are able to recover deep and important properties of a general wealth process Y defined by (WE) and (27) equivalently. We shall summarize these techniques into:

(G1)	D is almost surely unique solution to $dD = D \, d\left(I^W(G^\top)\right)$, $D_0 \overset{as}{=} 1$.
(G2)	D is a P-martingale such that $ED_t = 1$ for all $t \geq 0$.
(G3)	For arbitrary $T \geq 0$ the probability measure $Q_T \sim P$ defined by $dQ_T = D_T \, dP$ is such that $\widehat{W}^{(T)}$ is an $(Q_T, \mathcal{F}_t)$-Wiener process.
(G4)	For arbitrary $T \geq 0$, N is a Q_T-(local) martingale iff $D^T N$ is a P-(local) martingale.
(G5)	For $S \leq T$ and $F_S \in \mathcal{F}_S$ arbitrary, $Q_T(F_S) = Q_S(F_S)$ holds.

The measure Q_T will be referred to as T-**Girsanov measure** for the Black-Scholes model (BS1), (BS2) and (BS3).

In (G2), (G3) and (G4) (local) P, Q_T-martingales and P, Q_T-Wiener processes, respectively, are understood as (local) $(P, \mathcal{F}_t)$, $(Q_T, \mathcal{F}_t)$-martingales and $(P, \mathcal{F}_t)$, $(Q_T, \mathcal{F}_t)$-Wiener processes, respectively. Agree also to denote by E_{Q_T} the (conditional) expectation operator associated by the Girsanov measure Q_T and keep the notation E for that associated with the original measure P.

Indeed, (G1) follows, for example, by 2.2.13. Hence, D is a nonnegative $(P, \mathcal{F}_t)$-local martingale. Its definition (42) and 2.4.7 imply that $ED_t \equiv 1$ since G is a bounded progressive process. Thus, (G2) is true according to 1.4.2 (d). (G3) is simply proved by (G2) and by Girsanov Theorem 2.4.4 itself.. Finally, Q_T is defined as a measure in the Girsanov set $\mathcal{G}(P)$ with the derivative process $D(Q_T|P)$ that equals almost surely to D^T. Accordingly, (G4) is exactly the statement (a) in 2.4.1. To prove (G5) apply (G2) and write

$$Q_S(F_S) = \int_{F_S} D_S \, dP = \int_{F_S} D_T \, dP = Q_T(F_S).$$

The above reasoning touched a **supermartingale calculus** to be found frequently in action in the financial mathematics. According to 1.4.2 (c) and 1.3.13 we have

(SP1)   Any nonnegative local martingale M is a supermartingale.
(SP2)   If M is a nonnegative supermartingale and τ_0 its bankruptcy time then outside a P-null set $\ \ t \geq \tau_0 \Rightarrow M(t) = 0$.

Note that we may apply 1.4.2 (c) as $\mathcal{F}_0$ is a trivial σ-algebra.

The following pair of assertions explains the rôle of our definitions:

3.2.1   **Lemma.** *For an arbitrary portfolio* $\pi = (\pi_1, \ldots, \pi_n)^\top$ *the formula*

(44) $$\int d \cdot \pi^\top \sigma \, d\widehat{W} \overset{as}{=} \int d \cdot \pi^\top \cdot (b - r \cdot 1_n) \, ds + \int d \cdot \pi^\top \cdot \sigma \, dW$$

holds.

3.2.2   **Theorem.** *Let* (π, c) *be a PC-strategy and* $y \geq 0$. *Denote by* Y *the wealth process generated by* (π, c, y). *Then*

(45)
$$M := d \cdot Y + \int d \cdot c \, ds \overset{as}{=} y + \int d \cdot \pi^\top \cdot (b - r \cdot 1_n) \, ds + \int d \cdot \pi^\top \cdot \sigma \, dW$$
$$\overset{as}{=} y + \int d \cdot \pi^\top \cdot \sigma \, d\widehat{W}$$

defines a process such that M^T *is a* $(Q_T, \mathcal{F}_t)$-*local martingale for arbitrary* $T \geq 0$.

The assertion, that in fact coincides with Girsanov theorem, provides the principal tool of the diffusion financial mathematics: Under the T-Girsanov measure Q_T the process

$$M = [\text{discounted current wealth}] + [\text{discounted cumulative consumption}]$$

has no drift. Under the original probability distribution P, this can happen only in very restricted BS-models with $b = r \cdot 1_n$, hence in models that provide no motivation to trade risky assets $1, \ldots, n$.

Proof of 3.2.1. Denoting the right hand side in (44) as U it follows by 2.2.19 that $U \overset{\text{as}}{=} \int d \cdot \pi^{\top} dZ$ where

$$Z := \int (b - r \cdot 1_n) \, ds + \int \sigma \, dW$$

is an n-dimensional Itô process with bounded coefficients. Further, by 2.2.18 and (BS3) we get $Z \overset{\text{as}}{=} \int \sigma \, dZ_{\text{ort}}$, where

$$Z_{\text{ort}} := \int \sigma^{-1} \, dZ = \int \sigma^{-1}(b - r \cdot 1_n) \, ds + W =: \widehat{W}.$$

Thus, $U \overset{\text{as}}{=} \int d \cdot \pi^{\top} dZ \overset{\text{as}}{=} \int d \cdot \pi^{\top} \cdot \sigma \, dZ_{\text{ort}}$ by 2.2.15.

Remark that the assertions 2.2.15, 2.2.18 and 2.2.19 may be applied as above, since all processes $r, b, \sigma, \sigma^{-1}$ are assumed to be bounded and progressive, the processes π, c are chosen in $\text{PM}_2(W)$ and the process d is continuous adapted. □

Proof of 3.2.2. The middle equality in (45) is just wealth equation (27), the last one coincides with (44). Fix a $T \geq 0$ and look at (45). According to 2.1.5 and 2.1.22,

$$M^T - y \overset{\text{as}}{=} \int I_{[0,T]} \cdot d \cdot \pi^{\top} \cdot \sigma \, d\widehat{W} \overset{\text{as}}{=} \int I_{[0,T]} \cdot d \cdot \pi^{\top} \cdot \sigma \, d\widehat{W}^{(T)}$$

holds as $\widehat{W} \overset{\text{as}}{=} \widehat{W}^{(T)}$ on $[0,T]$. Thus, M^T is a $(Q_T, \mathcal{F}_t)$-local martingale by (G3) and by 2.4.2 (c). □

The preceeding results enable to recognize a wealth process among continuous adapted processes. We recall the results made available by Section 2.5 on **Brownian representations calculus** on a unique existence of a solution φ to the equation

$$(46) \qquad H \overset{\text{as}}{=} H(0) + \int \varphi^{\top} \, dW, \qquad \varphi = (\varphi_1, \ldots, \varphi_n)^{\top} \in \text{PM}_2(W),$$

where H is a local P-martingale. Since $\mathcal{F}_t := (\mathcal{F}_t^W)^P$, 2.5.7 and 2.5.10, respectively, provide the following information:

(BR1)   Any P-martingale can be modified to a continuous P-martingale.

(BR2)   For any local P-martingale H the equation (46) has a solution φ that is unique on $\mathbb{R}^+ \times \Omega$ up to a $\lambda \otimes P$-null set.

3.2.3   Theorem. *Let c be a consumption process and M a continuous adapted process with $M(0) \overset{\text{as}}{=} y \geq 0$ such that DM is a local P-martingale. Then*

$$(47) \qquad Y := d^{-1} \cdot \left(M - \int d \cdot c \, ds \right)$$

defines a wealth process that is generated by (π, c, y) *where*

(48)
$$\pi^\top := D^{-1} \cdot (\varphi - DMG)^\top \cdot d^{-1} \cdot \sigma^{-1}$$

and $\varphi \in PM_2(W)$ *is defined* $\lambda \otimes P$-*uniquely on* $\mathbb{R}^+ \times \Omega$ *as a solution to the equation* $D \cdot M \overset{as}{=} y + \int \varphi^\top \, dW$ *according to (BR2).*

Proof. If $H := D \cdot M$, then $dH := \varphi^\top \, dW$. Apply (G1) and 2.2.15 to prove the relations

(49) $dD = D \cdot G^\top \, dW, \quad d\langle H, D \rangle = D \cdot \varphi^\top \cdot G \, ds \quad d\langle D \rangle = D^2 \cdot G^\top \cdot G \, ds.$

Then, the Itô formula 2.2.8 for $f(H, D) = \frac{H}{D} = M$ applied in $CSM(P)$ and (49) give

$$
\begin{aligned}
dM &= D^{-1} \, dH - \frac{H}{D^2} \, dD - \frac{1}{D^2} \, d\langle H, D \rangle + \frac{H}{D^3} \, d\langle D \rangle \\
&= D^{-1} \cdot \varphi^\top \, dW - \frac{H}{D^2} \cdot D \cdot G^\top \, dW - \frac{1}{D^2} D \cdot \varphi^\top \cdot G \, ds + \frac{H}{D^3} \cdot D^2 \cdot G^\top \cdot G \, ds \\
&= D^{-1} \cdot [\varphi - H \cdot G]^\top \, dW - D^{-1} \cdot [\varphi - H \cdot G]^\top \cdot G \, ds \\
&= D^{-1} \cdot [\varphi - H \cdot G]^\top \, \widehat{dW} = (\pi^\top \cdot d \cdot \sigma) \, \widehat{dW},
\end{aligned}
$$

where the last but one equality follows by 2.2.19 and π *is a portfolio given by (48). Hence, according to 2.1, the process* Y *defined by (47) is a wealth process generated by* (π, c, y). $\square$

For further purposes we will denote the space of all wealth processes by $\mathcal{W}$ and its sections in the following manner: If c is a consumption process, then

$$\mathcal{W}(c) := \{\, Y \text{ is a wealth process generated by a } (\pi, c, y), \quad \pi \in PM_2(W), y \geq 0 \,\};$$

if c is a consumption process and $y \geq 0$, then

$$\mathcal{W}(c, y) := \{\, Y \text{ is a wealth process generated by a } (\pi, c, y), \quad \pi \in PM_2(W) \,\}.$$

The genesis of wealth processes Y is described completely by the following statement:

3.2.4   Theorem. *Let* Y *be a continuous adapted process with* $Y(0) \overset{as}{=} y \geq 0$ *and* c *a consumption process. Denote*

(50)
$$M := d \cdot Y + \int d \cdot c \, ds.$$

Then the following statements are equivalent:

(a) $Y \in \mathcal{W}(c, y)$.

(b) M^T *is a* Q_T-*local martingale for all* $T \geq 0$.

(c) DM *is a local* P-*martingale.*

Moreover,

> (d)   If M^T is a Q_T-local martingale for a fixed $T > 0$, then there exists a wealth process $Y^w \in \mathcal{W}(c, y)$ such that $Y = Y^w$ holds on $[0, T]$ almost surely.

Proof. (a)$\Rightarrow$(b) is proved by 2.2. If M^T is a local Q_T-martingale for arbitrary $T \geq 0$, (G4) applies to prove that $(DM)^T = D^T M^T$ is a local P-martingale for arbitrary $T \geq 0$. This obviously implies (c). The implication (c)$\Rightarrow$(a) is a consequence of 2.3.

To prove (d) put

$$M^w := M \cdot I_{[0,T]} + D^{-1} \cdot D(T) \cdot M(T) \cdot I_{(T,\infty)}$$

to define a continuous adapted process such that $D \cdot M^w = (D \cdot M)^T$ holds. Thus, if M^T is a local Q_T-martingale, it follows by (G4) that $D \cdot M^w$ is a local P-martingale. Denoting $Y^w := d^{-1} \cdot (M^w - \int d \cdot c \, ds)$ it is a consequence of the implication (c)$\Rightarrow$(a) that Y^w is a wealth process in $\mathcal{W}(c, y)$, obviously such that $Y^w \overset{as}{=} Y$ on $[0, T]$. $\square$

3.2.5   Example. Let c be a consumption process and $y \geq 0$. Endow both $\mathcal{W}(c, y)$ and the space $y + \mathrm{CM}_{loc}(P)$ of all local martingales that start from y by the almost sure equality on $\mathbb{R}^+$. Since M is a continuous adapted process iff its transformation $D^{-1} \cdot M$ has the property, we may combine 2.3 and $(a) \iff (c)$-part of 2.4 to state that

$$\mathcal{H}(Y) = D \cdot \left[d \cdot Y + \int d \cdot c \, ds \right], \quad Y \in \mathcal{W}(c, y)$$

is a one to one map $\mathcal{W}(c, y) \to y + \mathrm{CM}_{loc}(P)$.

If we provide the space of all portfolios $\mathrm{PM}_2(W)$ with the almost everywhere equality on $\mathbb{R}^+ \times \Omega$ with respect to the measure $\lambda \otimes P$, we observe that the map $\mathcal{H}$ yields through (BR2) and 2.3 a dual one to one transformation $h : \mathrm{PM}_2(W) \to \mathrm{PM}_2(W)$ defined by

$$\mathcal{H}(Y) = y + \int h(\pi) \, dW \quad \text{or equivalently as} \quad h(\pi)^\top = d \cdot D \cdot \pi^\top \cdot \sigma + \mathcal{H}(Y) \cdot G^\top$$

for $\pi \in \mathrm{PM}_2(W)$, where Y is the wealth process generated by (π, c, y).

Now we are prepared to fulfil our promise made in Section 3.1 to prove that any admissible PC-strategy generates a wealth process Y that will never recover after its bankruptcy τ_0.

3.2.6   Admissible Strategies. Let Y be a wealth process generated by (π, c, y) and τ_0 its bankruptcy time; $T > 0$. Then (π, c) is an admissible strategy up to time T for the initial wealth y if and only if there is a P-null set N such that for $\omega \notin N$

$$(51) \qquad\qquad \tau_0(\omega) \leq T \quad \Rightarrow \quad Y(\omega) = 0 \quad \text{on} \quad [\tau_0(\omega), T].$$

Hence, a PC-strategy (π, c) is admissible for a $y \geq 0$ iff $Y^{\tau_0} \overset{as}{=} Y$ holds for the wealth process Y generated by (π, c, y).

Proof. Assume that $Y(t) \geq 0$ almost surely for $0 \leq t \leq T$. If M is the process defined by (45) then M^T is, without loss of generality, a nonnegative local Q_T-martingale by 2.2. Thus, M^T is a Q_T-supermartingale according to (SP1). Since $\int d \cdot c\, ds$ is a non decreasing process, we conclude that $(d \cdot Y)^T$ is a Q_T-supermartingale, too. The processes Y^T and $(d \cdot Y)^T$ have obviously the same bankruptcy times and therefore the implication $(\Rightarrow)$ in (51) follows by (SP2). The rest of 2.6 is obvious. $\square$

Having a finite trading expiration time T the investor will, beyond any doubt, judge a plausibility of any single PC-strategy according to the final capital $Y(T)$, that may be attained by its application, compared with an initial endowment $Y(0) = y$. More precisely, claiming a final wealth w_T he searches for a wealth process Y that would respect a given consumption c, whose value at time T would be equal to the claim w_T and that, at the same time, would require as small initial endowment $Y(0) = y$ as possible.

In this respect our definitions will be as follows: Let $T > 0$ be a trading expiration time.

- A nonnegative random variable $w_T \in \mathcal{F}_T$ will be called a **claim** at time $T > 0$. Saying that w_T is a claim we shall always mean that T is a positive and finite time and that w_T is a claim in $\mathcal{F}_T$.

- A wealth process Y will be called a **valuation of a claim** w_T if $Y(T) \overset{as}{=} w_T$ holds. We may also say that Y valuates the claim w_T.

- A valuation of a claim w_T will be said an **admissible valuation of the claim** if $Y(t) \geq 0$ holds almost surely for all $0 \leq t \leq T$.

- The **minimal price** of a claim w_T and of a consumption c, or simply of (c, w_T), is defined as the minimum of all $y \in \mathbb{R}^+$ such that there exists an admissible valuation $Y \in \mathcal{W}(c)$ of w_T with the initial endowment y. If it exists, the minimal price will be denoted as $y(c, w_T)$.

In symbols,

$$y(c, w_T) := \min \left\{ y \in [0, \infty) : y \text{ is the initial endowment of an } Y \text{ in } \mathcal{W}^+(c, w_T) \right\},$$

where

$$\mathcal{W}^+(c, w_T) := \{ Y \in \mathcal{W}(c), \ Y \text{ is an admissible valuation of } w_T \}.$$

- If the minimal price $y(c, w_T)$ exists, then any process $Y \in \mathcal{W}^+(c, w_T)$ with the initial endowment $y(c, w_T)$ will be called a **minimal admissible valuation** of the claim w_T and of the consumption c or simply a minimal admissible valuation of (c, w_T).

Next three statements are the fundamental achievements of diffusion financial mathematics.

3.2.7   Valuations. *Consider a consumption process c and a claim w_T. Assume that there exists a valuation $Y \in \mathcal{W}(c)$ of w_T such that*

$$(52) \qquad d(t) \cdot Y(t) + \int_0^t d \cdot c\, ds \geq 0 \quad \text{holds almost surely for all} \quad 0 \leq t \leq T.$$

Then

$$(53) \qquad q := q(c, w_T) := E_{Q_T}\left(d(T) \cdot w_T + \int_0^T d \cdot c\, ds\right) < \infty,$$

in particular, w_T and $\int_0^T d \cdot c\, ds$ are Q_T-integrable random variables. Further denote

$$(54) \quad V(t) := V(t, c, w_T) := E_{Q_T}^{\mathcal{F}_t}\left[d(T) \cdot d(t)^{-1} \cdot w_T + \int_t^T d(s) \cdot d(t)^{-1} \cdot c(s)\, ds\right],$$

for $t \geq 0$ and assume that $Y \in \mathcal{W}(c)$ is a valuation of w_T that satisfies (52). Then

$$Y(t) \geq V(t) \quad \text{holds almost surely for all } 0 \leq t \leq T, \quad [\Rightarrow Y(0) \geq q \text{ a.s.}]$$

and $Y(t) \overset{as}{=} V(t)$ for all $t \in [0, T]$ if and only if $Y(0) \overset{as}{=} q$.

Note that $d(s) \cdot d(t)^{-1} = \exp\{-\int_t^s r\, du\}$ if $s \geq t$ and that the conditional expectation (54) exists by (53) as d^{-1} is a bounded process on $[0, T]$.

Proof. Note first that $V(0) \overset{as}{=} q$, and $V(T) \overset{as}{=} w_T$ and that

$$(55) \qquad d(t) \cdot V(t) + \int_0^t d \cdot c\, ds \overset{as}{=} E_{Q_T}^{\mathcal{F}_t}\left[d(T) \cdot w_T + \int_0^T d \cdot c\, ds\right]$$

holds for arbitrary $t \geq 0$ according to 1.2.7 (c). Assume that $Y \in \mathcal{W}(c)$ is a valuation of w_T such that (52) is satisfied. Since $(d \cdot Y + \int d \cdot c\, ds)^T$ is in this case a nonnegative local Q_T-martingale by 2.2, it is a Q_T-supermartingale by (SP1), especially (53) is valid. Hence,

$$(56) \qquad d(t) \cdot Y(t) + \int_0^t d \cdot c\, ds \geq E_{Q_T}^{\mathcal{F}_t}\left[d(T) \cdot Y(T) + \int_0^T d \cdot c\, ds\right]$$

$$= E_{Q_T}^{\mathcal{F}_t}\left[d(T) \cdot w_T + \int_0^T d \cdot c\, ds\right] = d(t) \cdot V(t) + \int_0^t d \cdot c\, ds$$

holds almost surely for all $0 \leq t \leq T$ by (55). We conclude that for any $t \in [0, T]$ the inequality $Y(t) \geq V(t)$ is valid almost surely.

If $Y(0) \overset{as}{=} q$ then $E_{Q_T}\left(d(t) \cdot Y(t) + \int_0^t d \cdot c\, ds\right) = q$ for both $t = 0$ and $t = T$ and therefore the Q_T-supermartingale $(d \cdot Y + \int d \cdot c\, ds)^T$ turns to be a Q_T-martingale and we get (56) as equality that holds almost surely for all $0 \leq t \leq T$. $\square$

The theorem sends an important message:

3.2.8 Admissible Valuations. *Let w_T be a claim, c a consumption process and $Y \in \mathcal{W}(c)$ a valuation of w_T. Then*

(a) *Y is an admissible valuation of w_T if and only if condition (52) is satisfied.*

If $Y \in \mathcal{W}(c)$ is an admissible valuation of w_T with an initial endowment $y \geq 0$, then $y \geq q(c, w_T)$. Moreover,

    (b) *$y = q$ if and only if M^T is a Q_T- martingale and if and only if Y^T is a modification of the process V^T.*

    (c) *$y > q$ holds if and only if M^T is an integrable local Q_T-martingale and it is not a true Q_T-martingale.*

The processes M and $V := V(c, w_T)$ in (a) and (b) are defined by (50) and (54), respectively and $q := q(c, w_T) \in \mathbb{R}^+$ is the number specified by (53).

Proof. If a wealth process $Y \in \mathcal{W}(c)$ satisfies (52) and valuates the claim w_T, then $Y(t) \geq V(t)$ almost surely for $0 \leq t \leq T$ by 2.7. Hence, Y is an admissible valuation of w_T.

If Y is an admissible valuation of w_T and $Y(0) \stackrel{as}{=} y \geq 0$, then $Y(t) \geq V(t)$ almost surely for $0 \leq t \leq T$ by 2.7 and therefore $y \geq q(c, w_T)$.

If $y = q(c, w_T)$, then $Y(t) \stackrel{as}{=} V(t)$ for $t \leq T$ according to 2.7, again, consequently

$$(57) \qquad M(t) \stackrel{as}{=} E_{Q_T}^{\mathcal{F}_t}\left(d(T) \cdot w_T + \int_0^T d \cdot c\, ds\right), \quad 0 \leq t \leq T$$

by (55) and M^T is seen to be a Q_T-martingale.

If M^T is a Q_T-martingale then (57) holds because $w_T \stackrel{as}{=} Y(T)$. Putting $t = 0$ into (57) we get $y = q(c, w_T)$. We have proved (b).

To prove (c) note that M^T is a nonnegative local Q_T-martingale by 2.4, consequently an integrable stochastic process by (SP1). This obviously implies that (b) and (c) are equivalent statements. $\square$

This result is substantially complemented by the following deep theorem:

3.2.9 Minimal Admissible Valuations. *Let w_T be a claim and c a consumption process. Then the minimal price $y(c, w_T)$ exists if and only if (53) holds.*

If the condition (53) is satisfied and Y is a minimal admissible valuation of (c, w_T) then

$$y(c, w_T) = q(c, w_T) \quad \text{and} \quad Y(t) \stackrel{as}{=} V(t, c, w_T), \quad 0 \leq t \leq T$$

hold.

In particular, if Y and Y^* are minimal admissible valuations of (c, w_T) generated by a (π, c, y) and by a (π^*, c, y), respectively, then $Y \stackrel{as}{=} Y^*$ on $[0, T]$ and $\pi = \pi^*$ almost everywhere on $[0, T] \times \Omega$ with respect to the measure $\lambda \otimes P$.

Proof. If the minimal price $y(c, w_T)$ exists then $y(c, w_T) < \infty$ by definition. Further, $y(c, w_T) \geq q(c, w_T)$ by 2.8, hence the existence of $y(c, w_T)$ implies (53).

Assume (53): According to (55) the equality

$$N(t) := d(t) \cdot V(t) + \int_0^t d \cdot c \, ds \overset{\text{as}}{=} E_{Q_T}^{\mathcal{F}_t} \left[d(T) \cdot w_T + \int_0^T d \cdot c \, ds \right], t \geq 0$$

is valid and therefore N is a Q_T-martingale. It follows by (G4) that $D^T \cdot N$ is a P-martingale, hence N has a continuous adapted modification, say M, by (BR1). Define a continuous adapted process Y as a solution to the equation (50). Note that $Y(0) \overset{\text{as}}{=} q(c, w_T)$, $Y(T) \overset{\text{as}}{=} w_T$ and that M^T is a Q_T-martingale as M^T is a modification of the process N^T. We further reason as follows:

According 2.4 (d) there is a wealth process $Y^w \in \mathcal{W}(c, q(c, w_T))$ such that $Y^w \overset{\text{as}}{=} Y$ on $[0, T]$ holds. Consequently,

$$d(t) \cdot Y^w(t) + \int_0^t d \cdot c \, ds \overset{\text{as}}{=} N(t) \geq 0 \quad \text{is true for all } 0 \leq t \leq T$$

and $Y^w \in \mathcal{W}(c, q(c, w_T))$ is an admissible valuation of w_T by 2.8 (a).

Since, according to 2.8, $q(c, w_T)$ is a lower bound of all $y \geq 0$ such that there is an admissible valuation of w_T with $Y(0) \overset{\text{as}}{=} y$, we have proved that the minimal price $y(c, w_T)$ exists and is equal to $q(c, w_T)$. If Y is an arbitrary minimal admissible valuation of (c, w_T), then $Y(0) \overset{\text{as}}{=} y(c, w_T) = q(c, w_T)$ and $Y(t) \overset{\text{as}}{=} V(t)$ for $0 \leq t \leq T$ by the last statement of 2.7.

The uniqueness part follows by 1.8. $\square$

The statements 2.7, 2.8 and 2.9 combined may be summarized to provide a check for a process to be a minimal admissible valuation and a method how to compute the minimal price. Also the term *the minimal admissible valuation* finds its justification:

3.2.10   A Summary. *Let w_T and c be a claim and a consumption process, respectively:*

(a)   *The minimal price $y(c, w_T)$ exists if and only if $q(c, w_T) < \infty$, $y(c, w_T) = q(c, w_T)$ in this case.*

(b)   *A valuation $Y \in \mathcal{W}(c)$ of w_T is a minimal admissible valuation of (c, w_T) if and only if $d \cdot Y + \int d \cdot c \, ds$ is a Q_T-martingale on $[0, T]$.*

Let $Y \in \mathcal{W}(c)$ be an admissible valuation of w_T. Then

(c) $Y(t) \geq V(t, c, w_T)$ *almost surely for $0 \leq t \leq T$.*

(d)   *Y is a minimal admissible valuation of (c, w_T) if and only if $Y(0) \overset{\text{as}}{=} q(c, w_T)$ and if and only if $Y(t) \overset{\text{as}}{=} V(t, c, w_T)$ for $0 \leq t \leq T$.*

(e) $Y(0) > q(c, w_T)$ *almost surely if and only if $[d \cdot Y + \int d \cdot c \, ds]^T$ is a local Q_T-martingale and not a true Q_T-martingale.*

Make it an exercise to prove the statements included to our summary.

The equivalence **(e)** in 2.10 creates through a mathematical pathology a possibility for a really suicide trading, indeed.

3.2.11 Example and Exercise. A Safe Way to the Bankruptcy. Let us consider a market where only one bond and one stock are traded until the expiration time $T = 1$ and fix a consumption process c.

Since the process $\widehat{W}^{(1)}$ defined by (43) is a $(Q_1, \mathcal{F}_t)$-Wiener process, the construction performed in 2.3.7 exhibits a Markov time $\tau < 1$ such that outside a P-null set

$$(58) \qquad \int F \, d\widehat{W}^{(1)} \geq -1, \quad \text{on } \mathbb{R}^+ \quad \text{and} \quad \int F \, d\widehat{W}^{(1)} = -1 \quad \text{on } [\tau, \infty),$$

$$\text{where} \quad F(s) := \frac{1}{1-s} \cdot \mathrm{I}_{[0,\tau]}(s), \quad s \geq 0$$

holds to be true. The integral $\int F \, d\widehat{W}^{(1)}$, being an integrable local Q_1-martingale and not a true Q_1-martingale, enters through (c) in 2.8 to arrange for the bankruptcy with probability one:

For an arbitrary $K \geq 0$ define a portfolio π^K by

$$(59) \qquad\qquad \pi^K(s) := K \cdot \frac{F(s)}{d(s) \cdot \sigma(s)}, \quad s \geq 0.$$

Consider an initial endowment $y \geq 0$ and let Y be the wealth process generated by (π^K, c, y). Verify the following statements:

(a) The wealth equation for Y is

$$(60) \qquad\qquad d \cdot Y + \int d \cdot c \, ds \overset{\text{as}}{=} y + K \cdot \int F \, d\widehat{W}^{(1)}$$

and therefore

$$d \cdot Y + \int c \cdot d \, ds \geq y - K \quad \text{on } \mathbb{R}^+ \quad \text{and} \quad d \cdot Y + \int d \cdot c \, ds = y - K \quad \text{on } [\tau, \infty)$$

hold outside a P-null set.

(b) (π^K, c) is an admissible strategy up to time 1 for an initial endowment $y \geq 0$ (briefly admissible) iff $\int_0^1 d \cdot c \, ds \leq y - K$ is valid almost surely.

In particular, if $y = K$ and (π^K, c) is an admissible strategy, then $c \overset{\text{ae}}{=} 0$ on $[0, 1]$ and (π^K, c) is not admissible for $y < K$. In other words, if $y < K$, then there is no consumption process c that could be paired with the portfolio π^K to make a PC-strategy that would be admissible up to time 1 with the initial endowment y.

Apply **(a)** and **(b)** to construct a really bad financial operation:

(c) Choosing to trade without any consumption, i.e. with $c = 0$, then for arbitrary initial investment $K \geq 0$ the portfolio π^K is admissible up to time 1 for the initial endowment K and forces the investor to go bankrupt with probability one not later than at time $T = 1$. According to **(a)**, the bankruptcy time τ_0 of the wealth process Y generated by $(\pi^K, 0, K)$ equals to the Markov time τ that enters the definition of the process F.

Thus, the portfolio (59) provides a very safe way how to loose arbitrarily large amount of money. Look how this is accomplished if for simplicity we assume that the interest rate and the volatility are given as $r = 0$ and $\sigma = 1$, respectively:

You need only to keep stubbornly increasing the capital invested to the stock with the rate $K \cdot (1 - t)^{-2}$ at any time t before you go bankrupt without any regards to the dynamics of the price process $X_1(t)$. Obviously, this trading can not be performed without a to infinity increasing negative position for the bond given according to (a) almost surely as

$$\phi_0(t) = \frac{Y(t) - \pi^K(t)}{X_0(t)} = \frac{K}{x_0} \cdot \left(\int_0^t F \, d\widehat{W}^{(1)} - \frac{t}{1-t} \right) \to \frac{K}{x_0(\tau - 1)}$$

if $t \to \tau^-$. Thus, a huge short sale of shares of the bond is necessarily involved.

We continue to assume that $r = 0$, $\sigma = 1$ and $c = 0$ and note that (π^K, c, K) generates the wealth process $Y = K + K \cdot \int F \, d\widehat{W}^{(1)} \in \mathcal{W}(c)$ that is an admissible valuation of the claim $w_1 = 0$ with the initial endowment $y = K$ and is not a Q_1-martingale on $[0,1]$. Trivially, $y(c, w_1) = 0$ and you would do much more better to skip your investments completely until $T = 1$ advised by the fact that the minimal admissible valuation of (c, w_1) is a wealth process whose trajectories are constant zero on $[0,1]$ generated by (π, c, y) where $\pi \cdot I_{[0,1]} = 0$. In this case the initial wealth K could be better invested elsewhere rather than to get it wasted by the suicide portfolio π^K.

According to 2.10 the problem of the minimal price of (c, w_T) when $c = 0$ reduces itself to the problem of finding an admissible valuation Y such that the discounted process $d \cdot Y$ is a Q_T-martingale on $[0,T]$. The important aspect of the BS-price model is that the price process X_i meets the requirement for arbitrary risky security $1 \le i \le n$.

3.2.12. *If only one bond and one stock are traded and $T > 0$ is arbitrary, then*

(a) $D \cdot d \cdot X_1 \overset{\text{as}}{=} x_1 \cdot \mathcal{E}\left(I^W((r - b)\sigma^{-1} + \sigma) \right)$ *is a P-martingale.*

(b) $d \cdot X_1$ *is a Q_T-martingale on $[0,T]$.*

(c) X_1^T *is a process in $CSM(Q_T)$ such that*

$$X_1(t) \overset{\text{as}}{=} x_1 + \int_0^t r \cdot X_1 \, ds + \int \sigma \cdot X_1 \, dW^{(T)}$$

holds almost surely for all $0 \le t \le T$.

Proof. The formula (a) follows by a direct computation. The right hand side is a P-martingale by 2.4.6 as the processes r, b and σ are bounded. It follows by (G4) that $d \cdot X_1$ is a Q_T-martingale on $[0,T]$ for arbitrary $T > 0$. It follows by (6) that X_1 is a continuous P-semimartingale with the stochastic differential $dX_1 = (b \cdot X_1) \, ds + (\sigma \cdot X_1) \, dW$ and therefore by (44) in 2.1

$$dX_1 = (r \cdot X_1) \, ds + (\sigma \cdot X_1) \, d\widehat{W}$$

holds. Since $\widehat{W} \stackrel{\text{as}}{=} \widehat{W}^{(T)}$ on $[0,T]$, we arrive to the formula

$$X_1^T \stackrel{\text{as}}{=} x_1 + \int rX_1 \cdot \mathrm{I}_{[0,T]}\, ds + \int \sigma X_1 \cdot \mathrm{I}_{[0,T]}\, dW^{(T)}$$

that is equivalent to **(c)** because $W^{(T)}$ is a Q_T-Wiener process by **(G3)**. □

More generally we may prove:

3.2.13. *For any stock* $1 \leq i \leq n$ *and* $T > 0$

$$(d \cdot X_i)^T \stackrel{\text{as}}{=} x_i \cdot \mathcal{E}\left(I^{\widehat{W}^{(T)}}(\sigma_i \cdot \mathrm{I}_{[0,T]})\right), \quad \sigma_i := (\sigma_{i1}, \ldots, \sigma_{in}).$$

Hence, by 2.4.7, $d \cdot X_i$ is a Q_T-martingale on $[0,T]$ as σ_i is a bounded process.

Proof. As follows from 1.11, X_i is a wealth process generated by $(\pi^{(i)}, c, y)$ where $\pi^{(i)} := X_i \cdot e_i$ and e_i the i-th unit vector, $c = 0$, $y = x_i$. Consequently, according to 2.2, **(G3)** and by 2.4.2 **(c)**, the corresponding wealth equation is given on the interval $[0,T]$ as

$$(d \cdot X_i)^T \stackrel{\text{as}}{=} x_i + \int (d \cdot X_i)^T \cdot \sigma_i^\top \cdot \mathrm{I}_{[0,T]}\, d\widehat{W}^{(T)},$$

where $\widehat{W}^{(T)}$ is understood as a Q_T-Wiener process. Hence,

$$(d \cdot X_i)^T \stackrel{\text{as}}{=} x_i + \int (d \cdot X_i)^T\, dL \text{ where } L := \int \sigma_i^\top \cdot \mathrm{I}_{[0,T]}\, d\widehat{W}^{(T)},$$

holds by 2.2.15. An application of 2.2.13 shows that $(d \cdot X_i)^T \stackrel{\text{as}}{=} x_i \cdot \mathcal{E}(L)$. □

3.2.14 Example and Exercise. Consider the Black-Scholes model for one bond and one stock with

$$r = 0, \quad x_0 = 1, \quad x_1 = 1, \quad \sigma = 1, \quad b \in \mathbb{R} \quad \text{a constant,}$$

$K > 0$ and further assume no consumption, i.e., $c = 0$. Imagine that you are offered to sign a very exotic contract to pay at time $t = 0$ an agreed price $y \geq 0$ to collect at time $T > 0$ a total given as

$$w_T := 1 + K \cdot (1 - X_1(\tau_0 \wedge T)) \text{ where } \tau_0 := \inf\left\{t \geq 0 : X_1(t) = 1 + K^{-1}\right\}.$$

In other words, for the price y paid at $t = 0$ you will get nothing if the price $X_1(t)$ enters $1+K^{-1}$ some time before T and you will collect $0 < 1+K \cdot (1-X_1(T)) \leq 1+K$ if not. What would be a fair price y to pay for the claim or rather option w_T? The answer is simple and involves the concepts of the minimal price and that of the minimal admissible valuation that exist in this case, since $E_{Q_T} w_T < \infty$:

If $y < y(0, w_T)$ and if you are allowed to trade only in the framework of admissible portfolios you would not hesitate to sign the contract, since there is no legal possibility how to valuate w_T by a wealth process $Y \in \mathcal{W}^+(0, w_T)$ with initial endowment y. If $y > y(0, w_T)$ the contract should be dismissed because the minimal admissible valuation of $(0, w_T)$ is less expensive and the rest $y - y(0, w_T)$ of the required price might be invested elsewhere to replicate the total profit.

Prove that $y(0, w_T) = 1$ and that the minimal admissible valuation of $(0, w_T)$ is given as Y^{τ_0} where

$$Y(t) := 1 + K - KX_1(t) = 1 + K - K \exp\left\{\left(b - \frac{1}{2}\right)t + W(t)\right\}, \quad t \geq 0.$$

Recall 1.14 to see that Y is a wealth process whose bankruptcy time coincides with τ_0 and $Y^{\tau_0} \in \mathcal{W}(0)$ is an admissible valuation of $(0, w_T)$. You will also find there a formula for $P[\tau_0 \leq T]$ that provides an information about the risk born by the contract.

The interpretation of the minimal price as a fair price for a claim or option will be thoroughly revisited in the next section.

Via 2.10 we have also solved the problem of the existence of an admissible PC-strategy up to a given finite expiration time for a given consumption process and a given initial capital.

3.2.15 The Existence of Admissible Strategies. *Let c be a consumption process, $T > 0$ and $y \geq 0$. Then*

(61)
$$E_{Q_T} \int_0^T d(s) \cdot c(s)\, ds \leq y$$

is a necessary and sufficient condition for the existence of a PC-strategy (π, c) admissible up to time T for the initial endowment y.

Proof. Assume first that (π, c) is an admissible strategy up to T for y and denote by Y the wealth process generated by (π, c, y). Consider $q := q(c, w_T)$ and $V := V(c, w_T)$ defined by (53) and (54), respectively, where $w_T := Y(T)$. Then, according to 2.10 (c), $E_{Q_T} \int_0^T d \cdot c\, ds \leq q \leq y < \infty$ holds.

On the contrary assume (61) and put $w_T := d(T)^{-1} \cdot (y - E_{Q_T} \int_0^T d \cdot c\, ds)$. Thus,

$$E_{Q_T}\left(d(T) \cdot w_T + \int_0^T d \cdot c\, ds\right) = y < \infty.$$

According to 2.10 (a), there exists an admissible valuation $Y \in \mathcal{W}(c)$ of the claim w_T with the initial endowment y. Consequently, an admissible PC-strategy up to time T for the initial capital y is proved to exist. □

PC-strategies admissible up to infinity may frequently cease to exist.

3.2.16 Exercise. Let c be a consumption process and $y \geq 0$. Then an admissible strategy (π, c) for the initial endowment y exists iff there is a process L with the following properties:

$$(62) \qquad L \in y + \mathrm{CM}_{loc}(P) \quad \text{and} \quad L \geq D \cdot \int d \cdot c \, ds \quad \text{almost surely.}$$

Prove:

    (a) If d and c are deterministic processes such that $\int_0^\infty d \cdot c \, ds < \infty$ and $y \geq 0$ arbitrary, then there is an admissible strategy (π, c) for y.

    (b) If an admissible strategy (π, c) for y exists then $E(D_t \cdot \int_0^t d \cdot c \, ds) \leq y$ holds for all $t \geq 0$.

    (c) If there is an $\epsilon > 0$ such that $d \cdot c \geq \epsilon$ holds on $\mathbb{R}^+$ with a positive probability, then there is no admissible strategy (π, c) whatever large initial capital y we may have available.

3.2.17 Exercise. Assume (61) for a consumption process c and a $T > 0$, put $w_T := d(T)^{-1}(y - E_{Q_T} \int_0^T d \cdot c \, ds)$ as in the proof of 2.15. Compute explicitly the minimal valuation Y of (c, w_T).

3.2.18 Exercise. Consider a utility function $u : \mathbb{R}^+ \times \mathbb{R} \to \mathbb{R}^+$ that is Borel measurable and such that

$$u(t, a) \leq u(t, b) \quad \text{for all} \quad t \geq 0, \quad a \leq b.$$

For $y \geq 0$ and $T > 0$ define

$$U(y) := \sup \left\{ E \int_0^T u(s, c(s)) \, ds, \quad c \in \mathcal{C}(y) \right\},$$

where $\mathcal{C}(y)$ is the set of all consumption processes c such that a portfolio π exists to form a strategy (π, c) that is admissible up to T for the initial endowment y. Prove that the maximal possible utility $U(y)$ is given as

$$U(y) = \sup \left\{ E \int_0^T u(s, c(s)) \, ds, \quad c \in \mathcal{D}(y) \right\},$$

where $\mathcal{D}(y)$ is the set of all consumption processes c with $E_{Q_T} \int_0^T d \cdot c \, ds = y$.

See, [115], [63] (pp. 160-161) and [95] (pp. 379-387) for constructions of a consumption process c^* that is optimal in the sense of $U(y) = E \int_0^T u(s, c^*(s)) \, ds$ where $U(s, x) = d(s) \cdot u(x)$ and $u(x)$ is a smooth concave function on $\mathbb{R}^+$.

In 2.3, 2.4, 2.5, 2.6 and in 2.16 we succeeded in formulating the corresponding results in terms of the original probability distribution P rather than to recall an associated Girsanov measure $Q_T \sim P$. The reader might also prefer the principal results on the minimal admissible valuations and the minimal price of a claim to be handled analogously. We shall do it now adding to our reasoning an integrability requirement on the claim under consideration.

3.2.19 Integrable Claims. *Let w_T be a claim and c a consumption process such that for some $\alpha > 1$*

$$(63) \qquad E\left[w_T + \int_0^T c\, ds \right]^\alpha < \infty$$

holds. Then

$$(64) \qquad E\left\{ \frac{D_T}{D_t} \cdot \left[d(T) \cdot w_T + \int_0^T d \cdot c\, ds \right] \right\} < \infty, \quad 0 \le t \le T$$

and the minimal price of (c, w_T) is

$$(65) \qquad y(c, w_T) = E\left\{ D_T \cdot \left(d(T) \cdot w_T + \int_0^T d \cdot c\, ds \right) \right\}.$$

If Y is a minimal admissible valuation of (c, w_T), then $Y(t) \overset{as}{=} V(t, c, w_T)$ for each $0 \le t \le T$ by 2.10 (d), and

$$(66) \quad V(t, c, w_T) \overset{as}{=} E^{\mathcal{F}_t}\left\{ \frac{D_T}{D_t \cdot d(t)} \cdot \left[d(T) \cdot w_T + \int_0^T d \cdot c\, ds \right] \right\} - \frac{1}{d(t)} \int_0^t d \cdot c\, ds$$

holds for all $0 \le t \le T$.

We shall need the following

3.2.20 Lemma. *For any $0 \le t \le T < \infty$ and $\beta \in \mathbb{R}$ the random variable $D_T^\beta \cdot D_t^{-\beta}$ has a finite expectation.*

Proof. Put $F := \beta \cdot I_{(t, \infty)} \cdot G \in PM_2(W)$ and apply 2.1.9 to compute that

$$D_T^\beta \cdot D_t^{-\beta} = \exp\left\{ \beta \cdot \int_t^T G^\top\, dW - \frac{\beta}{2} \int_t^T \|G^\top\|^2\, ds \right\}$$

$$= \exp\left\{ \int_0^T F^\top\, dW - \frac{1}{2} \int_0^T \|F^\top\|^2\, ds + \frac{\beta^2}{2} \cdot \int_t^T \|G^\top\|^2\, ds - \frac{\beta}{2} \cdot \int_t^T \|G^\top\|\, ds \right\}$$

$$= \mathcal{E}_T\left(I^W(F^\top) \right) \cdot \exp\left\{ \frac{\beta \cdot (\beta - 1)}{2} \cdot \int_t^T \|G^\top\|^2\, ds \right\} \le \mathcal{E}_T\left(I^W(F^\top) \right) \cdot K$$

holds almost surely for a constant $K < \infty$ because G is a bounded process. Since $\langle I^W(F^\top) \rangle(T) = \beta^2 \cdot \int_t^T \|G^\top\|^2\, ds$, it follows by 2.4.7 that $E\mathcal{E}_T\left(I^W(F^\top) \right) = 1$. The integrability of $D_T^\beta \cdot D_t^{-\beta}$ is proved. $\square$

Proof of 2.19. Denote

$$v_T := d(T) \cdot w_T + \int_0^T d \cdot c\, ds$$

and define $\beta > 0$ by $\frac{1}{\alpha} + \frac{1}{\beta} = 1$. Consequently, by Hölder inequality, 2.20 and (G3)

(67) $E\left(\dfrac{D_T}{D_t} \cdot v_T\right) \leq \left(E(D_T \cdot D_t^{-1})^\beta\right)^{\frac{1}{\beta}} \cdot \left(E(v_T)^\alpha\right)^{\frac{1}{\alpha}} < \infty, \quad 0 \leq t \leq T$

is seen to be true. Hence, (64) is proved. It follows that

$$q(c, w_T) := E_{Q_T} v_T = E D_T \cdot v_T < \infty.$$

The minimal price $y(c, w_T)$ exists and (65) is true by 2.10 (a).

Further, denoting $V := V(c, w_T)$, it follows by (55) that $d \cdot V + \int d \cdot c\, ds$ is a Q_T-martingale on $[0, T]$ and therefore $D \cdot \left(d \cdot V + \int d \cdot c\, ds\right)$ is a P-martingale on the interval by (G4). This yields

$$D_t\left(d(t) \cdot V(t) + \int_0^t d \cdot c\, ds\right) \overset{\text{as}}{=} E^{\mathcal{F}_t}\left(D_T \cdot \left[d(T) \cdot V(T) + \int_0^T d \cdot c\, ds\right]\right)$$

$$\overset{\text{as}}{=} E^{\mathcal{F}_t}\left(D_T \cdot \left[d(T) \cdot w_T + \int_0^T d \cdot c\, ds\right]\right).$$

According to 1.2.7 (c) we arrive at

(68) $d(t) \cdot V(t) + \displaystyle\int_0^t d \cdot c\, ds \overset{\text{as}}{=} E^{\mathcal{F}_t}\left(\dfrac{D_T}{D_t} \cdot \left[d(T) \cdot w_T + \int_0^T d \cdot c\, ds\right]\right),$

since (64) implies that the argument of the conditional expectation on the right hand side is an integrable random variable. Computing V as the solution to (68) with an assistance of 1.2.7 (c) again, we get (66). $\square$

Two Remarks. The Girsanov calculus enters the financial mathematics principally through the probability measures $Q_T \sim P$ whose purpose is to remove the drift that is present in the definition of a general wealth process. There is no harm as far as a finite expiration time $T > 0$ is concerned, as Q_T and the original distribution P bear the same random events of probability one and the Radon-Nikodym Theorem provides a reliable bridge between the Q_T and P- stochastic worlds.

The problem of the time horizon $T = \infty$ can not be easily overcome. Even though we are able in some special cases (see 2.4.9) to define uniquely a probability measure Q that coincides with any Q_T on the σ-algebra $\mathcal{F}_T^W$, we will not get a very neat mathematical model. The reasons are that it may happen that Q and P are singular measures, see 2.4.10, and therefore two different filtrations $(\mathcal{F}_t^W)^Q$ and $(\mathcal{F}_t^W)^P$ may enter our model with no purpose for financial modeling.

Compared with the standard terminology we have slightly simplified the definition of the contingent claim that is generally understood as an financial instrument (w_T, h) where w_T, called also a **terminal payoff**, is a claim, $h \geq 0$ a stochastic process, referred to as a **payoff rate**, whose mathematical definition coincides with that of a consumption process. In this context a PC-strategy (π, c) is called an admissible hedging strategy against a contingent claim (w_T, h) for an initial endowment $y \geq 0$ if $c(t) \overset{\text{as}}{=} h(t)$ for $0 \leq t \leq T$ and (π, c, y) generates an admissible valuation of the claim w_T. Thus, our definition of a claim is that standard one of a contingent claim plus (hopefully) an innocent identification of the concept of a payoff and that of a consumption.

3.3   Market Regulations and Option Pricing

We shall continue our presentation of the diffusion financial mathematics in the framework of a fixed Black-Scholes prices model $(X_0, X_1, \ldots, X_n)$ given by (BS1), (BS2) and (BS3) and assume no consumption being present in our trading model. Our notations and terminology will be modified in the following obvious manner:

If π is a portfolio and $y \geq 0$ an initial endowment, then a wealth process

$$(69) \qquad Y = d^{-1} \cdot \left\{ y + \int d \cdot \pi^\top (b - r \cdot 1_n) \, ds + \int d \cdot \pi^\top \cdot \sigma \, dW \right\}$$

generated by $(\pi, 0, y)$ will be simply referred to as the wealth process generated by (π, y). A portfolio π will be said admissible up to time $T > 0$ for an initial endowment $y \geq 0$, if $Y(t) \geq 0$ almost surely for all $0 \leq t \leq T$, where Y is defined by (69).

If w_T is a claim, then $y(w_T) := y(0, w_T)$ will be called the minimal price of w_T and a minimal admissible valuation of $(0, w_T)$ shortly as a minimal admissible valuation of w_T.

The mathematical technology developed in Section 2 and summarized by 2.10 makes possible a modest discussion of **market regulations** as are or as could be imposed with all regards to the need to have a market that is both safe and interesting for investors. One more concept contrasting the suicide admissible portfolios introduced by 2.11 is of considerable importance to make the discussion profitable.

We shall say that a portfolio π is an **arbitrage opportunity** at time $T > 0$ if $(\pi, 0)$ generates a wealth process Y such that

$$P[Y(T) \geq 0] = 1 \quad \text{and} \quad P[Y(T) > 0] > 0$$

holds.

Obviously, no responsible market regulator would allow such a portfolio to be operated. Even if he would, the following example shows that both the arbitrage opportunities and suicide portfolios (such as constructed in 2.11) are disqualified automatically by their spectacular and lasting need of a credit. The example also demonstrates that the arbitrage and suicide portfolios are in a sense dual concepts.

3.3.1   Example. Arbitrage Opportunities Exist. Consider $K > 0$ arbitrary, define an admissible portfolio π^K by (59) in 2.11 and generate wealth processes Y_s and Y_a by (π^K, K) and by $(-\pi^K, 0)$, respectively. The portfolio π^K is constructed so that the processes Y_s and Y_a behave as follows:

$$Y_s(0) = K, \qquad Y_s(1) \overset{\text{as}}{=} 0, \qquad Y_a(0) = 0, \qquad Y_a(1) \overset{\text{as}}{=} K \cdot e^{\int_0^t r \, ds}.$$

Thus, the portfolio $-\pi^K$ is not only an arbitrage opportunity at $T = 1$, it transforms a void initial capital into a positive wealth in a finite trading expiration time, that is $T = 1$ in our case.

As we have seen in 2.11, the suicide process Y_s requires a huge increasing negative position for the bond and therefore the arbitrage opportunity process Y_a may be operated only by means of increasing short sales of stock. A common necessary condition for such a behavior of an investor is an access to an almost unlimited credit. Even a very liberal lending practice rules out such a credit and in fact prohibits implicitly to operate such portfolios as π^K and $-\pi^K$ are.

3.3.2   Exercise. Let π be an arbitrage opportunity at $T > 0$, $y \geq 0$ and Y_y the wealth process generated by (π, y). Denote by Z_y the wealth process generated by $(0, y)$. Then

$$P[Y_y(T) \geq Z_y(T)] = 1 \quad \text{and} \quad P[Y_y(T) > Z_y(T)] > 0 \quad \text{holds.}$$

Thus, to operate an arbitrage opportunity portfolio starting with an arbitrary initial capital $y \geq 0$ means that you are sharply better than to invest the initial capital y to the bond and wait for the expiration time T.

As expected, neither true martingales nor non-negative wealth processes provide opportunities for arbitrage. More precisely:

3.3.3   No Arbitrage Portfolios. *Let π be a portfolio, denote by Y the wealth process generated by $(\pi, 0)$ and consider $T > 0$. If π is a portfolio admissible up to T for $y = 0$ or if $d \cdot Y$ is a Q_T-martingale on $[0, T]$ then π is not an arbitrage opportunity at T.*

Proof. If π is a portfolio admissible up to T for $y = 0$ then $d \cdot Y$ is a Q_T-supermartingale on $[0, T]$ by 2.2 and (SP1). Thus, $d \cdot Y$ is a Q_T-supermartingale on $[0, T]$ in both cases. Consequently, $E_{Q_T}(d(T) \cdot Y(T)) \leq 0$. On the other hand, if π is an arbitrage opportunity at T, then

$$P[Y(T) > 0] > 0 \quad \text{implies that} \quad Q_T[Y(T) > 0] > 0$$

as Q_T and P are equivalent measures. Hence, $E_{Q_T}(d(T) \cdot Y(T)) > 0$ and a contradiction is exhibited. $\square$

The above assertion is a simple but useful statement: In the set of all portfolios that are admissible up to T for a fixed initial endowment $y \geq 0$ you will find no portfolio which would provide an arbitrage opportunity at some time $0 < t \leq T$. On the other hand, according to 3.1, you will find there suicide portfolios that generate wealth processes Y such that $E_{Q_T} d(T) \cdot Y(T) < y$ holds. It follows by 2.10 (e) that the trouble making portfolios are exactly those that generate jointly with y wealth processes that are local Q_T-martingales and that are not true Q_T-martingales. Thus, it seems that the admissibility itself can not, at least mathematically, guarantee a safe and decent market and we are advised to restrict our trading to the set

$$\mathcal{P}(T) \subset \mathrm{PM}_2(W) \times \mathbb{R}^+ \text{ of all trading strategies } (\pi, y) \text{ such that}$$

$$\pi \text{ is a portfolio admissible up to } T \text{ for } y \text{ and } (\pi, y) \text{ generates}$$

$$\text{a wealth process } Y \text{ for which } d \cdot Y \text{ is a } Q_T\text{-martingale on } [0, T].$$

We may summarize the reasons in favour of a market defined by the set $\mathcal{P}(T)$ that deserve a consideration on the part of a market modeller: The investor restricted to $\mathcal{P}(T)$ is denied of all arbitrage opportunities and lives without any hope to overcome a possible bankruptcy according to 2.6. Not so good for him. On the other hand, he is rewarded a security in that sense that no suicide portfolio can be met in such a market which may be considered as a favourable compensation.

The Girsanov measure Q_T is designed to remove the drift. A growth of the wealth is present even under Q_T though $d \cdot Y$ becomes a (local) martingale in the Q_T-world, hence, basically a process whose expectation is constant on $[0, T]$.

3.3.4   Lemma. *Assume that the interest rate r is a non-negative process. Let M be an adapted process such that $d \cdot M$ is a Q_T-martingale on $[0, T]$ for a $T > 0$. Also let $g : \mathbb{R} \to \mathbb{R}$ be a convex function with $g(0) = 0$ such that $g(M(t))$ is a Q_T-integrable random variable for any $0 \le t \le T$. Then, $g(d \cdot M)$ and $d \cdot g(M)$ are both Q_T-submartingales on $[0, T]$.*

Proof. Indeed, if $r \ge 0$, then $0 < d \le 1$ is a non-increasing process and

(70)
$$g(\alpha x) \le \alpha g(x), \quad 0 \le \alpha \le 1, \quad x \in \mathbb{R}$$

holds. Hence, $g\left(d(t)M(t)\right)$ is a Q_T-integrable random variable for $0 \le t \le T$ and $g(d \cdot M)$ is a Q_T-submartingale according to 1.2.9 (d).

Having $0 \le s \le t \le T$, it follows by 1.2.7 (c), (70) and 1.2.7 (b) that outside a Q_T-null set

$$E^{\mathcal{F}_s}\left[d(t)g(M(t))\right] = d(s) \cdot E^{\mathcal{F}_s}\left[d(t)d(s)^{-1}g(M(t))\right]$$
$$\ge d(s) \cdot E^{\mathcal{F}_s}g\left(d(t)d(s)^{-1}M(t)\right)$$
$$\ge d(s)g\left(E^{\mathcal{F}_s}\left[d(t)d(s)^{-1}M(t)\right]\right) = d(s) \cdot g\left(d(s)^{-1}d(s)M(s)\right) = d(s)g(M(s))$$

holds. Hence, $d \cdot g(M)$ is a Q_T-submartingale by definition. $\square$

As a direct corollary to 3.4, 2.13 and 2.10 (b) we get

3.3.5   Submartingale Growth Under Q_T. *Assume that $r \ge 0$, let $g : \mathbb{R} \to \mathbb{R}$ be a convex function with $g(0) = 0$ and $T > 0$.*

If $g(X_1(t)) \in L_1(Q_T)$ for $0 \le t \le T$, then both $g(d \cdot X_1)$ and $d \cdot g(X_1)$ are Q_T-submartingales on $[0, T]$.

If Y is a minimal admissible valuation of a claim w_T and $g(Y(t)) \in L_1(Q_T)$ for $0 \le t \le T$, then also both $g(d \cdot Y)$ and $d \cdot g(Y)$ are Q_T-sumbartingales on $[0, T]$.

Thus, if $r \ge 0$ and g is a **convex utility function** with $g(0) = 0$ such that $g(Y)$ and $g(X_1)$ are Q_T-integrable processes on $[0, T]$, then

$$t \to E_{Q_T}\{d(t)g(Y(t))\} \quad \text{and} \quad t \to E_{Q_T}\{d(t)g(X_1(t))\}$$

are non decreasing functions on $[0, T]$.

Claims w_T whose minimal valuations we studied in Section 3.2 are most importantly constructed as a betting on the price of a stock i, say $i = 1$, at a specified time $T > 0$. We shall start with a notorious stock exchange game called a **European call** option that is a claim given as $w_T := (X_1(T) - K)^+$ where $T > 0$ is called the **exercise time** of the option (also the maturity or expiration time), and $K > 0$ its **exercise price**.

Obviously, the game goes as follows: The investor signs a contract that makes possible for him to buy at time T one share of the stock $i = 1$ at a price K if $X_1(T) > K$ and, obviously, to make a profit $X_1(T) - K$ if the share is sold immediately for $X_1(T)$. This option is, of course, not free of charge and we ask what would be its fair price $y > 0$ to be paid at time $t = 0$? The answer is simple, it is the minimal price of $w_T = (X_1(T) - K)^+$, since nobody is prepared to pay for the option a price $y > y(w_T)$ when there is a trading strategy $(\pi, y(w_T)) \in \mathcal{P}(T)$ that valuates w_T at time T and the rest $y - y(w_T)$ may be invested elsewhere.

More generally, having a Borel function $g : \mathbb{R}^+ \to \mathbb{R}^+$ and a $T > 0$ such that $E_{Q_T} g(X_1(T)) < \infty$, the claim $w_T := g(X_1(T))$ will be called the g-**option** and T referred to as its exercise time. Any European call is a g-option with $g(x) := (x - K)^+$ for a $K > 0$. The minimal price of a g-option exists and is equal to

$$q := E_{Q_T} (d(T) \cdot g(X_1(T))) < \infty$$

according to 2.10 (a). It will be called the **value of the g-option** $g(X_1(T))$. Obviously, it might not be enough to know only the value, we need to visualize a legal valuation Y of the claim $g(X_1(T))$ with $Y(0) \overset{\text{as}}{=} q$ to support the interpretation of q as that of the fair price of the option. Denoting

$$(71) \qquad V(t) := E_{Q_T}^{\mathcal{F}_t} \left[d(T)d(t)^{-1} \cdot g(X_1(T)) \right], \quad 0 \le t \le T$$

we call $V(t)$ the **value of the g-option** $g(X_1(T))$ at time t and stress that the process V is exactly what we need because any minimal admissible valuation of $g(X_1(T))$ is a continuous modification of V on the interval $[0, T]$ by 2.10 (d).

However, no explicit formula for the values $V(t)$ can by itself provide an exact control of our investments to attain the option terminal value $g(X_1(T))$. By 2.10 (b) there exists a portfolio $\pi \in \mathcal{P}(T)$ such that

$$(72) \qquad d(t) \cdot V(t) \overset{\text{as}}{=} q + \int_0^t d \cdot \pi^\top (b - r \cdot 1_n) \, ds + \int_0^t d \cdot \pi^\top \cdot \sigma \, dW$$

holds for $0 \le t \le T$. We are justified to call such a portfolio a **hedging strategy against the g-option** as such a portfolio π is determined uniquely $\lambda \otimes P$-almost everywhere on $[0, T] \times \Omega$ by 1.8.

The valuation of European calls is the highlight of the continuous time financial mathematics, indeed. The **Black-Scholes** Nobel price explicit **formula** for its value, see [23], was derived by means of solving a partial differential equation specified below.

3.3.6   Exercise. Let r, σ, T and K be positive real numbers. Denote

$$f(t, x) := x \cdot \Phi(\alpha(T - t, x)) - K \cdot e^{-r(T-t)} \Phi(\beta(T - t, x)), \quad x > 0, \quad 0 \le t < T,$$
$$:= (x - K)^+, \qquad\qquad\qquad\qquad\qquad\qquad\qquad\qquad x > 0, \quad t \ge T,$$

where Φ denotes the distribution function of $N(0, 1)$, $\Phi(-\infty) := 0$, $\Phi(+\infty) := 1$ and

$$\alpha(t, x) := \frac{1}{\sigma\sqrt{t}} \left[\ln\left(\frac{x}{K}\right) + t\left(r + \frac{\sigma^2}{2}\right) \right],$$
$$\beta(t, x) := \frac{1}{\sigma\sqrt{t}} \left[\ln\left(\frac{x}{K}\right) + t\left(r - \frac{\sigma^2}{2}\right) \right]$$

for $x > 0$ and $t > 0$.

Check that f is continuous on $\mathbb{R}^+ \times (0, \infty)$ $(\lim_{t \to T-} f(t, x) = (x - K)^+)$ and that the derivatives

$$f' := \frac{\partial f}{\partial t}, \quad f_x := \frac{\partial f}{\partial x} \quad \text{and} \quad f_{xx} := \frac{\partial^2 f}{\partial x^2}$$

exist continuous on $[0, T) \times (0, \infty)$. Also

(73) $\qquad \dfrac{\partial f}{\partial t} + rx \dfrac{\partial f}{\partial x} + \dfrac{1}{2} \sigma^2 x^2 \dfrac{\partial^2 f}{\partial x^2} = rf, \quad$ holds on $[0, T) \times (0, \infty)$.

In other words f solves the Cauchy problem with the boundary $f(x, T) = (x - K)^+$.

3.3.7 The Value of European Call.
Consider BS-model with one stock where the interest rate r and the volatility σ are positive constants. Also consider $T > 0$ and $K > 0$. Then

(74) $\quad V(t) \overset{as}{=} f(t, X_1(t)), \quad 0 \le t \le T, \quad$ and $\quad q = y((X_1(T) - K)^+) = f(0, x_1)$

where f is defined in 3.6, are the values of the European call with the exercise time T and the exercise price K.

Moreover, if π_1 is a hedging portfolio against $(X_1(T) - K)^+$, then

(75) $\quad \pi_1(t, \omega) = f_x(t, X_1(t, \omega)) \cdot X_1(t, \omega) \quad \lambda \otimes P - \text{almost everywhere on } [0, T) \times \Omega.$

3.3.8 Exercise.
We are able to determine the trading strategy ϕ that corresponds to the hedging portfolio π_1 and to the initial endowment $f(0, x_1)$ on the interval $[0, T]$. In the setting of 3.7 prove:

Let $Y(\phi)$ be a wealth process associated with a trading strategy $\phi = (\phi_0, \phi_1)$. Then $Y(\phi)$ is a minimal admissible valuation of $(X_1(T) - K)^+$ iff

$$\phi_0(t, \omega) = \frac{e^{-rt}}{rx_0} \cdot f'(t, X_1(t, \omega)) + \frac{e^{-rt}}{2rx_0} \cdot f_{xx}(t, X_1(t, \omega)) \cdot (X_1(t, \omega))^2$$
$$\text{and} \quad \phi_1(t, \omega) = f_x(t, X_1(t, \omega)) \quad \text{holds } \lambda \otimes P\text{-a.e. on } [0, T) \times \Omega.$$

Note that the values of European calls are independent of the rate of return process b but depend heavily on the interest rate r and the volatility σ. This is an important feature of the Black-Scholes formula (74), since while there is a chance to estimate the latter coefficients, it seems to be a problem in the case of b because of its basically stochastic or even diffusion character.

Proof of 3.7. The process $(d \cdot X_1)^T$ is a Q_T-martingale by 2.12. We shall prove that it is an (L_2, Q_T)-martingale: It follows by 2.13 that for $t \le T$

$$(d(t) \cdot X_1(t))^2 = x_1^2 \exp \left\{ 2 \cdot \left(\sigma \widehat{W}^{(T)}(t) - \frac{\sigma^2}{2} t \right) \right\} \le x_1^2 \exp \left\{ 2\sigma \widehat{W}^{(T)}(t) \right\}$$

is true and therefore, again for $t \leq T$, we get

$$E_{Q_T} (d(t) \cdot X_1(t))^2 \leq x_1^2 E_{Q_T} \exp\{2\sigma\sqrt{t}Z\} = x_1^2 \exp\{2\sigma^2 t\} < \infty,$$

where Z is a random variable with $\mathcal{L}(Z|Q_T) = N(0,1)$. Hence, $d \cdot X_1$ is an (L_2, Q_T)-martingale on $[0, T]$.

Further, define two continuous adapted processes Y and Z by

$$Y(t) := f(t, X_1(t)) \quad \text{and by} \quad Z(t) := d(t) \cdot Y(t), \quad t \geq 0.$$

The definition of $f(t, x)$ yields the inequality

$$\max_{t \leq T} |Z(t)|^2 \leq \max_{t \leq T} (d(t) \cdot X_1(t))^2 + d(T)^2 \cdot (X_1(T) - K)^2.$$

Apply 1.2.10 (b) to the (L_2, Q_T)-martingale $(d \cdot X_1)^T$ to see that

$$(76) \qquad E_{Q_T} \max_{t \leq T} |Z(t)|^2 \leq 4 E_{Q_T} (d(T) \cdot X_1(T))^2 + E_{Q_T} (X_1(T) - K)^2 < \infty.$$

Imagine that we have already proved that

$$(77) \qquad\qquad Z^U \text{ is a local } Q_T\text{-martingale for arbitrary } 0 \leq U < T.$$

If it is so, we may let $U_m \uparrow T$ and see that (76) implies that

$$(78) \qquad\qquad |Z^{U_m}| \leq \max_{t \leq T} |Z(t)| \in L_1(Q_T) \quad \text{holds for all } m \in \mathbb{N}.$$

Consequently, each Z^{U_m} is a Q_T-martingale by 1.4.2 (a) and $Z^T = \lim_{m \to \infty} Z^{U_m}$ is also easily seen to be a Q_T-martingale as an application of (78) and 1.2.7 (f). Hence, $E_{Q_T}^{\mathcal{F}_t} Z(T) \overset{as}{=} Z(t)$ for all $0 \leq t \leq T$ which is a statement equivalent to (74) due to (71) and 1.2.7 (c).

We shall prove (77) for a fixed $0 < U < T$. It follows by Itô formula 2.2.9 and 2.12 (c) that for all $t \leq U$

$$Y(t) \overset{as}{=} f(0, x_1) + \int_0^t \left(f'(X_1) + \frac{1}{2} f_{xx}(X_1) \cdot \sigma^2 \cdot X_1^2 + f_x(X_1) \cdot r \cdot X_1 \right) ds$$
$$+ \int_0^t f_x(X_1) \cdot \sigma \cdot X_1 \, d\widehat{W}^{(T)}$$

is true if the integrands are meant as $f'(X_1)(t) := f'(t, X_1(t)) \cdot I_{[0,U]}(t)$ and

$$f_x(X_1)(t) := f_x(t, X_1(t)) \cdot I_{[0,U]}(t), \quad f_{xx}(X_1)(t) := f_{xx}(t, X_1(t)) \cdot I_{[0,U]}(t).$$

Since $d' = -r \cdot d$ holds, we integrate per partes to get

$$Z(t) - Z(0) \overset{as}{=} \int_0^t d \cdot \left(f'(X_1) + \frac{1}{2} \cdot f_{xx}(X_1) \cdot \sigma^2 \cdot X_1^2 + f_x(X_1) \cdot r \cdot X_1 \right) ds$$
$$+ \int_0^t d \cdot f_x(X_1) \cdot \sigma \cdot X_1 \, d\widehat{W}^{(T)}(s) - \int_0^t d \cdot r \cdot Y \, ds, \quad 0 \leq t \leq U$$

and finally, because f is the solution to the Cauchy problem (73), we get

$$(79) \qquad Z(t) \overset{as}{=} f(0, x_1) + I_t^{\widehat{W}^{(T)}} \left(d \cdot f_x(X_1) \cdot \sigma \cdot X_1 \right), \quad 0 \le t \le U$$

and Z^U as a local Q_T-martingale as stated by (77).

It remains to verify (75): Fix $U < T$ and apply (79) and (44) to prove that $Z(t)$ equals almost surely to

$$
f(0, x_1) + \int_0^t d \cdot f_x(X_1) \cdot X_1 \cdot \sigma \, d\widehat{W}^{(T)} \overset{as}{=} f(0, x_1) + \int_0^t d \cdot f_x(X_1) \cdot X_1 \cdot \sigma \, d\widehat{W}
$$
$$
\overset{as}{=} f(0, x_1) + \int_0^t d \cdot f_x(X_1) \cdot X_1 \cdot (b - r) \, ds + \int_0^t d \cdot f_x(X_1) \cdot X_1 \cdot \sigma \, dW
$$

for all $t \le U$. If π_1 is a hedging portfolio against $(X_1(T) - K)^+$, then according to (72) we get that

$$
\int_0^t d \cdot \pi_1 \cdot (b - r) \, ds + \int_0^t d \cdot \pi_1 \cdot \sigma \, dW
$$
$$
\overset{as}{=} \int_0^t d \cdot f_x(X_1) \cdot X_1 \cdot (b - r) \, ds + \int_0^t d \cdot f_x(X_1) \cdot X_1 \cdot \sigma \, dW
$$

holds for all $0 \le t < U$ denoting again $f_x(X_1)(t) := f_x(t, X_1(t)) \cdot I_{[0,U]}(t)$.

It follows by 1.8 that $\pi_1 = f_x(X_1) \cdot X_1$ almost everywhere on $[0, U] \times \Omega$ w.r.t $\lambda \otimes P$ for all $U < T$. Hence (75) holds. $\square$

More generally, having positive constants r, σ, T and a continuous function $g : \mathbb{R}^+ \to \mathbb{R}^+$, we shall say that

$$ f : \mathbb{R}^+ \times (0, \infty) \to \mathbb{R} \text{ continuous, } f', f_x, f_{xx} \text{ continuous on } [0, T) \times (0, \infty) $$

solves the (r, σ, T)-**Cauchy problem** with the boundary condition g if (73) holds and $f(t, x) = g(x)$ for all $t \ge T$ and $x > 0$.

It presents no problem to transfer 3.8 to more general g-options:

3.3.9   The Value of g-Option. *Assume a BS-model as in 2.7 and consider a continuous function* $g : \mathbb{R}^+ \to \mathbb{R}^+$. *Let f be a solution to the (r, σ, T)-Cauchy problem with boundary g such that*

$$(80) \qquad \max_{t \le T} |f(t, x)| \le C \cdot (1 + |x|) \quad \text{holds for } x \in (0, \infty),$$

where $C < \infty$. Then $g(X_1(T))$ is the g-option with values given as $V(t) \overset{as}{=} f(t, X_1(t))$ for $0 \le t \le T$ and with a hedging portfolio defined by (75).

Indeed, $g(X_1(T))$ is the g-option, since

$$ E_{Q_T} \{d(T) \cdot g(X_1(T))\} = E_{Q_T} \{d(T) \cdot f(T, X_1(T))\} \le C \cdot E_{Q_T} (1 + X_1(T)) < \infty $$

holds.

The proof of 3.9 goes exactly as that of 3.7 the only difference being that $\max_{t \leq T} |Z(t)|^2$ is estimated by means of (80) as

$$\max_{t \leq T} |Z(t)|^2 = \max_{t \leq T} \{d(t) \cdot f(t, X_1(t))\}^2 \leq C^2 \cdot \max_{t \leq T} \{d(t) \cdot (1 + X_1(t))\}^2$$

and (76) then reads

$$E_{Q_T} \max_{t \leq T} |Z(t)|^2 \leq 2C^2 \cdot \left\{ 1 + E_{Q_T} \max_{t \leq T} \{d(t) \cdot X_1(t)\}^2 \right\}$$
$$\leq 2C^2 \cdot \left\{ 1 + 4E_{Q_T} \{d(T) \cdot X_1(T)\}^2 \right\} < \infty.$$

Surprisingly, we may also prove, as a corollary to 3.9, a uniqueness result for the Cauchy problem which is something very far off our present topic:

3.3.10   Exercise. Consider an (r, σ, T)-Cauchy problem with a boundary g : $\mathbb{R}^+ \to \mathbb{R}^+$. If f and h are its solutions that satisfy the linear growth requirement (80), then $f = h$ on $\mathbb{R}^+ \times (0, \infty)$.

In 3.9 and 3.10 we touched, if only slightly, the recent research on stochastic representations of Cauchy-Dirichlet problems. See, 5.7 Section in [95] and Section 15 in [151].

Choosing $g(X_1(T)) := (X(T) - K)^-$ we define a g-option called the **European put** option with an exercise price $K > 0$ and an exercise time $T > 0$. It is easy to see that the European put is a dual to the European call that suggests a game where a bet is placed on a small value $X_1(T)$ of the stock at the exercise time T. By buying a European put option we receive the right (not the obligation) to sell at time T one share of the stock at a price K.

3.3.11   Example and Exercise. The Value of European Put. Consider a BS-model as in 3.7, note that

$$(X_1(T) - K)^- = K + (X_1(T) - K)^+ - X_1(T),$$

and denote by $V(t)$ the value of the European put at time $0 \leq t \leq T$. It follows by 3.7 and by 2.12 (b) that

$$d(t) \cdot V(t) = E_{Q_T}^{\mathcal{F}_t} \{d(T) \cdot (X_1(T) - K)^-\} \overset{\text{as}}{=} d(T) \cdot K + d(T) \cdot f(t, X_1(t)) - d(t) \cdot X_1(t)$$

holds for $0 \leq t \leq T$, where f is the function defined in 3.6. Thus,

$$(81) \qquad V(t) \overset{\text{as}}{=} e^{-r(T-t)} \cdot K + f(t, X_1(t)) - X_1(t), \quad 0 \leq t \leq T$$

and therefore,

$$h(t, x) := K \cdot e^{-r(T-t)} \cdot \Phi(-\beta(T - t, x)) - x \cdot \Phi(-\alpha(T - t, x)), \quad x > 0, \quad t \leq T,$$
$$:= (x - K)^-, \qquad\qquad\qquad\qquad\qquad\qquad\qquad x > 0, \quad t \geq T$$

defines a function h such that the value of the European put at $0 \le t \le T$ is given by

$$V(t) \overset{\text{as}}{=} h(t, X_1(t)), \text{ in particular, its value is } q = K \cdot e^{-rT} + f(0, x_1) - x_1,$$

where $f(0, x_1)$ is the value of the European call.

Observe that the function h is the only solution to the (r, σ, T)-Cauchy problem with the boundary $g(x) = (x - K)^-$ that satisfies the requirement (80).

Prove that the hedging portfolio π_1 against the European put $(X_1(T) - K)^-$ is such that

$$(82) \qquad \pi_1(t, \omega) = X_1(t, \omega) \cdot (f_x(t, X_1(t)) - 1)$$

holds almost everywhere on $[0, T) \times \Omega$ with respect to $\lambda \otimes P$.

The concept of an option makes its appearance both in the market practice and in corresponding theoretical models even in a more general form.

For example, an **exotic option** with an expiration time $T > 0$ is a claim $w_T := g(X_1(s), s \le T)$ where $g : C[0, T] \to [-\infty, +\infty]$ is a Borel function with the following properties:

$$g(X_1(s), s \le T) \ge 0, \quad \text{and} \quad E_{Q_T} g(X_1(s), s \le T) < \infty.$$

Putting

$$q = E_{Q_T} \{d(T) \cdot g(X_1(s), s \le T)\}, \quad V(t) = E_{Q_T}^{\mathcal{F}_t} \{d(T) \cdot d(t)^{-1} \cdot g(X_1(s), s \le T)\}$$

for $0 \le t \le T$ again, we define the **value** and the **value at time** t of the exotic option $g(X_1(s), s \le T)$. Stress again that q is the minimal price of $w_T := g(X_1(s), s \le T)$ and V is the process such that any minimal admissible valuation of w_T is a continuous modification of V on the interval $[0, T]$.

The final value of an exotic option may depend on the history of the price of the stock $(X_1(s), s \le T)$ as one may wish or may invent: We may choose

$$g(x) := \frac{1}{T} \int_0^T x(s)\, ds, \quad x \in C[0, T] \quad \text{to get} \quad w_T := \frac{1}{T} \int_0^T X_1(s)\, ds,$$
$$g(x) := \max_{s \le T} x(s), \quad x \in C[0, T] \quad \text{to get} \quad w_T := \max_{s \le T} X_1(s)$$

and derive a variety of exotic options that are attractive both to the investors and mathematicians. The examples are such as follows:

$$\left(X_1(T) - \frac{1}{T} \int_0^T X_1(s)\, ds\right)^+, \quad L \cdot I_{[\max_{s \le T} X_1(s) \ge K]}, \quad L > 0, K > 1,$$

the former being called the **Asian option**, the latter the **exotic binary option**. We might even be able in some cases to establish the values of exotic options.

Both in 3.12 and 3.13 we shall assume that a BS-model with one stock is operated and consider the interest rate r and the volatility σ to be some constants in $(0, \infty)$.

3.3.12 Example. The Value of $T^{-1} \int_0^T X_1(s)\,ds$. The random variable $w_T := \frac{1}{T} \int_0^T X_1(s)\,ds$ is obviously a Q_T-integrable claim, hence the value of $\frac{1}{T} \int_0^T X_1(s)\,ds$ is given as

$$q = E_{Q_T} \left\{ \frac{d(T)}{T} \cdot \int_0^T X_1(s)\,ds \right\} ds = \frac{x_1 \cdot e^{-rT}}{T} \cdot \int_0^T e^{-sr}\,ds = \frac{x_1}{T \cdot r} \cdot \left(1 - e^{-rT} \right),$$

since $d \cdot X_1$ is a Q_T-martingale by 2.12 (b). Applying the property in a more sophisticated manner we get

$$E_{Q_T}^{\mathcal{F}_t} X_1(s) = d(s)^{-1} \cdot E_{Q_T}^{\mathcal{F}_t} \left\{ d(s) \cdot X_1(s) \right\} = d(s)^{-1} \cdot d(t) \cdot X_1(t), \quad s \geq t$$

and consequently, having fixed $0 \leq t \leq T$ and $F \in \mathcal{F}_t$, we get by Fubini Theorem

$$E_{Q_T} \left\{ I_F \cdot \int_t^T X_1(s)\,ds \right\} = \int_t^T E_{Q_T} \left\{ I_F \cdot X_1(s) \right\} ds$$

$$= \int_t^T E_{Q_T} \left\{ I_F \cdot d(s) \cdot X_1(s) \cdot d(s)^{-1} \right\} ds = E_{Q_T} \left\{ I_F \cdot e^{-rt} \cdot \frac{e^{rT} - e^{rt}}{r} \cdot X_1(t) \right\}$$

and therefore $E_{Q_T}^{\mathcal{F}_t} \int_t^T X_1(s)\,ds \overset{\text{as}}{=} \frac{e^{r(T-t)}-1}{r} \cdot X_1(t)$. Thus we have computed the value of w_T at time $t \leq T$ as

$$V(t) = E_{Q_T}^{\mathcal{F}_t} \left\{ d(T) \cdot d(t)^{-1} \cdot w_T \right\} = \frac{e^{-r(T-t)}}{T} \cdot \left\{ t \cdot w_t + \frac{e^{r(T-t)} - 1}{r} \cdot X_1(t) \right\},$$

where $w_t := t^{-1} \int_0^t X_1(s)\,ds$.

3.3.13 Example. The Value of Exotic Binary Option. Denote $w_T := \max_{s \leq T} \ln X_1(s)$ and assume that $x_1 = 1$. It follows by 1.1 that for all $t \leq T$

$$\ln X_1(t) = \int_0^t \left(b - \frac{\sigma^2}{2} \right) ds + \int_0^t \sigma\,dW$$

$$= \int_0^t (b - r)\,ds + \int_0^t \sigma\,dW + \int_0^t \left(r - \frac{\sigma^2}{2} \right) ds$$

$$= \sigma \left(W(t) - \int_0^t \frac{r - b}{\sigma}\,ds \right) + \int_0^t \left(r - \frac{\sigma^2}{2} \right) ds$$

$$= \sigma \left(\widehat{W}^{(T)}(t) + \mu t \right)$$

holds, if $\mu := \frac{r}{\sigma} - \frac{\sigma}{2}$. Since $\widehat{W}^{(T)}$ is a Q_T-Wiener process, it follows by 2.4.11 that $w_T \geq 0$ almost surely and that

$$E_{Q_T} w_T \leq \sigma E_{Q_T} \left\{ \max_{s \leq T} \widehat{W}^{(T)}(s) \right\} + \mu^+ \sigma T < \infty$$

is true. Hence, w_T is an exotic option (we may assume that $w_T \geq 0$ everywhere on Ω without loss of generality) and 2.4.11 also provides its Q_T-probability distribution as

$$Q_T[w_T \geq y] = \frac{y}{\sigma\sqrt{2\pi}} \int_0^T s^{-\frac{3}{2}} \exp\left\{-\frac{\left(\frac{y}{\sigma} - \mu s\right)^2}{2s}\right\} ds =: h(y,T), \quad y > 0.$$

Finally,

$$q = E_{Q_T} d(T) \cdot w_T = e^{-rT} \int_0^\infty h(y,T)\, dy$$

is the value of w_T at time $t = 0$.

Consider the exotic binary option and denote

$$v_T := L \cdot I_{[\max_{s \leq T} X_1(s) \geq K]}, \quad L > 0, \quad K > 1.$$

The minimal price of v_T, i.e., the value of the corresponding exotic g-option is computed as follows:

$$E_{Q_T} d(T) \cdot v_T = e^{-rT} \cdot L \cdot Q_T[\max_{s \leq T} X_1(s) \geq K]$$

$$= e^{-rT} \cdot L \cdot Q_T[\max_{s \leq T} \ln X_1(s) \geq \ln K] = e^{rT} \cdot L \cdot h(\ln K, T).$$

Up to now we have observed only options with a fixed exercise time $T > 0$. Let us now look at T as the final exercise time, an arbitrary earlier exercise time $0 \leq S \leq T$ being also made possible by the option contract. For a g-option $g(X_1(T))$, it means that the contractor chooses an $0 \leq S \leq T$ to receive $g(X_1(S))$ at time S not having been forced by the contract to specify S. It seems that a fair definition of a price $q^A(g,T)$ for such a contract should be

$$(83) \qquad q^A(g,T) := \sup_{S \leq T} q(g,S), \quad \text{where} \quad q(g,S) := E_{Q_S} \{d(S) \cdot g(X_1(S))\},$$

is the value of $g(X_1(S))$ at time $t = 0$. Since, $Q_T|\mathcal{F}_S = Q_S|\mathcal{F}_S$ for all $S \leq T$ by (G5) and the random variable $d(S) \cdot g(X_1(S))$ is $\mathcal{F}_S$-measurable, we also may write

$$(84) \quad q(g,S) = E_{Q_T} \{d(S) \cdot g(X_1(S))\}, \quad q^A(g,T) = \sup_{S \leq T} E_{Q_T} \{d(S) \cdot g(X_1(S))\}.$$

Agree to call $q^A(g,T) \in [0, \infty]$ the **American value of a g-option** $g(X_1(T))$.

If we consider a European call $(X_1(T) - K)^+$ and a European put $(X_1(T) - K)^-$ valued at time $t = 0$ by $q^A(g,T)$, where we denote either $g(x) := (x - K)^+$ or $g(x) := (x - K)^-$, respectively, we refer to them as to an **American call** option and an **American put** option, respectively, having on mind the option contract that makes possible an arbitrary earlier exercise time $S \leq T$. There are at least two items that should be clarified:

(1) A choice of an early exercise time $0 \leq S \leq T$ should not be expected deterministic, the contractor picks up S or waits on a later opportunity considering the price history $(X_1(s), s \le S)$ and therefore, in fact, his choice is a Markov time with values in the interval $[0, T]$. It seems that the American value of a g-option should be defined rather as

$$q^M(g, T) := \sup_{\tau \in M(T)} E_{Q_T} \{d(\tau) \cdot g(X_1(\tau))\},$$

where $M(T)$ denotes the set of all $\mathcal{F}_t$-Markov times $\tau \le T$. Recall that $d(\tau) \cdot g(X_1(\tau))$ is an $\mathcal{F}_\tau \subset \mathcal{F}_T$-measurable variable according to 1.3.9 (c) and 1.3.1 (e).

(2)   The definition of the American value of a g-option is based on a heuristics that seems to be fair only to the seller of the option. The contractor who is offered to buy the option for the American price value should deserve to know in which sense the offered price is minimal. Denote

$$\mathcal{M}(g, T) := \{Y \text{ is a wealth process} : Y(t) \ge g(X_1(t)) \text{ a.s. } \forall 0 \le t \le T\}$$

and

(85) $q^m(g, T) := \inf \left\{ y \ge 0 \text{ such that } Y(0) \overset{as}{=} y \text{ for some } Y \in \mathcal{M}(g, T) \right\}.$

The alternative definition of a fair price of the American option by $q^m(g, T)$ respects the interests of the investor. Indeed, there is no way how to accumulate by a trading the final profit that would be equivalent to that coming from entering the option contract, than to trade along a wealth process $Y \in \mathcal{M}(g, T)$.

If an American call $(X_1(T) - K)^+$ and **not** an American put $(X_1(T) - K)^-$ is considered everything goes smoothly:

3.3.14 The Value of American Option. *Assume that the interest rate r is a non negative process. Let $g : \mathbb{R}^+ \to \mathbb{R}^+$ be a convex function such that $g(0) = 0$ and $E_{Q_T} g(X_1(T)) < \infty$ hold. Then*

(86) $q(g, T) = q^A(g, T) = q^M(g, T) = q^m(g, T) < \infty$

and $q^m(g, T)$ defined by (85) is the minimum of the right-hand side in (85). If Y is a minimal admissible valuation of $g(X_1(T))$, then $Y \in \mathcal{M}(g, T)$.

Thus, for example, if $g(x) := (x - K)^+$ for a $K > 0$, then any of the American values q^A, q^M, q^m, we have suggested, is finite and equals to the European value $q(g, T) = E_{Q_T} \{d(T) \cdot (X_1(T) - K)^+\}$ of the option.

Nothing like that may be true for American puts and a very different behavior may be expected from the exotic options.

3.3.15 Example and Exercise. Denote $g_1(x) := (x - K)^+$ and $g_2(x) := (x - K)^-$ for a $K > 0$, assume that only one stock is traded and further simplify the setting by putting $x_1 = r = \sigma = K = 1$. Then

$$q(g_1, T) = \Phi(\tfrac{3}{2}\sqrt{T}) - e^{-T} \cdot \Phi(-\tfrac{1}{2}\sqrt{T}), \quad 0 = q(g_1, 0^+) < q(g_1, +\infty) = 1,$$

$$q(g_2, T) = e^{-T} + \Phi(\tfrac{3}{2}\sqrt{T}) - e^{-T} \cdot \Phi(-\tfrac{1}{2}\sqrt{T}) - 1, \quad q(g_2, 0^+) = 0 = q(g_2, +\infty)$$

is the value of the European call and put, respectively. Prove that $q(g_1, T)$ is increasing on $(0, \infty)$ and that there is a $T > 0$ for which $q^A(g_2, T) > q(g_2, T)$ holds. Specify the set of all T's with the latter property.

3.3.16   Exercise. Denote by $q(g, T)$ the value of the exotic option given as $g(X_1(s)), s \leq T) = T^{-1} \int_0^T X_1(s) \, ds$ and treated in 3.12 assuming there that r is positive constant. Prove that $q(g, T)$ is a decreasing function on $(0, +\infty)$ and $x_1 = q(g, 0^+) > q(g, +\infty) = 0$.

Proof of 3.14. Obviously, $q(g, T) \leq q^A(g, T) \leq q^M(g, T)$ holds for arbitrary g. If $\tau \in M(T)$ then

$$E_{Q_T} \{d(\tau)) \cdot g(X_1(\tau))\} \leq E_{Q_T} \{d(T) \cdot g(X_1(T))\} < \infty$$

by 1.3.10, since $d \cdot g(X_1)$ is a Q_T-submartingale by 3.5. The first three equalities in (86) are proved.

If Y is a process in $\mathcal{M}(g, T)$, then Y is an admissible valuation of the claim $v_T := Y(T)$ such that $v_T \geq g(X_1(T))$ holds almost surely. According to 2.7 we have $E_{Q_T} v_T < \infty$ and if $Y(0) \overset{\text{as}}{=} y$, then

$$y \geq E_{Q_T} \{d(T) \cdot v_T\} \geq E_{Q_T} \{d(T) \cdot g(X_1(T))\}$$

is true as $E_{Q_T} \{d(T) \cdot v_T\}$ is the minimal price of the claim v_T by 2.10 (a). Hence, $q(g, T) \leq \inf \{Y(0) : Y \in \mathcal{M}(g, T)\} =: q^m(g, T)$.

On the other hand, if Y is a minimal admissible valuation of the claim $g(X_1(T))$, then 3.5 yields that

$$d(t) \cdot Y(t) = d(t) \cdot V(t) = E_{Q_T}^{\mathcal{F}_t} \{d(T) \cdot g(X_1(T))\} \geq d(t) \cdot g(X_1(t))$$

holds almost surely for any $0 \leq t \leq T$ if $V(t)$ denotes the value of the option $g(X_1(T))$ at time t. It follows that $Y \in \mathcal{M}(g, T)$ and because $Y(0) \overset{\text{as}}{=} q(g, T)$ we get that $q^m(g, T) \leq q(g, T)$ and $q^m(g, T)$ is defined as the minimum of the right-hand side set in (85). □

3.3.17   Exercise. Let $r \geq 0$ and consider a g such as in 3.14. Prove that

$$\mathcal{M}^m(g, T) := \left\{ Y \in \mathcal{M}(g, T) : Y(0) \overset{\text{as}}{=} q^m(g, T) \right\}$$

is exactly the set of all minimal admissible valuations of $g(X_1(T))$.

We may summarize the message on American calls and puts provided by 3.14, 3.15 and 3.17 as follows. Let g_1 and g_2 be such as in 3.15:

If we do not want to buy the American call $g_1(X_1)$, we must trade along a wealth process $Y \in \mathcal{M}(g_1, T)$. However, to trade with the minimal possible initial endowment available in the set we are sent back to a minimal admissible valuation of the European call $g_1(X_1)$ as the only trading strategy for such a purpose. The values of the American and European call $g_1(X_1)$ are the same in all respects, indeed.

This is not true for the American and European put $g_2(X_1)$. However, the equality $q^M(g_2, T) = q^m(g_2, T)$ remains to be valid if one stock is traded and r, b and σ are constants. See [102], Theorem 3.39, p.3.5.

3.4   Helps to Some Exercises

1.4   Fix i. Since the process b_i is progressive and bounded, we apply the Fundamental Theorem of Calculus to verify that outside a $(\lambda \otimes P)$-null set N

$$\lim_{h \to 0+} h^{-1} \int_t^{t+h} b_i(s, \omega)\, ds = b_i(t, \omega)$$

holds. Thus, λ-almost everywhere $E|h^{-1} \int_t^{t+h} b_i(s)\, ds - b_i(t)|^2 \to 0$. Hence, the limit in 1.4 is true, since we easily prove that

$$\lim_{h \to 0+} h^{-2} \sum_{j=1}^d E \int_t^{t+h} \sigma_{ij}^2(s)\, ds = 0$$

holds, observing that the processes σ_{ij} are bounded.

1.10   The equation (35) reads as $Y = y + \int Y\, dV$ where the process V is defined by (37). According to 2.2.19, the equality (37) defines V as an Itô process whose quadratic variation is computed as $\langle V \rangle = \int p^\top \cdot \sigma \cdot \sigma^\top \cdot p\, ds$. It follows by 2.2.13 that (35) has a unique solution Y given by (36).

1.11   Note that (4) defines the price process $X_i(t)$ and at the same time says that $(\phi^{(i)}, 0)$ is a self-financing strategy.

Alternatively, as in 1.9, you may to construct the wealth process generated by $\pi := Y \cdot e_i$, $c := 0$ and by $y := x_i$. The wealth equation for Y reads then as

$$Y = x_i + \int Y \cdot b_i\, ds + \sum_{j=1}^d \int Y \cdot \sigma_{ij}\, dW_j.$$

Hence, $Y \stackrel{\text{as}}{=} X_i$ according to 1.1.

1.12   Denote by Y^* the wealth process generated by (π^*, c^*, y) and apply 2.1.5 to verify that $d^{\tau_0} \cdot Y^{\tau_0} \stackrel{\text{as}}{=} d \cdot Y^*$. Thus outside a P-null set $Y^{\tau_0} = Y^*$ holds on $[0, \tau_0)$. On the other hand, we have $Y^{\tau_0} = 0$ on $[\tau_0, \infty)$ and so $Y^{\tau_0} = Y^* = 0$ on $[\tau_0, \infty)$.

1.13   Express $X_i(t)$ as in (10), apply 1.2.3 (c) to prove that $\lim_{t \to \infty} X_i(t) \stackrel{\text{as}}{=} 0$.

1.14   In **(b)** and **(c)** apply 1.2.3 (c) and (b), respectively, to compute $\lim Y(t)$. In **(d)** note that $E \exp\{W(t) - \frac{t}{2}\} = 1$ for all t by 1.2.6., for example.

2.10 (a)   is proved by 2.9.

2.10 (b)   For a $Y \in \mathcal{W}(c)$ the reasoning goes as follows:

Y is a m.a.v. of (c, w_T) $\overset{2.9}{\Rightarrow}$ $Y(0) \stackrel{\text{as}}{=} V(0) \stackrel{\text{as}}{=} q(c, w_T)$

$\overset{2.8\ (b)}{\Rightarrow} [d \cdot Y + \int d \cdot c\, ds]^T$ is a Q_T-martingale $\Rightarrow Y(t) \stackrel{\text{as}}{=} V(t)$ for $0 \le t \le T$

$\Rightarrow$ Y is an admissible valuation with $Y(0) \stackrel{\text{as}}{=} q(c, w_T)$

$\overset{(a)}{\Rightarrow}$ Y is a m.a.v. of (c, w_T), where a m.a.v. is a shorthand for a minimal admissible valuation.

2.10 (c)   is proved by 2.7.

2.10 (d)   If Y is a m.a.v. of (c, w_T), then $Y(0) \overset{as}{=} q(c, w_T)$, by **(a)**. If Y is an admissible valuation of w_T with $Y(0) \overset{as}{=} q(c, w_T)$ then $d \cdot Y + \int d \cdot c \, ds$ is a Q_T-martingale on $[0, T]$ by 2.8 (b) and it follows by **(b)** that Y is a m.a.v. of (c, w_T).

2.10 (e)   is proved by 2.8 (b).

2.11 (a)   If processes $\widehat{W}$ and $\widehat{W}^{(1)}$ are defined by (43), it follows by 2.2 that

$$d \cdot Y + \int d \cdot c \, ds \overset{as}{=} y + \int K \cdot F \, d\widehat{W} \overset{as}{=} y + K \cdot \int F \, d\widehat{W}^{(1)},$$

where $\int F \, d\widehat{W}^{(1)}$ is a P-stochastic integral with respect to the P-semimartingale $\widehat{W}^{(1)}$ because $\widehat{W} \overset{as}{=} \widehat{W}^{(1)}$ on $[0, 1]$ and $F = 0$ on $[1, \infty)$. According to (G3) and 2.4.2 (c), $\int F \, d\widehat{W}^{(1)}$ is also a Q_1-stochastic integral, hence the equality (60) holds as stated. The rest of **(a)** follows by (60) and (58).

2.11 (b)   If Y is the wealth process generated by (π^K, c, y), then it is defined by (60) and we have

$$Y(1) \overset{as}{=} \frac{y - K - \int_0^1 d \cdot c \, ds}{d(1)}.$$

Hence the admissibility of (π^K, c) implies that $\int_0^1 d \cdot c \, ds \le y - K$. On the other hand, if $\int_0^1 d \cdot c \, ds \le y - K$ then $y \ge K$ and $Y(1) \in \mathcal{F}_1$ is a claim. It follows by 2.8 (a) that $Y \in \mathcal{W}(c)$ is an admissible valuation of $Y(1)$ iff (52) holds. But the inequality stated by (a) implies (52), as we have already proved that $y \ge K$.

2.14 Y^{τ_0} is an admissible valuation of $(0, w_T)$ and it is a Q_T-martingale on $[0, T]$ by 2.12 (b) and by 1.3.11. Consequently, Y^{τ_0} with $Y^{\tau_0}(0) \overset{as}{=} 1$ is a minimal admissible valuation of $(0, w_T)$ by 2.10 (b).

2.16   Having fixed a consumption c and a $y \ge 0$, consider the bijective map $\mathcal{H} : \mathcal{W}(c, y) \to y + \mathrm{CM}_{loc}(P)$ defined in 2.5 Then, a wealth process Y in $\mathcal{W}(c, y)$ is a nonnegative process iff $L := \mathcal{H}(Y) \ge D \cdot \int d \cdot c \, ds$ holds almost surely on $\mathbb{R}^+$. Hence, an admissible strategy (π, c) for y exists iff there is a process L such as in (62).

In **(a)** observe $L := y + D \cdot \int_0^\infty d \cdot c \, ds$ which is, according to (G2), a process with properties (62).

In **(b)** consider a process L that has the properties (62). Since L is a supermartingale, we get $E D_t \cdot \int_0^t d \cdot c \, ds \le E L(t) \le y$ to be true for all t.

2.18   Note that according to 2.15

$$\mathcal{D}(y) \subset \left\{ c \text{ a consumption} : E_{Q_T} \int_0^T d \cdot c \, ds \le y \right\} = \mathcal{C}(y).$$

Thus, it remains to prove that for any $c \in \mathcal{C}(y)$ there exists a $c^* \in \mathcal{D}(y)$ such that

$$c^* \geq c \qquad \lambda \otimes P\text{-almost everywhere on} \quad [0,T] \times \Omega.$$

For any $c \in \mathcal{C}(y)$ we have $x := E_{Q_T} \int_0^T d \cdot c \, ds \leq y$ by 2.15. If $x > 0$ put $c^* = \frac{y}{x} \cdot c$ and if $x = 0$ choose $c^* = \frac{y}{T} \cdot d^{-1}$ to construct such a consumption process.

3.2 Note that $Y_y = d^{-1} \cdot y + Y_0 = Z_y + Y_0$.

3.8 $Y(\phi)$ is a minimal admissible valuation of $(X_1(T) - K)^+$ iff

$$\phi_0(t) \cdot X_0(t) \overset{as}{=} V(t) - \pi_1(t), \quad 0 \leq t < T.$$

3.10 Fix $0 < t \leq T$. According to 3.9 we have

$$f(t, X_1(t)) \overset{as}{=} V(t) \overset{as}{=} h(t, X_1(t)),$$

where $V(t)$ is the value of the g-option $g(X_1(T))$ at time t. Hence, $f(t,x) = h(t,x)$ for $x \notin N \in \mathcal{B}(\mathbb{R})$ such that $P[X_1(t) \in N] = 0$. Since $\mathcal{L}(X_1(t)) \sim \lambda|(0,\infty)$, it follows by the continuity of f and h that $f(t,x) = h(t,x)$ for all $x > 0$ and finally that $f = h$ on $\mathbb{R}^+ \times (0,\infty)$.

3.11 Apply (81), (75) and 1.13 to prove that for arbitrary $U < T$ and arbitrary $t \leq U$

$$d(t) \cdot V(t) \overset{as}{=} q + \int_0^t d \cdot (f_x(X_1) \cdot X_1 - X_1) \cdot (b-r) \, ds + \int_0^t d \cdot (f_x(X_1) \cdot X_1 - X_1) \cdot \sigma \, dW$$

holds, denoting again $f_x(X_1)(s) := f_x(s, X_1(s)) \cdot I_{[0,U]}(s)$. This proves (82).

3.17 If Y is a minimal admissible valuation of $g(X_1(T))$, then it is a process in $\mathcal{M}^m(g,T)$ by 3.14. Let Y be a process in $\mathcal{M}^m(g,T)$. Then Y is an admissible valuation of $Y(T)$ and $Y(T) \in L_1(Q_T)$ by 2.7. Further,

$$E_{Q_T} \{d(T) \cdot Y(T)\} \leq Y(0) = q^m(g,T) = q(g,T) = E_{Q_T} \{d(T) \cdot g(X_1(T))\}$$

because the left hand term is exactly the minimal price of $Y(T)$. Since $Y(T) \geq g(X_1(T))$ holds almost surely, we conclude that $Y(T) \overset{as}{=} g(X_1(T))$. On the other hand $Y(0) = q^m(g,T) = q(g,T)$ and so the process Y is proved to be a minimal admissible valuation of $g(X_1(T))$.

III.4 BIBLIOGRAPHICAL NOTES

Not too many results in Part III of the present volume are new and therefore the main purpose of the following notes is to refer to the sources that inspired our presentation and proofs. Also, at times, some literature on recent developments is suggested for additional reading.

Generally, we expect our reader to be a mathematician familiar with the classical standards of measure and probability theory (discrete-time martingales included) as excellently summarized, for example, in Chapter II of [135] by L.C.G. ROGERS AND D. WILLIAMS or in [6] by H. BAUER. Czech speaking readers may substitute the above references by [152]. We also refer to the monograph [93] written by O. KALLENBERG for a comprehensive treatise on almost all important achievements of modern probability.

Martingales. The material covered in this chapter is a tiny section of that made available by such rich sources as those provided by K. ITÔ AND M. McKEAN in [86], D. REVUZ AND M. YOR in [131] and by I. KARATZAS AND D.E. SHREVE in [95], for example. The reader may find useful [32] by Y.S. CHOW AND H. TEICHER as a source of information on discrete time martingales. We have made also use of [37] by C. DELLACHERIE AND P.A. MEYER.

Our proof of Density Theorem 1.5.7 goes in essence along the lines suggested in [95] (proof of 2.8, p.137). The proof of Doob-Meyer decomposition 1.6.1 heavily depends on the idea employed in [131] to prove (1.3) Theorem, p. 115.

We refer to [135] and [136] by L.C.G ROGERS AND D. WILLIAMS for a more complex information about continuous local martingales and for recent developments in the theory of local martingales that need not be continuous.

Stochastic integration. We define the stochastic integral by (4) in III.2 as a stochastic version of Pettis integral and prove its existence and properties simultaneously. The idea is, of course, not an entirely new one, it is very clearly suggested by *Kunita-Watanabe Theorem*, see 15.12, p.282 in [93], for example. We have made use of many sources, name at the very least the books [86], [95], [131], [136] and [156].

Our proof of Itô formula relies on *Weierstrass Theorem* as in [93] or [136]. We boast to prove it indeed for a dimension $d > 1$ having found a fitting and proved Weierstrass theorem in [142] by L. SCHWARTZ. We prove the celebrated P. Lévy's martingale characterization of Brownian motion similarly as in [131] or [95] by an application of Itô formula. See [93] for another and really elegant proof. The important example 2.3.7 of an integrable local martingale that is not a true martingale is basically Exercise 2. in [93], p. 293.

The Girsanov Theory and Brownian representations are presented to respect needs of the chapter on diffusion financial mathematics. The treatment of this material is inspired mostly by [131]. Thus, the proof of Girsanov theorem itself is based on the characterization 2.3.8 that is, in fact, (3.14) Exercise in [131], p. 143 and the global Girsanov theorem 2.4.9 is a one dimensional refinement of (1.13) Proposition in [131], p. 317. The proof of 2.4.7, that in our text substitutes Novikov theorem, is borrowed from [136], see (37.8) Theorem on p.77. For our construction of Brownian representations by means of extremal martingale measures through the decomposition result 2.5.2 we are in debt to the Remark made on p. 202 in [131]. The results 2.5.7 and 2.5.11 are due to M. YOR, see V.4 Section in [131] whose origins come from [175]. H.V. WEIZSÄCKER AND G. WINKLER in [168] provide further exiting developments in the field that connects Choquet theory and stochastic analysis. See also [154] and [155].

There are many other topics in stochastic analysis that deserve a close attention from any financial probabilist: Time changed martingales, local times, strong and weak theory of stochastic differential equations, stochastic control are perhaps the most important items. We refer briefly again to [95], [131], [93], [136] and newly to [103] for a complete treatment of the above topics.

Diffusion Financial mathematics. Among the sources that influenced this chapter are undoubtedly [75] by J. HARRISON AND S.R. PLISKA as a deep and ingenuously clear initiation to continuous stochastic finance (we borrowed 3.1.14 Example there) and [123] by M. MUSSIELA AND M. RUTKOWSKI, a treasure reference text in the field. We also owe a debt to T. ROLSKI, H. SCHMIDLI AND J. TEUGELS [137] and R. KORN AND E. KORN [102] for precise mathematical definitions of the financial concepts that go hand in hand with thorough and elegant clarifications of their role in Economics. Our treatment of American options is based in part on the enjoyable monograph [151] of J.M. STEELE. We borrowed there 3.3.4 Lemma, some sophisticated observations on the meaning of various market regulations and share 3.2.11 Example of a suicide trading strategy.

Our treatment of Black-Scholes and Girsanov calculus is a modest development of the original results of I. KARATZAS AND D.E. SHREVE summarized in [95] and [96]. Namely, the cornerstone result on the minimal price of a claim 3.2.10 is a *compactification* of the information that is available in 5.8.A Section in [95].

As we started to write this part of the volume with a strong determination not to include the results that we could not honestly prove, however attractive they might be, we had to skip many important topics. Among them are:

The problem of the *optimal control* of portfolio and consumption processes that we contacted in 3.2.18 is perhaps the most important one. We refer our reader to [94] and [102] to obtain an up to date information on the topic. More *general models for stock prices*, that go beyond Black-Scholes Gaussian limitations, offer further possibilities to exploit the Girsanov Calculus and the properties of martingale measures, we introduced in 2.5 Section. For a more deep treatment of the concept of arbitrage see [123] and [151], for example. To get a more realistic trading model it is necessary to consider the size of transaction costs. See [149] and [170] for recent developments in the field.

REFERENCES

[1] Abaffy, J. et al., *A nonparametric model for analysis of the Euro bond market*, J. Econ. Dynamics & Control (to appear).

[2] Abaffy, J. et al., *On generating scenarios for bond portfolios*, Bulletin of the Czech Econometric Society **11** (2000), 3–27.

[3] Anari, A. and Kolari, J., *Stock prices and inflation*, J. of Financial Research (2001), To appear.

[4] Bachelier, L., *Théorie de la spéculation*, Ann. Sci. École Norm. Sup. **17** (1900), 21-86.

[5] Bank, B. et al., *Non-Linear Parametric Optimization*, Akademie-Verlag, Berlin, 1982.

[6] Bauer, H., *Probability Theory*, Walter de Gruyter, Berlin, New York, 1996.

[7] Bazaraa, M. S., Jarvis, J. J., and Sherali, H. D., *Linear Programming and Network Flows*, 2nd edition, Wiley, New York, 1990.

[8] Bazaraa, M. S., Sherali, H. D., and Shetty, C. M., *Nonlinear Programming: Theory and Algorithms*, 2nd edition, Wiley, New York, 1993.

[9] Benninga, S. and Wiener, Z., *Binomial option pricing, the Black-Scholes option pricing formula, and exotic options*, Mathematica in Education and Research **6** (1997), no. 4, 1–4.

[10] Benninga, S. and Wiener, Z., *The binomial option pricing model*, Mathematica in Education and Research **6** (1997), no. 3, 1–8.

[11] Benninga, S. and Wiener, Z., *Binomial term structure models*, Mathematica in Education and Research **7** (1998), no. 3, 1–9.

[12] Benninga, S. and Wiener, Z., *Dynamic hedging strategies*, Mathematica in Education and Research **7** (1998), no. 1, 1–5.

[13] Benninga, S. and Wiener, Z., *Term structure of interest rates*, Mathematica in Education and Research **7** (1998), no. 2, 1–9.

[14] Benninga, S. and Wiener, Z., *Value-at-Risk*, Mathematica in Education and Research **7** (1998), no. 4, 1–7.

[15] Bertocchi, M., Dupačová, J., and Moriggia, V., *Sensitivity of bond portfolio's behavior with respect to random movements in yield curve: A simulation study*, Annals of Oper. Res. **99** (2000), 267–286.

[16] Best, M. J. and Grauer, R. R., *Sensitivity analysis for mean-variance portfolio problems*, Management Science **37** (1991), 980–989.

[17] Bhatti, M. A., *Practical Optimization Methods with Mathematica® Applications*, TELOS Springer-Verlag, New York, 2000.

[18] Bílý, J., *Appendix D. Index numbers*, Mathematical Economics (Allen, R. G. D.), Academia, Praha, 1971, pp. 761–765. (Czech)

[19] Birge, J. R. and Louveaux, F., *Introduction to Stochastic Programming*, Springer Series in Operations Research, Springer, New York, 1997.

[20] Birge, J. R. et al., *A standard input format for multiperiod stochastic linear programs*, COAL Newsletter **17** (1987), 1–20.

[21] Bisschop, J. J. and Roelofs, M., *AIMMS The User's Guide*, Paragon Decision Technology B. V., Haarlem, 1999.

[22] Black, F., Derman, E., and Toy, W., *A one-factor model of interest rates and its application to treasury bond options*, Financial Analysts Journal (1990), no. Jan./Feb., 33–39.

[23] Black, F. and Scholes, M., *Pricing of Options and Corporate Liabilities*, J. Political Econ. **81** (1973), 637-654.

[24] Bradley, S. P. and Crane, D. B., *A dynamic model for bond portfolio management*, Management Science **19** (1972), 139–151.

[25] Brigham, E. F., *Fundamentals of Financial Management*, 6th ed., The Dryden Press, Forth Worth, 1992.

[26] Cambell, J. Y., Lo, A. W., and MacKinlay, A. C., *The Econometrics of Financial Markets*, Princeton University Press, Princeton, 1997.

[27] Cariño, D. R., Myers, D. H., and Ziemba, W. T., *Concepts, technical issues and uses of the Russell – Yasuda Kassai financial planning model*, Oper. Res. **46** (1998), 450–462.

[28] Cariño, D. R. and Ziemba, W. T., *Formulation of the Russell – Yasuda Kassai financial planning model*, Oper. Res. **46** (1998), 433–449.

[29] Casimir, R. J., *The newsboy and the flower-girl*, Omega **18** (1990), 395–398.

[30] Censor, Y. and Zenios, S.A., *Parallel Optimization: Theory, Algorithms, and Applications*, Oxford University Press, New York, 1997.

[31] Chopra, W. K. and Ziemba, W. T., *The effect of errors in means, variances and covariances on optimal portfolio choice*, J. of Portfolio Management **19** (1993), 6–11.

[32] Chow, Y. S. and Teicher, H., *Probability Theory: Independence, Interchangeability, Martingales*, 2nd edition, Springer Verlag, New York, 1988.

[33] Cipra, T., *Mathematics of Securities*, HZ, Praha, 2000. (Czech)

[34] Consigli, G., *Dynamic Stochastic Programming for Asset and Liability Management*, PhD Dissertation, Univ. of Essex, Colchester, 1997.

[35] Consigli, G. and Dempster, M. A. H., *Dynamic stochastic programming for asset-liability management*, Annals of Oper. Res. **81** (1998), 131–161.

[36] Damodaran, A., *Investment Valuation*, Wiley, New York, 1996.

[37] Dellacherie, C. and Meyer, P.A., *Probabilités et Potential*, Hermann, Paris, 1975-1993.

[38] Dembo, R. S., *Scenario immunization*, Financial Optimization (Zenios, S. A., ed.), Cambridge University Press, 1993, pp. 290–308.

[39] Dempster, M. A. H., ed., *Stochastic Programming*, Academic Press, London, 1980.

[40] Dempster, M. A. H. and Ireland, A. M., *A financial expert decision system*, Math. Models for Decision Support (Mitra, B. , ed.), NATO ASI Series, Vol. F48, Springer, 1988, pp. 415–440.

[41] Doob, J. L., *Stochastic Processes*, Willey, New York, 1953.

[42] Dowd, K., *Beyond Value at Risk*, Wiley, Chichester, 1998.

[43] Duffie, D., *Dynamic Asset Pricing Theory*, 2nd ed., Princeton University Press, Princeton, 1996.

[44] Dupačová, J., *Stochastic programming models in banking*, (see also Tutorial paper, IIASA Laxenburg, 1991), Ekonomicko - Matematický Obzor **27** (1991), 201–234. (Czech)

[45] Dupačová, J., *Multistage stochastic programs: The state-of-the-art and selected bibliography*, Kybernetika **31** (1995), 151–174.

[46] Dupačová, J., *Scenario based stochastic programs: Resistance with respect to sample*, Annals of Oper. Res. **64** (1996), 21–38.

[47] Dupačová, J., *Uncertainty about input data in portfolio management*, Modelling Techniques for Financial Markets and Bank Management (Bertocchi, M. et al., eds.), Physica Verlag, Heidelberg, 1996, pp. 17–33.

[48] Dupačová, J., *Stochastic programming: Approximation via scenarios*, Apportaciones Matematicás, Ser. Communicationes **24** (1998), 77–94.

[49] Dupačová, J., *Portfolio optimization via stochastic programming: Methods of output analysis*, MMOR **50** (1999), 245–270.

[50] Dupačová, J., *Stability properties of a bond portfolio management problem*, Annals of Oper. Res. **99** (2000), 251–265.

[51] Dupačová, J. and Bertocchi, M., *From data to model and back to data: A bond portfolio management problem*, European J. of Oper. Res. **134** (2001), 261–278.

[52] Dupačová, J., Bertocchi, M., and Moriggia, V., *Postoptimality for scenario based financial planning models with an application to bond portfolio management*, World Wide Asset and Liability Modeling (Ziemba, W.T. and Mulvey, J., eds.), Cambridge Univ. Press, 1998, pp. 263–285.

[53] Dupačová, J., Consigli, G., and Wallace, S. W., *Scenarios for multistage stochastic programs*, Annals of Oper. Res. **100** (2000), 25–53.

[54] Dupačová, J. and Sladký, K., *Comparison of multistage stochastic programs with recourse and stochastic dynamic programs with discrete time*, ZAMM (to appear).

[55] Dupačová, J. et al., *On estimating the yield and volatility curves*, Kybernetika **33** (1997), 659–673.

[56] Einstein, A., *On the theory of Brownian motion*, Investigations on the Theory of the Brownian Movement, Reprint ed. 1956, Dover, New York, originally published in 1906.

[57] Elton, E. J. and Gruber, M. J., *Modern Portfolio Theory and Investment Analysis*, 4th ed., Wiley, New York, 1991.

[58] Ermoliev, Y. M. and Wets, R. J.-B., *Numerical Techniques for Stochastic Optimization Problems*, Springer Series in Computational Mathematics 10, Springer, Berlin, 1988.

[59] Escudero, L. F. et al., *Production planning via scenario modeling*, Annals of Oper. Res. **43** (1993), 311–335.

[60] Fabozzi, F. J., ed., *The Handbook of Fixed Income Securities*, 5th ed., McGraw-Hill, New York, 1997.

[61] Fabozzi, F. J. and Kole, S., eds., *Selected Topics in Investment Management for Financial Planning*, Dow Jones-Irwin, Homewood (IL), 1985.

[62] Fiacco, A. V., *Introduction to Sensitivity and Stability Analysis in Nonlinear Programming*, Academic Press, New York,, 1993.

[63] Fleming, W. H. and Rischel, R. W., *Deterministic and Stochastic Optimal Control*, Springer Verlag, New York, 1975.

[64] Föllmer, H. and Leukert, P., *Quantile hedging*, Unpublished manuscript (1998), Humboldt-Universität zu Berlin.

[65] Fourer, R., Gay, D. M., and Kernighan, B. W., *AMPL — A Modeling Language for Mathematical Programming*, Duxbury Press/Brooks/Cole Publishing Company, New York, 1993.

[66] GAMS D.C., *GAMS 2.5 Documentation*, GAMS, Washington, 2000.

[67] Gänssler, P. and Stute, W., *Wahrscheinlichkeitstheorie*, Springer Verlag, New York, 1977.

[68] Gassmann, H. I., *MSLiP: A computer code for the multi-stage stochastic programming problem*, Math. Progr. **47** (1990), 407–423.

[69] Gol'shtein, E. G., *Vypukloje Programmirovanije. Elementy Teoriji*, [Theory of Convex Programming, Translations of Mathematical Monographs 36; American Mathematical Society, Providence (RI), 1972], Nauka, Moscow, 1970. (Russian)

[70] Golub, B. et al., *Stochastic programming models for portfolio optimization with mortgage-backed securities*, European J. of Oper. Res. **82** (1995), 282–296.

[71] Grinold, R. C., *Infinite horizon stochastic programs*, SIAM J. Control and Optimization **24** (1986), 1246–1260.

[72] Guddat, J. et al., *Multiobjective and Stochastic Optimization Based on Parametric Optimization*, Mathematical Research – Mathematische Forschung, Band 26, Akademie-Verlag, Berlin, 1985.

[73] Hadley, G., *Nonlinear and Dynamic Programming*, Addison-Wesley, Reading (MA), 1964.

[74] Hansen, P. and Jaumard, B., *Cluster analysis and mathematical programming*, Math. Progr. **79** (1997), 191–215.

[75] Harrison, J. and Pliska, S. R., *Martingales and stochastic integrals in the theory of continuous trading*, Stochastic Process. Appl. **11** (1981), 215-260.

[76] Higgins, R. C., *Financial Management*, Science Research Associates, Chicago, 1977.

[77] Higle, J. and Sen, S., *Stochastic Decomposition. A Statistical Method for Large Scale Stochastic Linear Programming*, Kluwer, Dordrecht, 1996.

[78] Hillermeier, C., *Nonlinear Multiobjective Optimization. A Generalized Homotopy Approach*, International Series of Numerical Mathematics 135, Birkhäuser, Basel, 2001.

[79] Hirschleifer, J., *Investment decision under uncertainty-choice theoretic approaches*, The Quarterly J. of Economics **74** (1965), 509–536.

[80] Høyland, K. and Wallace, S. W., *Generating scenario trees for multistage problems*, Management Science **47** (2001), 295–307.

[81] Hull, J. C., *Options, Futures, and Other Derivative Securities*, 2nd ed., Prentice-Hall, Englewood Cliffs, 1993.

[82] Hunt, P. J. and Kennedy, J. E., *Financial Derivatives in Theory and Practice*, Wiley, Chichester, 2000.

[83] Hurt, J. and Koch, R., *Mathematical methods of risk control*, Proceedings of the Seminar on Actuarial Sciences, Czech Society of Actuaries 2000/2001, Praha, 2001, pp. 53–67. (Czech)

[84] Infanger, G., *Planning under Uncertainty: Solving Large-Scale Stochastic Linear Programs*, Boyd and Fraser, Danvers, 1994.

[85] Ingersoll, J. E. Jr., *Theory of Financial Decision Making*, Rowman & Littlefield, Savage, 1987.

[86] Itô, K. and McKean, M., *Diffusion Processes and Their Sample Paths*, 2nd printing, Springer Verlag, New York, 1974.

[87] Jamshidian, F. and Yu Zhu, *Scenario simulation: Theory and methodology*, Finance and Stochastics **1** (1997), 43–67.

[88] Jarrow, R. A., Maksimovic, V., and Ziemba, W. T., eds., *Finance*, Handbooks in Operations Research and Management Science Volume 9, Elsevier, Amsterdam, 1995.

[89] Jorion, P., *Value at Risk*, McGraw-Hill, New York, 1997.

[90] Kahn, R. N., *Fixed income risk modelling*, The Handbook of Fixed Income Securities (Fabozzi, F., ed.), 3rd Edition, Irwin, New York, 1991, pp. 1307–1319.

[91] Kall, P. and Wallace, S. W., *Stochastic Programming*, Wiley-Interscience Series in Systems and Optimization, Wiley, Chichester, 1994.

[92] Kallberg, J. G. and Ziemba, W. T., *An algorithm for portfolio revision: Theory, computation, algorithm and empirical results*, Applications of Management Science **1** (1981), 267–291.

[93] Kallenberg, O., *Foundations of Modern Probability*, Springer Verlag, New York, 1997.

[94] Karatzas, I., *Optimization problems in the theory of continuous trading*, SIAM J. Control and Optimization **27** (1989), 1221-1259.

[95] Karatzas, I. and Shreve, D. E., *Brownian Motion and Stochastic Calculus*, Springer Verlag, New York, 1991.

[96] Karatzas, I. and Shreve, D. E., *Methods of Mathematical Finance*, Springer Verlag, New York, 1998.

[97] Kataoka, S., *A stochastic programming model*, Econometrica **31** (1963), 181–196.

[98] Kira, D. and Ziemba, W. T., *Equivalence among alternative portfolio selection criteria*, Financial Decision Making under Uncertainty (Levy, H. and Sarnat, M., eds.), Academic Press, New York, 1977, pp. 151–161.

[99] Klein Haneveld, W. H. and van der Vlerk, M., *Stochastic integer programming: General models and algorithms*, Annals of Oper. Res. **85** (1999), 39–57.

[100] Kolari, J. W., *An analytical model of risky yield curves*, J. of Financial Research **X** (1987), 295–303.

[101] Konno, H. and Yamazaki, H., *Mean-absolute deviation portfolio optimization model and its applications to Tokyo stock market*, Management Science **37** (1991), 519–531.

[102] Korn, R. and Korn, E., *Option Pricing and Portfolio Optimization*, American Mathematical Society, Providence (RI), 2000.

[103] Krylov, N., *Controlled Diffusion Processes*, Springer Verlag, New York, 1980.

[104] Kusy, M. I. and Ziemba, W. T., *A bank asset and liability management model*, Oper. Res. **34** (1986), 356–376.

[105] Kwok, Y. K., *Mathematical Models of Financial Derivatives*, Springer, Singapore, 1998.

[106] Lane, M. and Hutchinson, P., *A model for managing a certificate of deposit portfolio under uncertainty*, Stochastic Programming (Dempster, M. A. H., ed.), Academic Press, London, 1980, pp. 473–495.

[107] Lewis, A. L., *Option Valuation under Stochastic Volatility with Mathematica Code*, Finance Press, Newport Beach (CA), 2000.

[108] Louveaux, F. V. and Smeers, Y., *Optimal investments for electricity generation: A stochastic model and a test problem*, Numerical Techniques for Stochastic Optimization Problems (Ermoliev, Yu., Wets, R. J.-B., eds.), Springer, pp. 445–453.

[109] Madalla, G. S. and Rao, C. R., eds., *Statistical Methods in Finance*, Handbook of Statistics Volume 14, Elsevier, Amsterdam, 1996.

[110] Madansky, A., *Bounds on the expectation of a convex function of a multivariate random variable*, AMS **30** (1959), 743–746.

[111] Mandl, P., Lánská, V., and Vrkoč, I., *Exercises in stochastic analysis*, Supplement, Kybernetika **14** (1978), 36 pp.

[112] Markowitz, H. M., *Portfolio selection*, J. of Finance **6** (1952), 77–91.

[113] Mayer, J., *Stochastic Linear Programming Algorithms: A Comparison Based on a Model Management System*, Gordon and Beach Science Publ., Amsterdam, 1998.

[114] McCutcheon, J. J. and Scott, W. F., *An Introduction to the Mathematics of Finance*, Butterworth-Heinemann, Oxford, 1986.

[115] Merton, R. C., *Optimum consumption and portfolio rules in a continuous-time model*, J. Econom. Theory **3** (1971), 373-413.

[116] Merton, R. C., *Continuous-Time Finance*, Blackwell, Cambridge (MA), 1992.

[117] Moré, J. J. and Wright, S. J., *Optimization Software Guide*, SIAM, Philadelphia, 1993.

[118] Morgan, J. P., Reuters, *RiskMetrics^{TM} - Technical Document*, 4th ed., Morgan Guaranty Trust Company, New York, 1996.

[119] Mulvey, J. M., Rosenbaum, D. P., and Shetty, B., *Strategic financial risk management and operations research*, European J. of Oper. Res. **97** (1997), 1–16.

[120] Mulvey, J. M., Vanderbei, R. J., and Zenios, S. A., *Robust optimization of large scale systems*, Oper. Res. **43** (1995), 264–281.

[121] Mulvey, J. M. and Vladimirou, H., *Stochastic network optimization models for investment planning*, Annals of Oper. Res. **20** (1989), 187–217.

[122] Mulvey, J. M. and Vladimirou, H., *Applying the progressive hedging algorithm to stochastic generalized networks*, Annals of Oper. Res. **31** (1991), 399–424.

[123] Musiela, M. and Rutkowski, M., *Martingale Methods in Financial Modelling*, Springer Verlag, New York, 1997.

[124] Pflug, G. Ch., *Scenario tree generation for multiperiod financial optimization by optimal discretization*, Mathematical Programming **89** (2001), 251–271.

[125] Pflug, G. Ch. and Swietanowski, A., *Dynamic asset allocation under uncertainty for pension fund management*, Cybernetics and Control **28** (1999), 755–777.

[126] Popela, P., *An Object-Oriented Approach to Multistage Stochastic Programming: Models and Algorithms*, PhD Dissertation, UK MFF, Prague, 1998.

[127] Prášková, Z., *Volatility estimation and bootstrap*, Prob. Theory and Math. Stat. (Grigelionis B. et al., eds.), VSP, Zeist, 1999, pp. 589–598.

[128] Prékopa, A., ed., *Studies in Applied Stochastic Programming*, MTASzTAKI, Budapest, 1978.

[129] Prékopa, A., *Network planning using two stage stochastic programming under uncertainty*, Recent Results in Stochastic Programming (Kall, P. and Prékopa, A., eds.), LNIEMS 179, Springer, Berlin, 1980, pp. 215–237.

[130] Prékopa, A., *Stochastic Programming*, Kluwer and Akademiai Kiadó, Dordrecht and Budapest, 1995.

[131] Revuz, D. and Yor, M., *Continuous Martingales and Brownian Motion*, Springer Verlag, New York, 1994.

[132] Robinson, S. M., *A characterization of stability in linear programming*, Oper. Res. **25** (1976), 435–447.

[133] Rockafellar, R. T. and Wets, R. J.-B., *Nonanticipativity and $\mathcal{L}^1$-martingales in stochastic optimization problems*, Mathematical Programming Study **6** (1976), 170–187.

[134] Rockafellar, R. T. and Wets, R. J.-B., *Scenarios and policy aggregation in optimization under uncertainty*, Mathematics of Oper. Res. **15** (1991), 119–147.

[135] Rogers, L. C. G. and Williams, D., *Diffusions, Markov Processes and Martingales. Volume 1: Foundations*, Cambridge University Press, Cambridge, 2000.

[136] Rogers, L. C. G. and Williams, D., *Diffusions, Markov Processes and Martingales. Volume 2: Itô Calculus*, Cambridge University Press, Cambridge, 2000.

[137] Rolski, T. et al., *Stochastic Processes for Insurance and Finance*, Wiley, Chichester, 1999.

[138] Rose, P. S., Kolari, J., and Fraser, D. R., *Financial Institutions*, 4th ed., IRWIN, Homewood (IL), 1993.

[139] Roy, A. D., *Safety-first and the holding of assets*, Econometrica **20** (1952), 431–449.

[140] Saaty, T. L., *The Analytic Hierarchy Process*, RWS Publications, Pittsburgh, 1990.

[141] Samuels, J. M., Wilkes, F. M., and Brayshaw, R. E., *Management of Company Finance*, 5th ed., Chapman & Hall, London, 1990.

[142] Schwartz, L., *Analyse Mathematique I*, Hermann, Paris, 1967.

[143] Sears, R. S. and Trennepohl, G. L., *Investment Management*, The Dryden Press, Forth Worth, 1993.

[144] Serfling, R. J., *Approximation Theorems in Mathematical Statistics*, Wiley, New York, 1980.

[145] Sharpe, W. F., *Mean-absolute-deviation characteristic lines for securities and portfolios*, Manag. Sci. **18** (1971), B1–B13.

[146] Sharpe, W. F., *Investments*, 3rd ed., Prentice-Hall, Englewood Cliffs, 1985.

[147] Shaw, W. T., *Modelling Financial Derivatives with Mathematica®*, Cambridge University Press, Cambridge, 1998.

[148] Shaw, W. T. and Tigg, J., *Applied Mathematica®*, Addison-Wesley, Reading (MA), 1994.

[149] Shreve, S. E. and Soner, H. M., *Optimal investment and consumption with transaction costs*, The Annals of Applied Probability **4** (1994), 609-692.

[150] Späth, H., *Algorithmen für Elementare Ausgleichs-modelle*, R. Oldenburg Verlag, Munich, 1973.

[151] Steele, J.M., *Stochastic Calculus and Financial Applications*, Springer Verlag, New York, 2001.

[152] Štěpán, J., *Probability Theory.Mathematical Foundations*, Academia, Prague, 1987. (Czech)

[153] Štěpán, J. and Hlubinka, D., *One dimensional observation of two dimensional diffusion process*, ESES'96 Proceedings, Proceedings the Satellite Meeting of 4th World Congress of the Bernoulli Society (Horová, I., Jurečková, J., and Vosmanský, J., eds.), Masaryk University, Brno, 1996, pp. 149-156.

[154] Štěpán, J. and Ševčík, P., *Convex analysis for sets of local martingale measures*, Acta Universitatis Carolinae, Mathematica et Physica **41** (2000), 57-80.

[155] Štěpán, J. and Ševčík, P., *Local martingale measures*, Acta Universitatis Carolinae, Mathematica et Physica **41** (2000), 37-55.

[156] Stroock, D. W. and Varadhan, S. R. S., *Multidimensional Diffusion Processes*, Springer Verlag, New York, 1979.

[157] Takriti, S., Birge, J. R., and Long, E., *A stochastic model for the unit commitment problem*, IEEE Trans. on Power Systems **11** (1996), 1497-1508.

[158] Takriti, S., Krasenbrink, B., and Wu, L. S.-Y., *Incorporating fuel constraints and electricity spot prices into the stochastic unit commitment problem*, Oper. Res. **48** (2000), 268-280.

[159] Taylor, S. J., *Modelling Financial Time Series*, Wiley, Chichester, 1986.

[160] Telser, L. G., *Safety first and hedging*, Review of Economic Studies **23** (1955), 1-16.

[161] Van Horne, J. C., *Financial Market Rates and Flows*, Prentice-Hall, Englewood Clifs, 1978.

[162] Varian, H. R., ed., *Economic and Financial Modeling with Mathematica®*, TELOS Springer Verlag, New York, 1993.

[163] Varian, H. R., ed., *Computational Economics and Finance*, TELOS Springer Verlag, New York, 1996.

[164] Vašíček, O., *An equilibrium characterization of the term structure*, J. of Financial Economics **5** (1977), 177-188.

[165] Vetzal, K. R., *Stochastic volatility, movements in short term interest rates, and bond option valuation*, J. of Banking & Finance **21** (1997), 169-196.

[166] Ville, J., *Étude Critique de la Notion du Collectif*, Gauthier-Villars, Paris, 1939.

[167] Vladimirou, H. and Zenios, S. A., *Stochastic linear programs with restricted recourse*, European J. of Oper. Res. **101** (1997).

[168] von Weizsäcker, H. and Winkler, G., *Non-compact extremal integral representations: some probabilistic aspects*, Functional Analysis: Surveys and recent results II, Mathematics Studies 38, North Holland, Amsterdam, 1980, pp. 115-148.

[169] Wets, R. J.-B. and Ziemba, W. T., eds., *Stochastic Programming: The State of the Art 1998*, Annals of Oper. Res. 85, Baltzer, Bussum, 1999.

[170] Whalley, A. and Wilmott, P., *An asymptotic analysis of an optimal hedging model for option pricing with transaction costs*, Mathematical Finance **7** (1997), 307-322.

[171] Wiener, N., *Differential space*, J. Math. Phys. **2** (1923), 131-174.

[172] Wilmott, P., Howison, S., and Dewynne, J., *The Mathematics of Financial Derivatives*, A Student Introduction, Cambridge University Press, Cambridge, 1995.

[173] Wise, A. J., *A theoretical analysis of the matching of assets to liabilities*, J. of the Institute of Actuaries **III** (1984), 375-402.

[174] Wolfram, S., *The Mathematica® Book*, 4th ed., Wolfram Media/Cambridge University Press, Champaign/Cambridge, 1999.

[175] Yor, M., *Quelques Résultats sur Certaines Measures Extrémales à la Representation des Martingales*, Springer Lecture Notes in Mathematics 695, Springer Verlag, New York, 1979.

[176] Zenios, S. A., ed., *Financial Optimization*, Cambridge University Press, Cambridge, 1993.

[177] Zenios, S. A. and Shtilman, M. S., *Constructing optimal samples from a binomial lattice*, J. of Information & Optimization Sciences **14** (1993), 125-147.

[178] Ziemba, W. T. and Mulvey, J. M., eds., *World Wide Asset and Liability Modeling*, Cambridge University Press, Cambridge, 1998.

[179] Ziemba, W. T. and Vickson, R. G., eds., *Stochastic Optimization Models in Finance*, Academic Press, New York, 1975.

[180] Zvára K., *Regression Analysis*, Academia, Prague, 1989. (Czech)

INDEX